Electrical

Level One

Trainee Guide
2008 *NEC*® Revision

PEARSON
Prentice
Hall

Upper Saddle River, New Jersey
Columbus, Ohio

National Center for Construction Education and Research
President: Don Whyte
Director of Product Development: Daniele Stacey
Electrical Project Manager: Daniele Stacey
Production Manager: Tim Davis
Quality Assurance Coordinator: Debie Ness
Desktop Publishing Coordinator: James McKay
Editors: Rob Richardson, Matt Tischler, Brendan Coote

Writing and development services provided by Topaz Publications, Liverpool, NY
Lead Writer/Project Manager: Veronica Westfall
Desktop Publisher: Joanne Hart
Art Director: Megan Paye
Permissions Editors: Andrea LaBarge and Jackie Vidler
Writers: Tom Burke, Gerald Shannon, Nancy Brown, Charles Rogers

Pearson Education, Inc.
Product Manager: Lori Cowen
Product Development Editor: Janet Ryerson
Project Managers: Stephen C. Robb, Christina M. Taylor
AV Project Manager: Janet Portisch
Senior Operations Supervisor: Pat Tonneman
Art Director: Diane Y. Ernsberger
Interior Design: Kristina D. Holmes
Cover Designer: Kristina D. Holmes
Cover Photo: Tim Davis
Director of Marketing: David Gesell
Executive Marketing Manager: Derril Trakalo
Senior Marketing Coordinator: Alicia Dysert
Copyeditor: Sheryl Rose
Proofreader: Bret Workman

This book was set in Palatino and Helvetica by S4Carlisle Publishing Services and was printed and bound by Courier Kendallville, Inc. The cover was printed by Phoenix Color Corp.

Pearson Education Ltd., London
Pearson Education Singapore Pte. Ltd.
Pearson Education Canada, Inc.
Pearson Education—Japan

Pearson Education Australia Pty. Limited
Pearson Education North Asia Ltd., Hong Kong
Pearson Educación de Mexico, S.A. de C.V.
Pearson Education Malaysia Pte. Ltd.

10 9 8 7 6 5 4 3 2
ISBN-13: 978-0-13-604460-4
ISBN-10: 0-13-604460-3

Preface

Electricity powers the applications that make our daily lives more productive and efficient. Thirst for electricity has led to vast job opportunities in the electrical field. Electricians constitute one of the largest construction occupations in the United States, and they are among the highest-paid workers in the construction industry. According to the U.S. Bureau of Labor Statistics, job opportunities for electricians are expected to be excellent as the demand for skilled craftspeople is projected to outpace the supply of trained electricians.

Electricians install electrical systems in structures. They install wiring and other electrical components, such as circuit breaker panels, switches, and light fixtures. Electricians follow blueprints, the *National Electrical Code®*, and state and local codes. They use specialized tools and testing equipment, such as ammeters, ohmmeters, and voltmeters. Electricians learn their trade through craft and apprenticeship programs. These programs provide classroom instruction and on-the-job training with experienced electricians.

We wish you success as you embark on your first year of training in the electrical craft and hope that you will continue your training beyond this textbook. There are more than a half-million people employed in electrical work in the United States, and, as most of them can tell you, there are many opportunities awaiting those with the skills and desire to move forward in the construction industry.

NEW WITH *ELECTRICAL LEVEL ONE*

NCCER and Pearson Education, Inc. are pleased to present the sixth edition of *Electrical Level One*. In addition to being fully updated to the 2008 *National Electrical Code®*, this edition features a new "Orientation to the Electrical Trade" module and a completely rewritten "Electrical Test Equipment" module. Other updates include a feature called "Going Green" which is used to illustrate how decisions made by electricians can promote more efficient use of energy and lead to a healthier planet. To see what kinds of work you might be doing in the electrical trade, check out the opening pages of each of the twelve modules in this textbook. From being safe on the job and understanding how electricity

works to reading electrical blueprints and bending conduit, the knowledge that you need to get started in a rewarding career as an electrician is in this edition of *Electrical Level One*.

We invite you to visit the NCCER website at www.nccer.org for the latest releases, training information, newsletter, and much more. You can also reference the Contren® product catalog online at that site. Your feedback is welcome. Email your comments to curriculum@nccer.org or send general comments and inquiries to info@nccer.org.

CONTREN® LEARNING SERIES

The National Center for Construction Education and Research (NCCER) is a not-for-profit 501(c)(3) education foundation established in 1996 by the world's largest and most progressive construction companies and national construction associations. It was founded to address the severe workforce shortage facing the industry and to develop a standardized training process and curricula. Today, NCCER is supported by hundreds of leading construction and maintenance companies, manufacturers, and national associations. The Contren® Learning Series was developed by NCCER in partnership with Pearson Education, Inc., the world's largest educational publisher.

Some features of NCCER's Contren® Learning Series are as follows:

- An industry-proven record of success
- Curricula developed by the industry for the industry
- National standardization providing portability of learned job skills and educational credits
- Compliance with the Office of Apprenticeship requirements for related classroom training *(CFR 29:29)*
- Well-illustrated, up-to-date, and practical information

NCCER also maintains a National Registry that provides transcripts, certificates, and wallet cards to individuals who have successfully completed modules of NCCER's Contren® Learning Series. *Training programs must be delivered by an NCCER Accredited Training Sponsor in order to receive these credentials.*

Special Features of This Book

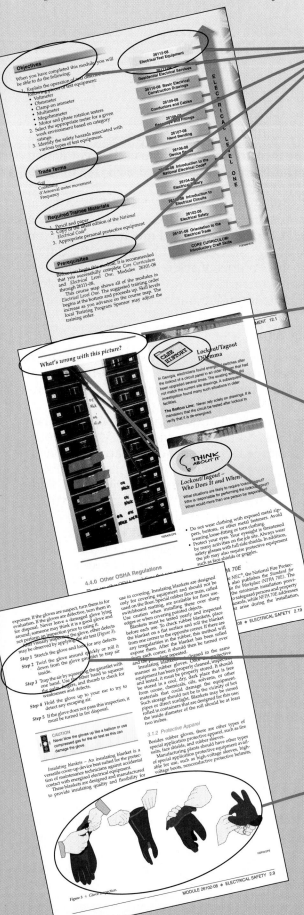

Introduction Page

This page is found at the beginning of each module and lists the Objectives, Trade Terms, Required Trainee Materials, Prerequisites, and Course Map for that module. The Objectives list the skills and knowledge you will need in order to complete the module successfully. The list of Trade Terms identifies important terms you will need to know by the end of the module. Required Trainee Materials list the materials and supplies needed for the module. The Prerequisites for the module are listed and illustrated in the Course Map. The Course Map also gives a visual overview of the entire course and a suggested learning sequence for you to follow.

What's wrong with this picture?

What's wrong with this picture? features include photos of actual code violations for identification and encourage you to approach each installation with a critical eye.

Case History

Case History features emphasize the importance of safety by citing examples of the costly (and often devastating) consequences of ignoring *National Electrical Code®* or OSHA regulations.

Think About It

Think About It features use "What if?" questions to help you apply theory to real-world experiences and put your ideas into action.

Step-by-Step Instructions

Step-by-step instructions are used throughout to guide you through technical procedures and tasks from start to finish. These steps show you not only how to perform a task but also how to do it safely and efficiently.

Color Illustrations and Photographs

Full-color illustrations and photographs are used throughout each module to provide vivid detail. These figures highlight important concepts from the text and provide clarity for complex instructions. Each figure reference is denoted in the text in *italic type* for easy reference.

Inside Track

Inside Track features provide a head start for those entering the electrical field by presenting technical tips and professional practices from master electricians in a variety of disciplines. Inside Tracks often include real-life scenarios similar to those you might encounter on the job site.

Trade Terms

Each module presents a list of Trade Terms that are discussed within the text, defined in the Glossary at the end of the module, and reinforced with a Trade Terms Quiz. These terms are denoted in the text with blue bold type upon their first occurrence. To make searches for key information easier, a comprehensive Glossary of Trade Terms from all modules is located at the back of this book.

Notes, Cautions, and Warnings

Safety features are set off from the main text in highlighted boxes and organized into three categories based on the potential danger of the issue being addressed. Notes simply provide additional information on the topic area. Cautions alert you of a danger that does not present potential injury but may cause damage to equipment. Warnings stress a potentially dangerous situation that may cause injury to you or a co-worker.

Going Green

Going Green looks at ways to preserve the environment, save energy, and make good choices regarding the health of the planet. Through the introduction of new construction practices and products, you will see how the "greening of America" has already taken root.

Profile in Success

Profiles in Success share the apprenticeship and career experiences of and advice from successful professionals in the electrical field. From contractors and consultants to construction engineers and corporate training managers to association executives, all of these professionals got their start in the electrical trade.

Review Questions

Review Questions are provided to reinforce the knowledge you have gained. This makes them a useful tool for measuring what you have learned.

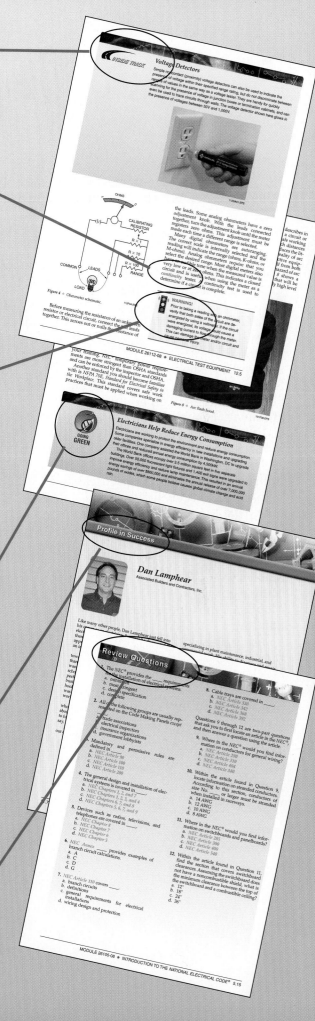

Contren® Curricula

NCCER's training programs comprise over 50 construction, maintenance, and pipeline areas and include skills assessments, safety training, and management education.

Boilermaking
Cabinetmaking
Carpentry
Concrete Finishing
Construction Craft Laborer
Construction Technology
Core Curriculum: Introductory Craft Skills
Drywall
Electrical
Electronic Systems Technician
Heating, Ventilating, and Air Conditioning
Heavy Equipment Operations
Highway/Heavy Construction
Hydroblasting
Industrial Coating and Lining Application
   Specialist
Industrial Maintenance Electrical and
   Instrumentation Technician
Industrial Maintenance Mechanic
Instrumentation
Insulating
Ironworking
Masonry
Millwright
Mobile Crane Operations
Painting
Painting, Industrial
Pipefitting
Pipelayer
Plumbing
Reinforcing Ironwork
Rigging
Scaffolding
Sheet Metal
Site Layout
Sprinkler Fitting
Welding

Pipeline
Control Center Operations, Liquid
Corrosion Control
Electrical and Instrumentation
Field Operations, Liquid
Field Operations, Gas
Maintenance
Mechanical

Safety
Field Safety
Safety Orientation
Safety Technology

Management
Introductory Skills for the Crew Leader
Project Management
Project Supervision

Spanish Translations
Andamios
Currículo Básico: Habilidades Introductorias del
   Oficio
Instalación de Rociadores Nivel Uno
Introducción a la Carpinteria
Orientación de Seguridad
Seguridad de Campo

Supplemental Titles
Applied Construction Math
Careers in Construction

Acknowledgments

This curriculum was revised as a result of the farsightedness and leadership of the following sponsors:

ABC of Iowa
ABC of New Mexico
ABC Pelican Chapter Southwest, Westlake, LA
Baker Electric
Beacon Electric Company
Cuyahoga Valley Career Center
M.C. Dean Inc.
Duck Creek Engineering
Hamilton Electric Construction Company
IMTI of New York and Connecticut

Lamphear Electric
Madison Comprehensive High School/Central Ohio ABC
Pumba Electric LLC
Putnam Career & Technical Center
Rust Constructors Inc.
TIC Industrial
Tri-City Electrical Contractors, Inc.
Trident Technical College
Vector Electric and Controls Inc.

This curriculum would not exist were it not for the dedication and unselfish energy of those volunteers who served on the Authoring Team. A sincere thanks is extended to the following:

John S. Autrey
Clarence "Ed" Cockrell
Scott Davis
Tim Dean
Gary Edgington
Tim Ely
Al Hamilton
William (Billy) Hussey
E. L. Jarrell
Dan Lamphear

Leonard R. "Skip" Layne
L. J. LeBlanc
David Lewis
Neil Matthes
Jim Mitchem
Christine Porter
Michael J. Powers
Wayne Stratton
Marcel Veronneau
Irene Ward

A final note: This book is the result of a collaborative effort involving the production, editorial, and development staff at Pearson Education, Inc., and the National Center for Construction Education and Research. Thanks to all of the dedicated people involved in the many stages of this project.

PARTNERING ASSOCIATIONS

ABC Texas Gulf Coast Chapter
American Fire Sprinkler Association
Associated Builders & Contractors, Inc.
Associated General Contractors of America
Association for Career and Technical Education
Association for Skilled & Technical Sciences
Carolinas AGC, Inc.
Carolinas Electrical Contractors Association
Center for the Improvement of Construction Management and Processes
Construction Industry Institute
Construction Users Roundtable
Design Build Institute of America
Electronic Systems Industry Consortium
Merit Contractors Association of Canada
Metal Building Manufacturers Association
NACE International

National Association of Minority Contractors
National Association of Women in Construction
National Insulation Association
National Ready Mixed Concrete Association
National Systems Contractors Association
National Technical Honor Society
National Utility Contractors Association
NAWIC Education Foundation
North American Crane Bureau
North American Technician Excellence
Painting & Decorating Contractors of America
Portland Cement Association
SkillsUSA
Steel Erectors Association of America
U.S. Army Corps of Engineers
University of Florida
Women Construction Owners & Executives, USA

Contents

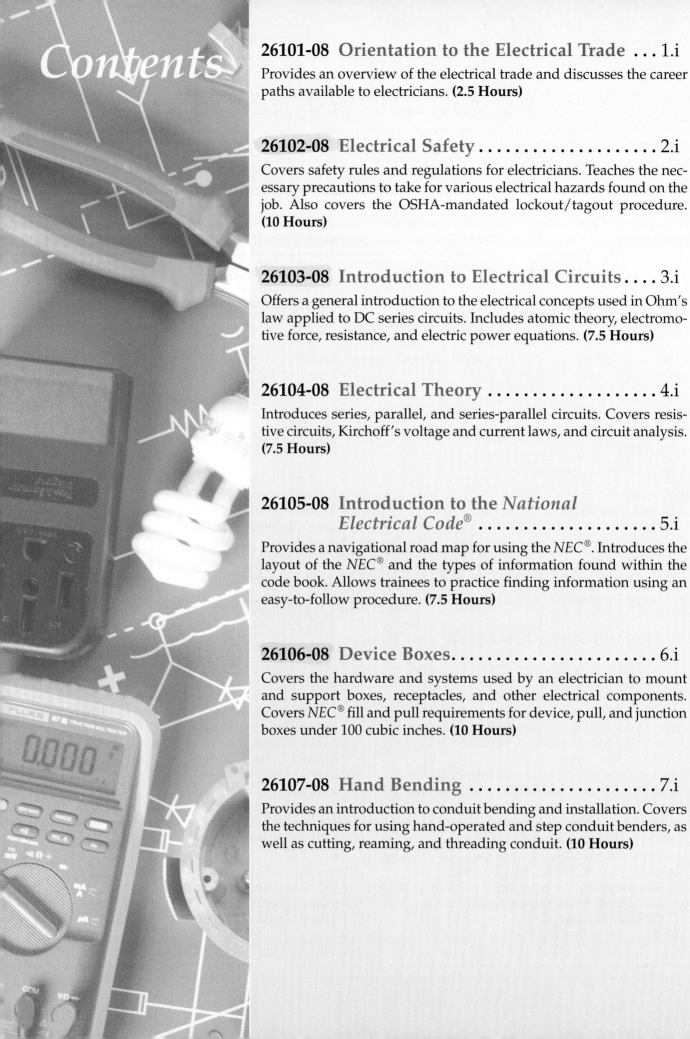

Orientation to the Electrical Trade

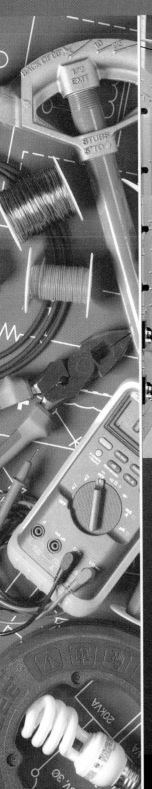

Pet Food Facility

With an extensive safety program and involvement in the project's pre-planning stages, Interstates Companies performed a complete electrical design-build, including all electrical engineering, construction, and automation programming, clocking 69,000 hours without a recordable incident.

26101-08

26101-08
Orientation to the Electrical Trade

Topics to be presented in this module include:

Overview

The electrical trade offers numerous job opportunities in residential, commercial, and industrial construction. Required skills include blueprint reading, job planning, selecting materials, selecting and using the correct tools, installing the components, testing the system, and troubleshooting. Electricians install electrical services, conductors, and fixtures.

Most of the knowledge and skills required for residential electricians are also required for commercial electricians. Commercial electricians require additional skills related to conduit bending, load calculations, and exposure to higher voltage levels.

Industrial electricians must also know how to install various types of conduit, large conductors, motors, and controls. They must possess keen testing and troubleshooting abilities. Maintenance electricians are responsible for keeping the electrical or higher-voltage systems and equipment in productive operating condition.

Objectives

When you have completed this module, you will be able to do the following:

1. Describe the apprenticeship/training process for electricians.
2. Describe various career paths/opportunities one might follow in the electrical trade.
3. Define the various sectors of the electrical industry.
4. State the tasks typically performed by an electrician.
5. Explain the responsibilities and aptitudes of an electrician.

Trade Terms

Electrical service
Occupational Safety and Health Administration (OSHA)
On-the-job training (OJT)
Raceway systems
Rough-in
Trim-out

Required Trainee Materials

1. Paper and pencil
2. Copy of the latest edition of the *National Electrical Code®*

Prerequisites

Before you begin this module, it is recommended that you successfully complete *Core Curriculum*.

This course map shows all of the modules in *Electrical Level One*. The suggested training order begins at the bottom and proceeds up. Skill levels increase as you advance on the course map. The local Training Program Sponsor may adjust the training order.

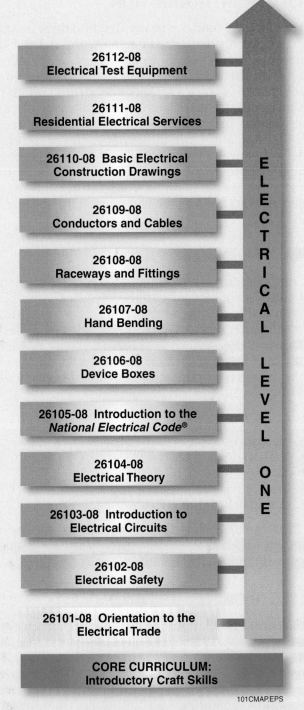

26112-08
Electrical Test Equipment

26111-08
Residential Electrical Services

26110-08 Basic Electrical Construction Drawings

26109-08
Conductors and Cables

26108-08
Raceways and Fittings

26107-08
Hand Bending

26106-08
Device Boxes

26105-08 Introduction to the *National Electrical Code®*

26104-08
Electrical Theory

26103-08 Introduction to Electrical Circuits

26102-08
Electrical Safety

26101-08 Orientation to the Electrical Trade

ELECTRICAL LEVEL ONE

CORE CURRICULUM:
Introductory Craft Skills

101CMAP.EPS

1.0.0 ◆ INTRODUCTION

The electrical field can be divided into three broad categories: residential, commercial, and industrial systems. As you study to become an electrician, you will reach a point at which you must decide which area of electrical work you want to pursue. Many skilled electricians become comfortable in residential or commercial wiring, while others feel at home in large industrial facilities such as petrochemical plants, installing or maintaining huge electrical systems including motors and control devices. Later on in your career, you may decide to start your own company or teach others the trade.

1.1.0 Residential Wiring

Components of a residential electrical system include an electrical supply, electrical service, nonmetallic-sheathed cable, nail-on device boxes, panelboards, and fixtures. Phases of residential electrical wiring include rough-in, trim-out, testing, and troubleshooting. *Figure 1* shows examples of some primary components of residential wiring, including:

- Pad-mounted transformer
- Electrical service
- Nail-on device box
- Nonmetallic-sheathed cable
- Interior panel (subpanel)
- Luminaire (lighting fixture)

Interior panel enclosures, such as the one shown in *Figure 1(E)*, are typically installed and partially terminated during the rough-in stage.

1.2.0 Commercial Wiring

Electrical installations in commercial structures contain many of the same elements as residential installations. One major exception, however, is that in commercial and industrial electrical installations, conductors are typically installed in metal raceways, requiring the installing electricians to be skilled in conduit bending. A well-trained electrician has the ability to install a metal raceway system with little or no waste in conduit, while an inexperienced beginner will typically go through several pieces of conduit before acquiring the necessary bend. Practice makes perfect in conduit bending.

Figure 2 shows some elements that make up a commercial electrical system. These include a pad-mounted commercial transformer, the electrical service, conduit, a fire alarm system, and lighting.

1.3.0 Industrial Wiring

Because of the hazardous materials that exist in many industrial facilities, the installation and maintenance of electrical systems in these volatile environments must follow rigid requirements as governed by the *National Electrical Code® (NEC®)*. For similar reasons, commercial and residential installations also have strict code requirements that must be followed.

Conduit systems in volatile environments must be sealed to outside vapors and gases, and any potential sparking or arcing device must be enclosed within a special enclosure or casing to prevent the ignition of hazardous vapors that might be present.

Industrial electricians are generally split into two groups: installers and maintenance personnel. In many large industrial facilities, electrical systems are typically installed by contract electricians who do not work directly for the facility in which they are working, but for a contractor hired by the facility. It is the responsibility of these electricians to install conduit systems, conductors, motors, and equipment. They then turn the completed system over to maintenance electricians who work directly for the facility. These electricians maintain the system once it is energized and operating. In smaller facilities, industrial plant electricians may both install and maintain the electrical equipment.

Figure 3 illustrates some of the electrical equipment that may be found in industrial facilities, including distribution switchgear, rigid metallic conduit (RMC), and a motor control center.

(A) PAD-MOUNTED TRANSFORMER

(B) RESIDENTIAL ELECTRICAL SERVICE

(C) DEVICE BOX

(D) NONMETALLIC-SHEATHED CABLE

(E) INTERIOR PANEL (SUBPANEL)

(F) LUMINAIRE

101F01.EPS

Figure 1 ◆ Primary components of residential wiring.

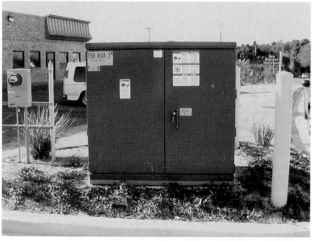

(A) PAD-MOUNTED COMMERCIAL TRANSFORMER

(B) COMMERCIAL ELECTRICAL SERVICE

(C) CONDUIT SYSTEM

(D) FIRE ALARM SYSTEM

(E) OFFICE LIGHTING

(F) OUTDOOR LIGHTING

101F02.EPS

Figure 2 ◆ Commercial electrical system.

(A) DISTRIBUTION SWITCHGEAR

(B) RIGID CONDUIT SYSTEM

(C) MOTOR CONTROL CENTER

101F03.EPS

Figure 3 ◆ Typical industrial electrical equipment.

2.0.0 ◆ CAREER OPPORTUNITIES IN THE ELECTRICAL FIELD

Try to visualize anything operating without electricity. We live in a world of electrical dependency, with most of us taking its availability for granted. We have all experienced the unexpected power failures and power outages due to weather conditions, brownouts, or even blackouts.

It takes a small army of electrically skilled individuals to generate, transmit, distribute, and maintain electrical systems and equipment in order for us to have the convenience of continuous and quality electrical energy at our fingertips. Examples of electrical occupations include residential electrician, commercial electrician, industrial electrician, and electrical maintenance technician.

2.1.0 What You Can Do with Electrical Training and Experience

Wherever you might go, there is a very good chance that you will come in contact with the effects of electricity. Driving through the city and suburbs you see buildings, stores, and signs lighted by electricity. The buildings you see have various kinds of electrical systems installed for cooling, heating, lighting, and for powering all the equipment in the building. This is true of a small home as well as the largest buildings in the world. You will find it difficult to do anything in today's world without having electricity involved. Getting ready for work, cooking, shopping, seeing a movie, surfing the Web, using power tools, and most other things you might do during a typical day require electricity.

Lighting the New York Skyline

The first searchlight on top of the Empire State Building heralded the election of Franklin D. Roosevelt in 1932. A series of floodlights were installed in 1964 to illuminate the top 30 floors of the building. Today, the color of the lights is changed to mark various events. For example, yellow marks the U.S. Open and red, white, and blue marks Independence Day.

The building is lit from the 72nd floor to the base of the TV antenna by 204 metal halide lamps and 310 fluorescent lamps. In 1984, a color-changing apparatus was added in the uppermost mooring mast. There are 880 vertical and 220 horizontal fluorescent lights. The colors can be changed with the flick of a switch.

101SA01.EPS

Once you begin to look, you will see the effects of electricity everywhere. All of the electrical systems in the United States and all over the world must be installed and maintained by someone qualified to do the work. That means that you have a very big opportunity for work that is rewarding both personally and financially.

Growth in the industry, new technology, upgrading and retrofitting existing equipment, and retirement of current workers create opportunities for trained and skilled electricians.

The demand for skilled electricians is high. There is a huge amount of construction work underway. New homes, schools, office buildings,

malls, airports, industrial plants, and many other types of structures are being constructed every day. According to the Bureau of Labor Statistics, this new construction creates a very large demand for new workers every year.

2.1.1 Earn While You Learn

While you are learning the trade, typically through an **on-the-job (OJT)** training program, you are making money. You are getting paid to learn the trade. Compare that with a typical college student who may be paying a large tuition bill while attending class. Another advantage of learning on the job is the hands-on experience you get. There is no substitute in any job for hands-on experience.

Pay in the construction industry is very good, and the pay for electricians is close to the top of the scale for all occupations.

2.1.2 A Diversified and Interesting Career

There are many ways to increase your skills and grow professionally in construction. There are many opportunities to try different types of jobs and earn more money within the electrical trade. You will find that electrical work is demanding but fulfilling. There is a large variety of work to be done. You may be indoors installing boxes or hooking up motor controllers, or you may be outside climbing a ladder or running conduit on a pipe rack high in the air.

Electricity and electrical equipment is needed everywhere. This means that jobs for electricians are also needed everywhere.

2.2.0 Residential Electrician

The primary goal of a residential electrician is to provide a complete electrical system in a residential structure. Elements of a residential wiring installation include installing the electrical service entrance equipment, branch circuit conductors, device boxes, panel enclosures, overcurrent protective devices (circuit breakers), and the fixtures such as lighting and smoke detectors.

Residential electrical contractors are the primary employers of residential electricians. Various methods of employment are available, depending

on the policies of the contractor. Residential electricians may work directly for the contractor, or may be hired as individual contractors who are responsible for filing their own taxes and providing their own insurance, as well as supplying their own equipment. These types of electricians are generally paid by piece work, meaning that they are paid a fixed price for each house they complete. Tract homes, similar to those shown in *Figure 4*, are frequently wired by residential electricians.

2.3.0 Commercial Electrician

Commercial electricians install power, light, and control wiring in a variety of locations including apartment buildings, stores, offices, service stations, and hospitals.

Employers of commercial electricians are generally electrical contractors who work as subcontractors for a general contractor. Many electrical contractors install both residential and commercial wiring, such as in metal frame commercial buildings, as shown in *Figure 5*. However, they often rely on electricians that are specialists in one or the other type of wiring.

2.4.0 Industrial Electrician

Electricians who specialize in installing electrical systems in industrial facilities require additional training due to the amount of specialty equipment that must be installed and tested. Electricians working in hazardous locations must understand the special code requirements associated with these locations. They must differentiate between

101F05.EPS

Figure 5 ◆ Metal frame commercial building.

the hazardous classes and divisions, and know the requirements for each class and division. They must be familiar with three-phase power, motors, and motor control systems. Industrial electricians may also be responsible for installing the conduit and wiring for process control instrumentation, such as the installation shown in *Figure 6*. They must be able to troubleshoot any of these systems should they fail during initial testing.

Contractors who specialize in designing and building industrial facilities are typically the employers of industrial electricians.

101F04.EPS

Figure 4 ◆ Tract homes (subdivisions).

Figure 6 ◆ Instrumentation installation.

101F06.EPS

101F07.EPS

Figure 7 ◆ Magnetic motor starter.

2.5.0 Electrical Maintenance Technician

Electrical maintenance technicians can be found in commercial and industrial facilities. They normally work directly for the owner or management of the facility, and in large facilities are members of a maintenance group supervised by a maintenance manager or supervisor.

In industrial facilities, maintenance electricians are frequently responsible for both the electrical and instrumentation systems and equipment, and are referred to as E&I technicians. Instrumentation expertise is a trade of its own, and requires additional training over and above the electrical skills training found in this course.

Electrical maintenance electricians are usually employees of the facility; however, there are contract maintenance groups that provide maintenance personnel who work side-by-side with full-time plant personnel in the maintenance of the electrical and instrumentation systems.

One common component that all electrical maintenance personnel must be familiar with in industrial facilities is the magnetic motor starter, as illustrated in *Figure 7*. Industrial maintenance electricians should be able to disassemble and reassemble, troubleshoot, and repair magnetic motor starters, as these components frequently fail in these environments.

3.0.0 ◆ YOUR TRAINING PROGRAM

The Department of Labor's Office of Apprenticeship sets the minimum standards for training programs across the country. These programs rely on mandatory classroom instruction and on-the-job training (OJT). They require at least 144 hours of classroom instruction per year and 2,000 hours of OJT per year. In a typical electrical apprenticeship program, trainees spend a total of at least 576 hours in classroom instruction and 8,000 hours in OJT before receiving journeyman certificates issued by registered apprenticeship programs.

To address the training needs of the professional communities, the National Center for Construction Education and Research (NCCER) developed a four-year electrical training program. NCCER uses the minimum Department of Labor standards as a foundation for comprehensive curricula that provide trainees with in-depth classroom and OJT experience.

This NCCER curriculum provides trainees with industry-driven training and education. It adopts a purely competency-based teaching philosophy. This means that trainees must demonstrate to the instructor that they possess the understanding and the skills necessary to perform the hands-on tasks that are covered in each module before they can advance to the next stage of the curriculum.

When the instructor is satisfied that a trainee has successfully demonstrated the required knowledge and skills for a particular module, that information is sent to NCCER and kept in the National Registry. The National Registry can then confirm training and skills for workers as they move from state to state, company to company, or even within a company (see the *Appendix* for samples of NCCER credential documents).

Whether you enroll in an NCCER program or another apprenticeship program, make sure that you work for an employer or sponsor who supports a nationally standardized training program that includes credentials to confirm your skill development.

3.1.0 Apprenticeship

Apprentice training goes back thousands of years, and its basic principles have not changed over time. First, it is a means for individuals entering the craft to learn from those who have mastered the craft. Second, it focuses on learning by doing; real skills versus theory. Although some theory is presented in the classroom, it is always presented in a way that helps the trainee understand the purpose behind the skill that is to be learned.

3.1.1 Apprenticeship Standards

All apprenticeship standards prescribe certain work-related or on-the-job training. This on-the-job training is broken down into specific tasks in which the apprentice receives hands-on training. In addition, a specified number of hours is required in each task. The total number of on-the-job training hours for the electrical apprenticeship program is traditionally 8,000, which amounts to four years of training. In a competency-based program, it may be possible to shorten this time by testing out of specific tasks through a series of performance exams.

In a traditional program, the required on-the-job training may be acquired in increments of 2,000 hours per year.

The apprentice must log all work time and turn it in to the apprenticeship committee so that accurate time control can be maintained. After each 1,000 hours of related work, the apprentice will receive a pay increase as prescribed by the apprenticeship standards.

Informal on-the-job training provided by employers is usually less thorough than that provided through a formal apprenticeship program. The degree of training and supervision in this type of program often depends on the size of the employing firm. A small contractor may provide training in only one area, while a large company may be able to provide training in several areas.

For those entering an apprenticeship program, a high school or technical school education is desirable, as are courses in shop, mechanical drawing, and general mathematics. Manual dexterity, good physical condition, and quick reflexes are important. The ability to solve problems quickly and accurately and to work closely with others is essential. You must also have a high concern for safety.

The prospective apprentice must submit certain information to the apprenticeship committee. This may include the following:

- Aptitude test (General Aptitude Test Battery or GATB Form Test) results (usually administered by the local Employment Security Commission)
- Proof of educational background (candidate should have school transcripts sent to the committee)
- Letters of reference from past employers and friends
- Proof of age
- If the candidate is a veteran, a copy of Form DD214
- A record of technical training received that relates to the construction industry and/or a record of any pre-apprenticeship training

NOTE

Some companies have physical activity requirements that must be met by apprentices. These requirements vary from company to company.

The apprentice must do the following:

- Wear proper safety equipment on the job
- Purchase and maintain tools of the trade as needed and required by the contractor
- Submit a monthly on-the-job training report to the committee
- Report to the committee if a change in employment status occurs
- Attend classroom-related instruction and adhere to all classroom regulations such as attendance requirements

3.1.2 Youth Apprenticeship Program

A Youth Apprenticeship Program is also available that allows students to begin their apprentice training while still in high school. A student entering the program in eleventh grade may complete as much as two years of the NCCER Standardized Craft Training four-year program by high school graduation. In addition, the program, in cooperation with local craft employers, allows students to work in the trade and earn money while still in school. Upon graduation, the student can enter the industry at a higher level and with more pay than someone just starting the apprenticeship program.

This training program is similar to the one used by NCCER learning centers, contractors, and colleges across the country. Students are recognized through official transcripts and can enter the next year of the program wherever it is offered. They may also have the option of applying the credits at a two-year or four-year college that offers degree or certification programs in the construction trades.

3.1.3 Licensing

After you complete your training, you will probably want to take your state or local licensing exam. The purpose of licensing is to provide assurance that you are qualified to install and/or maintain electrical systems. You will be able to work independently and earn a higher income. As a licensed electrician, you are not only responsible for your work, but are liable for that work as well. If someone is working for you, then you are also responsible and liable for that person's work.

Licensing requirements vary from state to state, and may also vary by municipality. Contact your local building department for the requirements in your area. After you receive your license, your state or locality may require continuing education in order to renew your license.

4.0.0 ◆ RESPONSIBILITIES OF THE EMPLOYEE

In order to be successful, you must be able to use current trade materials, tools, and equipment to finish the task quickly and efficiently. You must keep abreast of technical advancements and continually gain the skills to use them. A professional never takes chances with regard to personal safety or the safety of others.

4.1.0 Professionalism

The word professionalism is a broad term that describes the desired overall behavior and attitude expected in the workplace. Professionalism

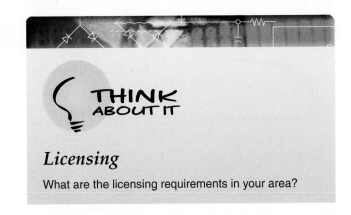

THINK ABOUT IT

Licensing

What are the licensing requirements in your area?

is too often absent from the construction site and the various trades. Most people would argue that professionalism must start at the top in order to be successful. It is true that management support of professionalism is important to its success in the workplace, but it is just as important that individuals recognize their own responsibility for professionalism.

Professionalism includes honesty, productivity, safety, civility, cooperation, teamwork, clear and concise communication, being on time, and coming prepared to work. It can be demonstrated in a variety of ways every minute you are in the workplace.

Professionalism is a benefit to both the employer and the employee. It is a personal responsibility. Our industry is what each individual chooses to make of it; choose professionalism and the industry image will follow.

4.2.0 Honesty

Honesty and personal integrity are important traits of the successful professional. Professionals pride themselves on performing a job well, and on being punctual and dependable. Each job is completed in a professional way, never by cutting corners or reducing materials. A valued professional maintains work attitudes and ethics that protect property such as tools and materials belonging to employers, customers, and other trades from damage or theft at the shop or job site.

Honesty and success go hand-in-hand for both the employer and the professional electrician. It is not simply a choice between good and bad, but a choice between success and failure. Dishonesty will always catch up with you. Whether you are stealing materials, tools, or equipment from the job site or simply lying about your work, it will not take long for your employer to find out. Of course, you can always go and find another employer, but this option will ultimately run out on you.

If you plan to be successful and enjoy continuous employment, consistency of earnings, and

Electricians Are Key Players for the NBA and NHL

Electricians are critical players in building modern sports complexes for professional sports franchises. The Pepsi Center was built in Denver, Colorado in 1999 for the NBA's Denver Nuggets, the NHL's Colorado Avalanche, and several other teams. During the two-year construction cycle, electricians laid over 120 miles of conduit and 569 miles of wire, and installed 13,000 fixtures and 280 panels. The lighting systems are computer controlled and can be preset for basketball games and concerts. In addition to quality sound for concerts, the state-of-the-art sound and security system includes CCTV monitoring and card access security controls.

101SA02.EPS

being sought after as opposed to seeking employment, then start out with the basic understanding of honesty in the workplace and you will reap the benefits.

Honesty means more, however, than just not taking things that do not belong to you. It also means giving a fair day's work for a fair day's pay. Employers place a high value on employees who display honesty.

4.3.0 Loyalty

Employees expect employers to look out for their interests, to provide them with steady employment, and to promote them to better jobs as openings occur. Employers feel that they, too, have a right to expect their employees to be loyal to them—to keep their interests in mind, to speak well of them to others, to keep any minor troubles strictly within the plant or office, and to keep absolutely confidential all matters that pertain to the business. Both employers and employees should keep in mind that

loyalty is not something to be demanded; rather, it is something to be earned.

4.4.0 Willingness to Learn

Every company and job site has its own way of doing things. Employers expect their workers to be willing to learn these ways. You must be willing to adapt to change and learn new methods and procedures as quickly as possible. Sometimes, a change in safety regulations or the purchase of new equipment makes it necessary for even experienced employees to learn new methods and operations. Successful people take every opportunity to learn more about their trade.

4.5.0 Willingness to Take Responsibility

Most employers expect their employees to see what needs to be done, then go ahead and do it. Once an assignment is received and the procedure and safety guidelines are fully understood, you

should assume the responsibility for that task without further reminders.

4.6.0 Willingness to Cooperate

To cooperate means to work together. In our modern business world, cooperation is the key to getting things done. Learn to work as a member of a team with your employer, supervisor, and fellow workers in a common effort to get the work done efficiently, safely, and on time.

4.7.0 Rules and Regulations

People can work together well only if there is some understanding about what work is to be done, when and how it will be done, and who will do it. Rules and regulations are a necessity in any work situation and must be followed by all employees.

4.8.0 Tardiness and Absenteeism

Tardiness means being late for work, and absenteeism means being off the job for one reason or another. Consistent tardiness and frequent absences are an indication of poor work habits, unprofessional conduct, and a lack of commitment.

Although workers don't get paid when they are absent or tardy, there is still a cost to the employer. For example, the worker's health care insurance must still be paid, even though the worker is not

on site. Also, jobs are bid and scheduled based on a certain work force size. If you are not there, work is not getting done and schedules are not being met. It is important for you to be at work, on time, every day.

If it is necessary to stay home, phone the office early in the morning so that your supervisor can find another worker for the day.

5.0.0 ◆ RESPONSIBILITIES OF THE EMPLOYER

Just as the employee has responsibilities on the job, the employer also has responsibilities. These are set out in the *Occupational Safety and Health Act* of 1970. The job of the **Occupational Safety and Health Administration (OSHA)** is to set occupational safety and health standards for all places of employment, enforce these standards, ensure that employers provide and maintain a safe workplace for all employees, and provide research and educational programs to support safe working practices.

OSHA was adopted with the stated purpose to assure as far as possible every working man and woman in the nation safe and healthful working conditions and to preserve our human resources.

OSHA requires each employer to provide a safe and hazard-free working environment. OSHA also requires that employees comply with OSHA rules and regulations that relate to their conduct on the job. To gain compliance, OSHA can perform spot inspections of job sites, impose fines for violations, and even stop any more work from proceeding until the job site is safe.

According to OSHA standards, you are entitled to on-the-job safety training. Your employer must do the following:

• Show you how to do each job safely
• Provide you with the required personal protective equipment
• Warn you about specific hazards
• Supervise you for safety while performing the work

The enforcement for this act of Congress is provided by the federal and state safety inspectors, who have the legal authority to impose fines for safety violations. The law allows states to have their own safety regulations and agencies to enforce them, but they must first be approved by the U.S. Secretary of Labor. For states that do not develop such regulations and agencies, federal OSHA standards must be obeyed.

These standards are listed in *OSHA Safety and Health Standards for the Construction Industry (29 CFR, Part 1926)*, sometimes called *OSHA Standards 1926*. Other safety standards that apply to construction are published in *OSHA Safety and Health Standards for General Industry (29 CFR, Parts 1900 to 1910)*, in *NFPA 70E*, and the *NEC®*.

The most important general requirements that OSHA places on employers in the construction industry are as follows:

• The employer must post, in an easily seen area, signs informing employees of their rights and responsibilities.
• The employer must ensure that there are no serious hazards on the job site, and make sure that the workplace is in compliance with OSHA rules and regulations.
• Warning signs, posters, and labels must be posted in all required areas.
• The employer must perform frequent and regular job site inspections of equipment.
• The employer must instruct all employees to recognize and avoid unsafe conditions, and to know the regulations that pertain to the job so they may control or eliminate any hazards.
• No one may use any tools, equipment, machines, or materials that do not comply with *OSHA Standards 1926*.
• The employer must ensure that only qualified individuals operate tools, equipment, and machines.
• The employer must provide medical training and examinations when required by OSHA.
• Employers with greater than 10 employees must keep records of work-related injuries and illnesses. These records must be available to employees.
• Employers must not discriminate against employees who are exercising their rights under OSHA regulations.

Additional employer responsibilities are described in the *Americans with Disabilities Act of 1990 (ADA)*. If a worker has a disability but is qualified to do the job, that worker has equal rights to employment. The U.S. Equal Employment Opportunity Commission, along with state and local civil rights agencies, enforce ADA regulations.

6.0.0 ◆ SAFETY

In exchange for the benefits of your employment and your own well-being, you are obligated to work safely. You are also obligated to make sure anyone you supervise or work with is working safely. Your employer is obligated to maintain a safe workplace for all employees. Safety is everyone's responsibility.

You have a responsibility to maintain a safe working environment. This means two things:

- Follow your company's rules for proper working procedures and practices.
- Report any unsafe equipment and conditions directly to your supervisor.

On the job, if you see something that is not safe, report it! Do not ignore it. It will not correct itself. In the long run, even if you do not think an unsafe condition affects you, it does. Always report unsafe conditions. Do not think your employer will be angry because your productivity suffers while the condition is being reported. On the contrary, your employer will be more likely to criticize you for not reporting a problem.

Your employer knows that the short time lost in making conditions safe again is nothing compared with shutting down the whole job because of a major disaster. If that happens, you are out of work anyway. In fact, OSHA regulations require you to report hazardous conditions.

This applies to every part of the construction industry. Whether you work for a large contractor or a small contractor, you are obligated to report unsafe conditions.

In addition to the OSHA standards, there are specific standards related to electrical systems and devices. The *National Electrical Code® (NEC®)* sets the minimum standards for the safe installation of electrical systems. You will become very familiar with the *NEC®* as you progress through your training. *NEC®* temporary power requirements are more stringent than OSHA standards and can be enforced by the inspector and OSHA.

Another standard you should become familiar with is *NFPA 70E, Standard for Electrical Safety in the Workplace.* This standard covers safe work practices that must be applied when working on or near exposed energized parts, and describes in detail the steps required for putting a circuit or electrical system into an electrically safe working condition. It also covers safe approach distances to exposed energized parts, and introduces the little understood but very dangerous reality of arc flash hazard. Proper personal protective equipment (PPE) required to protect oneself from both the hazard of electrical shock and the hazard of arc flash is extensively covered. *Figure 8* shows a higher level of PPE (arc flash hood) that will be required when exposed to a potentially high level of arc flash.

101F08.EPS

Figure 8 ◆ Arc flash hood.

Electricians Help Reduce Energy Consumption

GOING GREEN

Electricians are working to protect the environment and reduce energy consumption. Some companies specialize in energy efficiency in new installations and upgrading older facilities. One company assisted the World Bank in Washington, DC to upgrade their offices and reduced annual energy consumption by 4,500kW.

The World Bank offices occupy over 3.5 million square feet in five separate buildings. Over 50,000 fluorescent light fixtures and 1,400 exit signs were upgraded to improve energy efficiency and reduce lamp maintenance. This resulted in an annual energy savings of over $800,000 and eliminates the annual release of over 7,000,000 pounds of oxides, which some people believe causes global climate change and acid rain.

1. Interior panels in residential wiring are typically installed during which phase of work?
   a. Rough-in
   b. Trim-out
   c. Service installation
   d. Planning

2. A residential wiring system typically uses _____ cable.
   a. medium-voltage
   b. tray
   c. flat conductor
   d. nonmetallic-sheathed CABLE

3. Commercial wiring is typically installed in _____.
   a. pairs
   b. air ducting
   c. metal raceways
   d. PVC raceways

4. Which of the following would most likely require special knowledge of hazardous locations?
   a. Residential wiring
   b. Industrial wiring
   c. Commercial wiring
   d. Service work

5. RMC is a type of _____.
   a. electrical service
   b. conduit RIGID METAL CONDUIT
   c. transformer
   d. motor control

6. Minimum standards for apprentice training programs are established by _____.
   a. OSHA
   b. the DOL
   c. NCCER
   d. your employer

7. Once you have finished your apprenticeship, your training is over.
   a. True
   b. False

8. It is okay to be a little late for work as long as you make up the time.
   a. True
   b. False

9. The primary mission of OSHA is to _____.
   a. inspect job sites for safety violations
   b. fine companies that violate safety regulations
   c. distribute safety equipment to workers
   d. ensure that employers maintain a safe workplace

10. If you see a safety violation at your job site, you should ignore it unless it affects you directly.
    a. True
    b. False

Summary

The electrical trade can be divided into three main areas of expertise: residential, commercial, and industrial. Each of these areas requires specific skills and training.

Regardless of the area of expertise, all electricians must understand the fundamentals of electricity, the hazards associated with electrical shock and arc flash, and the practices that must be applied in order to work safely.

Job opportunities in the electrical field can provide an above-average income and a rewarding yet challenging career with many levels of advancement possible.

Notes

Trade Terms Quiz

1. _____OSHA_____ is the federal government agency established to ensure a safe and healthy environment in the workplace.

2. Job-related learning acquired while working is known as _____OJT_____ .

3. The main panelboard enclosure would probably be installed in the _ROUGH-IN_ stage.

4. The _ELECTRICAL SERV_ connects the commercial power to the premises wiring system.

5. Devices and fixtures would be installed during _TRIM-OUT_ .

6. Conduit is part of the _RACEWAY SYS_

Trade Terms

Electrical service
Occupational Safety and Health Administration (OSHA)
On-the-job training (OJT)
Raceway systems
Rough-in
Trim-out

Jared Garber

Four-Time Gold Medal Winner: State SkillsUSA, National SkillsUSA, ABC State, and ABC National Craft Training Competitions

Jared Garber is a winner. At the age of 20, with less than four years under his belt as an electrician, he's already worked his way up to Assistant Superintendent and owns his own home. He knew from an early age that he wanted a career in the electrical industry, and he didn't wait for it to come to him. He started working toward it in his early teens and received the education and training he needed to succeed. In his junior year of high school, Jared won both the State and National SkillsUSA Electrical Competitions. Two years later, he won the ABC State and National Craft Training Competitions.

How did you choose a career in the electrical field?
I job-shadowed my cousin in eighth grade. He was an electrician at a large performing arts center. The work seemed interesting and challenging, so I decided to pursue it.

What types of training have you been through?
I went through two years of training in my junior and senior years of high school, and then continued my education after I started working. I just finished up my fourth year of ABC training.

Tell us about your work experience.
I started out working for a small contractor. I worked there for about a year before I joined Beacon Electric Company. We do all types of work, from residential to large commercial projects. Right now, I'm heading up a 1.5-million-dollar project at a new elementary school.

What was your favorite part of the ABC National Competition?
My company paid to send my whole family to Las Vegas for the competition. We really enjoyed the trip. Winning was great—I received a cash prize and tools. My company surprised me with a pay raise when I won the state competition and then another when I won the national competition.

What factors have contributed most to your success?
Attention to detail, pride in my work, determination, and a work ethic instilled in me by my parents at an early age. I was raised on a 1,500-acre grain farm, and

I still work there in the evenings after I finish my regular job. On a farm, there are no days off and no excuses. I guess that work ethic has stayed with me.

Tell us about winning the ABC National Craft Training Competition.
You have to complete several commercial and residential electrical tasks. I started out with the commercial work, and for some reason, I couldn't get the start-stop switch to work. I spent too much time on it, and it threw me off. By lunchtime, when I should have had half the job done, I only had about a quarter of the job done. I could have given up, but I didn't. You were allowed one person to have lunch with you and I invited my fiancée Rachel, who calmed me down and encouraged me. That got me back on track. After lunch, I skipped the rest of the commercial part and knocked off the residential part, then went back to the commercial side and did everything else, and finally went back and tackled the start-stop switch. It was good knowing that I had gotten everything else out of the way. I finished with 45 minutes to spare. Later on, I found out that I was the only one who ever finished the competition and all of the devices worked, and I also had the best overall hands-on score ever. That was a great feeling.

What advice would you give to those new to the electrical field?
Pay attention to experienced electricians and watch their technique. You can learn a lot from them. Prioritize your work every day.

Trade Terms
Introduced in This Module

Electrical service: The electrical components that are used to connect the commercial power to the premises wiring system.

Occupational Safety and Health Administration (OSHA): The federal government agency established to ensure a safe and healthy environment in the workplace.

On-the-job training (OJT): Job-related learning acquired while working.

Raceway systems: Conduit, fittings, boxes, and enclosures that house the conductors in an electrical system.

Rough-in: The installation of the raceway system (including conduit, boxes, and enclosures), wiring, or cable.

Trim-out: After rough-in, the installation and termination of devices and fixtures.

Samples of NCCER
Apprentice Training Recognition

NATIONAL CENTER FOR CONSTRUCTION EDUCATION AND RESEARCH
BUILDING TOMORROW'S WORKFORCE

October 5, 2007

Student4 Sample
NCCER
3600 NW 43rd St Bldg G
Gainesville, FL 32614-1104

Dear Student4,

On behalf of the National Center for Construction Education and Research, I congratulate you for successfully completing NCCER's Contren® Learning Series program. I also congratulate you for choosing construction as a career.

You are now a valuable member of one of our nation's largest industries. The skills you have acquired will not only enhance your career opportunities, but will help build America.

Enclosed are your credentials from the National Registry. These industry-recognized credentials give you flexibility in planning your career and ensure your achievements follow you wherever you go.

To access your training accomplishments through the Automated National Registry, follow these instructions:
1. Go to www.nccer-anr.org.
2. Click the "Individuals" button.
3. Enter the NCCER card number, located on front of your wallet card or transcript, and your PIN.
    Note: The default PIN is the last four digits of your SSN. You may change your PIN after you login.
4. First-time users will be directed to answer a few security questions upon initial login.
5. Contact the registry department with any questions.

NCCER applauds your dedication and wishes you the best in your future endeavors.

Sincerely,

Donald E. Whyte

Donald E. Whyte
President, NCCER

Enc.

3600 NW 43rd St, Bldg G ○ Gainesville, FL 32606 ○ P 352.334.0911 ○ F 352.334.0932 ○ www.nccer.org
Affiliated with the University of Florida

101A01.EPS

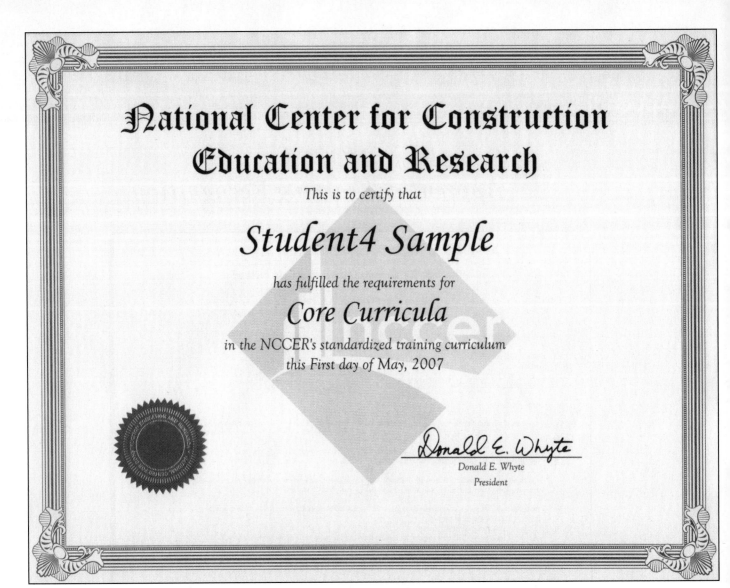

National Center for Construction Education and Research

This is to certify that

Student4 Sample

has fulfilled the requirements for

Core Curricula

in the NCCER's standardized training curriculum
this First day of May, 2007

Donald E. Whyte

Donald E. Whyte
President

101A02.EPS

NATIONAL CENTER FOR CONSTRUCTION EDUCATION AND RESEARCH

BUILDING TOMORROW'S WORKFORCE

3600 NW 43rd St, Bldg G ∘ Gainesville, FL 32606
P 352.334.0911 ∘ F 352.334.0932 ∘ www.nccer.org

Affiliated with the University of Florida

October 5, 2007

Official Transcript

Student4 Sample
NCCER
3600 NW 43rd St Bldg G
Gainesville. FL 32614-1104

Current Employer:

Card #: 4671443

Course / Description	Instructor	Training Location	Date Compl.
00101-04 Basic Safety	Don D Whyte	NCCER	5/1/2007
00102-04 Introduction to Construction Math	Don D Whyte	NCCER	5/1/2007
00103-04 Introduction to Hand Tools	Don D Whyte	NCCER	5/1/2007
00104-04 Introduction to Power Tools	Don D Whyte	NCCER	5/1/2007
00105-04 Introduction to Blueprints	Don D Whyte	NCCER	5/1/2007
00106-04 Basic Rigging	Don D Whyte	NCCER	5/1/2007

NO ENTRIES BELOW THIS LINE

Page 1

Donald E. Whyte
President, NCCER

101A03.EPS

To access the Automated National Registry (ANR), visit

http://www.nccer-anr.org

enter the number on the front of this card.

If found, drop in mailbox. Postage paid by
NCCER
3600 NW 43rd Street Building G
Gainesville FL 32606

monster·
today's the day·

Build your success... find jobs at
http://nccer.monster.com

21890

NATIONAL CENTER
FOR CONSTRUCTION
EDUCATION AND RESEARCH

nccer

Sample Student

2781481

101A04.EPS

Additional Resources

This module is intended to present thorough resources for task training. The following reference work is suggested for further study. This is optional material for continued education rather than for task training.

National Electrical Code® Handbook, Latest Edition. Quincy, MA: National Fire Protection Association.

NCCER makes every effort to keep these textbooks up-to-date and free of technical errors. We appreciate your help in this process. If you have an idea for improving this textbook, or if you find an error, a typographical mistake, or an inaccuracy in NCCER's Contren® textbooks, please write us, using this form or a photocopy. Be sure to include the exact module number, page number, a detailed description, and the correction, if applicable. Your input will be brought to the attention of the Technical Review Committee. Thank you for your assistance.

Instructors – If you found that additional materials were necessary in order to teach this module effectively, please let us know so that we may include them in the Equipment/Materials list in the Annotated Instructor's Guide.

Write: Product Development and Revision
National Center for Construction Education and Research
3600 NW 43rd St., Bldg. G, Gainesville, FL 32606

Fax: 352-334-0932

E-mail: curriculum@nccer.org

Craft _____ Module Name _____

Copyright Date _____ Module Number _____ Page Number(s) _____

Description _____

_____

_____

_____

(Optional) Correction _____

_____

_____

(Optional) Your Name and Address _____

_____

_____

Electrical Safety

Ponnequin Wind Farm

The Ponnequin Wind Farm generates electrical power from the wind and is located on the plains of eastern Colorado just south of the Wyoming border. It consists of 44 wind turbines and can generate up to 30 megawatts of electricity. Each turbine weighs nearly 100 tons and stands 181 feet tall.

26102-08

26102-08
Electrical Safety

Overview

Electricians work in all areas of a job site. They are exposed to safety hazards that other workers encounter, including hazards that can cause fall-related injuries, crushing injuries in excavations, electrical shock, injuries from being struck by falling objects, cuts, burns, punctures, chemical exposure, and other injuries. Electricians are exposed to the risk of electrical shock more often than other workers, which puts them at a higher risk for electrical burns and arc burns.

Safety regulations and company policies are designed to protect those working in the electrical field, but these regulations are only effective if the worker recognizes and understands the hazards that may be present and takes the proper precautions to avoid them. For that reason, the proper use of personal protective equipment and other safety gear is a critical element of the electrician's job.

In order to protect yourself and those around you from injury and possible death, you must become familiar with the various hazards on the job site, follow established safety procedures, and always keep safe work practices foremost in your mind.

Note: *National Electrical Code*® and *NEC*® are registered trademarks of the National Fire Protection Association, Inc., Quincy, MA 02269.
All *National Electrical Code*® and *NEC*® references in this module refer to the 2008 edition of the *National Electrical Code*®.

Objectives

When you have completed this module, you will be able to do the following:

1. Recognize safe working practices in the construction environment.
2. Explain the purpose of OSHA and how it promotes safety on the job.
3. Identify electrical hazards and how to avoid or minimize them in the workplace.
4. Explain safety issues concerning lockout/tagout procedures, confined space entry, respiratory protection, and fall protection systems.
5. Develop a task plan and a hazard assessment for a given task and select the appropriate PPE and work methods to safely perform the task.

Trade Terms

Double-insulated/ungrounded tool
Fibrillation
Grounded tool
Ground fault circuit interrupter (GFCI)
Polychlorinated biphenyls (PCBs)

Required Trainee Materials

1. Paper and pencil
2. Copy of the latest edition of the *National Electrical Code®*
3. Appropriate personal protective equipment

Prerequisites

Before you begin this module, it is recommended that you successfully complete *Core Curriculum*; and *Electrical Level One*, Module 26101-08.

This course map shows all of the modules in *Electrical Level One*. The suggested training order begins at the bottom and proceeds up. Skill levels increase as you advance on the course map. The local Training Program Sponsor may adjust the training order.

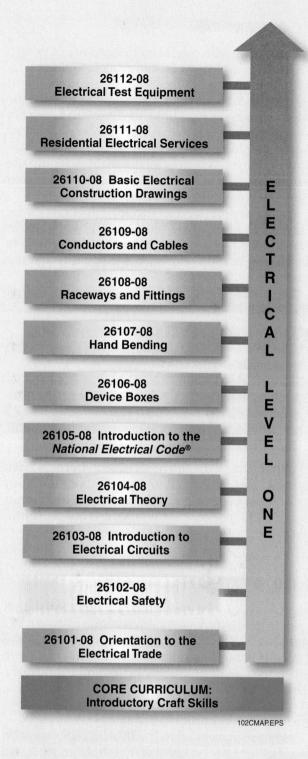

26112-08
Electrical Test Equipment

26111-08
Residential Electrical Services

26110-08 Basic Electrical Construction Drawings

26109-08
Conductors and Cables

26108-08
Raceways and Fittings

26107-08
Hand Bending

26106-08
Device Boxes

26105-08 Introduction to the *National Electrical Code®*

26104-08
Electrical Theory

26103-08 Introduction to Electrical Circuits

26102-08
Electrical Safety

26101-08 Orientation to the Electrical Trade

CORE CURRICULUM:
Introductory Craft Skills

ELECTRICAL LEVEL ONE

102CMAP.EPS

1.0.0 ◆ INTRODUCTION

In order to be safe, you need to be aware of potential hazards and stay constantly alert to these hazards. You must take the proper precautions and practice the basic rules of safety. You must be safety-conscious at all times and report any unsafe conditions to your supervisor and co-workers. Safety should become a habit. Keeping a safe attitude on the job will go a long way in reducing the number and severity of accidents. Remember that your safety is up to you.

As an apprentice electrician, you need to be especially careful. You should only work under the direction of experienced personnel who are familiar with the various job site hazards and the means of avoiding them.

The most life-threatening hazards on a construction site are:

- Falls when you are working in high places
- The possibility of being crushed by falling materials or equipment
- Electric shock and arc-related burns caused by coming into contact with live electrical circuits
- The possibility of being struck by flying objects or moving equipment/vehicles such as trucks, forklifts, and construction equipment

Other hazards include cuts, burns, back sprains, and getting chemicals or objects in your eyes. Most injuries, both those that are life-threatening and those that are less severe, are preventable if the proper precautions are taken.

2.0.0 ◆ ELECTRICAL SHOCK

Electricity can be described as a potential that results in the movement of electrons in a conductor. This movement of electrons is called electrical current. Some substances, such as silver, copper, steel, and aluminum, are excellent conductors. The human body is also a conductor. The conductivity of the human body greatly increases when the skin is wet or moistened with perspiration.

Electrical current flows along any path in which the voltage can overcome the resistance. If the human body contacts an electrically energized point and is also in contact with the ground or another point in the circuit, the human body becomes a path for the current. *Table 1* shows the effects of current passing through the human body. One mA is one milliamp, or one one-thousandth of an ampere.

What's wrong with this picture?

102SA01.EPS

Table 1   Current Level Effects on the Human Body

Current Value	Typical Effects
1mA	Perception level. Slight tingling sensation.
5mA	Slight shock. Involuntary reactions can result in serious injuries such as falls from elevations.
6 to 30mA	Painful shock, loss of muscular control.
50 to 150mA	Extreme pain, respiratory arrest, severe muscular contractions. Death possible.
1000mA to 4300mA	Ventricular fibrillation, severe muscular contractions, nerve damage. Typically results in death.

Source: Occupational Safety and Health Administration

102T01.EPS

INSIDE TRACK

Electrical Safety in the Workplace

Each year in the U.S., there are approximately 20,000 electricity-related accidents at home and in the workplace. In a recent year, these accidents resulted in 700 deaths. Electrical accidents are the third leading cause of death in the workplace.

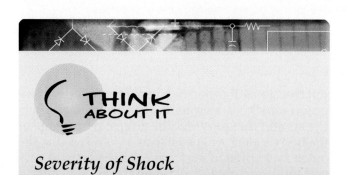

THINK ABOUT IT

Severity of Shock

In *Table 1*, how many milliamps separate a mild shock from a potentially fatal one? What is the fractional equivalent of this in amps? How many amps are drawn by a 60W light bulb?

A primary cause of death from electrical shock is when the heart's rhythm is overcome by an electrical current. Normally, the heart's operation uses a very low-level electrical signal to cause the heart to contract and pump blood. When an abnormal electrical signal, such as current from an electrical shock, reaches the heart, the low-level heartbeat signals are overcome. The heart begins twitching in an irregular manner and goes out of rhythm with the pulse. This twitching is called **fibrillation**. The use of cardiopulmonary resuscitation (CPR) can keep oxygen flowing to the body, but unless the normal heartbeat is restored using special defibrillation equipment (paddles), the individual will die. Other effects of electrical shock may include immediate heart stoppage and burns. In addition, the body's reaction to the shock can cause a fall or other accident. Delayed internal problems can also result. This is why it is critical for you to have a medical exam if you receive even a minor shock.

2.1.0 The Effect of Current

The amount of current measured in amperes that passes through a body determines the outcome of an electrical shock. The higher the voltage, the greater the chance for a fatal shock. In a one-year study in California, the following results were observed by the State Division of Industry Safety:

- Thirty percent of all electrical accidents were caused by contact with conductors. Of these accidents, 66% involved low-voltage conductors (those carrying 600 volts [V] or less).

NOTE

Electric shocks or burns are a major cause of accidents in the construction industry. According to the National Institute for Occupational Safety and Health, workers in the construction industry are four times more likely to be electrocuted at work than all other industries combined.

- Portable, electrically operated hand tools made up the second largest number of injuries (15%). Almost 70% of these injuries happened when the frame or case of the tool became energized. These injuries could have been prevented by following proper safety practices, using properly maintained grounded or **double-insulated/ungrounded tools**, and using **ground fault circuit interrupter (GFCI)** protection.

In one ten-year study, investigators found 9,765 electrical injuries in the U.S. A little more than 13% of the high-voltage injuries (over 600V) resulted in death. These high-voltage totals included limited-amperage contacts, which are often found on electronic equipment. When tools or equipment touch high-voltage overhead lines, the chance that a resulting injury will be fatal climbs to 28%. Of the low-voltage injuries, 1.4% were fatal.

CAUTION

High voltage, defined as 600V or more, is almost ten times as likely to kill as low voltage. However, on the job you spend most of your time working on or near lower voltages. Due to the frequency of contact, most electrocution deaths actually occur at low voltages. Attitude about the harmlessness of lower voltages undoubtedly contributes to this statistic.

These statistics have been included to help you gain respect for the environment where you work and to stress how important safe working habits really are.

2.1.1 Body Resistance

Electricity travels in closed circuits, and its normal route is through a conductor. Shock occurs when the body becomes part of the electric circuit (*Figure 1*). The current must enter the body at one point and leave at another. Shock normally occurs in one of three ways: the person must come in contact with both wires of the electric circuit; one wire of the electric circuit and the ground; or a metallic part that has become live by being in contact with an energized wire while the person is also in contact with the ground.

To fully understand the harm done by electrical shock, we need to understand something about the physiology of certain body parts: the skin, the heart, and muscles.

Skin covers the body and is made up of three layers. The most important layer, as far as electric shock is concerned, is the outer layer of dead cells

referred to as the horny layer. This layer is composed mostly of a protein called keratin, and it is the keratin that provides the largest percentage of the body's electrical resistance. When it is dry, the outer layer of skin may have a resistance of several thousand ohms, but when it is moist, there is a radical drop in resistance, as is also the case if there is a cut or abrasion that pierces the horny layer. The amount of resistance provided by the skin will vary widely from individual to individual. A worker with a thick horny layer will have a much higher resistance than a child. The resistance will also vary widely at different parts of the body. For instance, the worker with high-resistance hands may have low-resistance skin on the back of his calf.

The heart is the pump that sends life-sustaining blood to all parts of the body. The blood flow is caused by the contractions of the heart muscle, which is controlled by electrical impulses. The electrical impulses are delivered by an intricate system of nerve tissue with built-in timing mechanisms, which make the chambers of the heart contract at exactly the right time. An outside electric current of as little as 75 milliamperes can upset the rhythmic, coordinated beating of the heart by disturbing the nerve impulses. When this happens, the heart is said to be in fibrillation, and the pumping action stops. Death will occur quickly if the normal beat is not restored. Remarkable as it may seem, what is needed to defibrillate the heart is a shock of an even higher intensity.

The other muscles of the body are also controlled by electrical impulses delivered by nerves. Electric shock can cause loss of muscular control, resulting in the inability to let go of an electrical conductor. Electric shock can also cause injuries of

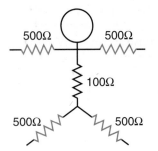

- HAND TO HAND 1000Ω
- 120 VOLT
- FORMULA: I = E/R
- 120/1000 = 0.120 AMPS OR 120 MILLIAMPS

102F01.EPS

Figure 1 ◆ Body resistance.

an indirect nature in which involuntary muscle reaction from the electric shock can cause bruises, fractures, and even death resulting from collisions or falls.

The severity of shock received when a person becomes a part of an electric circuit is affected by three primary factors: the amount of current flowing through the body (measured in amperes), the path of the current through the body, and the length of time the body is in the circuit. Other factors that may affect the severity of the shock are the frequency of the current, the phase of the heart cycle when shock occurs, and the general health of the person prior to the shock. Effects can range from a barely perceptible tingle to immediate cardiac arrest. Although there are no absolute limits, or even known values that show the exact injury at any given amperage range, *Table 1* lists the general effects of electric current on the body for different current levels. As this table illustrates, a difference of only 100 milliamperes exists between a current that is barely perceptible and one that is likely to kill you.

A severe shock can cause considerably more damage to the body than is visible. For example, a person may suffer internal hemorrhages and destruction of tissues, nerves, and muscle. In addition, shock is often only the beginning in a chain of events. The final injury may well be from a fall, cuts, burns, or broken bones.

2.1.2 Burns

The most common shock-related injury is a burn. Burns suffered in electrical accidents may be of three types: electrical burns, arc burns, and thermal contact burns.

Electrical burns are the result of electric current flowing through the tissues or bones. Tissue damage is caused by the heat generated by the current flow through the body. An electrical burn is one of the most serious injuries you can receive, and should be given immediate attention. Since the most severe burning is likely to be internal, what may appear at first to be a small surface wound could, in fact, be an indication of severe internal burns.

Arc burns make up a substantial portion of the injuries from electrical malfunctions. The electric arc between metals can be up to 35,000°F, which is about four times hotter than the surface of the sun. Workers several feet from the source of the arc can receive severe or fatal burns. Since most electrical safety guidelines recommend safe working distances based on shock considerations, workers can be following these guidelines and still be at risk from arc. Electric arcs can occur due to poor

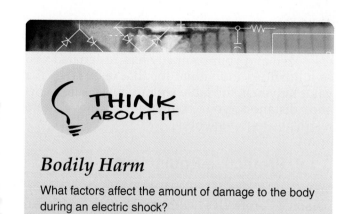

THINK ABOUT IT

Bodily Harm

What factors affect the amount of damage to the body during an electric shock?

electrical contact or failed insulation. Electrical arcing is caused by the passage of substantial amounts of current through the vaporized terminal material (usually metal or carbon).

CAUTION

Since the heat of the arc is dependent on the short circuit current available at the arcing point, arcs generated by low-voltage systems can be just as dangerous as those generated at 13,000V.

The third type of burn is a thermal contact burn. It is caused by contact with objects thrown during the blast associated with an electric arc. This blast comes from the pressure developed by the near-instantaneous heating of the air surrounding the arc, and from the expansion of the metal as it is vaporized. (Copper expands by a factor in excess of 65,000 times in boiling.) The pressure wave can be great enough to hurl people, switchgear, and cabinets considerable distances. It can stop your heart, impale you with shrapnel, blow off limbs, cause deafness, and cause you to inhale vaporized metal. Another hazard associated with the blast is the hurling of molten metal droplets, which can also cause thermal contact burns and associated damage.

3.0.0 ◆ REDUCING YOUR RISK

There are many things that can be done to greatly reduce the chance of receiving an electrical shock. Always comply with your company's safety policy and all applicable rules and regulations, including job site rules. In addition, the Occupational Safety and Health Administration (OSHA) publishes the *Code of Federal Regulations (CFR)*. *CFR Part 1910* covers the OSHA standards for general industry and *CFR Part 1926* covers the OSHA standards for the construction industry.

Do not approach any electrical conductors closer than indicated in *Table 2* unless they are de-energized and your company has designated you as a qualified individual for that task. Also, the values given in the table are minimum safe clearance distances; your company may have more restrictive requirements.

3.1.0 Protective Equipment

You should also become familiar with common personal protective equipment. In particular, know the voltage rating of each piece of equipment. Rubber gloves are used to prevent the skin from coming into contact with energized circuits. A separate leather cover protects the rubber glove from punctures and other damage (see *Figure 2*). The leather protectors also provide the wearer with a certain level of protection against arc flash. OSHA addresses the use of protective equipment, apparel, and tools in *CFR 1910.335(a)*. This article is divided into two sections: *Personal Protective Equipment* and *General Protective Equipment and Tools*.

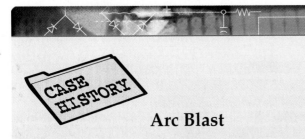

Arc Blast

An electrician in Louisville, KY was cleaning a high-voltage switch cabinet. He removed the padlock securing the switch enclosure and opened the door. He used a voltage meter to verify absence of voltage on the three load phases at the rear of the cabinet. However, he did not test all potentially energized parts within the cabinet, and some components were still energized. He was wearing standard work boots and safety glasses, but was not wearing protective rubber gloves or an arc suit. As he used a paintbrush to clean the switch, an arc blast occurred, which lasted approximately one-sixth of a second. The electrician was knocked down by the blast. He suffered third-degree burns and required skin grafts on his arms and hands. The investigation determined that the blast was caused by debris, such as a cobweb, falling across the open switch.

The Bottom Line: Always wear the appropriate personal protective equipment and test all components for voltage before working in or around electrical devices.

Table 2 Limited Approach Boundaries to Live Parts

Nominal System Voltage Range (Phase-to-Phase)	Limited Approach Boundary
50 to 300	3 ft 6 in
301 to 750	3 ft 6 in
751 to 15kV	5 ft 0 in
15.1kV to 36kV	6 ft 0 in
36.1kV to 46kV	8 ft 0 in
46.1kV to 72.5kV	8 ft 0 in
72.6kV to 121kV	8 ft 0 in
138kV to 145kV	10 ft 0 in
161kV to 169kV	11 ft 8 in
230kV to 242kV	13 ft 0 in
345kV to 362kV	15 ft 4 in
500kV to 550kV	19 ft 0 in
765kV to 800kV	23 ft 9 in

102T02.EPS

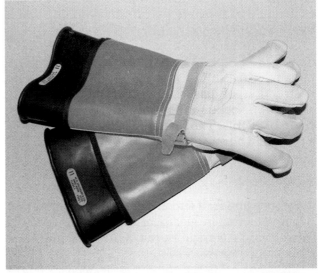

102F02.EPS

Figure 2 ◆ Rubber gloves and leather protectors.

The first section, *Personal Protective Equipment*, includes the following requirements:

- Employees working in areas where there are potential electrical hazards shall be provided with, and shall use, electrical protective equipment that is appropriate for the specific parts of the body to be protected and for the work to be performed.
- Protective equipment shall be maintained in a safe, reliable condition and shall be periodically inspected or tested, as required by *CFR 1910.137/1926.95*.
- If the insulating capability of protective equipment may be subject to damage during use, the insulating material shall be protected.

- Employees shall wear nonconductive head protection wherever there is a danger of head injury from electric shock or burns due to contact with exposed energized parts.
- Employees shall wear protective equipment for the eyes and face wherever there is danger of injury to the eyes or face from electric arcs or flashes or from flying objects resulting from an electrical explosion.

The second section, *General Protective Equipment and Tools,* includes the following requirements:

- When working near exposed energized conductors or circuit parts, each employee shall use insulated tools or handling equipment if the tools or handling equipment might make contact with such conductors or parts. If the insulating capability of insulated tools or handling equipment is subject to damage, the insulating material shall be protected.
- Fuse pullers, insulated for the circuit voltage, shall be used to remove or install fuses.

WARNING!

Fuses should never be installed or removed when energized. This could cause an electrical arc, which can result in death or a serious injury.

- Ropes and handlines used near exposed energized parts shall be nonconductive.
- Protective shields, protective barriers, or insulating materials shall be used to protect each employee from shock, burns, or other electrically related injuries while that employee is working near exposed energized parts that might be accidentally contacted or where dangerous electric heating or arcing might occur. When normally enclosed live parts are exposed for maintenance or repair, they shall be guarded to protect unqualified persons from contact with the live parts.

The types of electrical safety equipment, protective apparel, and protective tools available for use are quite varied. This module will discuss the most common types of safety equipment. These include the following:

- Currently tested rubber protective equipment, including gloves and blankets
- Protective apparel
- Natural fiber clothing
- Hot sticks
- Fuse pullers
- Shorting probes
- Safety glasses
- Face shields

3.1.1 Rubber Protective Equipment

All electrical workers may be exposed to energized circuits or equipment. Two of the most important articles of protection for electrical workers are insulated rubber gloves and rubber blankets, which must be matched to the voltage rating for the circuit or equipment. Rubber protective equipment is designed for the protection of the user. If it fails during use, a serious injury could occur.

Rubber protective equipment is available in two types. Type 1 designates rubber protective equipment that is manufactured of natural or synthetic rubber that is properly vulcanized, and Type 2 designates equipment that is ozone resistant, made from any elastomer or combination of elastomeric compounds. Ozone is a form of oxygen that is produced from electricity and is present in the air surrounding a conductor under high voltages. Normally, ozone is found at voltages of 10kV and higher, such as those found in electric utility transmission and distribution systems. Type 1 protective equipment can be damaged by corona cutting, which is the cutting action of ozone on natural rubber when it is under mechanical stress. Type 1 rubber protective equipment can also be damaged by ultraviolet rays. However, it is very important that the rubber protective equipment in use today be made of natural rubber or Type 1 equipment. Type 2 rubber protective equipment is very stiff and is not as easily worn as Type 1 equipment.

Various classes – The American National Standards Institute (ANSI) and the American Society for Testing and Materials International (ASTM) have designated a specific classification system for rubber protective equipment. The maximum AC use voltage and equipment tag colors are as follows:

- Class 00 (beige tag)   500V
- Class 0 (red tag)   1,000V
- Class 1 (white tag)   7,500V
- Class 2 (yellow tag)   17,000V
- Class 3 (green tag)   26,500V
- Class 4 (orange tag)   36,000V

Referring back to *Figure 2,* note that the gloves shown in the picture have a yellow tag and are therefore Class 2 equipment.

Inspection of protective equipment – Before rubber protective equipment can be worn by personnel in the field, all equipment must have a current test date stenciled on the equipment. Insulating gloves must be inspected each day by the user before they can be used. They must also be electrically tested every six months and any time the insulating value is in question. Because rubber protective equipment is used for personal protection and serious injury could result from its misuse or failure, it is important that an adequate safety factor be provided between the voltage on which it is to be used and the voltage at which it was tested.

All rubber protective equipment must be marked with the appropriate voltage rating and last inspection date. The markings that are required to be on rubber protective equipment must be applied in a manner that will not interfere with the protection that is provided by the equipment.

 WARNING!
Never work on anything energized without direct instruction from your employer.

Gloves – Both high- and low-voltage rubber gloves are of the gauntlet type and are available in various sizes. To get the best possible protection and service life, here are a few general rules that apply whenever they are used in electrical work:

- Always wear leather protectors over your gloves as they provide burn protection not provided by the gloves themselves. Any direct contact with sharp or pointed objects may cut, snag, or puncture the gloves and take away the protection you are depending on.
- Always wear rubber gloves right side out (serial number and size to the outside).
- Always keep the gauntlets up. Rolling them down sacrifices a valuable area of protection. Tuck the sleeves of your shirt or protective suit under the glove cuffs to prevent an arc blast from coming inside your clothing.
- Always inspect and field check gloves before using them. Always check the inside for any debris.
- Use light amounts of manufacturer-approved glove dust or cotton liners with the rubber gloves. This gives the user more comfort, and it also helps to absorb some of the perspiration that can damage the gloves over years of use.
- Wash the rubber gloves in lukewarm, clean, fresh water after each use. Dry the gloves inside and out prior to returning to storage. Never use any type of cleaning solution on the gloves.
- Once the gloves have been properly cleaned, inspected, and tested, they must be properly stored. Store them in a cool, dry, dark place that is free from ozone, chemicals, oils, solvents, or other materials that could damage the gloves.

Do not store gloves near hot pipes or in direct sunlight. Store both gloves and sleeves in their natural shape in a bag or box inside their leather protectors. They should be undistorted, right side out, and unfolded.

- Gloves can be damaged by many different chemicals, especially petroleum-based products such as oils, gasoline, hydraulic fluid inhibitors, hand creams, pastes, and salves. If contact is made with these or other petroleum-based products, the contaminant should be wiped off immediately. If any signs of physical damage or chemical deterioration are found (e.g., swelling, softness, hardening, stickiness, ozone deterioration, or sun checking), the protective equipment must not be used.
- Never wear watches or rings while wearing rubber gloves; this can cause damage from the inside out and defeats the purpose of using rubber gloves. Never wear anything conductive.
- Rubber gloves must be electrically tested every six months by a certified testing laboratory. Always check the inspection date before using gloves.
- Use rubber gloves only for their intended purpose, not for handling chemicals or other work. This also applies to the leather protectors.

Before rubber gloves are used, a visual inspection and an air test should be made. This should be done prior to use and as many times during the day as you feel necessary. To perform a visual inspection, stretch a small area of the glove, checking to see that no defects exist, such as:

- Embedded foreign material
- Deep scratches
- Pinholes or punctures
- Snags or cuts

Gloves and sleeves can be inspected by rolling the outside and inside of the protective equipment between the hands. This can be done by squeezing together the inside of the gloves or sleeves to bend the outside area and create enough stress to the inside surface to expose any cracks, cuts, or other defects. When the entire surface has been checked in this manner, the equipment is then turned inside out, and the procedure is repeated. It is very important not to leave the rubber protective equipment inside out as that places stress on the preformed rubber.

Remember, any damage at all reduces the insulating ability of the rubber glove. Look for signs of deterioration from age, such as hardening and slight cracking. Also, if the glove has been exposed to petroleum products, it should be considered suspect because deterioration can be caused by such exposure. If the gloves are suspect, turn them in for evaluation. If the gloves are defective, turn them in for disposal. Never leave a damaged glove lying around; someone may think it is a good glove and not perform an inspection prior to using it.

After visually inspecting the glove, other defects may be observed by applying the air test (*Figure 3*).

Step 1 Stretch the glove and look for any defects.

Step 2 Twirl the glove around quickly or roll it down from the glove gauntlet to trap air inside.

Step 3 Trap the air by squeezing the gauntlet with one hand. Use the other hand to squeeze the palm, fingers, and thumb to check for weaknesses and defects.

Step 4 Hold the glove up to your ear to try to detect any escaping air.

Step 5 If the glove does not pass this inspection, it must be turned in for disposal.

CAUTION

Never blow the gloves up like a balloon or use compressed gas for the air test as this can damage the glove.

Insulating blankets – An insulating blanket is a versatile cover-up device best suited for the protection of maintenance technicians against accidental contact with energized electrical equipment.

These blankets are designed and manufactured to provide insulating quality and flexibility for use in covering. Insulating blankets are designed only for covering equipment and should not be used on the floor. Special rubber floor mats, called switchboard matting, are available for floor use. Use caution when installing these over sharp edges or when covering pointed objects.

Blankets must be tested yearly and inspected before each use. To check rubber blankets, place the blanket on a flat surface and roll the blanket from one corner to the opposite corner. If there are any irregularities in the rubber, this method will expose them. After the blanket has been rolled from each corner, it should then be turned over and the procedure repeated.

Insulating blankets are cleaned in the same manner as rubber gloves. Once the protective equipment has been properly cleaned, inspected, and tested, it must be properly stored. It should be stored in a cool, dry, dark place that is free from ozone, chemicals, oils, solvents, or other materials that could damage the equipment. Such storage should not be in the vicinity of hot pipes or direct sunlight. Blankets may be stored rolled in containers that are designed for this use; the inside diameter of the roll should be at least two inches.

3.1.2 Protective Apparel

Besides rubber gloves, there are other types of special application protective apparel, such as fire suits, face shields, and rubber sleeves.

Manufacturing plants should have other types of special application protective equipment available for use, such as high-voltage sleeves, high-voltage boots, nonconductive protective helmets,

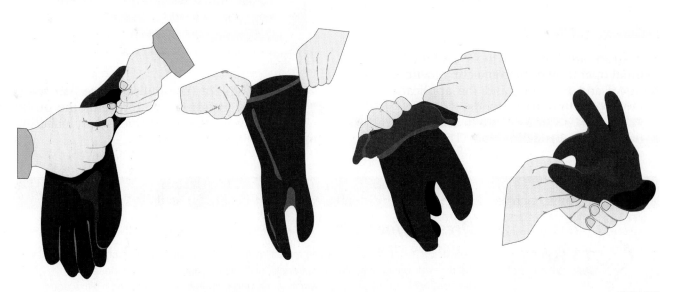

102F03.EPS

Figure 3 ◆ Glove inspection.

Field-Marked Flash Protection Signs

INSIDE TRACK

In other than dwelling units, *NEC Section 110.16* requires that all switchboards, panelboards, industrial control panels, meter socket enclosures, and motor control centers be clearly marked to warn qualified persons of potential electric arc flash hazards.

nonconductive eyewear and face protection, and switchboard blankets.

All equipment must be inspected before use and during use, as necessary. The equipment used and the extent of the precautions taken depend on each individual situation; however, it is better to be overprotected than underprotected when you are trying to prevent electric shock, arc blast, and burns.

When working with energized equipment, flash suits may be required in some applications.

Face shields must also be worn during all switching operations where arcs are a possibility. Always use a rated face shield.

3.1.3 Personal Clothing

Any individual who will perform work in an electrical environment or in plant substations should dress accordingly. Never wear synthetic-fiber clothing on a job site; these types of materials will melt when exposed to high temperatures and because they burn hotter, will actually increase the severity of a burn. Wear cotton clothing, fiberglass-toe boots or shoes, safety glasses, and hard hats. Use hearing protection where needed.

3.1.4 Hot Sticks

Hot sticks are insulated tools designed for the manual operation of disconnecting switches, fuse removal and insertion, and the application and removal of temporary grounds.

A hot stick is made up of two parts, the head or hood and the insulating rod. The head can be made of metal or hardened plastic, while the insulating section may be wood, plastic, laminated wood, or other effective insulating materials. There are also telescoping sticks available.

Most plants have hot sticks available for different purposes. Select a stick of the correct type and size for the application and inspect it before use. Look for signs of obvious damage, deep scratches, dust, or surface contaminants. Never use a damaged hot stick.

Storage of hot sticks is important. They should be hung up vertically on a wall to prevent any damage. They should also be stored away from direct sunlight and prevented from being exposed to petroleum products.

3.1.5 Fuse Pullers

Use the plastic or fiberglass style of fuse puller for removing and installing low-voltage cartridge fuses. All fuse pulling and replacement operations must be done using fuse pullers.

 WARNING!

Fuses should only be pulled when they are safely de-energized. Failure to do so can result in a serious injury or death.

The best type of fuse puller is one that has a spread guard installed. This prevents the puller from opening if resistance is met when installing fuses.

Dressing for Safety

THINK ABOUT IT

How could minor flaws in clothing cause harm? What about metal components, such as the rivets in jeans, or synthetic materials, such as polyester? How are these dangerous? What about rings, watches, earrings, or other body piercings? How does protective apparel prevent accidents?

3.1.6 Shorting Probes

Before working on de-energized circuits that have capacitors installed, you must discharge the capacitors using a safety shorting probe. This is a procedure that requires special training and may only be performed by qualified individuals.

3.1.7 Eye and Face Protection

NFPA 70E and OSHA require that you wear safety glasses at all times whenever you are working on or near energized circuits. Face protection is worn over your safety glasses, and is required whenever there is danger of electrical arcs or flashes, or from flying or falling objects resulting from an electrical explosion. *NFPA 70E* is discussed in detail later in this module.

3.2.0 Verify That Circuits Are De-energized

You should always assume that all the circuits are energized until you have verified that the circuit is de-energized. This is called a live-dead-live test. Follow these steps to verify that a circuit is de-energized:

Step 1 Ensure that the circuit is properly tagged and locked out *(CFR 1910.333/1926.417).*

Step 2 Verify the test instrument operation on a known source using the appropriately rated tester.

Step 3 Using the test instrument, check the circuit to be de-energized. The voltage should be zero.

Step 4 Verify the test instrument operation, once again on a known power source.

3.3.0 Other Precautions

There are several other precautions you can take to help make your job safer. For example:

- Always remove all jewelry (e.g., rings, watches, body piercings, bracelets, and necklaces) before working on electrical equipment. Most jewelry is made of conductive material and wearing it can result in a shock, as well as other injuries if the jewelry gets caught in moving components.
- When working on energized equipment, it is safer to work in pairs. In doing so, if one of the workers experiences a harmful electrical shock, the other worker can call for help.
- Plan each job before you begin it. Make sure you understand exactly what it is you are going to do. If you are not sure, ask your supervisor.

 WARNING!
The first time you perform a specific task, it cannot be done energized.

Using a Phasing Tester

Using a phasing tester or other voltage detection equipment can expose you to a considerable arc flash hazard. Always take the time to use the appropriate personal protective equipment. Remember, you only have one chance to protect yourself.

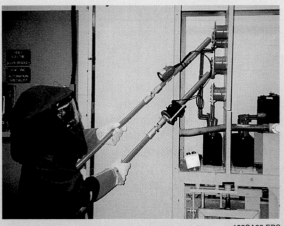

102SA02.EPS

- You will need to look over the appropriate prints and drawings to locate isolation devices and potential hazards. Never defeat safety interlocks. Remember to plan your escape route before starting work. Know where the nearest phone is and the emergency number to dial for assistance.
- If you realize that the work will go beyond the scope of what was planned, stop and get instructions from your supervisor before continuing. Do not attempt to plan as you go.
- It is critical that you stay alert. Workplaces are dynamic, and situations relative to safety are always changing. If you leave the work area to pick up material, take a break, or have lunch, reevaluate your surroundings when you return. Remember, plan ahead.

4.0.0 ◆ OSHA

The purpose of OSHA is "to ensure safe and healthful working conditions for working men and women." OSHA is authorized to enforce standards and assist and encourage the states in their efforts to ensure safe and healthful working conditions. OSHA assists states by providing for research, information, education, and training in the field of occupational safety and health.

The law that established OSHA specifies the duties of both the employer and employee with respect to safety. Some of the key requirements are outlined below. This list does not include everything, nor does it override the procedures called for by your employer.

- Employers shall provide a place of employment free from recognized hazards likely to cause death or serious injury.
- Employers shall comply with the standards of the act.
- Employers shall be subject to fines and other penalties for violation of those standards.

 WARNING!

OSHA states that employees have a duty to follow the safety rules laid down by the employer. Additionally, some states can reduce the amount of benefits paid to an injured employee if that employee was not following known, established safety rules. Your company may also terminate you if you violate an established safety rule.

INSIDE TRACK

Safety on the Job Site

Uncovered openings present several hazards:

- Workers may trip over them.
- If they are large enough, workers may fall through them.
- If there is a work area below, tools or other objects may fall through them, causing serious injury to workers below.

Any hole deeper than 6' and more than 2" in any direction must be protected with a cover or with guardrails, as shown here. A cover must be able to handle twice the anticipated load, be secured in place, and have the word "HOLE" or "COVER" on it.

102SA03.EPS

Working Around Open Pipes

Did you know that OSHA could cite your company for working around the open pipe in the photo shown here, even if you weren't responsible for creating it? If you are working near uncovered openings, you have only two options:

• Cover the opening properly (either you may do it or you may have the responsible contractor cover it).
• Stop work and leave the job site until the opening has been covered.

102SA04.EPS

4.1.0 Safety Standards

The OSHA standards are split into several sections. As discussed earlier, the two that affect you the most are *CFR 1926*, construction specific, and *CFR 1910*, which is the standard for general industry. Either or both may apply depending on where you are working and what you are doing. Your company may also have its own policies and procedures. In addition, you will be required to follow the safety procedures of any plant or facility where you are working.

4.2.0 Safety Philosophy and General Safety Precautions

The most important piece of safety equipment required when performing work in an electrical environment is common sense. All areas of electrical safety precautions and practices draw upon common sense and attention to detail. One of the most dangerous conditions in an electrical work area is a poor attitude toward safety.

 WARNING!
Only qualified individuals may work on or near electrical equipment. Your employer will determine who is qualified. Remember, your employer's safety rules must always be followed.

As stated in *CFR 1910.333(a)/1926.403*, safety-related work practices shall be employed to prevent electric shock or other injuries resulting from either direct or indirect electrical contact when work is performed near or on equipment or circuits that are or may be energized. The specific safety-related work practices shall be consistent with the nature and extent of the associated electrical hazards. The following are considered some of the basic and necessary attitudes and electrical safety precautions that lay the groundwork for a proper safety program. Before going on any electrical work assignment, these safety precautions should be reviewed and adhered to.

• *All work on electrical equipment should be done with circuits de-energized and cleared or grounded* – It is obvious that working on energized equipment is much more dangerous than working on equipment that is de-energized. Work on energized electrical equipment should be avoided if at all possible. *CFR 1910.333(a)(1)/1926.403* states that live parts to which an employee may be exposed shall be de-energized before the employee works on or near them, unless the employer can demonstrate that de-energizing introduces additional or increased hazards or is not possible because of equipment design or operational limitations. Live parts that operate at less than 50 volts to ground need not be de-energized if there will be no increased exposure to electrical burns or to explosion due to electric arcs.

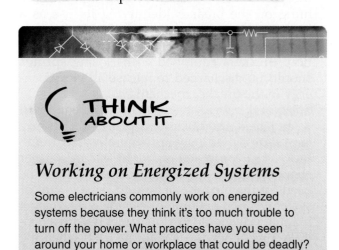

THINK ABOUT IT

Working on Energized Systems

Some electricians commonly work on energized systems because they think it's too much trouble to turn off the power. What practices have you seen around your home or workplace that could be deadly?

- *All conductors, buses, and connections should be considered energized until proven otherwise* – As stated in *1910.333(b)(1)/1926.417*, conductors and parts of electrical equipment that have not been locked out or tagged out in accordance with this section should be considered energized. Routine operation of the circuit breakers and disconnect switches contained in a power distribution system can be hazardous if not approached in the right manner. Several basic precautions that can be observed in switchgear operations are:
  - Wear proper clothing made of natural fiber or fire-resistant fabric.
  - Wear eye, face, and head protection.
  - Whenever operating circuit breakers in low-voltage or medium-voltage systems, always stand off to the side of the unit.
  - Always try to operate disconnect switches and circuit breakers under a no-load condition.
  - Never intentionally force an interlock on a system or circuit breaker.

Often, a circuit breaker or disconnect switch is used for providing lockout on an electrical system. To ensure that a lockout is not violated, perform the following procedures when using the device as a lockout point:

- Breakers must always be locked out and tagged as discussed previously whenever you are working on a circuit that is tied to an energized breaker. Breakers capable of being opened and racked out to the disconnected position should have this done. Afterward, approved safety locks must be installed. The breaker may be removed from its cubicle completely to prevent unexpected mishaps. Always follow the standard rack-out and removal procedures that were supplied with the switchgear. Once removed, a sign must be hung on the breaker identifying its use as a lockout point, and approved safety locks must be installed when the breaker is used for isolation. Breakers equipped with closing springs should be discharged to release all stored energy in the breaker mechanism.
- Some of the circuit breakers used are equipped with keyed interlocks for protection during operation. These locks are generally called kirklocks and are relied upon to ensure proper sequence of operation only. These are not to be used for the purpose of locking out a circuit or system. When opening or closing a disconnect manually, it should be done quickly with a positive force. Lockouts should be used when the disconnects are open.

- Whenever performing switching or fuse replacements, always use the protective equipment necessary to ensure personnel safety. Never make the assumption that because things have gone fine the last 999 times, they will not go wrong this time. Always prepare yourself for the worst case accident when performing switching.
- Whenever re-energizing circuits following maintenance or removal of a faulted component, use extreme care. Always verify that the equipment is in a condition to be re-energized safely. All connections should be insulated and all covers should be installed. Have all personnel stand clear of the area for the initial re-energization. Never assume everything is in perfect condition. Verify the conditions.

The following procedure is provided as a guideline for ensuring that equipment and systems will not be damaged by reclosing low-voltage circuit breakers into faults. If a low-voltage circuit breaker has opened for no apparent reason, perform the following:

Step 1 Verify that the equipment being supplied is not physically damaged and shows no obvious signs of overheating or fire.

Step 2 Make all appropriate tests to locate any faults.

Step 3 Reclose the feeder breaker. Stand off to the side when closing the breaker.

Step 4 If the circuit breaker trips again, do not attempt to reclose the breaker. In a plant environment, Electrical Engineering should be notified, and the cause of the trip must be isolated and repaired.

The same general procedure should be followed for fuse replacement, with the exception of transformer fuses. If a transformer fuse blows, the transformer and feeder cabling should be inspected and tested before re-energizing. A blown fuse to a transformer is very significant because it normally indicates an internal fault. Transformer failures are catastrophic in nature and can be extremely dangerous. If applicable, contact the in-plant Electrical Engineering Department prior to commencing any effort to re-energize a transformer.

Power must always be removed from a circuit when removing and installing fuses. The air break disconnects (or quick disconnects) provided on the upstream side of a large transformer must be opened prior to removing the transformer's fuses. Otherwise, severe arcing will occur as the fuse is removed. This arcing can result in personnel injury and equipment damage.

To replace fuses servicing circuits below 600 volts:

- Turn off the power to the disconnect.
- Verify that the fuses are de-energized.
- Remove the blown fuse.
- Install the new fuse. Push it in firmly and verify that it is seated properly.
- Turn the power back on.

When replacing fuses servicing systems above 600 volts:

- Open and lock out the disconnect switches.
- Unlock the fuse compartment.
- Verify that the fuses are de-energized.
- Attach the fuse removal hot stick to the fuse and remove it.

4.3.0 Electrical Regulations

OSHA has certain regulations that apply to job site electrical safety. These regulations include:

- All electrical work shall be in compliance with the latest *NEC*® and OSHA standards.
- The noncurrent-carrying metal parts of fixed, portable, and plug-connected equipment shall be grounded. Choose either grounded tools or double-insulated tools. *Figure 4* shows an example of a double-insulated tool.

- Extension cords shall be the three-wire type, shall be protected from damage, and if hung overhead, shall not be fastened with staples or bare wire or hung in a manner that could cause damage to the outer jacket or insulation. Never run an extension cord through a doorway or window that can pinch the cord. Also, never allow vehicles or equipment to drive over cords.
- Exposed lamps in temporary lights shall be guarded to prevent accidental contact, except where lamps are deeply recessed in the reflector. Temporary lights shall not be suspended, except in accordance with their listed labeling.
- Receptacles for attachment plugs shall be of an approved type and properly installed. Installation of the receptacle will be in accordance with the listing and labeling for each receptacle.
- *NEC Section 590.6(A)* states that all 125V, single-phase, 15A, 20A, and 30A receptacle outlets shall be ground fault protected.
- Each disconnecting means for motors and appliances and each service feeder or branch circuit at the point where it originates shall be legibly marked to indicate its purpose and voltage.
- Flexible cords shall be used in continuous lengths (no splices) and shall be of a type listed in *NEC Table 400.4*.
- Receptacles other than 125V, single-phase, 15A, 20A, and 30A should be ground fault protected if possible. If that is not possible, a written assured equipment grounding conductor program must be continuously enforced at the site for all cord sets per *NEC Section 590.6(B)*. *Figure 5* shows a typical ground fault circuit interrupter (GFCI).

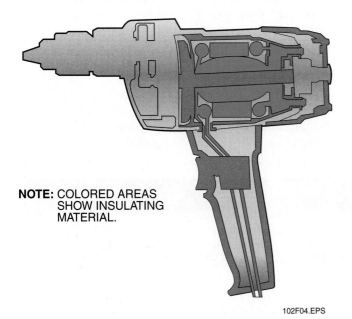

NOTE: COLORED AREAS SHOW INSULATING MATERIAL.

102F04.EPS

Figure 4 ◆ Double-insulated electric drill.

CASE HISTORY

Potential Hazards

A self-employed builder was using a metal cutting tool on a metal carport roof and was not using GFCI protection. The male and female plugs of his extension cord partially separated, and the active pin touched the metal roofing. When the builder grounded himself on the gutter of an adjacent roof, he received a fatal shock.

The Bottom Line: Always use GFCI protection and be on the lookout for potential hazards.

Figure 5 ◆ Typical GFCI receptacle.

102F05.EPS

4.3.1 OSHA Lockout/Tagout Rule

OSHA released the *29 CFR 1926* lockout/tagout rule in December 1991. This rule covers the specific procedure to be followed for the "servicing and maintenance of machines and equipment in which the unexpected energization or startup of the machines or equipment, or releases of stored energy, could cause injury to employees." This standard establishes minimum performance requirements for the control of such hazardous energy.

The first step to be completed before working on a circuit is to ensure that equipment is isolated from all potentially hazardous energy (for example, electrical, mechanical, hydraulic, chemical, or thermal), and tagged and locked out before employees perform any servicing or maintenance activities in which the unexpected energization, startup, or release of stored energy could cause injury. All employees shall be instructed in the lockout/tagout procedure.

 CAUTION

Although 99% of your work may be electrical, be aware that you may also need to lock out mechanical and other types of energy equipment.

The following is an example of a lockout/tagout procedure. Make sure to use the procedure that is specific to your employer or job site.

 WARNING!

This procedure is provided for your information only. The OSHA procedure provides only the minimum requirements for lockouts/tagouts. Consult the lockout/tagout procedure for your company and the plant or job site at which you are working. Remember that your life could depend on the lockout/tagout procedure. It is critical that you use the correct procedure for your site. The *NEC*® requires that remote-mounted motor disconnects be permanently equipped with a lockout feature.

I. *Introduction*
   A. This lockout/tagout procedure has been established for the protection of personnel from potential exposure to hazardous energy sources during construction, installation, service, and maintenance of electrical energy systems.
   B. This procedure applies to and must be followed by all personnel who may be potentially exposed to the unexpected startup or release of hazardous energy (e.g., electrical, mechanical, pneumatic, hydraulic, chemical, or thermal).

Exception: This procedure does not apply to process and/or utility equipment or systems with cord and plug power supply systems when the cord and plug are the only source of hazardous energy, are removed from the source, and remain under the exclusive control of the authorized employee.

Exception: This procedure does not apply to troubleshooting (diagnostic) procedures and installation of electrical equipment and systems when the energy source cannot be de-energized

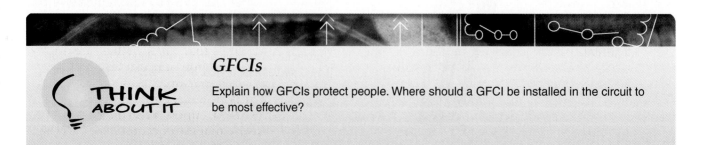

GFCIs

THINK ABOUT IT

Explain how GFCIs protect people. Where should a GFCI be installed in the circuit to be most effective?

because continuity of service is essential or shutdown of the system is impractical. Additional personal protective equipment for such work is required and the safe work practices identified for this work must be followed.

II. *Definitions*
- *Affected employee* – Any person working on or near equipment or machinery when maintenance or installation tasks are being performed by others during lockout/tagout conditions.
- *Appointed authorized employee* – Any person appointed by the job site supervisor to coordinate and maintain the security of a group lockout/tagout condition.
- *Authorized employee* – Any person authorized by the job site supervisor to use lockout/tagout procedures while working on electrical equipment.
- *Authorized supervisor* – The assigned job site supervisor who is in charge of coordination of procedures and maintenance of security of all lockout/tagout operations at the job site.
- *Energy isolation device* – An approved electrical disconnect switch capable of accepting approved lockout/tagout hardware for the purpose of isolating and securing a hazardous electrical source in an open or safe position.
- *Lockout/tagout hardware* – A combination of padlocks, danger tags, and other devices designed to attach to and secure electrical isolation devices.

III. *Training*
A. Each authorized supervisor, authorized employee, and appointed authorized employee shall receive initial and as-needed user-level training in lockout/tagout procedures.
B. Training is to include recognition of hazardous energy sources, the type and magnitude of energy sources in the workplace, and the procedures for energy isolation and control.
C. Retraining will be conducted on an as-needed basis whenever lockout/tagout procedures are changed or there is evidence that procedures are not being followed properly.

IV. *Protective Equipment and Hardware*
A. Lockout/tagout devices shall be used exclusively for controlling hazardous energy sources.
B. All padlocks must be numbered and assigned to one employee only.
C. No duplicate or master keys will be made available to anyone except the site supervisor.
D. A current list with the lock number and authorized employee's name must be maintained by the site supervisor.
E. Danger tags must be of the standard white, red, and black *DANGER—DO NOT OPERATE* design and shall include the authorized employee's name, the date, and the appropriate network company (use permanent markers).
F. Danger tags must be used in conjunction with padlocks, as shown in *Figure 6*.

V. *Procedures*
A. Preparation for lockout/tagout:
1. Check the procedures to ensure that no changes have been made since you last used a lockout/tagout.
2. Identify all authorized and affected employees involved with the pending lockout/tagout.
B. Sequence for lockout/tagout:
1. Notify all authorized and affected personnel that a lockout/tagout is to be used and explain the reason why.
2. Shut down the equipment or system using the normal OFF or STOP procedures.
3. Lock out energy sources and test disconnects to be sure they cannot be moved to the ON position and open the control cutout switch. If there is no cutout switch, block the magnet in the switch open position before working on electrically operated equipment/apparatus such as motors, relays, etc. Remove the control wire.
4. Lock and tag the required switches in the open position. Each authorized employee must affix a separate lock and tag. An example is shown in *Figure 7*.
5. Dissipate any stored energy by attaching the equipment or system to ground.
6. Verify that the test equipment is functional via a known power source.
7. Confirm that all switches are in the open position and use test equipment to verify that all parts are de-energized.
8. If it is necessary to temporarily leave the area, upon returning, retest to ensure that the equipment or system is still de-energized.

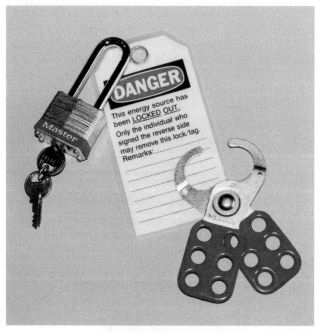

ELECTRICAL LOCKOUT

102F06A.EPS

PNEUMATIC LOCKOUT

102F06B.EPS

Figure 6 ◆ Lockout/tagout devices.

C. Restoration of energy:
   1. Confirm that all personnel and tools, including shorting probes, are accounted for and removed from the equipment or system.
   2. Completely reassemble and secure the equipment or system.
   3. Replace and/or reactivate all safety controls.

102F07.EPS

Figure 7 ◆ Multiple lockout/tagout device.

   4. Remove locks and tags from isolation switches. Authorized employees must remove their own locks and tags.
   5. Notify all affected personnel that the lockout/tagout has ended and the equipment or system is energized.
   6. Operate or close isolation switches to restore energy.

VI. *Emergency Removal Authorization*
   A. In the event a lockout/tagout device is left secured, and the authorized employee is absent, or the key is lost, the authorized supervisor can remove the lockout/tagout device.
   B. The authorized employee must be informed that the lockout/tagout device has been removed.
   C. Written verification of the action taken, including informing the authorized employee of the removal, must be recorded in the job journal.

What's wrong with this picture?

102SA05.EPS

4.4.0 Other OSHA Regulations

There are other OSHA regulations that you need to be aware of on the job site. For example:

- OSHA requires the posting of hard hat areas. Be alert to those areas and always wear your hard hat properly, with the bill in front. Hard hats should be worn whenever overhead hazards exist, or there is the risk of exposure to electric shock or burns.
- You should wear safety shoes on all job sites. Keep them in good condition.

Lockout/Tagout Dilemma

In Georgia, electricians found energized switches after the lockout of a circuit panel in an older system that had been upgraded several times. The existing wiring did not match the current site drawings. A subsequent investigation found many such situations in older facilities.

The Bottom Line: Never rely solely on drawings. It is mandatory that the circuit be tested after lockout to verify that it is de-energized.

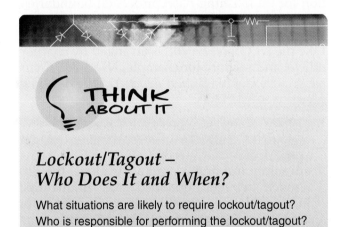

THINK ABOUT IT

Lockout/Tagout – Who Does It and When?

What situations are likely to require lockout/tagout? Who is responsible for performing the lockout/tagout? When would more than one person be responsible?

- Do not wear clothing with exposed metal zippers, buttons, or other metal fasteners. Avoid wearing loose-fitting or torn clothing.
- Protect your eyes. Your eyesight is threatened by many activities on the job site. Always wear safety glasses with full side shields. In addition, the job may also require protective equipment such as face shields or goggles.

5.0.0 ◆ NFPA 70E

In addition to the *NEC®*, the National Fire Protection Association also publishes the *Standard for Electrical Safety in the Workplace (NFPA 70E)*. The *NEC®* specifies the minimum installation provisions necessary to safeguard persons and property from electrical hazards, while *NFPA 70E* addresses the hazards that arise during the installation, operation, and maintenance of electrical equipment. In other words, the *NEC®* applies to installations, while *NFPA 70E* applies to workplaces. *NFPA 70E* is a national consensus standard, which means that it is a standard developed by the same people it affects and is then adopted by a nationally recognized institution. Other national consensus standards include those developed by ASTM International and the American National Standards Institute (ANSI).

The electrical safety requirements in *OSHA 1910.331* to *1910.335* are to use appropriate safe work practices, including de-energization of equipment, and the use of PPE appropriate to the hazard. *NFPA 70E* is really the first consensus standard to provide practical guidance in hazard assessment, training of qualified persons, and management of electrical hazards. For many, *NFPA 70E* provided the first functional understanding of the electrical arc flash hazard, direction for establishing flash protection boundaries, and levels of protection when qualified persons must work within the flash protection boundary. Like the *NEC®*, *NFPA 70E* is arranged by chapter, article, and section; for example, *NFPA 70E Chapter 4, Article 420, Section 420.1(A)*. *NFPA 70E* is organized as follows:

- *Chapter 1, Safety-Related Work Practices* – This chapter covers the basic safety requirements for working on or near electrical equipment. It covers safety training programs, explains the difference between qualified and unqualified individuals, and includes requirements for analysis of both the electrical shock hazard and the electrical arc flash/blast hazard for a given task. The analysis includes assessment of the presence and extent of hazardous voltage and the available fault current for arc flash. The end result of each hazard analysis is also selection of safe work practices and appropriate PPE to be used in performing the task.
- *Chapter 2, Safety-Related Maintenance Requirements* – This chapter covers the maintenance requirements for various types of electrical equipment. This includes premises wiring, various types of electric equipment and devices, and personal safety and protective equipment.
- *Chapter 3, Safety Requirements for Special Equipment* – This chapter covers electrolytic cells, batteries, lasers, and various types of power electronic equipment, such as radio and television transmitters, UPS systems, arc welding equipment, and lighting controllers.

- *Chapter 4, Installation Safety Requirements* – This chapter provides an abbreviated version of the requirements in the *NEC®*.
- *Annexes A through M* – The annexes provide various technical resources. These include safe approach distances, determination of flash protection boundaries and incident energy exposure from arc flash, a sample electrical lockout/tagout procedure, and other useful information.

6.0.0 ◆ LADDERS AND SCAFFOLDS

Ladders and scaffolds account for about half of the injuries from workplace electrocutions. The involuntary recoil that can occur when a person is shocked can cause the person to be thrown from a ladder or high place.

6.1.0 Ladders

Many job site accidents involve the misuse of ladders. Make sure to follow these general rules every time you use any ladder. Following these rules can prevent serious injury or even death.

- Before using any ladder, inspect it. Look for loose or missing rungs, cleats, bolts, or screws. Also check for cracked, bent, broken, or badly worn rungs, cleats, or side rails. See *Figure 8*.
- Before you climb a ladder, make sure you clear any debris from the base of the ladder so you do not trip over it when you descend.
- If you find a ladder in poor condition, do not use it. Report it and tag it for repair or disposal.
- Never modify a ladder by cutting it or weakening its parts.
- Do not set up ladders where they may be run into by others, such as in doorways or walkways. If it is absolutely necessary to set up a ladder in such a location, protect the ladder with barriers.
- Do not increase a ladder's reach by putting it in a mechanical lift or standing it on boxes, barrels, or anything other than a flat, solid surface.
- Check your shoes for grease, oil, or mud before climbing a ladder. These materials could make you slip.
- Always face the ladder and maintain three-point contact with the ladder (either have both feet and one hand on the ladder or both hands and one foot as you climb).
- Never lean out from the ladder. Keep your belt buckle centered between the rails. If something is out of reach, get down and move the ladder.

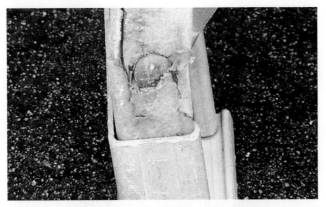

(A) CRUMBLING RAIL

(B) CRACKED STEPLADDER

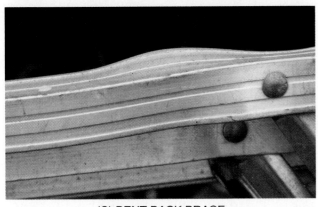

(C) BENT BACK BRACE

102F08.EPS

Figure 8 ◆ Types of ladder damage.

 WARNING!

When performing electrical work, always use ladders made of nonconductive material.

6.1.1 Straight and Extension Ladders

There are some specific rules to follow when working with straight and extension ladders:

- Always place a straight ladder at the proper angle. The horizontal distance from the ladder feet to the base of the wall or support should be about one-fourth the working height of the ladder. See *Figure 9*.
- Secure straight ladders to prevent slipping. Use ladder shoes or hooks at the top and bottom. Another method is to secure a board to the floor against the ladder feet. For brief jobs, someone can hold the straight ladder.
- Side rails should extend above the top support point by at least 36 inches.
- It takes two people to safely extend and raise an extension ladder. Extend the ladder only after it has been raised to an upright position.
- Never carry an extended ladder.
- Never use two ladders spliced together.
- Ladders should not be painted because paint can hide defects.

6.1.2 Step Ladders

There are also a few specific rules to use with a step ladder:

- Always open the step ladder all the way and lock the spreaders to avoid collapsing the ladder accidentally.

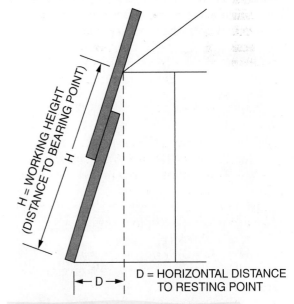

H = WORKING HEIGHT (DISTANCE TO BEARING POINT)

D = HORIZONTAL DISTANCE TO RESTING POINT

THE RATIO OF H TO D SHOULD BE 4 TO 1.

102F09.EPS

Figure 9 ◆ Straight ladder positioning.

- Use a step ladder that is high enough for the job so that you do not have to reach. Get someone to hold the ladder if it is more than 10 feet high.
- Never use a step ladder as a straight ladder.
- Never stand on or straddle the top two rungs of a step ladder.
- Ladders are not shelves.

 WARNING!
Do not leave tools or materials on a step ladder.

Sometimes you will need to move or remove protective equipment, guards, or guardrails to complete a task using a ladder. Remember, always replace what you moved or removed before leaving the area.

INSIDE TRACK

Fireman's Rule

To set up a straight ladder, place your feet against the side rails, stand straight up, and put your hands at right angles to your body directly in front of you. If you can grab the side rails, the ladder is at the correct angle. This is called the Fireman's Rule.

102SA06.EPS

What's wrong with this picture?

102SA07.EPS

6.2.0 Scaffolds

Working on scaffolds (*Figure 10*) also involves being safe and alert to hazards. In general, keep scaffold platforms clear of unnecessary material or scrap. These can become deadly tripping hazards or falling objects. Carefully inspect each part of the scaffold as it is erected. Your life may depend on it! Makeshift scaffolds have caused many injuries and deaths on job sites. Use only scaffolding and planking materials designed and marked for their specific use. When working on a scaffold, follow the established specific requirements set by OSHA for the use of fall protection. When appropriate, wear an approved harness with a lanyard properly anchored to the structure.

 NOTE
The following requirements represent a compilation of the more stringent requirements of both *CFR 1910* and *CFR 1926*.

The following are some of the basic OSHA rules for working safely on scaffolds:

- Scaffolds must be erected on sound, rigid footing (referred to as a mud sill) that can carry the

Scaffolds and Electrical Hazards

Remember that scaffolds are excellent conductors of electricity. Recently, a maintenance crew needed to move a scaffold and although time was allocated in the work order to dismantle and rebuild the scaffold, the crew decided to push it instead. They did not follow OSHA recommendations for scaffold clearance and did not perform a job site survey. During the move, the five-tier scaffold contacted a 12,000V overhead power line. All four members of the crew were killed and the crew chief received serious injuries.

The Bottom Line: Never take shortcuts when it comes to your safety and the safety of others. Trained safety personnel should survey each job site prior to the start of work to assess potential hazards. Safe working distances should be maintained between scaffolding and power lines.

maximum intended load. If the scaffolding is not erected on concrete or another firm surface, you can use 2 × 10 or 2 × 12 scaffold grade lumber to support the base plates of the scaffolding.

- Scaffolds must be erected straight and plumb with no bent or deformed pieces. Correctly erected scaffolding will be symmetrical with the same parts on both sides (unless it has an outrigger attachment).

- Guardrails and toe boards must be installed on the open sides and ends of platforms higher than ten feet above the ground or floor.
- There must be a screen of ½-inch maximum openings between the toe board and the midrail where persons are required to work or pass under the scaffold.
- Scaffold planks must extend over their end supports not less than 6 inches nor more than 12 inches and must be properly blocked.
- If the scaffold does not have built-in ladders that meet the standard, then it must have an attached ladder access.
- All employees must be trained to erect, dismantle, and use scaffold(s).
- Unless it is impossible, fall protection must be worn while building or dismantling all scaffolding.
- Work platforms must be completely decked for use by employees.
- Use trash containers or other similar means to keep debris from falling and never throw or sweep material from above.

7.0.0 ◆ LIFTS, HOISTS, AND CRANES

On the job, you may be working in the operating area of lifts, hoists, or cranes. The following safety rules are for those who are working in the area with overhead equipment but are not directly involved in its operation.

- Stay alert and pay attention to the warning signals from operators.
- Never stand or walk under a load, regardless of whether it is moving or stationary.

What's wrong with this picture?

102SA08.EPS

102F10.EPS

Figure 10 ◆ Scaffolding.

Vertical Towers

This electrician is installing a light fixture using a vertical tower or Genie® lift. This lift is designed to fold up small enough so that it can be maneuvered through house doorways.

102SA09.EPS

- Always warn others of moving or approaching overhead loads.
- Never attempt to distract signal persons or operators of overhead equipment.
- Obey warning signs.
- Do not use equipment that you are not qualified to operate.
- Cranes that are operated in areas with places in which a person can become trapped or pinched must have barricades placed around them to warn away workers.

- Hoists rigged in shafts or the outside of buildings must be secured to prevent them from being pulled down.
- Never overload a hoist, lift, or crane. Always follow lift ratings—there is no such thing as a safety factor that can be used to cheat a load.
- Cranes, lifts, and hoists must be inspected daily.
- Never lift a person using a crane, hoist, or material lift.
- Only people who have been trained in rigging should ever do rigging for a lift. It is extremely easy for a load to fall out of inappropriately rigged lifts.
- Personnel hoists require gates on every landing which can only be opened from the hoist side. This ensures that they are not opened onto a fall hazard.

8.0.0 ◆ LIFTING

Back injuries cause many lost working hours every year. That is in addition to the misery felt by the person with the hurt back! Learn how to lift properly and size up the load. To lift, first stand close to the load. Then, squat down and keep your back straight. Get a firm grip on the load and keep the load close to your body. Lift by straightening your legs. Make sure that you lift with your legs and not your back. Do not be afraid to ask for help if you feel the load is too heavy. See *Figure 11* for an example of proper lifting.

Keep the following precautions in mind when lifting:

- Make the lift smoothly and under control.
- Move your feet to pivot; do not twist or you may injure yourself.
- Constantly scan the path ahead for obstructions. If you cannot see your path over or around the object being carried, then you must have help to transport the object.
- Avoid lifting objects over your head.
- Never lift over the side or tailgate of a pickup truck.
- Don't twist your body when lifting up or setting an object down.
- Never reach over an obstacle to lift a load.
- Don't step over objects in your way.

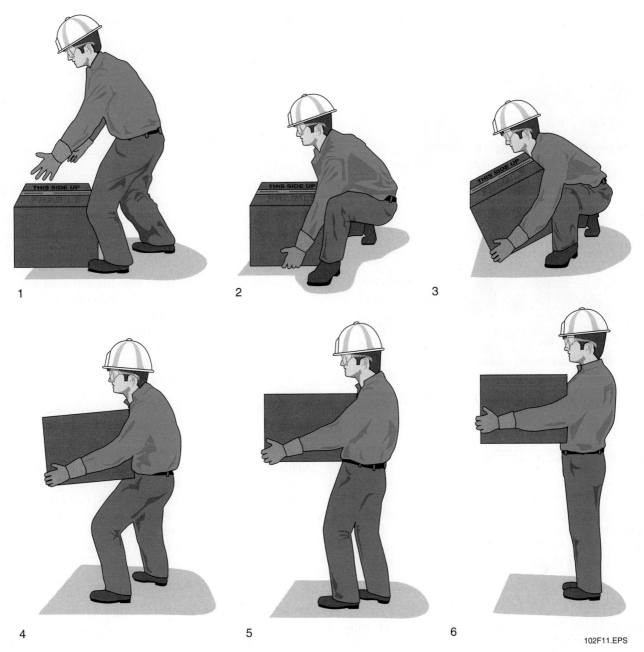

1   2   3

4   5   6

102F11.EPS

Figure 11 ◆ Proper lifting.

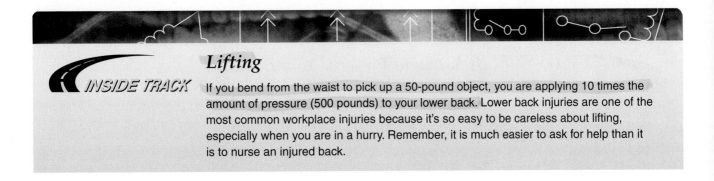

Lifting

If you bend from the waist to pick up a 50-pound object, you are applying 10 times the amount of pressure (500 pounds) to your lower back. Lower back injuries are one of the most common workplace injuries because it's so easy to be careless about lifting, especially when you are in a hurry. Remember, it is much easier to ask for help than it is to nurse an injured back.

What's wrong with this picture?

102SA10.EPS

9.0.0 ◆ BASIC TOOL SAFETY

When using any tools for the first time, read the operator's manual to learn the recommended safety precautions. If you are not certain about the operation of any tool, ask the advice of a more experienced worker. Before using a tool, you should know its function and how it works.

9.1.0 Hand Tool Safety

Hand tools are non-powered tools and may include anything from screwdrivers to cable strippers (*Figure 12*). Hand tools are dangerous if they are misused or improperly maintained.

Keep the following precautions in mind when using hand tools:

- Only use tools for their designated purpose.
- Always maintain your tools properly. If the wooden handle on a tool such as an ax or hammer is loose, splintered, or cracked, the head of the tool may fly off and strike the user or another person.
- Repair or replace damaged or worn tools. A wrench with its jaws sprung might easily slip, causing hand injuries. If the wrench flies, it may strike the user or another person.
- Impact tools such as chisels, wedges, and drift pins are unsafe if they have mushroomed heads. The heads might shatter when struck, sending sharp fragments flying.
- Use bladed tools with the blades and points aimed away from yourself and other people.
- Store bladed tools properly; use the sheath or protective covering if there is one.
- Keep blades sharp and inspect them regularly. Dull blades are difficult to use and control and can be far more dangerous than well-maintained blades.
- Never leave tools on top of ladders or scaffolding.

In general, your risks are greatly reduced by inspecting and maintaining tools regularly and always wearing appropriate personal protective equipment such as safety goggles, hard hats, and filtering masks. Also, keep floors dry and clean to prevent accidental slips that can result in injuries caused by the tools you may be using.

9.2.0 Power Tool Safety

Power tools can be hazardous when they are improperly used or not well maintained. Most of the risks associated with hand tools are also risks when using power tools. When you add a power source to a tool, however, the risk factors increase.

Lift Safely to Preserve Your B.A.C.K.

B – Balance: keep your stance wide, get a good grip on the object.
A – Alignment: keep your back relaxed and upright.
C – Contract and close: contract your stomach muscles and hold loads close.
K – Knees: make sure you bend them, not your waist.

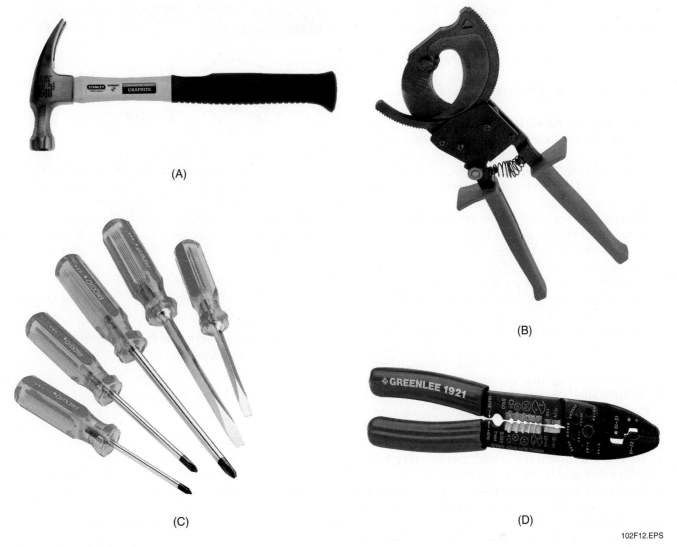

(A)

(B)

(C)

(D)

102F12.EPS

Figure 12 ◆ Hand tools.

Power tools are powered by different sources. Some examples of power sources for power tools include:

- Electricity
- Pneumatics (air pressure)
- Liquid fuel (gasoline or propane)
- Hydraulics (fluid pressure)

You must know the safety rules and proper operating procedures for each tool you use. Specific operating procedures and safety rules for using a tool are provided in the operator's/user's manual supplied by the manufacturer. Before operating any power tool for the first time, always read the manual to familiarize yourself with the tool. If the manual is missing, contact the manufacturer for a replacement.

 WARNING!
Never use a tool if you are unsure of how to use it correctly.

Follow these general guidelines in order to prevent accidents and injury:

- Inspect all tools for damage before use. Remove damaged tools from use and tag DO NOT USE.
- Never carry or lower a tool by the cord or hose.
- Keep cords and hoses away from heat, oil, and sharp edges.
- Do not attempt to operate any power tool before being checked out by your instructor on that particular tool.

- Always wear eye protection, a hard hat, and any other required personal protective equipment when operating power tools.
- Wear face and hearing protection when required.
- Wear proper respiratory equipment when necessary.
- Wear the appropriate clothing for the job being done. Wear close-fitting clothing that cannot become caught in moving tools. Roll up or button long sleeves, tuck in shirttails, and tie back long hair. Do not wear any jewelry, including watches or rings.
- Do not distract others or let anyone distract you while operating a power tool.
- Do not engage in horseplay.
- Do not run or throw objects.
- Consider the safety of others, as well as yourself. Observers should be kept at a safe distance away from the work area.
- Never leave a power tool running while it is unattended.
- Assume a safe and comfortable position before using a power tool. Be sure to maintain good footing and balance in order to respond to kickbacks, jumps, or sudden shifts.
- Secure work with clamps or a vise, freeing both hands to safely operate the tool.
- To avoid accidental starting, never carry a tool with your finger on the switch.
- Be sure that a power tool is properly grounded and connected to a ground fault circuit interrupter (GFCI) before using it.
- Ensure that power tools are disconnected before performing maintenance or changing accessories.
- Use a power tool only for its intended use.
- Keep your feet, fingers, and hair away from the blade and/or other moving parts of a power tool.
- Never use a power tool with guards or safety devices removed or disabled.
- Never operate a power tool if your hands or feet are wet.
- Keep the work area clean at all times.
- Become familiar with the correct operation and adjustments of a power tool before attempting to use it.
- Keep tools sharp and clean for the best performance.
- Follow the instructions in the user's manual for lubricating and changing accessories.
- Keep a firm grip on the power tool at all times.
- Use electric extension cords of sufficient size to service the particular power tool you are using.
- Do not run extension cords across walkways where they will pose a tripping hazard.

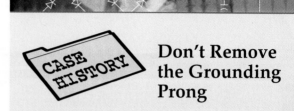

Don't Remove the Grounding Prong

CASE HISTORY

An employee was climbing a metal ladder to hand an electric drill to the journeyman installer on a scaffold about 5' above him. When the victim reached the third rung from the bottom of the ladder, he received an electric shock that killed him. The investigation revealed that the extension cord had a missing grounding prong and that a conductor on the green grounding wire was making intermittent contact with the energized black wire, thereby energizing the entire length of the grounding wire and the drill's frame. The drill was not double insulated.

The Bottom Line: Do not disable any safety device on a power tool. A ground fault can be deadly.

Source: The Occupational Safety and Health Administration (OSHA)

- Report unsafe conditions to your instructor or supervisor.
- Tools that shoot nails (*Figure 13*), rivets, or staples, and operate at pressures greater than 100 pounds per square inch (psi), must be equipped with a safety device that won't allow fasteners to be shot unless the muzzle is pressed against a work surface.
- Compressed-air guns should never be pointed toward anyone and the muzzle should never be pressed against a person.

102F13.EPS

Figure 13 ◆ Pneumatic nail gun.

- The use of powder-actuated tools requires special training and certification.
- Never use explosive or flammable materials around powder-actuated tools.
- Never point powder-actuated tools at anybody.
- Never pick up an unattended powder-actuated tool. Instead, tell your supervisor that a powder-actuated tool has been left unattended.
- Never play with powder-actuated tools. These tools are as dangerous as a loaded gun.

10.0.0 ◆ CONFINED SPACE ENTRY PROCEDURES

Occasionally, you may be required to do your work in a manhole or vault. If this is the case, there are some special safety considerations that you need to be aware of. For details on the subject of working in manholes and vaults, refer to *CFR 1910.146/1926.21(a)(6)(i) and (ii)*. The general precautions are listed in the following paragraphs.

10.1.0 General Guidelines

A confined space includes (but is not limited to) any of the following: a manhole (*Figure 14*), boiler, tank, trench (four feet or deeper), tunnel, hopper, bin, sewer, vat, pipeline, vault, pit, air duct, or vessel. A confined space is identified as follows:

- It has limited entry and exit.
- It is not intended for continued human occupancy.
- It has poor ventilation.
- It has the potential for entrapment/engulfment.
- It has the potential for accumulating a dangerous atmosphere.

Figure 14 ◆ Manhole.

Entry into a confined space occurs when any part of the body crosses the plane of entry. No employee shall enter a confined space unless the employee has been trained in confined space entry procedures. Other requirements for confined spaces include the following:

- All hazards must be eliminated or controlled before a confined space entry is made.
- The air quality in the confined space must be continually monitored.
- All appropriate personal protective equipment shall be worn at all times during confined space entry and work. The minimum required equipment includes a hard hat, safety glasses, full body harness, and life line.
- Ladders used for entry must be secured.
- A rescue retrieval system must be in use when entering confined spaces and while working in permit-required confined spaces (discussed later). Each employee must be capable of being rescued by the retrieval system.
- Only no-entry rescues will be performed by company personnel. Entry rescues will be performed by trained rescue personnel identified on the entry permit.
- The area outside the confined space must be properly barricaded, and appropriate warning signs must be posted.
- Entry permits can only be issued and signed by a qualified person such as the job site supervisor. Permits must be kept at the confined space while work is being conducted. At the end of the shift, the entry permits must be made part of the job journal and retained for one year.

10.2.0 Confined Space Hazard Review

Before determining the proper procedure for confined space entry, a hazard review shall be performed. The hazard review shall include, but not be limited to, the following conditions:

- The past and current uses of the confined space
- The physical characteristics of the space including size, shape, air circulation, etc.
- Proximity of the space to other hazards
- Existing or potential hazards in the confined space, such as:
  - Atmospheric conditions (oxygen levels, flammable/explosive levels, and/or toxic levels)
  - Presence/potential for liquids
  - Presence/potential for particulates
- Potential for mechanical/electrical hazards in the confined space (including work to be done)

Once the hazard review is completed, the supervisor, in consultation with the project managers and/or safety manager, shall classify the confined space as one of the following:

- A nonpermit confined space
- A permit-required confined space controlled by ventilation
- A permit-required confined space

Once the confined space has been properly classified, the appropriate entry and work procedures must be followed.

 WARNING!
Only qualified and trained individuals may enter a confined space.

11.0.0 ◆ FIRST AID

You should be prepared in case an accident does occur on the job site or anywhere else. First aid training that includes certification classes in CPR and artificial respiration could be the best insurance you and your fellow workers ever receive. Make sure that you know where first aid is available at your job site. Also, make sure you know the accident reporting procedure. Each job site should also have a first aid manual or booklet giving easy-to-find emergency treatment procedures for various types of injuries. Emergency first aid telephone numbers should be readily available to everyone on the job site. Refer to *CFR 1910.151/ 1926.23* and *1926.50* for specific requirements.

12.0.0 ◆ SOLVENTS AND TOXIC VAPORS

The solvents that are used by electricians may give off vapors that are toxic enough to make people temporarily ill or even cause permanent injury. Many solvents are skin and eye irritants. Solvents can also be systemic poisons when they are swallowed or absorbed through the skin.

Solvents in spray or aerosol form are dangerous in another way. Small aerosol particles or solvent vapors mix with air to form a combustible mixture with oxygen. The slightest spark could cause an explosion in a confined area because the mix is perfect for fast ignition. There are procedures and methods for using, storing, and disposing of most solvents and chemicals. These procedures are normally found in the material safety data sheets (MSDSs) available at your facility.

An MSDS is required for all materials that could be hazardous to personnel or equipment.

These sheets contain information on the material, such as the manufacturer and chemical makeup. As much information as possible is kept on the hazardous material to prevent a dangerous situation; or, in the event of a dangerous situation, the information is used to rectify the problem in as safe a manner as possible. See *Figure 15* for an example of MSDS information you may find on the job.

12.1.0 Precautions When Using Solvents

It is always best to use a nonflammable, nontoxic solvent whenever possible. However, any time solvents are used, it is essential that your work area be adequately ventilated and that you wear the appropriate personal protective equipment:

- Wear a chemical face shield with chemical goggles to protect the eyes and skin from sprays and splashes.
- Wear a chemical apron to protect your body from sprays and splashes. Remember that some solvents are acid-based. If they come into contact with your clothes, solvents can eat through your clothes to your skin.
- A paper filter mask does not stop vapors; it is used only for nuisance dust. In situations where a paper mask does not supply adequate protection, chemical cartridge respirators might be needed. These respirators can stop many vapors if the correct cartridge is selected. In areas where ventilation is a serious problem, a self-contained breathing apparatus (SCBA) must be used.
- Make sure that you have been given a full medical evaluation and that you are properly trained in using respirators at your site.

12.2.0 Respiratory Protection

The best respiratory protection is to avoid the hazard entirely. Off-shift work, ventilation, and/or rescheduled work schedules should always be used to eliminate the need for working in areas with poor air quality. For example, in an area where hazardous solvents are used, the electrical work can be done off schedule when the solvents are not being used. When this cannot be done, protection against high concentrations of dust, mist, fumes, vapors, and gases is provided by appropriate respirators.

Appropriate respiratory protective devices should be used for the hazardous material involved and the extent and nature of the work performed.

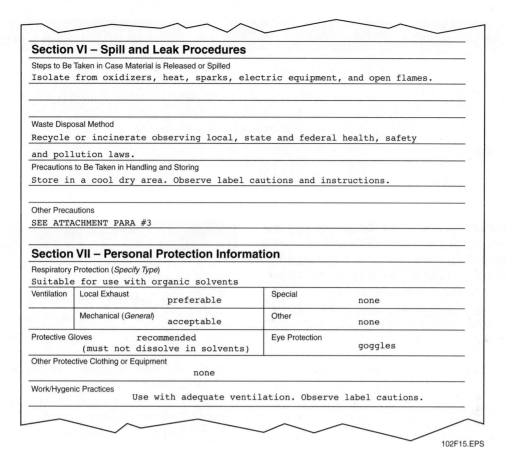

Section VI – Spill and Leak Procedures

Steps to Be Taken in Case Material is Released or Spilled
Isolate from oxidizers, heat, sparks, electric equipment, and open flames.

Waste Disposal Method
Recycle or incinerate observing local, state and federal health, safety

and pollution laws.
Precautions to Be Taken in Handling and Storing
Store in a cool dry area. Observe label cautions and instructions.

Other Precautions
SEE ATTACHMENT PARA #3

Section VII – Personal Protection Information

Respiratory Protection (*Specify Type*)
Suitable for use with organic solvents

Ventilation	Local Exhaust		Special	
		preferable		none
	Mechanical (*General*)		Other	
		acceptable		none
Protective Gloves	recommended		Eye Protection	
	(must not dissolve in solvents)			goggles

Other Protective Clothing or Equipment
none

Work/Hygenic Practices
Use with adequate ventilation. Observe label cautions.

102F15.EPS

Figure 15 ◆ Portion of an MSDS.

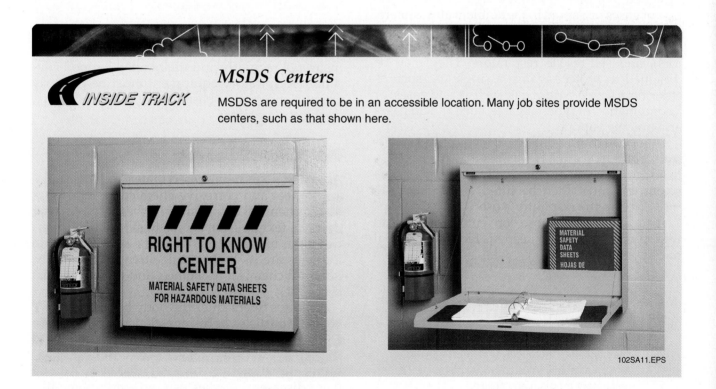

INSIDE TRACK

MSDS Centers

MSDSs are required to be in an accessible location. Many job sites provide MSDS centers, such as that shown here.

102SA11.EPS

Chemical Safety

The first line of defense with chemicals is to read and follow the directions found on the container. If you follow these instructions, you should be safe from chemical exposure. Be aware that everyone reacts differently to chemicals and you may be hypersensitive to a particular chemical that does not bother your co-workers. Leave the area at the first sign of an allergic reaction and seek medical attention.

An air-purifying respirator is, as its name implies, a respirator that removes contaminants from air inhaled by the wearer. The respirators may be divided into the following types: particulate-removing (mechanical filter), gas- and vapor-removing (chemical filter), and a combination of particulate-removing and gas- and vapor-removing.

Particulate-removing respirators are designed to protect the wearer against the inhalation of particulate matter in the ambient atmosphere. They may be designed to protect against a single type of particulate, such as pneumoconiosis-producing and nuisance dust, toxic dust, metal fumes or mist, or against various combinations of these types.

Gas- and vapor-removing respirators are designed to protect the wearer against the inhalation of gases or vapors in the ambient atmosphere. They are designated as gas masks, chemical cartridge respirators (nonemergency gas respirators), and self-rescue respirators. They may be designed to protect against a single gas such as chlorine; a single type of gas, such as acid gases; or a combination of types of gases, such as acid gases and organic vapors.

If you are required to use a respiratory protective device, you must be evaluated by a physician to ensure that you are physically fit to use a respirator. You must then be fitted and thoroughly instructed in the respirator's use.

Any employee whose job entails having to wear a respirator must keep his face free of facial hair in the seal area.

WARNING!

Do not use any respirator unless you have been fitted for it and thoroughly understand its use. As with all safety rules, follow your employer's respiratory program and policies.

Altered Respiratory Equipment

A self-employed man applied a solvent-based coating to the inside of a tank. Instead of wearing the proper respirator, he used nonstandard air supply hoses and altered the face mask. All joints and the exhalation slots were sealed with tape. He collapsed and was not discovered for several hours.

The Bottom Line: Never alter or improvise safety equipment.

Respiratory protective equipment must be inspected regularly and maintained in good condition. Respiratory equipment must be properly cleaned on a regular basis and stored in a sanitary, dustproof container.

13.0.0 ◆ ASBESTOS

Asbestos is a mineral-based material that is resistant to heat and corrosive chemicals. Depending on the chemical composition, asbestos fibers may range in texture from coarse to silky. The properties that make asbestos fibers so valuable to industry are its high tensile strength, flexibility, heat and chemical resistance, and good frictional properties.

Asbestos fibers enter the body by inhalation of airborne particles or by ingestion and can become embedded in the tissues of the respiratory or digestive systems. Exposure to asbestos can cause numerous disabling or fatal diseases. Among these diseases are asbestosis, an emphysema-like condition; lung cancer; mesothelioma, a cancerous tumor that spreads rapidly in the cells of membranes covering the lungs and body organs; and gastrointestinal cancer. The use of asbestos was banned in 1978.

Because asbestos was still in the manufacturing pipeline for a while after it was banned, you need to assume that any facility constructed before 1980 has asbestos in it. The owner must have a survey with any work rules needed to work safely around the asbestos. Common products that contain asbestos include thermal pipe insulation, mastic for ducts and insulation, spray-on fireproofing, floor tiles, ceiling tiles, roof insulation, exterior building sheathing, old wire insulation, and even pipe. As an electrician, you must not drill through or otherwise work with asbestos—you can only be trained to work around it when it can be done safely. Asbestos work is a trade that requires special training and protective equipment.

The following signs must be placed in areas containing asbestos.

DANGER
ASBESTOS
CANCER AND LUNG DISEASE HAZARD
AUTHORIZED PERSONNEL ONLY
RESPIRATORS AND PROTECTIVE CLOTHING ARE REQUIRED IN THIS AREA

DANGER
CONTAINS ASBESTOS FIBERS
AVOID CREATING DUST
CANCER AND LUNG DISEASE HAZARD

14.0.0 ◆ BATTERIES

Working around wet cell batteries can be dangerous if the proper precautions are not taken. Batteries often give off hydrogen gas as a byproduct. When hydrogen mixes with air, the mixture can be explosive in the proper concentration. For this reason, smoking is strictly prohibited in battery rooms, and only insulated tools should be used. Proper ventilation also reduces the chance of explosion in battery areas. Follow your company's procedures for working near batteries. Also, ensure that your company's procedures are followed for lifting heavy batteries.

WARNING!
Battery-powered scissor lifts may have unsealed batteries that require water levels to be checked and topped off. Never use a flame to look inside a battery—it can cause an explosion. Charging batteries with low water levels can damage them.

14.1.0 Acids

Batteries also contain acid, which will eat away human skin and many other materials. Personal protective equipment for battery work typically includes chemical aprons, sleeves, gloves, face shields, and goggles to prevent acid from contacting skin and eyes. Follow your site procedures for dealing with spills of these materials. Also, know the location of first aid when working with these chemicals.

14.2.0 Wash Stations

Because of the chance that battery acid may contact someone's eyes or skin, wash stations are located near battery rooms. Do not connect or disconnect batteries without proper supervision. Everyone who works in the area should know where the nearest wash station is and how to use it. Battery acid should be flushed from the skin and eyes with large amounts of water or with a neutralizing solution.

CAUTION
If you come in contact with battery acid, flush the affected area with water and report it immediately to your supervisor.

15.0.0 ◆ PCBs AND VAPOR LAMPS

Polychlorinated biphenyls (PCBs) are chemicals that were marketed under various trade names as a liquid insulator/cooler in older transformers. In addition to being used in older transformers, PCBs are also found in some large capacitors and in the small ballast transformers used in street lighting and ordinary fluorescent light fixtures. Disposal of these materials is regulated by the Environmental Protection Agency (EPA) and must be done through a regulated disposal company; use extreme caution and follow your facility procedures.

In addition, any vapor lamps, such as fluorescent, halide, or mercury vapor lamps, must be recycled. The tubes must be packaged and handled carefully to avoid breakage.

16.0.0 ◆ FALL PROTECTION

All employees must receive documented training before working in any area where there is the possibility of exposure to a fall of 6' or more. This training must be renewed annually and must include ladder safety. The 6' rule does not apply to ladders, scaffolds, and mechanical lifts, which have their own standards.

16.1.0 Fall Protection Procedures

Fall protection must be used when employees are on a walking or working surface that is 6' or more above a lower level and has an unprotected edge or side. The areas covered include, but are not limited to the following:

- Finished and unfinished floors or mezzanines
- Temporary or permanent walkways/ramps
- Finished or unfinished roof areas
- Elevator shafts and hoistways
- Floor, roof, or walkway holes
- Working 6' or more above dangerous equipment

Exception: If the dangerous equipment is unguarded, fall protection must be used at all heights regardless of the fall distance.

Figure 16 shows a simple temporary guardrail installed on the stairway of a residence under construction. *Figure 17* shows a complex guardrail system used during the construction of a department store. The large central opening will eventually house the building escalators.

NOTE

Walking/working surfaces do not include ladders, scaffolds, vehicles, or trailers. Also, an unprotected edge or side is an edge/side where there is no guardrail system at least 39" high.

102F16.EPS

Figure 16 ◆ Temporary guardrail on stairs.

102F17.EPS

Figure 17 ◆ Complex guardrail system.

According to OSHA, an employee can never be exposed to a fall of more than 6'. This is called 100% fall protection. The employee must be protected by one of the following in this order of preference:

1. Guardrail systems
2. Personal fall arrest systems (PFAS)
3. Controlled access zone or other administrative system

16.1.1 Guardrail Systems

Guardrail systems must be constructed as follows:

- The top rails must be 42" (±3") and must be capable of withstanding 200 lbs.
- The mid-rails must be at 21" (±3") and must be capable of withstanding 150 lbs.

What's wrong with this picture?

102SA12.EPS

- The toe board must be 4" tall and no more than ¼" above the floor for drainage.
- Guardrails can be made of 2 × 4s, pipes, chains, or cables.
- Chains and cables require flags every 6'. If cables are used, they must be secured to avoid deflection greater than 3" in, out, or down from the 42" requirement.
- Banding material is not allowed for guardrail construction. Cable guardrails require the use of cable clamps on the cable. Clamps must be forged and not malleable. They must be torqued and installed properly. Ensure that the clamp does not damage the loadbearing line.

NOTE

These ratings apply to all portions of the rail system such as anchors, anchorage material, and clamps. Guardrails cannot be used to secure a PFAS unless they are designed to support 5,000 lbs per person.

Warning lines, signs, or barricades must be installed back from the edge at least 6'. That way, even if someone falls over the barrier, they will not fall to the level below.

16.1.2 Personal Fall Arrest Systems (PFAS)

PFAS provide fall arrest after an employee falls. This equipment must be selected, inspected, donned, anchored, and maintained to be effective. The complete system usually consists of a full-body harness, lanyard, and anchorage device.

Full-body harnesses – Full-body harnesses are the only acceptable equipment to wear for PFAS. Select the appropriate harness based on size and gender. Inspect the equipment before use. Harnesses must be worn snug (but not tight) with all required straps attached. When properly applied, you should be able to slide two fingers under the straps with little difficulty. The D-ring in the back of the harness must be centered between the shoulder blades. After donning the harness, have a co-worker pull sharply up on the D-ring. You

What's wrong with this picture?

102SA13.EPS

TIE OFF SEPARATE LANYARDS TO
DIFFERENT D-RINGS

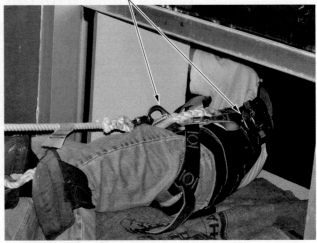

102F18.EPS

Figure 18 ◆ Electrician tied off to two lanyards.

should feel the grab around the thighs, chest, and buttocks. Jobs that require positioning must be accomplished using a full-body harness with side D-rings. Safety belts are not allowed.

Lanyards – Lanyards are used to connect the harness to the attachment point. As no employee can be exposed to a fall of more than 6', standard lanyards must be no longer than 6'. You can be exposed to 1,800 lbs of force in a properly worn harness. The use of shock absorbing lanyards or retractable lanyards can reduce that force to as low as 400 to 600 lbs. Shock absorbers work by slowing the employee to a stop by ripping stitches while elongating up to 42". All lanyards must have locking snap hooks. Never attach two locking snap hooks to the same D-ring as they can foul each other, causing the relatively weak gates to break. *Figure 18* shows an electrician making an adjustment to an outdoor wallwasher lighting fixture while hanging out of a 15th-story window. For extra protection, he is attached to two lanyards on two separate D-rings.

A twin-tailed lanyard is required when climbing. While climbing, you cannot unhook your lanyard to move it to another anchorage and still have 100% fall protection. Thus, with two lanyards, you can "walk" to where you are working.

Retractable lanyards come in a variety of sizes from 10' to over 150'. *Figure 19* shows a worker

RETRACTABLE
LANYARD

102F19.EPS

Figure 19 ◆ Proper fall protection on a boom lift.

tied off to a retractable lanyard when working on a boom lift. Retractable lanyards are used where movement or close proximity with the ground will render standard lanyards ineffective or inefficient. Retractable lanyards can be longer than 6' because when a fall occurs they grab hold within 2'. This quick reaction also eliminates the need for a shock absorber.

WARNING!
Do not put a shock absorber in line with a retractable lanyard because it may interfere with the quick response of the retractable lanyard.

Anchorage devices – Anchorage devices and points are the interface between the PFAS and the structure to which they are attached. This point must hold 5,000 lbs (this is the equivalent of a full-size extended cab pickup truck).

A lanyard cannot be wrapped around an anchorage and then attached to itself unless it is specially designed with cross arm straps. Cross arm straps are made of webbing, two inches wide, of any necessary length, with two different size D-rings. The cross arm straps are passed over whatever object you are going to attach them to and wrapped around to reduce the length of the lanyard. Beam clamps, wire hangers, trolleys, and other manufactured devices are also used for specific applications.

What's wrong with this picture?

102SA14.EPS

Equipment inspection process – All PFAS must be inspected when received and before each use. Check the manufacturer's tag for the manufacturer's inspection date. If the fall equipment has no date, it should be disposed of immediately. Carefully look over the webbing. If you observe any burns, ripped stitches, color marker threads, distorted grommets, bent or cracked buckle tongues, distorted D-rings, or bent, cracked, or pulled fabrics, remove the PFAS from service. Retractable lanyards must undergo the same inspection process as other PFAS. In addition, you must also pull out the entire lanyard for inspection, let it back in, and then pull out 2' to 4' of lanyard and give it a swift tug to see if it properly engages. If any of these inspections fail, the equipment must be destroyed immediately or tagged DO NOT USE and returned to the shop.

WARNING!
All fall protection equipment that is involved in a fall must be taken out of service and destroyed. Any employees involved in a fall must receive medical attention, even if they do not feel they have been injured. Falls can cause internal injuries that are not readily apparent to the victim.

Rescue – Never pull anyone up by their fall protection; always rescue them with ladders or equipment from below. If the standard equipment is not available to provide rescue, a plan must be created before work can proceed. Rescues must be accomplished from below using ladders, lifts, and/or scaffolds.

WARNING!
Unless a person is in immediate danger, never attempt to lift them up by their lanyard. This could cause an additional drop for the fallen worker and/or injure the rescuers.

Immediately summon the fire department to assist in the rescue effort unless you can rescue the person without assistance. Rescue must take place as quickly as possible, as hanging from a harness presents additional hazards. If you fall, continue to move your limbs while awaiting rescue. This will help maintain circulation in your lower extremities.

Putting It All Together

THINK ABOUT IT

This module has described a professional approach to electrical safety. How does this professional outlook differ from an everyday attitude? What do you think are the key features of a professional philosophy of safety?

 WARNING!

Any employee whose weight exceeds 310 pounds, including their tools, cannot wear fall protection, as that is the maximum weight for which it is designed.

PFAS selection – The type of system selected depends on the fall hazards associated with the work to be performed. First, a hazard analysis must be conducted by the job site supervisor prior to the start of work. Based on the hazard analysis, the job site supervisor and project manager, in consultation with the safety manager, will select the appropriate fall protection system. All employees must be instructed in the use of the fall protection system before starting work.

16.1.3 Controlled Access Zones

There are times when a guardrail cannot be attached to the building. In these cases, a controlled access zone must be installed. A controlled access zone may consist of guards, barricades, badge systems, or other administrative measures.

Figure 20 shows a controlled access zone on a roof. Note that the barricade is located 6' from the edge. This way, even if someone trips over the barricade, they won't fall off the roof.

SIX FEET FROM EDGE

102F20.EPS

Figure 20 ◆ Controlled access zone.

1. The most life-threatening hazards on a construction site typically include all of the following *except* _____.
   a. falls
   b. electric shock
   c. being crushed or struck by falling or flying objects
   d. chemical burns

2. If a person's heart begins to fibrillate due to an electrical shock, the solution is to _____.
   a. leave the person alone until the fibrillation stops
   b. immerse the person in ice water
   c. use the Heimlich maneuver
   d. have a qualified person use emergency defibrillation equipment

3. Low-voltage conductors rarely cause injuries.
   a. True
   b. False

4. Class 00 rubber gloves are used when working with voltages less than _____.
   a. 500 volts
   b. 1,000 volts
   c. 5,000 volts
   d. 7,500 volts

5. An important use of a hot stick is to _____.
   a. replace busbars
   b. test for voltage
   c. replace fuses
   d. test for continuity

6. Which of these statements correctly describes a double-insulated power tool?
   a. There is twice as much insulation on the power cord.
   b. It can safely be used in place of a grounded tool.
   c. It is made entirely of plastic or other nonconducting material.
   d. The entire tool is covered in rubber.

7. Which of the following applies in a lockout/tagout procedure?
   a. Only the supervisor can install lockout/tagout devices.
   b. If several employees are involved, the lockout/tagout equipment is applied only by the first employee to arrive at the disconnect.
   c. Lockout/tagout devices applied by one employee can be removed by another employee as long as it can be verified that the first employee has left for the day.
   d. Lockout/tagout devices are installed by every authorized employee involved in the work.

8. The *NEC*® provides requirements for safe electrical installations, while *NFPA 70E* provides guidance for establishing safe electrical workplaces.
   a. True
   b. False

9. What is the proper distance from the feet of a straight ladder to the wall?
   a. one-fourth the working height of the ladder
   b. one-half the height of the ladder
   c. three feet
   d. one-fourth of the square root of the height of the ladder

10. What are the minimum and maximum distances (in inches) that a scaffold plank can extend beyond its end support?
    a. 4; 8
    b. 6; 10
    c. 6; 12
    d. 8; 12

11. All of the following are considered confined spaces *except* a _____.
    a. 3 ft trench
    b. sewer
    c. pipeline
    d. manhole

12. The best way to protect yourself from solvent hazards is to _____.
a. always wear vinyl gloves and a paper filter mask
b. ask your supervisor or a co-worker
c. ask the supplier
d. read and follow all instructions on the product's MSDS

13. Asbestos was banned in _____; therefore, you must assume that any facility constructed before _____ has asbestos in it.
a. 1943; 1945
b. 1964; 1966
c. 1978; 1980
d. 1988; 1990

14. You may throw vapor lamps out with regular trash as long as you wrap them carefully to avoid breakage.
a. True
b. False

15. A PFAS anchorage point must be able to hold _____ lbs.
a. 250
b. 500
c. 1,000
d. 5,000

Summary

Safety must be your concern at all times so that you do not become either the victim of an accident or the cause of one. Safety requirements and safe work practices are provided by OSHA and your employer. It is essential that you adhere to all safety requirements and follow your employer's safe work practices and procedures. Also, you must be able to identify the potential safety hazards of your job site. The consequences of unsafe job site conduct can often be expensive, painful, or even deadly. Report any unsafe act or condition immediately to your supervisor. You should also report all work-related accidents, injuries, and illnesses to your supervisor immediately. Remember, proper construction techniques, common sense, and a good safety attitude will help to prevent accidents, injuries, and fatalities.

Notes

1. A life-threatening condition of the heart in which the muscle fibers contract irregularly is called _FIBRILLATION_

2. Any tool that has a case made of nonconductive material and that has been constructed so that the case is insulated from electrical energy is a _DOUBLE-INSULATED/UNGROUNDED TOOL_

3. _PCB'S_ are chemicals often found in liquids that are used to cool certain types of large transformers and capacitors.

4. A _GFCI_ will de-energize a circuit or a portion of it if the current to ground exceeds some predetermined value.

5. A _GROUNDED TOOL_ has a three-prong plug at the end of its power cord or some other means to ensure that stray current travels to ground without passing through the body of the operator.

Trade Terms

Double-insulated/ungrounded tool
Fibrillation
Grounded tool
Ground fault circuit interrupter (GFCI)
Polychlorinated biphenyls (PCBs)

Michael J. Powers
Tri-City Electrical Contractors, Inc.

How did you choose a career in the electrical field?
My father was an electrician and after I "burned out" with a career in fast-food management, I decided to choose a completely different field.

Tell us about your apprenticeship experience.
It was excellent! I worked under several very knowledgeable electricians and had a pretty good selection of teachers. Over my four-year apprenticeship, I was able to work on a variety of jobs, from photomats to kennels to colleges.

What positions have you held and how did they help you to get where you are now?
I have been an electrical apprentice, a licensed electrician, a job-site superintendent, a master electrician, and am currently a corporate safety and training director. The knowledge I acquired in electrical theory in apprenticeship school and preparing for my licensing exams, as well as the practical on-the-job experience over thirty years in the trade, were wonderful training for my current position.

I also serve on the authoring team for NCCER's Electrical Curricula, which has provided me with not only the opportunity to share what I have learned, but is also a great way to keep current in other areas by meeting with electricians from a variety of disciplines (we have commercial, residential, and industrial electricians on the team, as well as instructors).

What would you say was the single greatest factor that contributed to your success?
Choosing a company that recognized and rewarded competent, hard workers and provided them with the support and guidance to allow them to develop and succeed in the industry.

What does your current job entail?
I am responsible for safe work practices and procedures through the job-site management team at Tri-City Electrical Contractors, which currently has a workforce of over 1,100. I also assist in developing, delivering, and administering the training program, from apprenticeship to in-house to outsourced training.

What advice do you have for trainees?
Training in all its aspects is the key to your success and advancement in the industry. Any time you are given a training opportunity, take it, even if it might not appear relevant at the time. Eventually, all knowledge can be applied to some situation.

Most importantly, have fun! The construction industry is composed of good people. I firmly believe that construction workers, as a group, are much more honest and direct than any other comparable group. Wait—did I say comparable group? That's a misstatement—there is no comparable group. Construction workers build America!

Trade Terms
Introduced in This Module

Double-insulated/ungrounded tool: An electrical tool that is constructed so that the case is insulated from electrical energy. The case is made of a nonconductive material.

Fibrillation: Very rapid irregular contractions of the muscle fibers of the heart that result in the heartbeat and pulse going out of rhythm with each other.

Grounded tool: An electrical tool with a three-prong plug at the end of its power cord or some other means to ensure that stray current travels to ground without passing through the body of the user. The ground plug is bonded to the conductive frame of the tool.

Ground fault circuit interrupter (GFCI): A protective device that functions to de-energize a circuit or portion thereof within an established period of time when a current to ground exceeds some predetermined value. This value is less than that required to operate the overcurrent protective device of the supply circuit.

Polychlorinated biphenyls (PCBs): Toxic chemicals that may be contained in liquids used to cool certain types of large transformers and capacitors.

This module is intended to present thorough resources for task training. The following reference works are suggested for further study. These are optional materials for continued education rather than for task training.

29 CFR Parts 1900–1910, Standards for General Industry. Occupational Safety and Health Administration, U.S. Department of Labor.

29 CFR Part 1926, Standards for the Construction Industry. Occupational Safety and Health Administration, U.S. Department of Labor.

National Electrical Code® Handbook, Latest Edition. Quincy, MA: National Fire Protection Association.

Standard for Electrical Safety in the Workplace, Latest Edition. Quincy, MA: National Fire Protection Association.

NCCER makes every effort to keep these textbooks up-to-date and free of technical errors. We appreciate your help in this process. If you have an idea for improving this textbook, or if you find an error, a typographical mistake, or an inaccuracy in NCCER's Contren® textbooks, please write us, using this form or a photocopy. Be sure to include the exact module number, page number, a detailed description, and the correction, if applicable. Your input will be brought to the attention of the Technical Review Committee. Thank you for your assistance.

Instructors – If you found that additional materials were necessary in order to teach this module effectively, please let us know so that we may include them in the Equipment/Materials list in the Annotated Instructor's Guide.

Write: Product Development and Revision
National Center for Construction Education and Research
3600 NW 43rd St., Bldg. G, Gainesville, FL 32606

Fax: 352-334-0932

E-mail: curriculum@nccer.org

Craft _____ Module Name _____

Copyright Date _____ Module Number _____ Page Number(s) _____

Description _____

_____

_____

_____

(Optional) Correction _____

_____

_____

(Optional) Your Name and Address _____

_____

_____

Introduction to Electrical Circuits

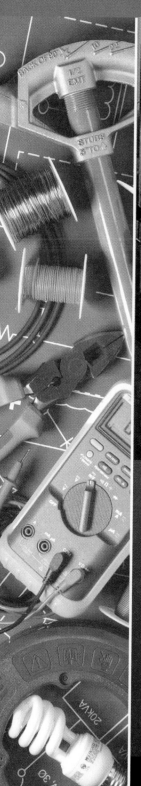

Reno Transportation Rail Access Corridor

The Reno Transportation Rail Access Corridor (ReTRAC) is the largest public works project ever undertaken in Nevada and includes the design and construction of a 2.1-mile long, 54-foot wide, 33-foot deep train trench running through downtown Reno. ReTRAC was built using the design-build method, resulting in shorter construction times, lessened traffic impacts, and lower costs.

26103-08

26103-08
Introduction to Electrical Circuits

Topics to be presented in this module include:

Overview

The foundation for successful and safe electrical installations and trouble-shooting is a sound understanding of electrical theory. Electrical theory involves the study of atoms, their reactions, and their involvement in electrical circuits. Electricians must understand electrical theory to fully understand the roles that voltage, current, and resistance play in electrical systems.

The primary action in any designed electrical circuit or system is the controlled flow of electrons. Electricians must know what electrons are, what makes them flow, how their flow is controlled, and how this flow is used to perform work. In addition, they must know what to expect if an unintentional or catastrophic flow of electrons should occur.

Having a solid understanding of electrical theory enables electricians to complete quality installations and troubleshoot a circuit or electrical system quickly and efficiently. These skills are fundamental for a successful career as an electrician.

Objectives

When you have completed this module, you will be able to do the following:

1. Define voltage and identify the ways in which it can be produced.
2. Explain the difference between conductors and insulators.
3. Define the units of measurement that are used to measure the properties of electricity.
4. Identify the meters used to measure voltage, current, and resistance.
5. Explain the basic characteristics of series and parallel circuits.

Trade Terms

Ammeter	Ohmmeter
Ampere (A)	Ohm's law
Atom	Power
Battery	Protons
Circuit	Relay
Conductor	Resistance
Coulomb	Resistor
Current	Schematic
Electron	Series circuit
Insulator	Solenoid
Joule (J)	Transformer
Kilo	Valence shell
Matter	Volt (V)
Mega	Voltage
Neutrons	Voltage drop
Nucleus	Voltmeter
Ohm (Ω)	Watt (W)

Required Trainee Materials

1. Paper and pencil
2. Appropriate personal protective equipment
3. Calculator

Prerequisites

Before you begin this module, it is recommended that you successfully complete *Core Curriculum* and *Electrical Level One*, Modules 26101-08 and 26102-08.

This course map shows all of the modules in *Electrical Level One*. The suggested training order begins at the bottom and proceeds up. Skill levels increase as you advance on the course map. The local Training Program Sponsor may adjust the training order.

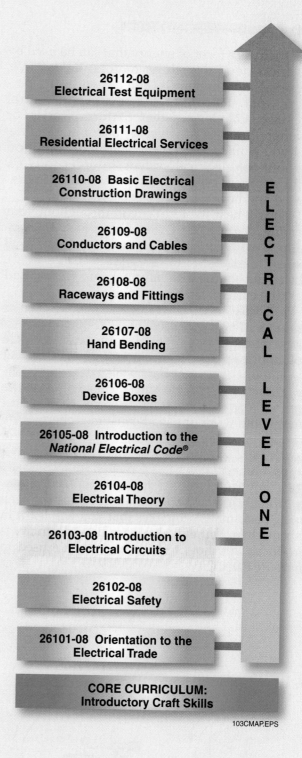

26112-08
Electrical Test Equipment

26111-08
Residential Electrical Services

26110-08 Basic Electrical
Construction Drawings

26109-08
Conductors and Cables

26108-08
Raceways and Fittings

26107-08
Hand Bending

26106-08
Device Boxes

26105-08 Introduction to the
National Electrical Code®

26104-08
Electrical Theory

26103-08 Introduction to
Electrical Circuits

26102-08
Electrical Safety

26101-08 Orientation to the
Electrical Trade

CORE CURRICULUM:
Introductory Craft Skills

ELECTRICAL LEVEL ONE

103CMAP.EPS

1.0.0 ◆ INTRODUCTION

Electricity is a form of energy that can be used by electrical devices such as motors, lights, TVs, heaters, and numerous other devices to perform work. Electricity is also used to control non-electrical devices that perform work. For example, your car is driven by a gasoline engine, but you wouldn't be able to start it or turn it off without the electrical system. In order to work with electricity, you need to know how it is produced and how it acts in electrical circuits.

You will hear the term circuit throughout your training. An electrical circuit contains, at minimum, a voltage source, a load, and conductors (wires) to carry the electrical current (*Figure 1*). The circuit should also have a means to stop and start the current, such as a switch.

Electricity is all about cause and effect. The presence of voltage (volts) in a closed circuit will cause current (amps) to flow. The more voltage you apply, the more current will flow. However, the amount of current flow is also determined by how much resistance (ohms) the load offers to the flow of current. In order to convert electrical energy into work, the load consumes energy. The amount of energy a device consumes is called power, and is expressed in watts (W). Volts (V), amps, ohms, and watts are related in such a way that if any one of them changes, the others are proportionally affected. This relationship can be seen using basic math principles that you will learn in this module. You will also learn how electricity is produced and how test instruments are used to measure electricity.

2.0.0 ◆ ATOMIC THEORY

In order to understand electrical theory, you must first understand the basic concepts of atomic theory. Atomic theory explains the construction and behavior of atoms, including the transfer of electrons that results in current flow.

2.1.0 The Atom

The atom is the smallest part of an element that enters into a chemical change, but it does so in the form of a charged particle. These charged particles are called ions, and are of two types—positive and negative. A positive ion may be defined as an atom that has become positively charged. A negative ion may be defined as an atom that has become negatively charged. One of the properties of charged ions is that ions of the same charge tend to repel one another, whereas ions of unlike charge will attract one another. The term charge can be taken to mean a quantity of electricity that is either positive or negative.

The structure of an atom is best explained by a detailed analysis of the simplest of all atoms, that of the element hydrogen. The hydrogen atom in *Figure 2* is composed of a nucleus containing one proton and a single orbiting electron. As the electron revolves around the nucleus, it is held in this orbit by two counteracting forces. One of these forces is called centrifugal force, which is the force that tends to cause the electron to fly outward as it travels around its circular orbit. The second force acting on the electron is electrostatic force. This force tends to pull the electron in toward the

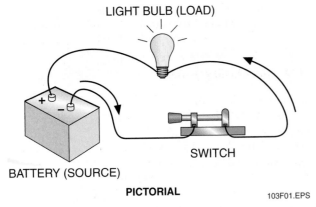

Figure 1 ◆ Basic electrical circuit.

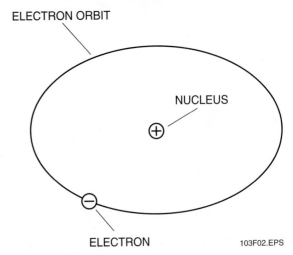

Figure 2 ◆ Hydrogen atom.

nucleus and is provided by the mutual attraction between the positive nucleus and the negative electron. At some given radius, the two forces will balance each other, providing a stable path for the electron.

- A proton (+) repels another proton (+).
- An electron (−) repels another electron (−).
- A proton (+) attracts an electron (−).

Basically, an atom contains three types of subatomic particles that are of interest in electricity: electrons, protons, and neutrons.

The protons and neutrons are located in the center, or nucleus, of the atom, and the electrons travel about the nucleus in orbits.

Because protons are relatively heavy, the repulsive force they exert on one another in the nucleus of an atom has little effect.

The attracting and repelling forces on charged materials occur because of the electrostatic lines of force that exist around the charged materials. In a negatively charged object, the lines of force of the excess electrons add to produce an electrostatic field that has lines of force coming into the object from all directions. In a positively charged object, the lines of force of the excess protons add to produce an electrostatic field that has lines of force going out of the object in all directions. The electrostatic fields either aid or oppose each other to attract or repel.

2.1.1 The Nucleus

The nucleus is the central part of the atom. It is made up of heavy particles called protons and neutrons. The proton is a charged particle containing the smallest known unit of positive electricity. The neutron has no electrical charge. The

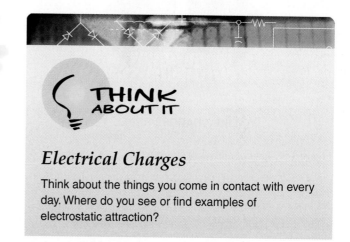

THINK ABOUT IT

Electrical Charges

Think about the things you come in contact with every day. Where do you see or find examples of electrostatic attraction?

number of protons in the nucleus determines how the atom of one element differs from the atom of another element.

Although a neutron is actually a particle by itself, it is generally thought of as an electron and proton combined and is electrically neutral. Since neutrons are electrically neutral, they are not considered important to the electrical nature of atoms.

2.1.2 Electrical Charges

The negative charge of an electron is equal but opposite to the positive charge of a proton. The charges of an electron and a proton are called electrostatic charges. The lines of force associated with each particle produce electrostatic fields. Because of the way these fields act together, charged particles can attract or repel one another. The Law of Electrical Charges states that particles with like charges repel each other and those with unlike charges attract each other. This is shown in *Figure 3*.

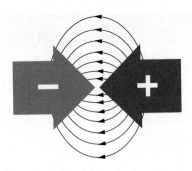

UNLIKE CHARGES ATTRACT

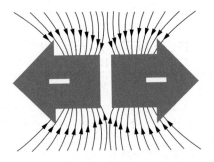

LIKE CHARGES REPEL

103F03.EPS

Figure 3 ◆ Law of electrical charges.

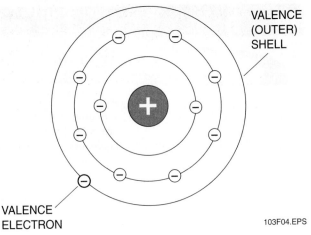

VALENCE
(OUTER)
SHELL

VALENCE
ELECTRON

103F04.EPS

Figure 4 ◆ Valence shell and electrons.

2.2.0 Conductors and Insulators

The difference between atoms, with respect to chemical activity and stability, depends on the number and position of the electrons included within the atom. In general, the electrons reside in groups of orbits called shells. The shells are arranged in steps that correspond to fixed energy levels.

The outer shell of an atom is called the **valence shell**, and the electrons contained in this shell are called valence electrons (*Figure 4*). The number of valence electrons determines an atom's ability to gain or lose an electron, which in turn determines the chemical and electrical properties of the atom. An atom that is lacking only one or two electrons from its outer shell will easily gain electrons to complete its shell, but a large amount of energy is required to free any of its electrons. An atom having a relatively small number of electrons in its outer shell in comparison to the number of electrons required to fill the shell will easily lose these valence electrons.

It is the valence electrons that we are most concerned with in electricity. These are the electrons that are easiest to break loose from their parent atom. Normally, a **conductor** has three or less

valence electrons, an **insulator** has five or more valence electrons, and semiconductors usually have four valence electrons.

All the elements of which **matter** is made may be placed into one of three categories: conductors, insulators, and semiconductors.

Conductors, for example, are elements such as copper and silver that will conduct a flow of electricity very readily. Because of their good conducting abilities, they are formed into wire and used whenever it is desired to transfer electrical energy from one point to another.

Insulators, on the other hand, do not conduct electricity to any great degree and are used when it is desirable to prevent the flow of electricity. Compounds such as porcelain and plastic are good insulators.

Materials such as germanium and silicon are not good conductors but cannot be used as insulators either, since their electrical characteristics fall between those of conductors and those of insulators. These in-between materials are classified as semiconductors. As you will learn later in your training, semiconductors play a crucial role in electronic circuits.

2.3.0 Magnetism

The operation of many electrical components relies on the power of magnetism. Motors, **relays**, **transformers**, and **solenoids** are examples. Magnetized iron generates a magnetic field consisting of magnetic lines of force, also known as magnetic flux lines (*Figure 5*). Magnetic objects within the field will be attracted or repelled by the magnetic field. The more powerful the magnet, the more powerful the magnetic field around

PERMANENT MAGNET

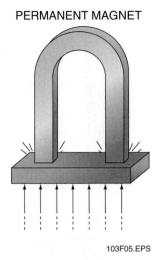

103F05.EPS

Figure 5 ◆ Magnetism.

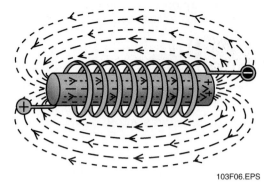

103F06.EPS

Figure 6 ◆ Electromagnet.

it. Each magnet has a north pole and a south pole. Opposing poles attract each other; like poles repel each other.

Electricity also produces magnetism. Current flowing through a conductor produces a small magnetic field around the conductor. If the conductor is coiled around an iron bar, the result is an electromagnet (*Figure 6*) that attracts and repels other magnetic objects just like an iron magnet. This is the basis on which electric motors and other components operate.

3.0.0 ◆ ELECTRICAL POWER GENERATION AND DISTRIBUTION

Electricity comes from electrical generating plants (*Figure 7*) operated by utilities like your local power company. Steam from coal-burning or nuclear power plants is used to power huge generators called turbines, which generate electricity. There are also hydroelectric power plants where water flowing through dams is used to drive turbines.

The electrical power that travels through long-distance transmission lines may be as high as 750,000 volts (V). Devices known as transformers

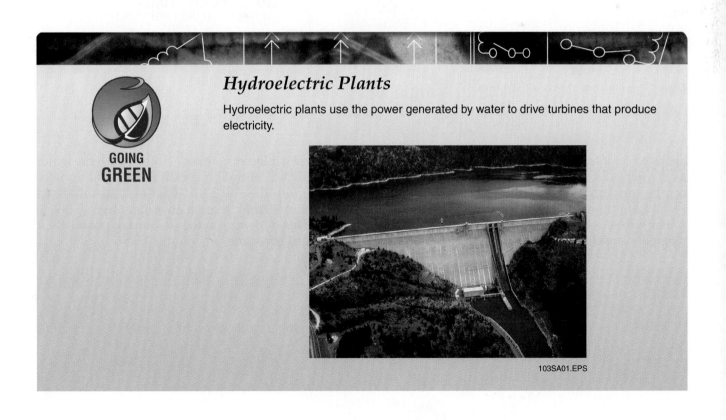

GOING GREEN

Hydroelectric Plants

Hydroelectric plants use the power generated by water to drive turbines that produce electricity.

103SA01.EPS

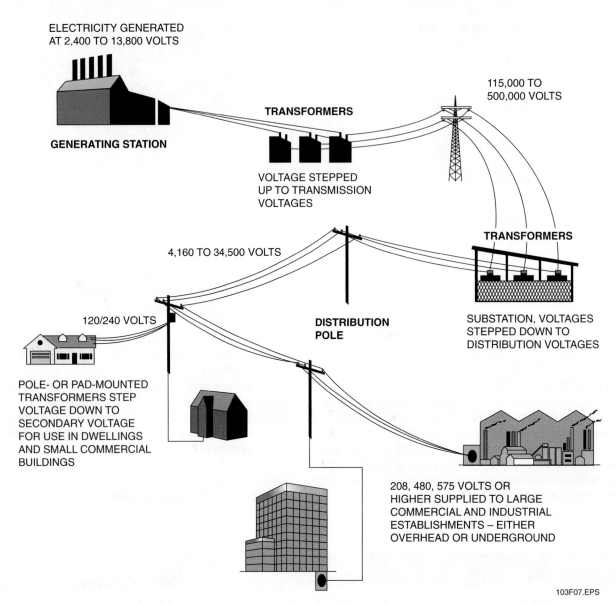

ELECTRICITY GENERATED
AT 2,400 TO 13,800 VOLTS

GENERATING STATION

TRANSFORMERS

VOLTAGE STEPPED
UP TO TRANSMISSION
VOLTAGES

115,000 TO
500,000 VOLTS

TRANSFORMERS

SUBSTATION, VOLTAGES
STEPPED DOWN TO
DISTRIBUTION VOLTAGES

4,160 TO 34,500 VOLTS

DISTRIBUTION
POLE

120/240 VOLTS

POLE- OR PAD-MOUNTED
TRANSFORMERS STEP
VOLTAGE DOWN TO
SECONDARY VOLTAGE
FOR USE IN DWELLINGS
AND SMALL COMMERCIAL
BUILDINGS

208, 480, 575 VOLTS OR
HIGHER SUPPLIED TO LARGE
COMMERCIAL AND INDUSTRIAL
ESTABLISHMENTS – EITHER
OVERHEAD OR UNDERGROUND

103F07.EPS

Figure 7 ◆ Electrical power distribution.

are used to step the voltage down to lower levels as it reaches electrical substations and eventually our homes, offices, and factories. The voltage we receive at home is usually about 240V. At the wall outlet where we plug in small appliances such as televisions and toasters, the voltage is about 120V (*Figure 8*). Electric stoves, clothes dryers, water heaters, and central air conditioning systems usually require the full 240V. Commercial buildings and factories may receive anywhere from 208V to 575V. This depends on the amount of power their machines consume.

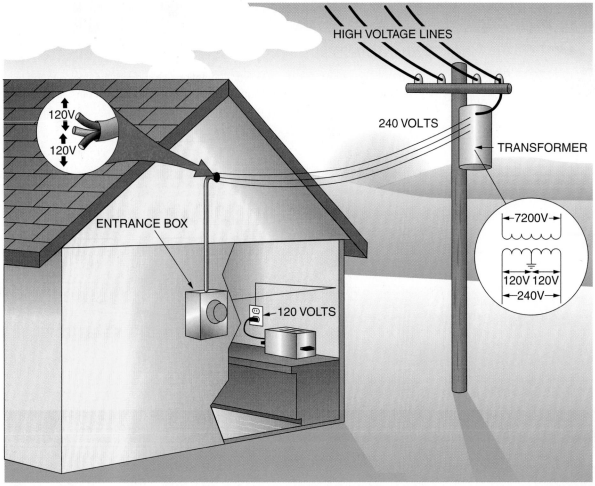

High voltage lines

120V
120V

240 VOLTS

TRANSFORMER

ENTRANCE BOX

120 VOLTS

7200V

120V 120V
240V

103F08.EPS

Figure 8 ◆ Internal power distribution.

4.0.0 ◆ ELECTRIC CHARGE AND CURRENT

An electric charge has the ability to do the work of moving another charge by attraction or repulsion. The ability of a charge to do work is called its potential. When one charge is different from another, there is a difference in potential between them. The sum of the difference of potential of all the charges in the electrostatic field is referred to as electromotive force (emf) or voltage. Voltage is frequently represented by the letter E.

Electric charge is measured in coulombs. An electron has 1.6×10^{-19} coulombs of charge. Therefore, it takes 6.25×10^{18} electrons to make up one coulomb of charge, as shown below.

$$\frac{1}{1.6 \times 10^{-19}} = 6.25 \times 10^{18} \text{ electrons}$$

If two particles, one having charge Q_1 and the other charge Q_2, are a distance (d) apart, then the force between them is given by Coulomb's law,

which states that the force is directly proportional to the product of the two charges and inversely proportional to the square of the distance between them:

$$\text{Force} = \frac{k \times Q_1 \times Q_2}{d^2}$$

If Q_1 and Q_2 are both positive or both negative, then the force is positive; it is repulsive. If Q_1 and Q_2 are of opposite charges, then the force is negative; it is attractive. The letter k equals a constant with a value of 10^9.

4.1.0 Current Flow

The movement of the flow of electrons is called current. To produce current, the electrons are moved by a potential difference. Current is represented by the letter I. The basic unit in which current is measured is the ampere (A), also called the amp. The symbol for the ampere is A. One ampere of current is defined as the movement of one

Transformers

Large distribution transformers at power substations step down the power to the level required for local distribution. Pole transformers like the one shown here step it down further to the voltages needed for homes and businesses.

103SA02.EPS

Units of Electricity and Volta

A disagreement with a fellow scientist over the twitching of a frog's leg eventually led 18th-century physicist Alessandro Volta to theorize that when certain objects and chemicals come into contact with each other, they produce an electric current. Believing that electricity came from contact between metals only, Volta coined the term metallic electricity. To demonstrate his theory, Volta placed two discs, one of silver and the other of zinc, into a weak acidic solution. When he linked the discs together with wire, electricity flowed through the wire. Thus, Volta introduced the world to the battery, also known as the Voltaic pile. Now Volta needed a term to measure the strength of the electric push or the flowing charge; the volt is that measure.

coulomb past any point of a conductor during one second of time. One coulomb is equal to 6.25×10^{18} electrons; therefore, one ampere is equal to 6.25×10^{18} electrons moving past any point of a conductor during one second of time.

The definition of current can be expressed as an equation:

$$I = \frac{Q}{T}$$

Where:

I = current (amperes)

Q = charge (coulombs)

T = time (seconds)

Charge differs from current in that charge (Q) is an accumulation of charge, while current (I) measures the intensity of moving charges.

In a conductor, such as copper wire, the free electrons are charges that can be forced to move with relative ease by a potential difference. If a potential difference is connected across two ends of a copper wire, as shown in *Figure 9,* the applied voltage forces the free electrons to move. This current is a flow of electrons from the point of negative charge (–) at one end of the wire, moving through the wire to the positive charge (+) at the other end. The direction of the electron flow is from the negative side of the **battery**, through the wire, and back to the positive side of the battery.

The direction of current flow is therefore from a point of negative potential to a point of positive potential.

ELECTRON FLOW

COPPER WIRE CONDUCTOR

FREE ELECTRONS IN MOTION

$-Q_1$ $+Q_2$

$-$ $+$

BATTERY CELL

103F09.EPS

Figure 9 ◆ Potential difference causing electric current.

4.2.0 Voltage

The force that causes electrons to move is called voltage, potential difference, or electromotive force (emf). One volt is the potential difference between two points for which one coulomb of electricity will do one **joule (J)** of work. A battery is one of several means of creating voltage. It chemically creates a large reserve of free electrons at the negative (–) terminal. The positive (+) terminal has electrons chemically removed and will therefore accept them if an external path is provided from the negative (–) terminal. When a battery is no longer able to chemically deposit electrons at the negative (–) terminal, it is said to be dead, or in need of recharging. Batteries are normally rated in volts. Large batteries are also rated in ampere-hours, where one ampere-hour is a current of one amp supplied for one hour.

Joule's Law

INSIDE TRACK

While other scientists of the 19th century were experimenting with batteries, cells, and circuits, James Joule was theorizing about the relationship between heat and energy. He discovered, contrary to popular belief, that work did not just move heat from one place to another; work, in fact, generated heat. Furthermore, he demonstrated that over time, a relationship existed between the temperature of water and electric current. These ideas formed the basis for the concept of energy. In his honor, the modern unit of energy was named the joule.

The Visual Language of Electricity

INSIDE TRACK

Learning to read circuit diagrams is like learning to read a book—first you learn to read the letters, then you learn to read the words, and before you know it, you are reading without paying attention to the individual letters anymore. Circuits are the same way— you will struggle at first with the individual pieces, but before you know it, you will be reading a circuit without even thinking about it. Studying the table below will help you to understand the fundamental language of electricity.

What's Measured	Unit of Measurement	Symbol	Ohm's Law Symbol
Amount of current	Amp	A	I
Electrical power	Watt	W	P
Force of current	Volt	V	E
Resistance to current	Ohm	Ω	R

4.3.0 Resistance

Resistance is directly related to the ability of a material to conduct electricity. All conductors have very low resistance; insulators have very high resistance.

4.3.1 Characteristics of Resistance

Resistance can be defined as the opposition to current flow. To add resistance to a circuit, electrical components called **resistors** are used. A resistor is a device whose resistance to current flow is a known, specified value. Resistance is measured in ohms and is represented by the symbol *R* in equations. One ohm is defined as the amount of resistance that will limit the current in a conductor to one ampere when the voltage applied to the conductor is one volt. The symbol for an ohm is Ω.

The resistance of a wire is proportional to the length of the wire, inversely proportional to the cross-sectional area of the wire, and dependent upon the kind of material of which the wire is made. The relationship for finding the resistance of a wire is:

$$R = \rho \frac{L}{A}$$

Where:

R = resistance (ohms)

L = length of wire (feet)

A = area of wire (circular mils, CM, or cm²)

ρ = specific resistance (ohm-CM/ft or microhm-CM)

A mil equals 0.001 inch; a circular mil is the cross-sectional area of a wire one mil in diameter.

The specific resistance is a constant that depends on the material of which the wire is made. *Table 1* shows the properties of various wire conductors.

Table 1 shows that at 75°F, a one-mil diameter, pure annealed copper wire that is one foot long has a resistance of 10.351 ohms; while a one-mil

Table 1 Conductor Properties

Metal	Specific Resistance (Resistance of 1 CM/ft in ohms)	
	32°F or 0°C	75°F or 23.8°C
Silver, pure annealed	8.831	9.674
Copper, pure annealed	9.390	10.351
Copper, annealed	9.590	10.505
Copper, hard-drawn	9.810	10.745
Gold	13.216	14.404
Aluminum	15.219	16.758
Zinc	34.595	37.957
Iron	54.529	62.643

103T01.EPS

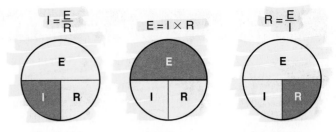

	LETTER SYMBOL	UNIT OF MEASUREMENT
CURRENT	I	AMPERES (A)
RESISTANCE	R	OHMS (Ω)
VOLTAGE	E	VOLTS (V)

103F10.EPS

Figure 10 ◆ Ohm's law circle.

diameter, one-foot-long aluminum wire has a resistance of 16.758 ohms. Temperature is important in determining the resistance of a wire. The hotter a wire, the greater its resistance.

5.0.0 ◆ OHM'S LAW

Ohm's law defines the relationship between current, voltage, and resistance. There are three ways to express Ohm's law mathematically.

- The current in a circuit is equal to the voltage applied to the circuit divided by the resistance of the circuit:

$$I = \frac{E}{R}$$

- The resistance of a circuit is equal to the voltage applied to the circuit divided by the current in the circuit:

$$R = \frac{E}{I}$$

- The applied voltage to a circuit is equal to the product of the current and the resistance of the circuit:

$$E = I \times R = IR$$

Where:

$$I = \text{current (amperes)}$$
$$R = \text{resistance (ohms)}$$
$$E = \text{voltage or emf (volts)}$$

If any two of the quantities E, I, or R are known, the third can be calculated.

The Ohm's law equations can be memorized and practiced effectively by using an Ohm's law circle, as shown in *Figure 10*. To find the equation for E, I, or R when two quantities are known, cover the unknown third quantity. The other two quantities in the circle will indicate how the covered quantity may be found.

Example 1:

Find I when E = 120V and R = 30Ω.

$$I = \frac{E}{R}$$
$$I = \frac{120V}{30Ω}$$
$$I = 4A$$

This formula shows that in a DC circuit, current (I) is directly proportional to voltage (E) and inversely proportional to resistance (R).

Example 2:

Find R when E = 240V and I = 20A.

$$R = \frac{E}{I}$$
$$R = \frac{240V}{20A}$$
$$R = 12Ω$$

Example 3:

Find E when I = 15A and R = 8Ω.

$$E = I \times R$$
$$E = 15A \times 8Ω$$
$$E = 120V$$

6.0.0 ◆ SCHEMATIC REPRESENTATION OF CIRCUIT ELEMENTS

The simple electric circuit shown earlier is shown in both pictorial and schematic forms in *Figure 11*. The schematic diagram is a shorthand way to draw an electric circuit, and circuits are usually represented in this way. In addition to the connecting wire, three components are shown symbolically: the battery, the switch, and the lamp. Note the positive (+) and negative (–) markings in both the pictorial and schematic representations of the battery. The schematic components represent the pictorial components in a simplified manner. A schematic diagram is one that shows, by means of graphic symbols, the electrical connections and functions of the different parts of a circuit.

The standard graphic symbols for commonly used electrical and electronic components are shown in *Figure 12*.

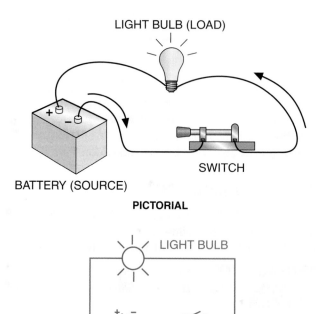

103F11.EPS

Figure 11 ◆ Electrical circuit.

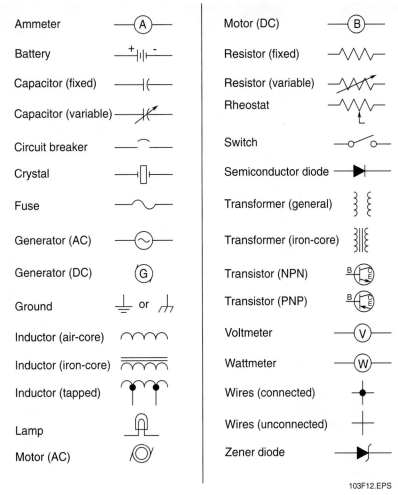

Ammeter	—(A)—	Motor (DC)	—(B)—	
Battery		Resistor (fixed)		
Capacitor (fixed)		Resistor (variable)		
Capacitor (variable)		Rheostat		
Circuit breaker		Switch		
Crystal		Semiconductor diode		
Fuse		Transformer (general)		
Generator (AC)		Transformer (iron-core)		
Generator (DC)		Transistor (NPN)		
Ground		Transistor (PNP)		
Inductor (air-core)		Voltmeter		
Inductor (iron-core)		Wattmeter		
Inductor (tapped)		Wires (connected)		
Lamp		Wires (unconnected)		
Motor (AC)		Zener diode		

103F12.EPS

Figure 12 ◆ Standard schematic symbols.

7.0.0 ◆ RESISTORS

The function of a resistor is to offer a particular resistance to current flow. For a given current and known resistance, the change in voltage across the component, or **voltage drop**, can be predicted using Ohm's law. Voltage drop refers to a specific amount of voltage used, or developed, by that component. An example is a very basic circuit of a 10V battery and a single resistor in a **series circuit**. The voltage drop across that resistor is 10V because it is the only component in the circuit and all voltage must be dropped across that resistor. Similarly, for a given applied voltage, the current that flows may be predetermined by selection of the resistor value. The required power dissipation largely dictates the construction and physical size of a resistor.

The two most common types of electronic resistors are wire-wound and carbon composition construction. A typical wire-wound resistor consists of a length of nickel wire wound on a ceramic tube and covered with porcelain. Low-resistance connecting wires are provided, and the resistance value is usually printed on the side of the component. *Figure 13* illustrates the construction of typical resistors. Carbon composition resistors are constructed by molding mixtures of powdered carbon and insulating materials into a cylindrical shape. An outer sheath of insulating material affords mechanical and electrical protection, and copper connecting wires are provided at each end. Carbon composition resistors are smaller and less expensive than the wire-wound type. However, the wire-wound type is the more rugged of the two and is able to survive much larger power dissipations than the carbon composition type.

Using Your Intuition

Learning the meanings of various electrical symbols may seem overwhelming, but if you take a moment to study *Figure 12*, you will see that most of them are intuitive—that is, they are shaped (in a symbolic way) to represent the actual object. For example, the battery shows + and –, just like an actual battery. The motor has two arms that suggest a spinning rotor. The transformer shows two coils. The resistor has a jagged edge to suggest pulling or resistance. Connected wires have a black dot that reminds you of solder. Unconnected wires simply cross. The fuse stretches out in both directions as though to provide extra slack in the line. The circuit breaker shows a line with a break in it. The capacitor shows a gap. The variable resistor has an arrow like a swinging compass needle. As you learn to read schematics, take the time to make mental connections between the symbol and the object it represents.

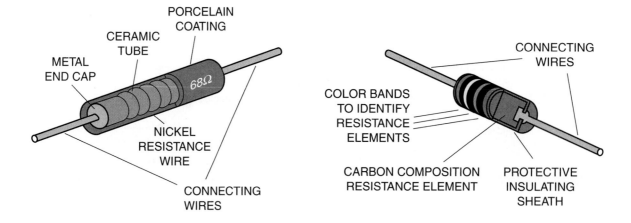

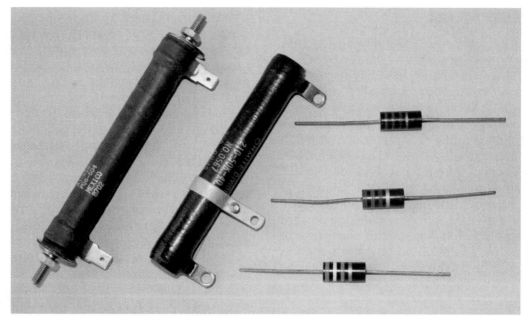

103F13.EPS

Figure 13 ◆ Common resistors.

Most resistors have standard fixed values, so they can be termed fixed resistors. Variable resistors, also known as adjustable resistors, are used a great deal in electronics. Two common symbols for a variable resistor are shown in *Figure 14*.

A variable resistor consists of a coil of closely wound insulated resistance wire formed into a partial circle. The coil has a low-resistance terminal at each end, and a third terminal is connected to a movable contact with a shaft adjustment facility. The movable contact can be set to any point on a connecting track that extends over one (uninsulated) edge of the coil.

Using the adjustable contact, the resistance from either end terminal to the center terminal may be adjusted from zero to the maximum coil resistance.

Another type of variable resistor is known as a decade resistance box. This is a laboratory component that contains precise values of switched series-connected resistors.

7.1.0 Resistor Color Codes

Because carbon composition resistors are physically small (some are less than 1 cm in length), it is not convenient to print the resistance value on the side. Instead, a color code in the form of colored bands is employed to identify the resistance value and tolerance. The color code is illustrated in *Figure 15*. Starting from one end of the resistor, the first two bands identify the first and second digits of the resistance value, and the third band indicates the number of zeros. An exception to this is when the third band is either silver or gold, which indicates a 0.01 or 0.1 multiplier, respectively. The fourth band is always either silver or gold, and in this position, silver indicates a ± 10% tolerance and gold indicates a ± 5% tolerance. Where no fourth band is present, the resistor tolerance is ± 20%.

We can put this information to practical use by determining the range of values for the carbon resistor in *Figure 16*.

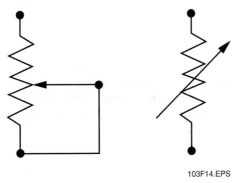

103F14.EPS

Figure 14 ◆ Symbols used for variable resistors.

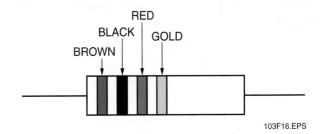

103F16.EPS

Figure 16 ◆ Sample color codes on a fixed resistor.

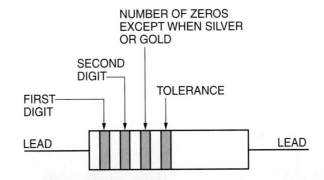

0	BLACK	7	VIOLET
1	BROWN	8	GRAY
2	RED	9	WHITE
3	ORANGE	0.1	GOLD
4	YELLOW	0.01	SILVER
5	GREEN	5%	GOLD – TOLERANCE
6	BLUE	10%	SILVER – TOLERANCE

103F15.EPS

Figure 15 ◆ Resistor color codes.

The color code for this resistor is as follows:

- Brown = 1, black = 0, red = 2, gold = a tolerance of ± 5%
- First digit = 1, second digit = 0, number of zeros (2) = 1,000Ω

Since this resistor has a value of 1,000W ± 5%, the resistor can range in value from 950W to 1,050Ω.

8.0.0 ◆ ELECTRICAL CIRCUITS

You will often hear the terms series circuit and parallel circuit during your training. When you hear these terms, keep in mind that they refer to the way loads are connected in the circuit.

8.1.0 Series Circuits

A series circuit provides only one path for current flow and is a voltage divider. The total resistance of the circuit is equal to the sum of the individual resistances. The 12V series circuit in *Figure 17* has two 30Ω loads. The total resistance is therefore 60Ω. The amount of current flowing in the circuit is 0.2A.

$$I = \frac{E}{R} = \frac{12V}{60\Omega} = 0.2A$$

If there were five 30Ω loads, the total resistance would be 150Ω. The current flow is the same through all the loads. The voltage measured across any load (voltage drop) depends on the resistance of that load. The sum of the voltage drops equals the total voltage applied to the circuit. Circuits containing loads in series are uncommon. An important trait of a series circuit is that if the circuit is open at any point, no current will flow. For example, if you have five light bulbs connected in series and one of them blows, all five lights will go off.

8.2.0 Parallel Circuits

In a parallel circuit, each load is connected directly to the voltage source; therefore, the voltage drop is the same through all loads and current is divided between the loads. The source sees the circuit as two or more individual circuits containing one load each. In the parallel circuit in *Figure 17*, the source sees three circuits, each containing a 30Ω load. The current flow through any load is determined by the resistance of that load. Thus the total current drawn by the circuit is the sum of the individual currents. The total resistance of a parallel circuit is calculated differently from that of a series circuit. In a parallel circuit, the total resistance is less than the smallest of the individual resistances.

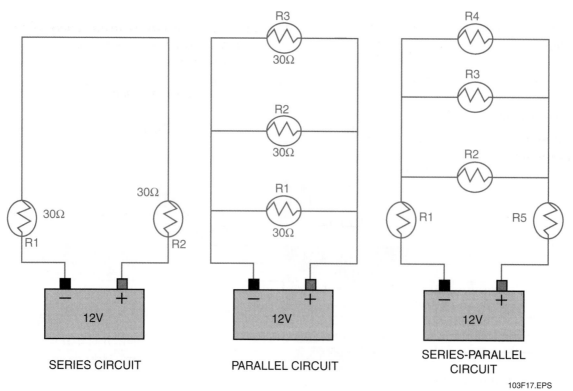

SERIES CIRCUIT   PARALLEL CIRCUIT   SERIES-PARALLEL CIRCUIT

103F17.EPS

Figure 17 ◆ Types of circuits.

Is It a Series Circuit?

When the term series circuit is used, it refers to the way the loads are connected. The same is true for parallel and series-parallel circuits. You will rarely, if ever, find loads connected in series, or in a series-parallel arrangement. The simple circuit shown here illustrates this point. At first glance, you might think it is a series-parallel circuit. On closer examination, you can see that there are only two loads—the relay and the contactor—and they are connected in parallel. Therefore, it is a parallel circuit. The control devices are wired in series with the loads, but only the loads are considered in determining the type of circuit.

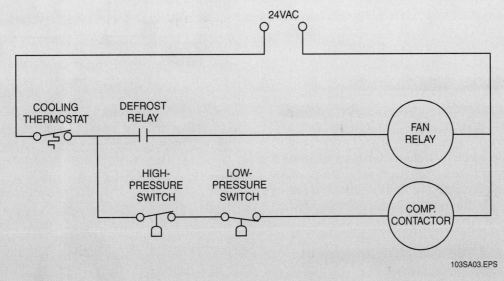

103SA03.EPS

For example, each of the 30Ω loads draws 0.4A at 12V; therefore, the total current is 1.2A:

$$I = \frac{E}{R} = \frac{12V}{30} = 0.4A \text{ per circuit}$$

0.4A per circuit × three circuits = 1.2A

Now, Ohm's law can be used again to calculate the total resistance:

$$R = \frac{E}{I} = \frac{12V}{1.2A} = 10\Omega$$

This one was simple because all the resistances were the same value. The process is the same when the resistances are different, but the current calculation has to be done for each load. The individual currents are added to get the total current.

Unlike series circuits, parallel circuits continue working even if one circuit opens. Household circuits are wired in parallel. In fact, almost all the load circuits you encounter will be parallel circuits.

Either of the following formulas can be used to convert parallel resistances to a single resistance

value. The first one is used when there are two resistances in parallel. The second is used when there are three or more.

$$\text{Total resistance} = \frac{R1 \times R2}{R1 + R2}$$

$$\text{Total resistance} = \frac{1}{\dfrac{1}{R1} + \dfrac{1}{R2} + \dfrac{1}{R3}}$$

Example:

1. The total resistance of the parallel circuit below is 6Ω.

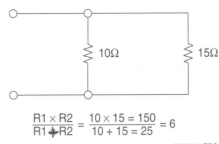

$$\frac{R1 \times R2}{R1 + R2} = \frac{10 \times 15 = 150}{10 + 15 = 25} = 6$$

UA0301.EPS

2. The total resistance of the parallel circuit below is 4.76Ω.

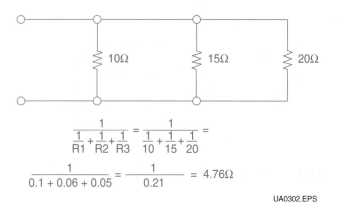

$$\frac{1}{\frac{1}{R1}+\frac{1}{R2}+\frac{1}{R3}} = \frac{1}{\frac{1}{10}+\frac{1}{15}+\frac{1}{20}} =$$

$$\frac{1}{0.1+0.06+0.05} = \frac{1}{0.21} = 4.76\Omega$$

UA0302.EPS

8.3.0 Series-Parallel Circuits

Electronic circuits often contain a hybrid arrangement known as a series-parallel circuit (*Figure 17*). It is unlikely, however, that you will ever have to determine the electrical characteristics of one of these circuits. If it becomes necessary, the parallel loads must be converted to their equivalent series resistance. Then the load resistances are added to determine total circuit resistance.

9.0.0 ◆ ELECTRICAL MEASURING INSTRUMENTS

Electricians frequently use test meters to measure voltage, current, and resistance. The most common test meter is the volt-ohm-milliammeter (VOM), also called a multimeter. *Figure 18* shows both digital and analog multimeters. The analog meter is so-called because the pointer moves in proportion to the value being measured. The person using the meter must then interpret the scale to determine the measured value. Digital meters display the result numerically on the screen.

Multimeters are commonly used to measure AC and DC voltage, DC current, and resistance. They can also be used to measure AC current in the milliamp range. For larger current values, it is usually necessary to use a clamp-on ammeter (*Figure 19*).

 WARNING!

Only qualified individuals may use these meters. Consult your company's safety policy for applicable rules.

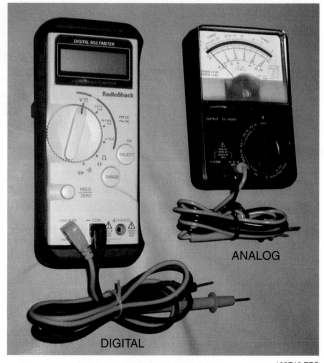

Figure 18 ◆ Digital and analog meters.

103F18.EPS

Figure 19 ◆ Clamp-on ammeter.

103F19.EPS

9.1.0 Measuring Current

A clamp-on ammeter (*Figure 20*) is used to measure current. The jaws of the ammeter are placed around a single conductor. Current flowing through the wire creates a magnetic field, which induces a proportional current in the ammeter

Figure 20 ◆ Clamp-on ammeter in use.

jaws. This current is read by the meter movement and appears as a direct readout or, on an analog meter, as a deflection of the meter needle.

In-line ammeters (*Figure 21*) are less common. This type of meter must be connected in series with the circuit, which means that the circuit must be opened.

Aside from following good safety practices, there are a few things to remember when measuring current:

- If the ammeter jaws are dirty or misaligned, a meter will not read correctly.
- When using an analog meter, always start at a high range and work down to avoid damaging the meter.
- Do not clamp the meter jaws around two different conductors at the same time, or an inaccurate reading will result.

9.2.0 Measuring Voltage

A **voltmeter** must be connected in parallel with (across) the component or circuit to be tested (*Figure 22*). If a circuit function is not operating, the voltmeter can be used to determine if the correct voltage is available to the circuit. Voltage must be checked with power applied.

9.3.0 Measuring Resistance

An **ohmmeter** contains an internal battery that acts as a voltage source. Therefore, resistance measurements are always made with the system

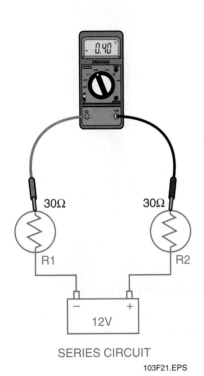

SERIES CIRCUIT

103F21.EPS

Figure 21 ◆ In-line ammeter test setup.

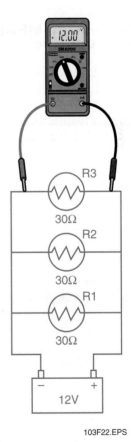

103F22.EPS

Figure 22 ◆ Voltmeter connection.

power shut off. Sometimes, an ohmmeter is used to measure resistance in a load; motor windings are a good example. More often, an ohmmeter is used to check continuity in a circuit. A wire or closed switch offers negligible resistance. With the ohmmeter connected as in *Figure 23*, and the three switches closed, the current produced by the ohmmeter battery will flow unopposed and the meter will show zero resistance. The circuit has continuity; that is, it is continuous. If a switch is open, however, there is no path for current and the meter will see infinite resistance; that is, a lack of continuity.

A continuity tester (*Figure 24*) is a simple device consisting mainly of a battery and either an audible or visual indicator. It can be used in place of an ohmmeter to test the continuity of a wire and to identify individual wires contained in a conduit or other raceway. To test the continuity of a wire, strip the insulation off the end of the wire to be tested at one end of the conduit run, then connect (short) the wire to the metal conduit. At the other end of the conduit run, clip the alligator clip lead of the tester to the conduit and touch the probe to the end of the wire under test. If the tester audible alarm sounds or the indicator light comes on, there is continuity. Note that this only indicates that there is continuity between the two points

being tested; it does not indicate the actual value of the resistance. If there is no indication, the wire is open.

To identify individual wires in a conduit run, touch the tester probe to the wires in the conduit one at a time until the tester audible alarm sounds or the indicator lights. Then, put matching identification tags on both ends of the wire. Continue this procedure until all the wires have been identified.

9.4.0 Voltage Testers

Figure 25 shows one of the wide variety of devices available for checking for the presence of voltage. It can be used as a troubleshooting tool and as a safety device to make sure the voltage is turned off before touching any terminals or conductors. When the probes are touched to the circuit, the light on the instrument will turn on if a voltage is present. Instruments like these are available in several voltage ranges, so it is important to know something about the circuit you are checking.

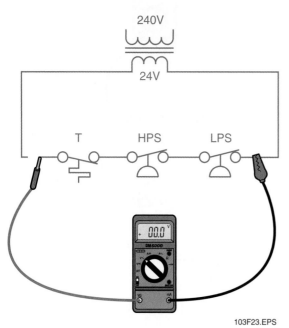

103F23.EPS

Figure 23 ◆ Ohmmeter connection for continuity testing.

103F24.EPS

Figure 24 ◆ Continuity tester.

Test Instruments—Old and New

Early electricians used individual meters to test circuit parameters. Today, those instruments seem primitive given the availability of all-purpose instruments like the combination multimeter and clamp-on ammeter with its direct digital readout shown here.

103SA04.EPS

103SA05.EPS

103SA06.EPS

Using an Ohmmeter

An ohmmeter has its own battery to test the resistance or continuity of a circuit. Therefore, the circuit must be de-energized because the ohmmeter is calibrated for its own power source.

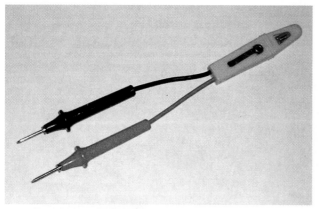

103F25.EPS

Figure 25 ◆ Voltage tester.

10.0.0 ◆ ELECTRICAL POWER

Power is defined as the rate of doing work. This is equivalent to the rate at which energy is used or dissipated. Electrons passing through a resistance dissipate energy in the form of heat. In electrical circuits, power is measured in units called watts (W). The power in watts equals the rate of energy conversion. One watt of power equals the work done in one second by one volt of potential difference in moving one coulomb of charge. One coulomb per second is an ampere;

therefore, power in watts equals the product of amperes times volts.

The work done in an electrical circuit can be useful work or it can be wasted work. In both cases, the rate at which the work is done is still measured in power. The turning of an electric motor is useful work. On the other hand, the heating of wires or resistors in a circuit is wasted work, since no useful function is performed by the heat.

The unit of electrical work is the joule. This is the amount of work done by one coulomb flowing through a potential difference of one volt. Thus, if five coulombs flow through a potential difference of one volt, five joules of work are done. The time it takes these coulombs to flow through the potential difference has no bearing on the amount of work done.

It is more convenient when working with circuits to think of amperes of current rather than coulombs. As previously discussed, one ampere equals one coulomb passing a point in one second. Using amperes, one joule of work is done in one second when one ampere moves through a potential difference of one volt. This rate of one joule of work in one second is the basic unit of power, and is called a watt. Therefore, a watt is the power used when one ampere of current flows through a potential difference of one volt, as shown in *Figure 26*.

Meter Applications

Different electrical systems may have differing levels of available current as well as different voltages. Keep in mind that the meter must be matched to the application. For example, a meter that has enough protective features to be used safely on interior branch circuits may not be safe to use on the service feeders.

Ohmmeter Batteries

When you use an ohmmeter to measure resistance or test continuity, the power to the circuit under test must be turned off. The internal multimeter battery provides a small DC current that is used to measure resistance. In order to ensure accurate readings, it is important to replace or recharge the battery in accordance with the meter manufacturer's instructions.

Power

We take electrical power for granted, never stopping to think how surprising it is that a flow of submicroscopic electrons can pump thousands of gallons of water or illuminate a skyscraper. Our lives now constantly rely on the ability of the electron to do work. Think about your day up to this moment. How has electrical power shaped your experience?

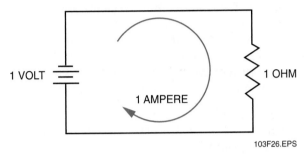

Figure 26 ◆ One watt.

Mechanical power is usually measured in units of horsepower (hp). To convert from horsepower to watts, multiply the number of horsepower by 746. To convert from watts to horsepower, divide the number of watts by 746. Conversions for common units of power are given in *Table 2*.

The kilowatt-hour (kWh) is commonly used for large amounts of electrical work or energy. (The prefix **kilo** means one thousand.) The amount is calculated simply as the product of the power in kilowatts multiplied by the time in hours during which the power is used. If a light bulb uses 300W or 0.3kW for 4 hours, the amount of energy is 0.3 × 4, which equals 1.2kWh.

Very large amounts of electrical work or energy are measured in megawatts (MW). (The prefix **mega** means one million.)

Electricity usage is figured in kilowatt-hours of energy. The power line voltage is fairly constant at 120V. Suppose the total load current in the main

line equals 20A. Then the power in watts from the 120V line is:

$$P = 120V \times 20A$$
$$P = 2{,}400W \text{ or } 2.4kW$$

If this power is used for five hours, then the energy of work supplied equals:

$$2.4 \times 5 = 12kWh$$

10.1.0 Power Equation

When one ampere flows through a difference of two volts, two watts must be used. In other words, the number of watts used is equal to the number of amperes of current times the potential difference. This is expressed in equation form as:

$$P = I \times E \text{ or } P = IE$$

Where:

$$P = \text{power used in watts}$$
$$I = \text{current in amperes}$$
$$E = \text{potential difference in volts}$$

The equation is sometimes called Ohm's law for power, because it is similar to Ohm's law. This equation is used to find the power consumed in a circuit or load when the values of current and voltage are known. The second form of the equation is used to find the voltage when the power and current are known:

$$E = \frac{P}{I}$$

The third form of the equation is used to find the current when the power and voltage are known:

$$I = \frac{P}{E}$$

Using these three equations, the power, voltage, or current in a circuit can be calculated whenever any two of the values are already known.

Table 2	Conversion Table
1,000 watts (W) = 1 kilowatt (kW)	
1,000,000 watts (W) = 1 megawatt (MW)	
1,000 kilowatts (kW) = 1 megawatt (MW)	
1 watt (W) = 0.00134 horsepower (hp)	
1 horsepower (hp) = 746 watts (W)	

103T02.EPS

Example 1:

Calculate the power in a circuit where the source of 100V produces 2A in a 50Ω resistance.

$$P = IE$$
$$P = 2 \times 100$$
$$P = 200W$$

This means the source generates 200W of power while the resistance dissipates 200W in the form of heat.

Example 2:

Calculate the source voltage in a circuit that consumes 1,200W at a current of 5A.

$$E = \frac{P}{I}$$
$$E = \frac{1,200}{5}$$
$$E = 240V$$

Example 3:

Calculate the current in a circuit that consumes 600W with a source voltage of 120V.

$$I = \frac{P}{E}$$
$$I = \frac{600}{120}$$
$$I = 5A$$

Components that use the power dissipated in their resistance are generally rated in terms of power. The power is rated at normal operating voltage, which is usually 120V. For instance, an appliance that draws 5A at 120V would dissipate 600W. The rating for the appliance would then be 600W/120V.

To calculate I or R for components rated in terms of power at a specified voltage, it may be convenient to use the power formula in different forms. There are actually three basic power formulas, but each can be rearranged into two other forms for a total of nine combinations:

$$P = IE \qquad P = I^2R \qquad P = \frac{E^2}{R}$$

$$I = \frac{P}{E} \qquad R = \frac{P}{I^2} \qquad R = \frac{E^2}{P}$$

$$E = \frac{P}{I} \qquad I = \sqrt{\frac{P}{R}} \qquad E = \sqrt{PR}$$

Note that all of these formulas are based on Ohm's law (E = IR) and the power formula (P = I × E). *Figure 27* shows all of the applicable power, voltage, resistance, and current equations.

10.2.0 Power Rating of Resistors

If too much current flows through a resistor, the heat caused by the current will damage or destroy the resistor. This heat is caused by I^2R heating, which is power loss expressed in watts. Therefore, every resistor is given a wattage, or power rating, to show how much I^2R heating it can take before it burns out. This means that a resistor with a power rating of one watt will burn out if it is used in a circuit where the current causes it to dissipate heat at a rate greater than one watt.

If the power rating of a resistor is known, the maximum current it can carry is found by using an equation derived from $P = I^2R$:

$$P = I^2R$$
$$I^2 = P/R$$
$$I = \sqrt{P/R}$$

Using this equation, find the maximum current that can be carried by a 1Ω resistor with a power rating of 4W:

$$I = \sqrt{P/R} = \sqrt{4/1} = 2 \text{ amperes}$$

If such a resistor conducts more than 2 amperes, it will dissipate more than its rated power and burn out.

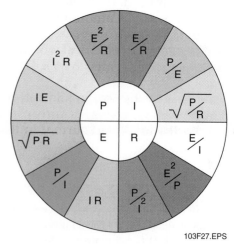

103F27.EPS

Figure 27 ◆ Expanded Ohm's law circle.

Power ratings assigned by resistor manufacturers are usually based on the resistors being mounted in an open location where there is free air circulation, and where the temperature is not higher than 104°F (40°C). Therefore, if a resistor is mounted in a small, crowded, enclosed space, or where the temperature is higher than 104°F, there is a good chance it will burn out even before its power rating is exceeded. Also, some resistors are designed to be attached to a chassis or frame that will carry away the heat.

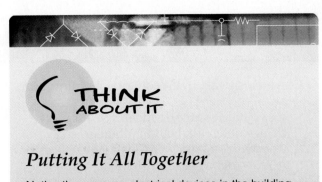

Putting It All Together

Notice the common electrical devices in the building you're in. What is their wattage rating? How much current do they draw? How would you test their voltage or amperage?

1. An electrical circuit contains, at minimum, a(n) _____.
   a. voltage source, load, and switch
   b. ammeter, load, and voltage source
   c. voltage source, load, and conductors
   d. conductor, switch, and load

2. A type of subatomic particle with a positive charge is a(n) _____.
   a. proton
   b. neutron
   c. electron
   d. nucleus

3. Which of the following substances is considered an insulator?
   a. Gold
   b. Copper
   c. Silver
   d. Porcelain

4. The voltage commonly supplied to a residence by the local utility is _____.
   a. 120V
   b. 240V
   c. 480V
   d. 208V

5. Another term used for voltage is _____.
   a. emf
   b. coulomb
   c. current
   d. joule

6. In order to calculate the current flowing in a circuit, you would multiply voltage by resistance.
   a. True
   b. False

7. The color band that represents tolerance on a resistor is the _____.
   a. 4th band
   b. 3rd band
   c. 2nd band
   d. 1st band

8. In a parallel circuit, the total resistance is _____ the smallest resistance.
   a. greater than
   b. equal to
   c. less than

9. Circuit continuity is checked using the _____ function of a multimeter.
   a. ammeter
   b. voltmeter
   c. ohmmeter
   d. wattmeter

10. The power in a circuit with 120 volts and 5 amps is _____.
    a. 24 watts
    b. 600 watts
    c. 6,000 watts
    d. ¼ watt

Summary

Electricians often test and troubleshoot electrical circuits. This work can be done safely and more effectively if you know the theory of electricity and the interrelationships of voltage, current, resistance, and power. The basic tool for understanding these relationships is Ohm's law.

Testing and troubleshooting of electrical circuits involves the use of test instruments such as the multimeter, or VOM. The multimeter combines the voltmeter, ammeter, and ohmmeter into a single instrument. In analog meters, a pointer moves across a scale in proportion to the current flowing though the meter. In a digital meter, the measured value is displayed directly on the screen of the meter.

Notes

Trade Terms Quiz

1. A(n) _AMMETER_ is an instrument for measuring electrical current.

2. Measured in amperes, _CURRENT_ is the flow of electrons in a circuit.

3. A(n) _AMPERE (A)_ is the force required to produce a current of one ampere through a resistance of one ohm.

4. Voltage is measured with a(n) _VOLTMETER_.

5. One volt applied across one ohm of resistance causes a current flow of one _VOLT (V)_.

6. One volt is the potential difference between two points for which one coulomb of electricity will do one _JOULE_ of work.

7. A(n) _COULOMB_ is the common unit used for specifying the size of a given charge.

8. _VOLTAGE_ is the driving force that makes current flow in a circuit.

9. The basic unit of measurement for electrical power is the _WATT (W)_

10. The _ATOM_ is the smallest particle of an element that will still retain the properties of that element.

11. The _NUCLEUS_ is the center of an atom.

12. Found in the nuclei of atoms, _PROTONS_ are electrically positive particles and _NEUTRONS_ are electrically neutral particles.

13. The outermost ring of electrons orbiting the nucleus of an atom is known as the _VALANCE SHELL_

14. A(n) _ELECTRON_ is a negatively charged particle that orbits the nucleus of an atom.

15. _MATTER_ is any substance that has mass and occupies space.

16. The prefix used to indicate one thousand is _KILO_.

17. The prefix used to indicate one million is _MW (MEGA WATTS)_

18. Consisting of two or more cells, _BATTERY_ convert chemical energy into electrical energy.

19. A(n) _CIRCUIT_ is a complete path for current flow.

20. A material through which it is relatively easy to maintain an electric current is a(n) _CONDUCTOR_.

21. A(n) _INSULATOR_ is a material through which it is difficult to conduct an electric current.

22. The basic unit of measurement for resistance is the _OHMS Ω_.

23. The instrument that is used to measure resistance is called a(n) _OHMMETER_.

24. _OHM'S LAW_ is a statement of the relationship between current, voltage, and resistance in an electrical circuit.

25. _POWER_ is the rate of doing work or the rate at which energy is used or dissipated.

26. Measured in ohms, _RESISTANCE_ is the electrical property that opposes the flow of current through a circuit.

27. A(n) _RESISTOR_ is a component that normally opposes current flow in a DC circuit.

28. A(n) _SCHEMATIC_ is a drawing in which symbols are used to represent the components in a system.

29. A(n) _SERIES_ circuit has only one route for current flow.

30. The change in voltage across a component is called _VOLTAGE DROP_

31. A(n) _RELAY_ is an electromechanical component used as a switching device.

32. A device containing one or more coils of wire wrapped around a common core is called a(n) _TRANSFORMER_

33. An electromagnetic device used to control a mechanical device such as a valve is called a(n) _SOLENOID_ .

Trade Terms

Ammeter
Ampere (A)
Atom
Battery
Circuit
Conductor
Coulomb
Current
Electron

Insulator
Joule (J)
Kilo
Matter
Mega
Neutrons
Nucleus
Ohm (Ω)
Ohmmeter

Ohm's law
Power
Protons
Relay
Resistance
Resistor
Schematic
Series circuit

Solenoid
Transformer
Valence shell
Volt (V)
Voltage
Voltage drop
Voltmeter
Watt (W)

E. L. Jarrell
Associated Builders and Contractors

Eurlin Layne (E. L.) Jarrell is another prime example of a master electrician giving back to the electrical community by teaching and mentoring.

After serving in the United States Army, E. L. went to work for Cities Services, now known as CITGO. He stayed at CITGO for 38 years, finally retiring in 1995. It was during his employment at CITGO that he first received apprenticeship training in the electrical field.

While at CITGO, E. L. worked as a process unit operator before moving to the electrical department. While there, he worked as a trainee electrician for three years until he became a first-class electrician. A few years later, he was promoted to temporary supervisor, planning and scheduling shut-down maintenance. In 1983, he took and passed the Block Master Electrician test for the City of Lake Charles, Louisiana. In 1997, E. L. became involved with Associated Builders and Contractors (ABC).

E. L. is currently the Electrical Department Head for the ABC Training Center, where he works in the lab, overseeing students doing hands-on electrical work. During his first semester teaching at the ABC

Training Center, it became clear to E. L. that many students simply had no time to study because they worked 10-hour days, drove over 100 miles to work, and had family obligations. In response, E. L. began an in-class study guide. He encouraged students to form study groups, and he gave students time to study in class.

E. L. was an instrumental member of NCCER's Technical Review Committee, which completely rewrote all four levels of NCCER's Electrical Curriculum. In addition, E. L. is currently a member of both NCCER's National Skills Assessment Written Test Committee and the Performance Verification Packet for Industrial Electricians Committee.

E. L. has decided to give back to the electrical community with his expertise and mentoring. Many of E. L.'s students have become his personal friends. He says, "At this point in my life, I just want to continue being the best electrical instructor that I can be and share some of my knowledge and experience with my students and hope that I can make a difference in their lives and careers."

Trade Terms Introduced in This Module

Ammeter: An instrument for measuring electrical current.

Ampere (A): A unit of electrical current. For example, one volt across one ohm of resistance causes a current flow of one ampere.

Atom: The smallest particle to which an element may be divided and still retain the properties of the element.

Battery: A DC voltage source consisting of two or more cells that convert chemical energy into electrical energy.

Circuit: A complete path for current flow.

Conductor: A material through which it is relatively easy to maintain an electric current.

Coulomb: An electrical charge equal to 6.25×10^{18} electrons or 6,250,000,000,000,000,000 electrons. A coulomb is the common unit of quantity used for specifying the size of a given charge.

Current: The movement, or flow, of electrons in a circuit. Current (I) is measured in amperes.

Electron: A negatively charged particle that orbits the nucleus of an atom.

Insulator: A material through which it is difficult to conduct an electric current.

Joule (J): A unit of measurement that represents one newton-meter (Nm), which is a unit of measure for doing work.

Kilo: A prefix used to indicate one thousand; for example, one kilowatt is equal to one thousand watts.

Matter: Any substance that has mass and occupies space.

Mega: A prefix used to indicate one million; for example, one megawatt is equal to one million watts.

Neutrons: Electrically neutral particles (neither positive nor negative) that have the same mass as a proton and are found in the nucleus of an atom.

Nucleus: The center of an atom. It contains the protons and neutrons of the atom.

Ohm (Ω): The basic unit of measurement for resistance.

Ohmmeter: An instrument used for measuring resistance.

Ohm's law: A statement of the relationships among current, voltage, and resistance in an electrical circuit: current (I) equals voltage (E) divided by resistance (R). Generally expressed as a mathematical formula: $I = E/R$.

Power: The rate of doing work or the rate at which energy is used or dissipated. Electrical power is the rate of doing electrical work. Electrical power is measured in watts.

Protons: The smallest positively charged particles of an atom. Protons are contained in the nucleus of an atom.

Relay: An electromechanical device consisting of a coil and one or more sets of contacts. Used as a switching device.

Resistance: An electrical property that opposes the flow of current through a circuit. Resistance (R) is measured in ohms.

Resistor: Any device in a circuit that resists the flow of electrons.

Schematic: A type of drawing in which symbols are used to represent the components in a system.

Series circuit: A circuit with only one path for current flow.

Solenoid: An electromagnetic coil used to control a mechanical device such as a valve.

Transformer: A device consisting of one or more coils of wire wrapped around a common core. It is commonly used to step voltage up or down.

Valence shell: The outermost ring of electrons that orbit about the nucleus of an atom.

Volt (V): The unit of measurement for voltage (electromotive force or difference of potential). One volt is equivalent to the force required to produce a current of one ampere through a resistance of one ohm.

Voltage: The driving force that makes current flow in a circuit. Voltage (E) is also referred to as electromotive force or difference of potential.

Voltage drop: The change in voltage across a component that is caused by the current flowing through it and the amount of resistance opposing it.

Voltmeter: An instrument for measuring voltage. The resistance of the voltmeter is fixed. When the voltmeter is connected to a circuit, the current passing through the meter will be directly proportional to the voltage at the connection points.

Watt (W): The basic unit of measurement for electrical power.

This module is intended to present thorough resources for task training. The following reference works are suggested for further study. These are optional materials for continued education rather than for task training.

American Electrician's Handbook, Terrell Croft, Wilfred Summers. New York: McGraw-Hill.

Electronics Fundamentals: Circuits, Devices, and Applications, Thomas L. Floyd. New York: Prentice Hall.

Principles of Electric Circuits, Thomas L. Floyd. New York: Prentice Hall.

NCCER makes every effort to keep these textbooks up-to-date and free of technical errors. We appreciate your help in this process. If you have an idea for improving this textbook, or if you find an error, a typographical mistake, or an inaccuracy in NCCER's Contren® textbooks, please write us, using this form or a photocopy. Be sure to include the exact module number, page number, a detailed description, and the correction, if applicable. Your input will be brought to the attention of the Technical Review Committee. Thank you for your assistance.

Instructors – If you found that additional materials were necessary in order to teach this module effectively, please let us know so that we may include them in the Equipment/Materials list in the Annotated Instructor's Guide.

Write: Product Development and Revision
National Center for Construction Education and Research
3600 NW 43rd St., Bldg. G, Gainesville, FL 32606

Fax: 352-334-0932

E-mail: curriculum@nccer.org

Craft _____   Module Name _____

Copyright Date _____   Module Number _____   Page Number(s) _____

Description _____

_____

_____

_____

(Optional) Correction _____

_____

_____

(Optional) Your Name and Address _____

_____

_____

Electrical Theory

GM Lansing Delta Township Assembly Complex

This project consisted of converting 375 acres of 1,100 acres of farm land into a campus-style automotive assembly facility totaling 2.4 million square feet. In 20 months, a greenfield site was transformed into a major complex where raw steel enters at one end and finished vehicles are shipped to dealers at the opposite end. This project was the first design-build/guaranteed maximum price new manufacturing facility that General Motors has created,

26104-08

26104-08
Electrical Theory

Topics to be presented in this module include:

Overview

Troubleshooting complex electrical or electronic circuitry requires an advanced understanding of electrical theory. When a complex circuit does not respond as designed, the electrician or technician must safely locate, determine, and repair the problem. To accomplish this task in a relatively short period and with reasonable success, the electrician must apply both fundamental concepts, such as Ohm's law, and more advanced concepts, such as Kirchhoff's laws. Studying advanced electrical theory is the only way to know when and where to apply such laws.

Objectives

When you have completed this module, you will be able to do the following:

1. Explain the basic characteristics of combination circuits.
2. Calculate, using Kirchhoff's voltage law, the voltage drop in series, parallel, and series-parallel circuits.
3. Calculate, using Kirchhoff's current law, the total current in parallel and series-parallel circuits.
4. Using Ohm's law, find the unknown parameters in series, parallel, and series-parallel circuits.

Trade Terms

Kirchhoff's current law
Kirchhoff's voltage law
Parallel circuits
Series circuits
Series-parallel circuits

Required Trainee Materials

1. Paper and pencil
2. Appropriate personal protective equipment

Prerequisites

Before you begin this module, it is recommended that you successfully complete *Core Curriculum* and *Electrical Level One*, Modules 26101-08 through 26103-08.

This course map shows all of the modules in *Electrical Level One*. The suggested training order begins at the bottom and proceeds up. Skill levels increase as you advance on the course map. The local Training Program Sponsor may adjust the training order.

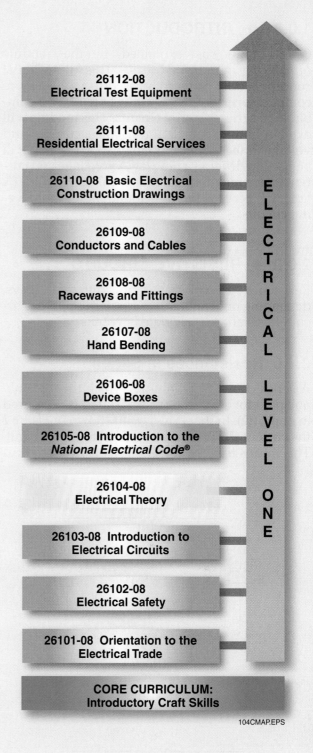

26112-08
Electrical Test Equipment

26111-08
Residential Electrical Services

26110-08 Basic Electrical Construction Drawings

26109-08
Conductors and Cables

26108-08
Raceways and Fittings

26107-08
Hand Bending

26106-08
Device Boxes

26105-08 Introduction to the *National Electrical Code®*

26104-08
Electrical Theory

26103-08 Introduction to Electrical Circuits

26102-08
Electrical Safety

26101-08 Orientation to the Electrical Trade

CORE CURRICULUM:
Introductory Craft Skills

ELECTRICAL LEVEL ONE

104CMAP.EPS

1.0.0 ◆ INTRODUCTION

Ohm's law was explained in the module *Introduction to Electrical Circuits.* This fundamental concept is now going to be used to analyze more complex **series circuits**, **parallel circuits** and combination **series-parallel circuits**. This module will explain how to calculate resistance, current, and voltage in these complex circuits. Ohm's law will be used to develop a new law for voltage and current determination. This law, called Kirchhoff's law, will become the new foundation for analyzing circuits.

2.0.0 ◆ RESISTIVE CIRCUITS

Resistance is calculated in different ways, depending on whether it is a series or parallel circuit. Resistance is calculated in ohms.

2.1.0 Resistances in Series

A series circuit is a circuit in which there is only one path for current flow. Resistance is measured in ohms (Ω). In the series circuit shown in *Figure 1* the current (I) is the same in all parts of the circuit. This means that the current flowing through R_1 is

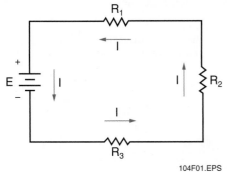

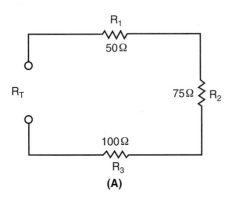

Figure 1 ◆ Series circuit.

Figure 2 ◆ Total resistance.

the same as the current flowing through R_2 and R_3, and it is also the same as the current supplied by the battery.

When resistances are connected in series as in this example, the total resistance in the circuit is equal to the sum of the resistances of all the parts of the circuit:

$$R_T = R_1 + R_2 + R_3$$

Where:

R_T = total resistance

$R_1 + R_2 + R_3$ = resistances in series

Example 1:

The circuit shown in *Figure 2(A)* has 50Ω, 75Ω, and 100Ω resistors in series. Find the total resistance of the circuit.

Add the values of the three resistors in series:

$$R_T = R_1 + R_2 + R_3 = 50 + 75 + 100 = 225\Omega$$

Example 2:

The circuit shown in *Figure 2(B)* has three lamps connected in series with the resistances shown. Find the total resistance of the circuit.

Add the values of the three lamp resistances in series:

$$R_T = R_1 + R_2 + R_3 = 20 + 40 + 60 = 120\Omega$$

2.2.0 Resistances in Parallel

The total resistance in a parallel resistive circuit is given by the formula:

$$R_T = \cfrac{1}{\cfrac{1}{R_1} + \cfrac{1}{R_2} + \cfrac{1}{R_3} + \cfrac{1}{R_n}}$$

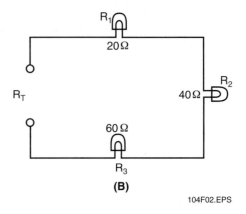

Series Circuits

Simple series circuits are seldom encountered in practical wiring. The only simple series circuit you may recognize is older strands of Christmas lights, in which the entire string went dead when one lamp burned out. Think about what the actual wiring of a series circuit would look like in household receptacles. How would the circuit physically be wired? What kind of illumination would you get if you wired your household receptacles in series and plugged half a dozen lamps into those receptacles?

Where:

R_T = total resistance in parallel

R_1, R_2, R_3, and R_n = branch resistances

Example 1:

Find the total resistance of the 2Ω, 4Ω, and 8Ω resistors in parallel shown in *Figure 3*.

Write the formula for the three resistances in parallel:

$$R_T = \frac{1}{\frac{1}{R_1} + \frac{1}{R_2} + \frac{1}{R_3}}$$

Substitute the resistance values:

$$R_T = \frac{1}{\frac{1}{2} + \frac{1}{4} + \frac{1}{8}}$$

$$R_T = \frac{1}{0.5 + 0.25 + 0.125}$$

$$R_T = \frac{1}{0.875}$$

$$R_T = 1.14\Omega$$

Note that when resistances are connected in parallel, the total resistance is always less than the resistance of any single branch.

In this case:

$$R_T = 1.14\Omega < R_1 = 2\Omega, R_2 = 4\Omega, \text{ and } R_3 = 8\Omega$$

Example 2:

Add a fourth parallel resistor of 2Ω to the circuit in *Figure 3*. What is the new total resistance, and what is the net effect of adding another resistance in parallel?

Write the formula for four resistances in parallel:

$$R_T = \frac{1}{\frac{1}{R_1} + \frac{1}{R_2} + \frac{1}{R_3} + \frac{1}{R_4}}$$

Substitute values:

$$R_T = \frac{1}{\frac{1}{2} + \frac{1}{4} + \frac{1}{8} + \frac{1}{2}}$$

$$R_T = \frac{1}{0.5 + 0.25 + 0.125 + 0.5}$$

$$R_T = \frac{1}{1.375}$$

$$R_T = 0.73\Omega$$

The net effect of adding another resistance in parallel is a reduction of the total resistance from 1.14Ω to 0.73Ω.

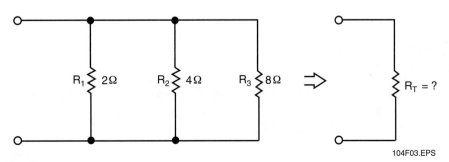

104F03.EPS

Figure 3 ◆ Parallel branch.

2.2.1 Simplified Formulas

The total resistance of *equal* resistors in parallel is equal to the resistance of one resistor divided by the number of resistors:

$$R_T = \frac{R}{N}$$

Where:

R_T = total resistance of equal resistors in parallel

R = resistance of one of the equal resistors

N = number of equal resistors

If two resistors with the same resistance are connected in parallel, the equivalent resistance is half of that value, as shown in *Figure 4*.

The two 200Ω resistors in parallel are the equivalent of one 100Ω resistor; the two 100Ω resistors are the equivalent of one 50Ω resistor; and the two 50Ω resistors are the equivalent of one 25Ω resistor.

When any two unequal resistors are in parallel, it is often easier to calculate the total resistance by multiplying the two resistances and then dividing the product by the sum of the resistances:

$$R_T = \frac{R_1 \times R_2}{R_1 + R_2}$$

Where:

R_T = total resistance of unequal resistors in parallel

R_1, R_2 = two unequal resistors in parallel

Example 1:

Find the total resistance of a 6Ω (R_1) resistor and an 18Ω (R_2) resistor in parallel:

$$R_T = \frac{R_1 \times R_2}{R_1 + R_2} = \frac{6 \times 18}{6 + 18} = \frac{108}{24} = 4.5\Omega$$

Example 2:

Find the total resistance of a 100Ω (R_1) resistor and a 150Ω (R_2) resistor in parallel:

$$R_T = \frac{R_1 \times R_2}{R_1 + R_2} = \frac{100 \times 150}{100 + 150} = \frac{15,000}{250} = 60\Omega$$

2.3.0 Series-Parallel Circuits

To find current, voltage, and resistance in series circuits and parallel circuits is fairly easy. When working with either type, use only the rules that apply to that type. In a series-parallel circuit, some parts of the circuit are series connected and other parts are parallel connected. Thus, in some parts the rules for series circuits apply, and in other parts, the rules for parallel circuits apply. To analyze or solve a problem involving a series-parallel circuit, it is necessary to recognize which parts of the circuit are series connected and which parts are parallel connected. This is obvious if the circuit is simple. Many times, however, the circuit must be redrawn, putting it into a form that is easier to recognize.

In a series circuit, the current is the same at all points. In a parallel circuit, there are one or more points where the current divides and flows in separate branches. In a series-parallel circuit, there are both separate branches and series loads. The easiest way to find out whether a circuit is a series, parallel, or series-parallel circuit is to start at the negative terminal of the power source and trace the path of current through the circuit back to the positive terminal of the power source. If the current does not divide anywhere, it is a series circuit.

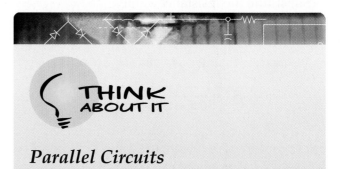

THINK ABOUT IT

Parallel Circuits

An interesting fact about circuits is the drop in resistance in a parallel circuit as more resistors are added. But this fact does not mean that you can add an endless number of devices, such as lamps, in a parallel circuit. Why not?

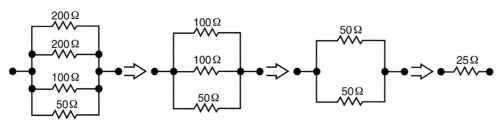

104F04.EPS

Figure 4 ◆ Equal resistances in a parallel circuit.

Parallel Circuits

Most practical circuits are wired in parallel, like the pole lights shown here.

104SA01.EPS

If the current divides into separate branches, but there are no series loads, it is a parallel circuit. If the current divides into separate branches and there are also series loads, it is a series-parallel circuit. *Figure 5* shows electric lamps connected in series, parallel, and series-parallel circuits.

After determining that a circuit is series-parallel, redraw the circuit so that the branches and the series loads are more easily recognized. This is especially helpful when computing the total resistance of the circuit. *Figure 6* shows resistors connected in a series-parallel circuit and the equivalent circuit redrawn to simplify it.

2.3.1 Reducing Series-Parallel Circuits

Very often, all that is known about a series-parallel circuit is the applied voltage and the values of the individual resistances. To find the voltage drop across any of the loads or the current in any of the branches, the total circuit current must usually be known. But to find the total current, the total resistance of the circuit must be known. To find the total resistance, reduce the circuit to its simplest form, which is usually one resistance that forms a series circuit with the voltage source. This simple series circuit has the equivalent resistance of the series-parallel circuit it was derived from, and also has the same total current. There are four basic steps in reducing a series-parallel circuit:

- If necessary, redraw the circuit so that all parallel combinations of resistances and series resistances are easily recognized.
- For each parallel combination of resistances, calculate its effective resistance.
- Replace each of the parallel combinations with one resistance whose value is equal to the effective resistance of that combination. This provides a circuit with all series loads.
- Find the total resistance of this circuit by adding the resistances of all the series loads.

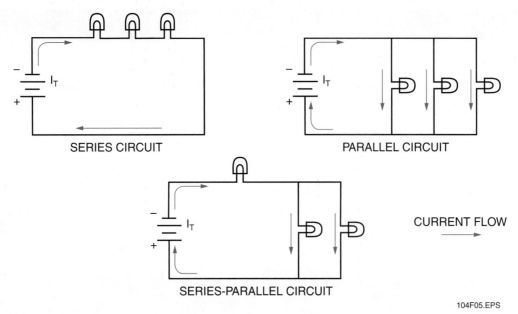

SERIES CIRCUIT

PARALLEL CIRCUIT

CURRENT FLOW

SERIES-PARALLEL CIRCUIT

104F05.EPS

Figure 5 ◆ Series, parallel, and series-parallel circuits.

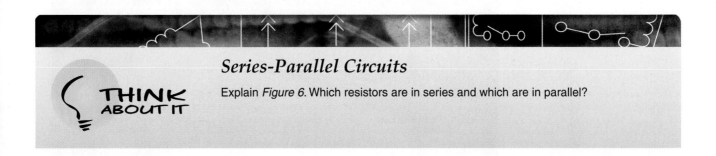

Series-Parallel Circuits

THINK ABOUT IT

Explain *Figure 6*. Which resistors are in series and which are in parallel?

Examine the series-parallel circuit shown in *Figure 7* and reduce it to an equivalent series circuit.

In this circuit, resistors R_2 and R_3 are connected in parallel, but resistor R_1 is in series with both the battery and the parallel combination of R_2 and R_3. The current I_T leaving the negative terminal of the voltage source travels through resistor R_1 before it is divided at the junction of resistors R_1, R_2, and R_3 (Point A) to go through the two branches formed by resistors R_2 and R_3.

Given the information in *Figure 7*, calculate the resistance of R_2 and R_3 in parallel and the total resistance of the circuit, R_T.

The total resistance of the circuit is the sum of R_1 and the equivalent resistance of R_2 and R_3 in parallel. To find R_T, first find the resistance of R_2 and R_3 in parallel. Because the two resistances have the same value of 20Ω, the resulting equivalent resistance is 10Ω. Therefore, the total resistance (R_T) is 15Ω ($5\Omega + 10\Omega$).

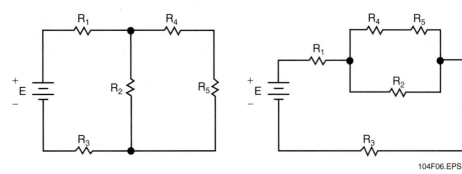

104F06.EPS

Figure 6 ◆ Redrawing a series-parallel circuit.

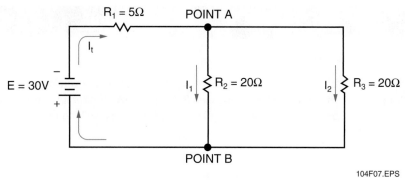

Figure 7 ◆ Reducing a series-parallel circuit.

104F07.EPS

2.4.0 Applying Ohm's Law

In resistive circuits, unknown circuit parameters can be found by using Ohm's law and the techniques for determining equivalent resistance.

2.4.1 Voltage and Current in Series Circuits

Ohm's law may be applied to an entire series circuit or to the individual parts of the circuit. When it is used on a particular part of a circuit, the voltage across that part is equal to the current in that part multiplied by the resistance of that part.

For example, given the information in *Figure 8*, calculate the total resistance (R_T) and the total current (I_T).

To find R_T:

$$R_T = R_1 + R_2 + R_3$$
$$R_T = 20 + 50 + 120$$
$$R_T = 190\Omega$$

To find I_T using Ohm's law:

$$I_T = \frac{E_T}{R_T}$$

$$I_T = \frac{95}{190}$$

$$I_T = 0.5A$$

Find the voltage across each resistor. In a series circuit, the current is the same; that is, $I = 0.5A$ through each resistor:

$$E_1 = IR_1 = 0.5(20) = 10V$$
$$E_2 = IR_2 = 0.5(50) = 25V$$
$$E_3 = IR_3 = 0.5(120) = 60V$$

The voltages E_1, E_2, and E_3 found for *Figure 8* are known as voltage drops or IR drops. Their effect is to reduce the voltage that is available to be applied across the rest of the components in the circuit. The sum of the voltage drops in any series circuit is always equal to the voltage that is

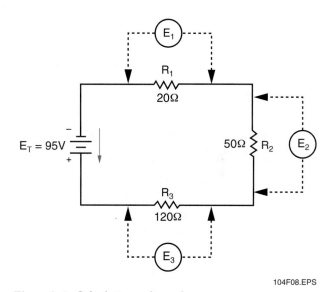

Figure 8 ◆ Calculating voltage drops.

104F08.EPS

applied to the circuit. The total voltage (E_T) is the same as the applied voltage and can be verified in this example ($E_T = 10 + 25 + 60$ or 95V).

2.4.2 Voltage and Current in Parallel Circuits

A parallel circuit is a circuit in which two or more components are connected across the same voltage source, as illustrated in *Figure 9*. The resistors R_1, R_2, and R_3 are in parallel with each other and with the battery. Each parallel path is then a branch with its own individual current. When the total current I_T leaves the voltage source E, part I_1 of the current I_T will flow through R_1, part I_2 will flow through R_2, and the remainder I_3 will flow through R_3. The branch currents I_1, I_2, and I_3 can be different. However, if a voltmeter is connected across R_1, R_2, and R_3, the respective voltages E_1, E_2, and E_3 will be equal to the source voltage E.

The total current I_T is equal to the sum of all branch currents.

Voltage Drops

THINK ABOUT IT

Calculating voltage drops is not just a schoolroom exercise. It is important to know the voltage drop when sizing circuit components. What would happen if you sized a component without accounting for a substantial voltage drop in the circuit?

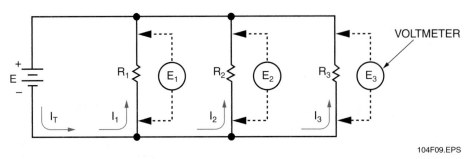

104F09.EPS

Figure 9 ◆ Parallel circuit.

This formula applies for any number of parallel branches whether the resistances are equal or unequal.

Using Ohm's law, each branch current equals the applied voltage divided by the resistance between the two points where the voltage is applied. Hence, for each branch in *Figure 9* we have the following equations:

$$\text{Branch 1: } I_1 = \frac{E_1}{R_1} = \frac{E}{R_1}$$

$$\text{Branch 2: } I_2 = \frac{E_2}{R_2} = \frac{E}{R_2}$$

$$\text{Branch 3: } I_3 = \frac{E_3}{R_3} = \frac{E}{R_3}$$

With the same applied voltage, any branch that has less resistance allows more current through it than a branch with higher resistance.

Example 1:

The two branches R_1 and R_2, shown in *Figure 10(A)*, across a 110V power line draw a total line current of 20A. Branch R_1 takes 12A. What is the current I_2 in branch R_2?

Transpose to find I_2 and then substitute given values:

$$I_T = I_1 + I_2$$
$$I_2 = I_T - I_1$$
$$I_2 = 20 - 12 = 8A$$

Example 2:

As shown in *Figure 10(B)*, the two branches R_1 and R_2 across a 240V power line draw a total line current of 35A. Branch R_2 takes 20A. What is the current I_1 in branch R_1?

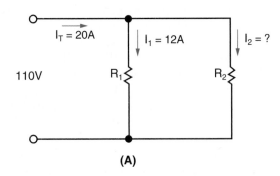

(A)

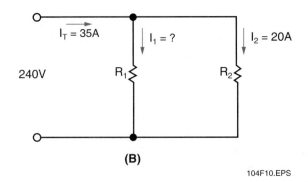

(B)

104F10.EPS

Figure 10 ◆ Solving for an unknown current.

Transpose to find I_1 and then substitute given values:

$$I_T = I_1 + I_2$$
$$I_1 = I_T - I_2$$
$$I_1 = 35 - 20 = 15A$$

2.4.3 Voltage and Current in Series-Parallel Circuits

Series-parallel circuits combine the elements and characteristics of both the series and parallel configurations. By properly applying the equations and methods previously discussed, the values of individual components of the circuit can be determined. *Figure 11* shows a simple series-parallel circuit with a 1.5V battery.

The current and voltage associated with each component can be determined by first simplifying the circuit to find the total current, and then working across the individual components.

This circuit can be broken into two components: the series resistances R_1 and R_2, and the parallel resistances R_3 and R_4.

R_1 and R_2 can be added together to form the equivalent series resistance R_{1+2}:

$$R_{1+2} = R_1 + R_2$$
$$R_{1+2} = 0.5k\Omega + 0.5k\Omega$$
$$R_{1+2} = 1k\Omega$$

R_3 and R_4 can be totaled using either the general reciprocal formula or, since there are two resistances in parallel, the product over sum method. Both methods are shown below.

$$R_{3+4} = \cfrac{1}{\cfrac{1}{R_3} + \cfrac{1}{R_4}} = \cfrac{1}{\cfrac{1}{1k\Omega} + \cfrac{1}{1k\Omega}}$$

$$= \cfrac{1}{\cfrac{2}{1,000\Omega}} = \frac{1}{0.002} = 0.5k\Omega$$

$$R_{3+4} = \frac{R_3 \times R_4}{R_3 + R_4} = \frac{1k\Omega \times 1k\Omega}{1k\Omega + 1k\Omega}$$

$$= \frac{1,000,000\Omega}{2,000\Omega} = 0.5k\Omega$$

The equivalent circuit containing the R_{1+2} resistance of 1kΩ and the R_{3+4} resistance of 0.5kΩ is shown in *Figure 12*.

Using the Ohm's law relationship that total current equals voltage divided by circuit resistance, the circuit current can be determined. First, however, total circuit resistance must be found. Since the simplified circuit consists of two resistances in series, they are simply added together to obtain total resistance.

$$R_T = R_{1+2} + R_{3+4}$$
$$R_T = 1k\Omega + 0.5k\Omega$$
$$R_T = 1.5k\Omega$$

Applying this to the current/voltage equation:

$$I_T = \frac{E_T}{R_T}$$

$$I_T = \frac{1.5V}{1.5k\Omega}$$

$$I_T = 1mA \text{ or } 0.001A$$

Now that the total current is known, voltage drops across individual components can be determined:

$$E_{R1} = I_T R_1 = 1mA \times 0.5K\Omega = 0.5V$$
$$E_{R2} = I_T R_2 = 1mA \times 0.5K\Omega = 0.5V$$

Since the total voltage equals the sum of all voltage drops, the voltage drop from A to B can be determined by subtraction:

$$E_T = E_{R1} + E_{R2} + E_{A+B}$$
$$E_T - E_{R1} - E_{R2} = E_{A+B}$$
$$1.5V - 0.5V - 0.5V = E_{A+B} = 0.5V$$

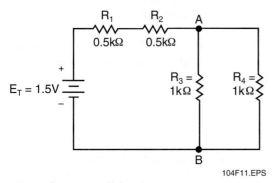

104F11.EPS

Figure 11 ◆ Series-parallel circuit.

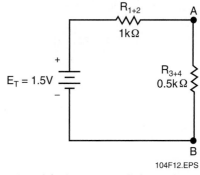

104F12.EPS

Figure 12 ◆ Simplified series-parallel circuit.

Since R_3 and R_4 are in parallel, some of the total current must pass through each resistor. R_3 and R_4 are equal, so the same current should flow through each branch. Using the relationship:

$$I = \frac{E}{R}$$

$$I_{R3} = \frac{E_{R3}}{R_3} \qquad I_{R4} = \frac{E_{R4}}{R_4}$$

$$I_{R3} = \frac{0.5V}{1k\Omega} \qquad I_{R4} = \frac{0.5V}{1k\Omega}$$

$$I_{R3} = 0.5mA \qquad I_{R4} = 0.5mA$$

$$0.5mA + 0.5mA = 1mA$$

Therefore, the total current for the circuit passes through R_1 and R_2 and is evenly divided between R_3 and R_4.

3.0.0 ◆ KIRCHHOFF'S LAWS

Kirchhoff's laws provide a simple, practical method of solving for unknown parameters in a circuit.

3.1.0 Kirchhoff's Current Law

In its most general form, **Kirchhoff's current law** states that at any point in a circuit, the total current entering that point must equal the total current leaving that point. For parallel circuits, this implies that the current in a parallel circuit is equal to the sum of the currents in each branch.

When using Kirchhoff's laws to solve circuits, it is necessary to adopt conventions that determine the algebraic signs for current and voltage terms. A convenient system for current is to consider all current flowing into a branch point as positive, and all current directed away from that point as negative.

As an example, in *Figure 13* the currents can be written as:

$$I_A + I_B - I_C = 0$$

or

$$5A + 3A - 8A = 0$$

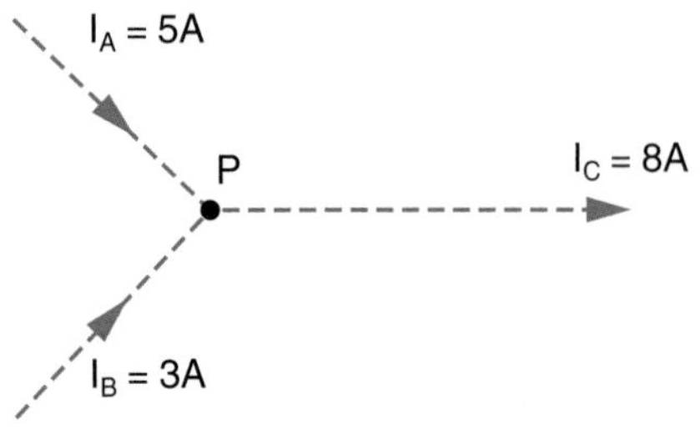

104F13.EPS

Figure 13 ◆ Kirchhoff's current law.

Currents I_A and I_B are positive terms because these currents flow into P, but I_C, directed out of P, is negative.

For a circuit application, refer to Point C at the top of the diagram in *Figure 14*. The 6A I_T into Point C divides into the 2A I_3 and 4A $I_{4/5}$, both directed out. Note that $I_{4/5}$ is the current through R_4 and R_5. The algebraic equation is:

$$I_T - I_3 - I_{4/5} = 0$$

Substituting the values for each current:

$$6A - 2A - 4A = 0$$

For the opposite direction, refer to Point D at the bottom of *Figure 14*. Here, the branch currents into Point D combine to equal the mainline current I_T returning to the voltage source. Now, I_T is directed out from Point D, with I_3 and $I_{4/5}$ directed in. The algebraic equation is:

$$I_3 + I_{4/5} - I_T = 0$$
$$2A + 4A - 6A = 0$$

Note that at either Point C or Point D, the sum of the 2A and 4A branch currents must equal the 6A total line current. Therefore, Kirchhoff's current law can also be stated as:

$$I_{IN} = I_{OUT}$$

For *Figure 14*, the equations for current can be written as shown below.

At Point C:

$$6A = 2A + 4A$$

At Point D:

$$2A + 4A = 6A$$

Kirchhoff's current law is really the basis for the practical rule in parallel circuits that the total line current must equal the sum of the branch currents.

3.2.0 Kirchhoff's Voltage Law

Kirchhoff's voltage law states that the algebraic sum of the voltages around any closed path is zero.

Referring to *Figure 15*, the sum of the voltage drops around the circuit must equal the voltage applied to the circuit:

$$E_A = E_1 + E_2 + E_3$$

Where:

E_A = voltage applied to the circuit

E_1, E_2, and E_3 = voltage drops in the circuit

Another way of stating this law is that the algebraic sum of the voltage rises and voltage drops

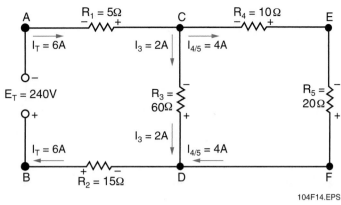

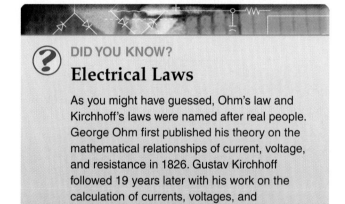

Figure 14 ◆ Application of Kirchhoff's current law.

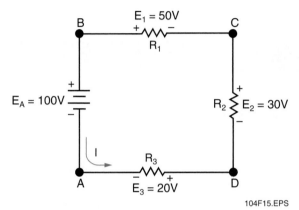

Figure 15 ◆ Kirchhoff's voltage law.

must be equal to zero. A voltage source is considered a voltage rise; a voltage across a resistor is a voltage drop. (For convenience in labeling, letter subscripts are shown for voltage sources and numerical subscripts are used for voltage drops.) This form of the law can be written by transposing the right members to the left side:

Voltage applied – sum of voltage drops = 0

Substitute letters:

$$E_A - E_1 - E_2 - E_3 = 0$$
$$E_A - (E_1 + E_2 + E_3) = 0$$

3.3.0 Loop Equations

Any closed path for current flow is called a loop. A loop equation specifies the voltages around the loop. Refer to *Figure 16*.

Consider the inside loop through A, C, D, and B. This includes the voltage drops E_1, E_3, and E_2, and the source E_T. In a clockwise direction, starting at Point A, the algebraic sum of the voltages is:

$$- E_1 - E_3 - E_2 + E_T = 0$$

or

$$- 30V - 120V - 90V + 240V = 0$$

Voltages E_1, E_3, and E_2 have a negative value, because there is a decrease in voltage seen across each of the resistors in a clockwise direction. However, the source E_T is a positive term because an increase in voltage is seen in that same direction.

For the opposite direction, going counterclockwise in the same loop from Point B, E_T is negative while E_1, E_2, and E_3 have positive values. Therefore:

$$- E_T + E_2 + E_3 + E_1 = 0$$

or

$$- 240V + 90V + 120V + 30V = 0$$

When the negative term is transposed, the equation becomes:

$$240V = 90V + 120V + 30V$$

In this form, the loop equation shows that Kirchhoff's voltage law is really the basis for the practical rule in series circuits that the sum of the voltage drops must equal the applied voltage.

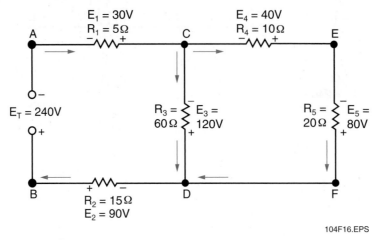

Figure 16 ◆ Loop equation.

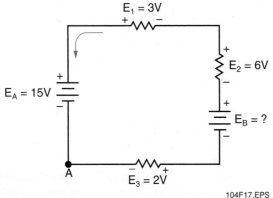

Figure 17 ◆ Applying Kirchhoff's voltage law.

For example, determine the voltage E_B for the circuit shown in *Figure 17*. The direction of the current flow is shown by the arrow. First mark the polarity of the voltage drops across the resistors and trace the circuit in the direction of the current flow starting at Point A. Then write the voltage equation around the circuit:

$$- E_3 - E_B - E_2 - E_1 + E_A = 0$$

Solve for E_B:

$$E_B = E_A - E_3 - E_2 - E_1$$
$$E_B = 15V - 2V - 6V - 3V$$
$$E_B = 4V$$

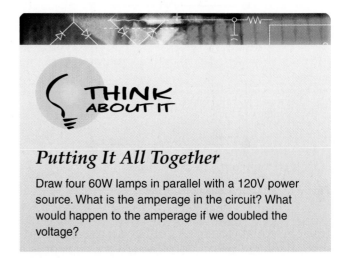

Putting It All Together

Draw four 60W lamps in parallel with a 120V power source. What is the amperage in the circuit? What would happen to the amperage if we doubled the voltage?

Since E_B was found to be positive, the assumed direction of current is in fact the actual direction of current.

In its most general form, Kirchhoff's voltage law states that the algebraic sum of all the potential differences in a closed loop is equal to zero. A closed loop means any completely closed path consisting of wire, resistors, batteries, or other components. For series circuits, this implies that the sum of the voltage drops around the circuit is equal to the applied voltage. For parallel circuits, this implies that the voltage drops across all branches are equal.

1. The formula for calculating the total resistance in a series circuit with three resistors is _____.
   a. $R_T = R_1 + R_2 + R_3$
   b. $R_T = R_1 - R_2 - R_3$
   c. $R_T = R_1 \times R_2 \times R_3$
   d. $R_T = \dfrac{1}{\dfrac{1}{R_1} + \dfrac{1}{R_2} + \dfrac{1}{R_3}}$

2. Find the total resistance in a series circuit with three resistances of 10Ω, 20Ω, and 30Ω.
   a. 1Ω
   b. 15Ω
   c. 20Ω
   d. 60Ω

3. The formula for calculating the total resistance in a parallel circuit with three resistors is _____.
   a. $R_T = R_1 + R_2 + R_3$
   b. $R_T = R_1 - R_2 - R_3$
   c. $R_T = R_1 \times R_2 \times R_3$
   d. $R_T = \dfrac{1}{\dfrac{1}{R_1} + \dfrac{1}{R_2} + \dfrac{1}{R_3}}$

4. The total resistance in *Figure 1* is _____.
   a. 100Ω
   b. 129Ω
   c. 157Ω
   d. $1,040\Omega$

5. In a parallel circuit, the voltage across each path is equal to the _____.
   a. total circuit resistance times path current
   b. source voltage minus path voltage
   c. path resistance times total current
   d. applied voltage

6. The value for total current in *Figure 2* is _____ amps.
   a. 1.25
   b. 2.50
   c. 5
   d. 10

7. A resistor of 32Ω is in parallel with a resistor of 36Ω, and a 54Ω resistor is in series with the pair. When 350V is applied to the combination, the current through the 54Ω resistor is _____ amps.
   a. 2.87
   b. 3.26
   c. 4.93
   d. 5.86

8. A 242Ω resistor is in parallel with a 180Ω resistor, and a 420Ω resistor is in series with the combination in a 50V circuit. A current of 22mA flows through the 242Ω resistor. The current through the 180Ω resistor is _____ mA.
   a. 36.4
   b. 59.4
   c. 60.3
   d. 73.6

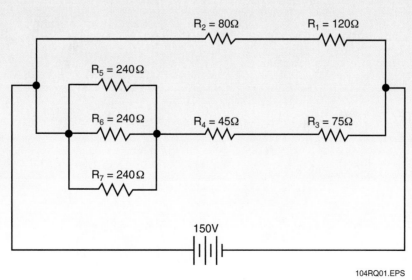

104RQ01.EPS

Figure 1

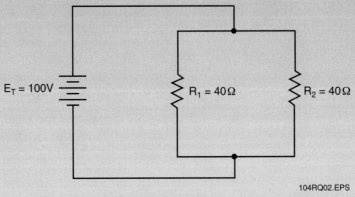

104RQ02.EPS

Figure 2

9. Kirchhoff's voltage law states that the algebraic sum of the voltages around any
   closed current path is _____.
   a. infinity
   b. zero
   c. twice the current
   d. always less than the individual voltages
      due to voltage drop

10. Two 24Ω resistors are in parallel, and a 42Ω
    resistor is in series with the combination.
    When 78V is applied to the three resistors,
    the voltage drop across the 42Ω resistor is
    about _____ volts.
    a. 49.8
    b. 55.8
    c. 60.5
    d. 65.3

Summary

The relationships among current, voltage, resistance, and power in Ohm's law are the same for both DC series and DC parallel circuits. Understanding and being able to apply these concepts is necessary for effective circuit analysis and troubleshooting. DC series-parallel circuits also have these fundamental relationships. Since DC series-parallel circuits are a combination of simple series and parallel circuits, Kirchhoff's voltage and current laws will apply. Calculating I, E, R, and P for series-parallel circuits is no more difficult than calculating these values for simple series or parallel circuits. However, for series-parallel circuits, these calculations require more careful circuit analysis in order to use Ohm's law correctly.

Notes

1. _____ states that the total amount of current flowing through a parallel circuit is equal to the sum of the amounts of current flowing through each current path.

2. _____ states that the sum of all the voltage drops in a circuit is equal to the source voltage of the circuit.

3. _____ contain both series and parallel current paths.

4. _____ contain two or more parallel paths through which current can flow.

5. _____ contain only one path for current flow.

Trade Terms

Kirchhoff's current law
Kirchhoff's voltage law
Parallel circuits
Series circuits
Series-parallel circuits

James Mitchem

TIC —The Industrial Company

Jim Mitchem serves as a troubleshooter for a large electrical contractor. During his career in the electrical industry, he worked his way up from apprentice to technical services manager.

How did you become an electrician?
Quite by accident. A couple of years after college, I was working as a relief operator in a plant when the lead electrician retired, creating a vacancy. I liked the idea that electricians were expected to use their knowledge and initiative to keep the place running. I applied and was accepted as a trainee.

How did you get your training?
I took an electrical apprenticeship course by correspondence, and I was fortunate enough to work with good people who helped me along. I worked in an environment that exposed me to a variety of equipment and applications, and just about everyone I've ever worked with has taught me something. Now I'm passing my knowledge on to others.

What kinds of work have you done in your career?
I've worked as an apprentice, journeyman, instrument and controls technician, instrument fitter, foreman, general foreman, superintendent, and startup engineer. Each of these positions required that I learn new skills, both technical and managerial. My experience in many disciplines and types of projects has given me a high level of credibility with my employer and our clients.

Now I act as a technical resource and troubleshooter for job sites and in-house functions such as safety, quality assurance, and training. I visit job sites to help solve problems and help out with commissioning and startup.

What factor or factors have contributed the most to your success?
There are several factors. Two very important ones have been a desire to learn and a willingness to do whatever is asked of me. I also keep an eye on the big picture. When I'm on a job, I'm not just pulling wire, I'm building a power plant or whatever the project is. I also think it has helped me to remain with the same employer for 18 years.

Any advice for apprentices just beginning their careers?
Keep learning! And don't depend on others to train you. Take the initiative to buy or borrow books and trade journals. Take licensing tests and do whatever is necessary to keep your licenses current. Finally, make sure you know your own personal and professional values and work with a company that shares those values.

Trade Terms
Introduced in This Module

Kirchhoff's current law: The statement that the total amount of current flowing through a parallel circuit is equal to the sum of the amounts of current flowing through each current path.

Kirchhoff's voltage law: The statement that the sum of all the voltage drops in a circuit is equal to the source voltage of the circuit.

Parallel circuits: Circuits containing two or more parallel paths through which current can flow.

Series circuits: Circuits with only one path for current flow.

Series-parallel circuits: Circuits that contain both series and parallel current paths.

This module is intended to present thorough resources for task training. The following reference works are suggested for further study. These are optional materials for continued education rather than for task training.

Electronics Fundamentals: Circuits, Devices, and Applications, Thomas L. Floyd. New York: Prentice Hall.

Principles of Electric Circuits, Thomas L. Floyd. New York: Prentice Hall.

NCCER makes every effort to keep these textbooks up-to-date and free of technical errors. We appreciate your help in this process. If you have an idea for improving this textbook, or if you find an error, a typographical mistake, or an inaccuracy in NCCER's Contren® textbooks, please write us, using this form or a photocopy. Be sure to include the exact module number, page number, a detailed description, and the correction, if applicable. Your input will be brought to the attention of the Technical Review Committee. Thank you for your assistance.

Instructors – If you found that additional materials were necessary in order to teach this module effectively, please let us know so that we may include them in the Equipment/Materials list in the Annotated Instructor's Guide.

Write: Product Development and Revision
National Center for Construction Education and Research
3600 NW 43rd St., Bldg. G, Gainesville, FL 32606

Fax: 352-334-0932

E-mail: curriculum@nccer.org

Craft _____ Module Name _____

Copyright Date _____ Module Number _____ Page Number(s) _____

Description _____

_____

_____

_____

(Optional) Correction _____

_____

_____

(Optional) Your Name and Address _____

_____

_____

Introduction to the
National Electrical Code®

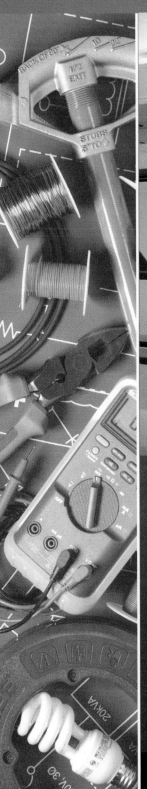

Birck Nanotechnology Center at Purdue University

The Gaylor Group's team of nearly 60 electricians followed strict cleanroom protocols to install electrical systems in this leading-edge nanotechnology facility while facing shared-space challenges and work restrictions.

26105-08

26105-08

Introduction to the National Electrical Code®

Topics to be presented in this module include:

Overview

The primary reference book for safe electrical installations is the *National Electrical Code®*, often referred to as the *NEC®*. The *NEC®* states that its primary purpose is "the practical safeguarding of persons and property from hazards arising from the use of electricity." It is not intended to be used as a book of installation standards. Every working electrician should include a copy of the current edition in his or her arsenal of tools.

The *NEC®* governs just about every task an electrician does. Therefore, it is important to understand the layout of the *NEC®* and be able to navigate through it quickly. The *NEC®* serves as the standard reference book in many electrical examinations, especially those related to the issuance of journeyman and master electrician licenses.

The *NEC®* is updated every three years by a panel of experts. It is important to keep up to date on the *NEC®*, because each revision contains changes that can affect your work.

Note: *National Electrical Code®* and *NEC®* are registered trademarks of the National Fire Protection Association, Inc., Quincy, MA 02269. All *National Electrical Code®* and *NEC®* references in this module refer to the 2008 edition of the *National Electrical Code®*.

Objectives

When you have completed this module, you will be able to do the following:

1. Explain the purpose and history of the *NEC*®.
2. Describe the layout of the *NEC*®.
3. Demonstrate how to navigate the *NEC*®.
4. Describe the purpose of the National Electrical Manufacturers Association and the NFPA.
5. Explain the role of nationally recognized testing laboratories.

Trade Terms

Articles
Chapters
Exceptions
Fine print note (FPN)
National Electrical Manufacturers Association (NEMA)

National Fire Protection Association (NFPA)
Nationally Recognized Testing Laboratories (NRTLs)
Parts
Sections

Required Trainee Materials

1. Paper and pencil
2. Copy of the latest edition of the *National Electrical Code*®

Prerequisites

Before you begin this module, it is recommended that you successfully complete *Core Curriculum* and *Electrical Level One*, Modules 26101-08 through 26104-08.

This course map shows all of the modules in *Electrical Level One*. The suggested training order begins at the bottom and proceeds up. Skill levels increase as you advance on the course map. The local Training Program Sponsor may adjust the training order.

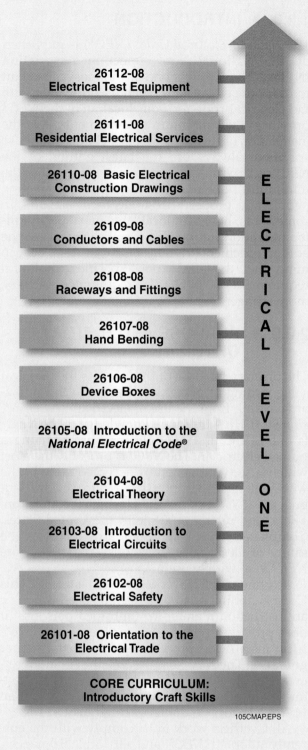

26112-08
Electrical Test Equipment

26111-08
Residential Electrical Services

26110-08 Basic Electrical Construction Drawings

26109-08
Conductors and Cables

26108-08
Raceways and Fittings

26107-08
Hand Bending

26106-08
Device Boxes

26105-08 Introduction to the *National Electrical Code*®

26104-08
Electrical Theory

26103-08 Introduction to Electrical Circuits

26102-08
Electrical Safety

26101-08 Orientation to the Electrical Trade

CORE CURRICULUM:
Introductory Craft Skills

ELECTRICAL LEVEL ONE

105CMAP.EPS

1.0.0 ◆ INTRODUCTION

The *National Electrical Code®* (*NEC®*) is published by the **National Fire Protection Association (NFPA)**. The *NEC®* is one of the most important tools for the electrician. When used together with the electrical code for your local area, the *NEC®* provides the minimum requirements for the installation of electrical systems. Unless otherwise specified, always use the latest edition of the *NEC®* as your on-the-job reference. It specifies the minimum provisions necessary for protecting people and property from electrical hazards. In some areas, however, local laws may specify different editions of the *NEC®*, so be sure to use the edition specified by your employer. Also, bear in mind that the *NEC®* only specifies minimum requirements, so local or job requirements may be more stringent.

2.0.0 ◆ PURPOSE AND HISTORY OF THE *NEC®*

The primary purpose of the *NEC®* is the practical safeguarding of persons and property from hazards arising from the use of electricity [*NEC Section 90.1(A)*]. A thorough knowledge of the *NEC®* is one of the first requirements for becoming a trained electrician. The *NEC®* is probably the most widely used and generally accepted code in the world. It has been translated into several languages. It is used as an electrical installation, safety, and reference guide in the United States. Compliance with *NEC®* standards increases the safety of electrical installations—the reason the *NEC®* is so widely used.

Although *NEC Section 90.1(C)* states, "This Code is not intended as a design specification or an instruction manual for untrained persons," it does provide a sound basis for the study of electrical installation procedures—under the proper guidance. The *NEC®* has become the standard reference of the electrical construction industry. Anyone involved in electrical work should obtain an up-to-date copy and refer to it frequently.

All electrical work must comply with the currently adopted *NEC®* and all local ordinances.

Like most laws, the *NEC®* is easier to work with once you understand the language and know where to look for the information you need.

> **NOTE**
>
> This module is not a substitute for the *NEC®*. You need to acquire a copy of the most recent edition and keep it handy at all times. The more you know about the *NEC®*, the better an electrician you will become.

2.1.0 History

In 1881, the National Association of Fire Engineers met in Richmond, Virginia. From this meeting came the idea to draft the first *National Electrical Code*. The first nationally recommended electrical code was published by the National Board of Fire Underwriters (now the American Insurance Association) in 1895.

In 1896, the National Electric Light Association (NELA) was working to make the requirements of the fire insurance organizations and electrical utilities fit together. NELA succeeded in promoting a conference that would result in producing a standard national code. The NELA code would serve the interests of the insurance industry, operating concerns, manufacturing, and industry.

The conference produced a set of requirements that was unanimously accepted. In 1897, the first edition of the *NEC®* was published, and the *NEC®* became the first cooperatively produced national code. The organization that produced the *NEC®* was known as the *National Conference on Standard Electrical Rules*. This group became a permanent organization, and its job was to develop the *NEC®*.

In 1911, the NFPA took over administration and control of the *NEC®*. However, the National Board of Fire Underwriters continued to publish the *NEC®* until 1962. From 1911 until now, the *NEC®* has experienced several major changes, as well as regular three-year updates. In 1923, the *NEC®* was rearranged and rewritten, and in 1937, it was edi-

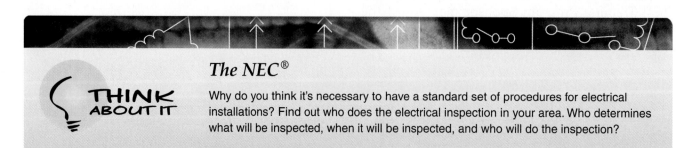

The *NEC®*

Why do you think it's necessary to have a standard set of procedures for electrical installations? Find out who does the electrical inspection in your area. Who determines what will be inspected, when it will be inspected, and who will do the inspection?

What's wrong with these pictures?

105SA01.EPS

105SA02.EPS

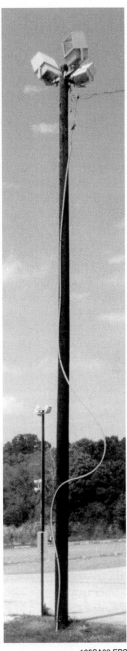

105SA03.EPS

torially revised. In 1959, the *NEC*® adopted a numbering system in which each **section** of an **article** was identified with an article/section number.

2.1.1 Who Is Involved?

The creation of a universally accepted set of rules is an involved and complicated process. Rules made by a committee have the advantage that they usually do not leave out the interests of any of the groups represented on the committee. However, since the rules must represent the interests and requirements of an assortment of groups, they are often quite complicated and wordy.

In 1949, the NFPA reorganized the *NEC*® into its present structure. The present structure consists of a Correlating Committee and Code Making Panels (CMPs). The Correlating Committee consists of principal voting members and alternates.

The principal function of the Correlating Committee is to ensure that:

- No conflict of requirements exists
- Correlation has been achieved
- NFPA regulations governing committee projects have been followed
- A practical schedule of revision and publication is established and maintained

Each of the CMPs have members who are experts on particular subjects and have been assigned certain articles to supervise and revise as required. Members of the CMPs represent special interest groups such as trade associations, electrical contractors, electrical designers and engineers, electrical inspectors, electrical manufacturers and suppliers, electrical testing laboratories, and insurance organizations.

Each panel is structured so that not more than one-third of its members are from a single interest group. The members of the *NEC®* Committee create or revise requirements for the *NEC®* through probing, debating, analyzing, weighing, and reviewing new input. Anyone, including you, can submit proposals to amend the *NEC®*. Sample forms for this purpose may be obtained from the

Secretary of the Standards Council at NFPA Headquarters. The *NEC®* describes the code-making process in detail.

The NFPA membership is drawn from the fields listed above. In addition to publishing the *NEC®*, the duties of the NFPA include the following:

- Developing, publishing, and distributing standards that are intended to minimize the possibility and effects of fire and explosion
- Conducting fire safety education programs for the general public
- Providing information on fire protection, prevention, and suppression
- Compiling annual statistics on causes and occupancies of fires, large-loss fires (over one million dollars), fire deaths, and firefighter casualties
- Providing field service by specialists on electricity, flammable liquids and gases, and marine fire problems
- Conducting research projects that apply statistical methods and operations research to develop computer models and data management systems

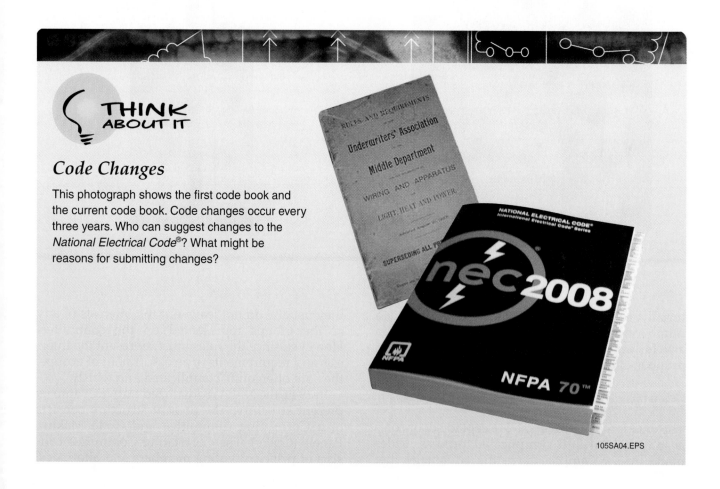

THINK ABOUT IT

Code Changes

This photograph shows the first code book and the current code book. Code changes occur every three years. Who can suggest changes to the *National Electrical Code®*? What might be reasons for submitting changes?

105SA04.EPS

3.0.0 ◆ THE LAYOUT OF THE *NEC*®

The *NEC*® begins with a brief history. *Figure 1* shows how the *NEC*® is organized.

3.1.0 Types of Rules

There are two basic types of rules in the *NEC*®: mandatory rules and permissive rules. It is important to understand these rules as they are defined in *NEC Section 90.5*. Mandatory rules contain the words "shall" or "shall not" and must be adhered to. Permissive rules identify actions that are allowed but not required and typically cover options or alternative methods. Permissive rules are indicated by the phrases "shall be permitted" or "shall not be required." Be aware that local ordinances may amend requirements of the *NEC*®. This means that a city or county may have additional requirements or prohibitions that must be followed in that jurisdiction.

3.2.0 *NEC*® Introduction

The main body of the text begins with an Introduction entitled *NEC Article 90*. This introduction gives you an overview of the *NEC*® (see *Figure 1*). Items included in this section are:

- Purpose of the *NEC*®
- Scope of the code book
- Code arrangement
- Enforcement
- Mandatory rules and explanatory material
- Formal interpretation
- Examination of equipment for safety
- Wiring planning
- Metric units of measurement

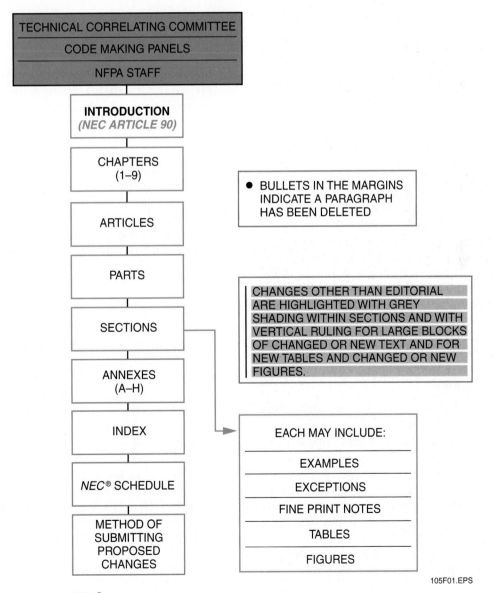

Figure 1 ◆ The layout of the *NEC*®.

105F01.EPS

3.3.0 The Body of the *NEC®*

The remainder of the code is organized into nine **chapters,** followed by annexes, the index, a schedule for the next code cycle, and information on submitting proposed changes.

NEC Chapters 1 through 8 are subdivided into articles. Each chapter focuses on a general category of electrical application, such as *NEC Chapter 2, Wiring and Protection.* Each article emphasizes a more specific **part** of that category, such as *NEC Article 210, Branch Circuits, Part I General Provisions.* Each section gives examples of a specific application of the *NEC®,* such as *NEC Section 210.4, Multiwire Branch Circuits. NEC Chapter 9* contains tables that are referenced by any of the articles in *NEC Chapters 1 through 8. Annexes A through H* provide informational material and examples that are helpful when applying *NEC®* requirements.

The chapters of the *NEC®* are organized into four major categories:

- *NEC Chapters 1, 2, 3, 4* – The first four chapters present the rules for the design and installation of electrical systems. They generally apply to all electrical installations.
- *NEC Chapters 5, 6, and 7* – These chapters are concerned with special occupancies, equipment, and conditions. Rules in these chapters may modify or amend those in the first four chapters.
- *NEC Chapter 8* – This chapter covers communications systems, such as the telephone and television receiving equipment. It may also reference other articles, such as the installation of grounding electrode conductor connections as covered in *NEC Section 250.52.*
- *NEC Chapter 9* – This chapter contains tables that are applicable when referenced by other chapters in the *NEC®.*
- *Annexes A through H* – These annexes contain helpful information that is not mandatory.
  - *Annex A* contains a list of product safety standards. These standards provide further references for requirements that are in addition to the *NEC®* requirements for the electrical components mentioned.
  - *Annex B* contains information for determining ampacities of conductors under engineering supervision.
  - *Annex C* contains the conduit fill tables for multiple conductors of the same size and type within the accepted raceways.
  - *Annex D* contains examples of calculations for branch circuits, feeders, and services as well as other loads such as motor circuits.
  - *Annex E* contains information on types of building construction.
  - *Annex F* is for informational purposes and covers critical operations power systems.
  - *Annex G* is for informational purposes and covers supervisory control and data acquisition (SCADA) systems.
  - *Annex H* contains the requirements for administration and enforcement of the *NEC®.*

3.4.0 Text in the *NEC*®

When you open the *NEC*®, you will notice several different types of text or printing used. Here is an explanation of each type of text:

- *Bold black letters* – Headings for each *NEC*® application are written in bold black letters.
- *Exceptions* – These explain the circumstances under which a specific part of the *NEC*® does not apply. Exceptions are written in *italics* under that part of the *NEC*® to which they pertain.
- *Fine print notes (FPNs)* – These explain something in an application, suggest other sections to read about the application, or provide tips about the application. These are defined in the text by the term (FPN) shown in parentheses before a paragraph in smaller print.
- *Figures* – These may be included with explanations to give you a picture of what your application may look like.
- *Tables* – These are often included when there is more than one possible application of the *NEC*®. You would use a table to look up the specifications of your application.

4.0.0 ◆ NAVIGATING THE *NEC*®

To locate information for a particular procedure being performed, use the following steps:

Step 1 Familiarize yourself with *NEC Articles 90, 100, and 110* to gain an understanding of the material covered in the *NEC*® and the definitions used in it.

Step 2 Turn to the *Table of Contents* at the beginning of the *NEC*®.

Step 3 Locate the chapter that focuses on the desired category.

Step 4 Find the article pertaining to your specific application.

Step 5 Turn to the page indicated. Each application will begin with a bold heading.

> **NOTE**
>
> An index is provided at the end of the *NEC*®. The index lists specific topics and provides a reference to the location of the material within the *NEC*®. The index is helpful when you are looking for a specific topic rather than a general category.

Once you are familiar with *NEC Articles 90, 100, and 110,* you can move on to the rest of the *NEC*®. There are several key sections used often in servicing electrical systems.

4.1.0 *NEC* Chapter 1, General

NEC Article 100 contains a list of common definitions used in the *NEC*®. Definitions specific to a single article are included in the second section of each article. For example, *NEC Section 240.2* contains definitions specific to overcurrent protection.

Two definitions from *NEC Article 100* you should become familiar with are:

- *Labeled* – "Equipment or materials to which has been attached a label, symbol, or other identifying mark of an organization that is acceptable to the authority having jurisdiction and concerned with product evaluation, that maintains periodic inspection of production of labeled equipment or materials, and by whose labeling the manufacturer indicates compliance with appropriate standards or performance in a specified manner."
- *Listed* – "Equipment, materials, or services included in a list published by an organization that is acceptable to the authority having jurisdiction and concerned with evaluation of products or services, that maintains periodic inspection of production of listed equipment or materials, or periodic evaluation of services, and whose listing states that the equipment, material, or services either meets appropriate designated standards or has been tested and found suitable for use for a specified purpose."

Besides installation rules, you will also have to be concerned with the type and quality of materials that are used in electrical wiring systems. Nationally recognized testing laboratories are product safety certification laboratories. Underwriters Laboratories, Inc., also called UL, is one such laboratory. These laboratories establish and operate product safety certification programs to make sure that items produced under the service are safeguarded against reasonable foreseeable risks. Some of these organizations maintain a worldwide network of field representatives who make unannounced visits to manufacturing facilities to counter-check products bearing their seal of approval. The UL label is shown in *Figure 2*.

NEC Article 110 gives the general requirements for electrical installations. It is important for you to be familiar with this general information and the definitions.

105F02.EPS

Figure 2 ♦ Underwriters Laboratories label.

4.2.0 *NEC Chapter 2, Wiring and Protection*

NEC Chapter 2 discusses wiring design and protection, the information electrical technicians need most often. It covers the use and identification of grounded conductors, branch circuits, feeders, calculations, services, overcurrent protection, and grounding. This is essential information for all types of electrical systems. If you run into a problem related to the design or installation of a conventional electrical system, you can probably find a solution for it in this chapter.

NEC Article 200 contains important information on the use and identification of grounded conductors.

NEC Articles 210 and 215 cover the provisions and requirements for branch circuits, their ratings and required outlets, and installation and overcurrent protection requirements for feeders.

NEC Article 220 contains the requirements for calculating branch circuit, feeder, and service loads, including the calculation methods for farm loads.

NEC Articles 225 and 230 cover outside branch circuit, feeder, and service installation requirements.

Underwriters Laboratories, Inc.

Underwriters Laboratories, Inc. was established in 1893. The Chicago World's Fair opened and thanks to Edison's introduction of the electric light bulb, it lit up the world. But all was not perfect—wires soon sputtered and crackled, and, ironically, the Palace of Electricity caught fire. The Fair's insurance company called in a troubleshooting engineer, who found faulty and brittle insulation, worn out and deteriorated wiring, bare wires, and overloaded circuits. He called for standards in the electrical industry, and then set up a testing laboratory above a Chicago firehouse to do just that. Hence, Underwriters Laboratories, Inc. (UL) was born.

NEC Article 240 covers the requirements for overcurrent protection and overcurrent protective devices, including the standard ratings for fuses and fixed-trip circuit breakers. *NEC Article 240, Part IX* covers overcurrent protection for installations over 600 volts.

NEC Article 250 covers the requirements for grounding and bonding electrical systems. It lists the systems required, permitted, and not permitted to be grounded and provides the requirements for grounding connection locations. It also covers accepted methods of grounding and bonding, including types and sizes of grounding and bonding conductors and electrodes.

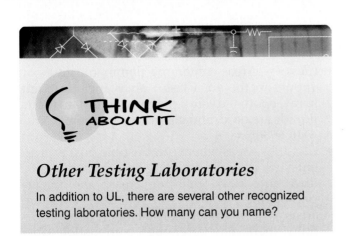

THINK ABOUT IT

Other Testing Laboratories

In addition to UL, there are several other recognized testing laboratories. How many can you name?

4.3.0 NEC Chapter 3, Wiring Methods and Materials

NEC Chapter 3 lists the rules on wiring methods and materials. The materials and procedures to use on a particular system depend on the type of building construction, the type of occupancy, the location of the wiring in the building, the type of atmosphere in the building or in the area surrounding the building, mechanical factors, and the relative costs of different wiring methods. The general requirements for conductors and wiring methods that form an integral part of manufactured equipment are not included in the requirements of the code per *NEC Article 300.1(B)*.

NEC Article 300 provides the general requirements for all wiring methods, including information such as minimum burial depths and permitted wiring methods for areas above suspended ceilings.

NEC Article 310 contains a description of acceptable conductors for the wiring methods contained in *NEC Chapter 3*.

NEC Articles 312 and 314 give rules for raceways, boxes, cabinets, and raceway fittings. Outlet boxes vary in size and shape, depending on their use, the size of the raceway, the number of conductors entering the box, the type of building construction, and the atmospheric conditions of the area. These articles should answer most questions on the selection and use of these items.

The *NEC®* does not describe in detail all types and sizes of outlet boxes. However, the manufacturers of outlet boxes provide excellent catalogs showing their products. Collect these catalogs, since these items are essential to your work.

NEC Articles 320 through 340 cover sheathed cables of two or more conductors, such as nonmetallic-sheathed and metal-clad cable.

NEC Articles 342 through 356 cover conduit wiring systems, such as rigid and flexible metal and nonmetallic conduit.

NEC Articles 358 through 362 cover tubing wiring methods, such as electrical metallic and nonmetallic tubing.

NEC Articles 366 through 390 cover other wiring methods, such as busways and wireways. Cable trays are a system of support for the wiring methods found not only in *NEC Article 392*, but also in *NEC Chapters 4, 7, and 8*.

4.4.0 NEC Chapter 4, Equipment for General Use

NEC Article 400 covers the use and installation of flexible cords and cables, including the trade name, type letter, wire size, number of conductors, conductor insulation, outer covering, and use of each. *NEC Article 402* covers fixture wires, again giving the trade name, type letter, and other important details.

NEC Article 404 covers the requirements for the uses and installation of switches, switching devices, and circuit breakers where used as switches.

NEC Article 406 gives the rules for the ratings, types, and installation of receptacles, cord connectors, and attachment plugs (cord caps).

NEC Article 408 covers the requirements for switchboards, panelboards, and distribution boards used to control light and power circuits. It also covers battery-charging panels supplied from light or power circuits.

NEC Article 410 on lighting fixtures (luminaires) is especially important. It gives installation procedures for fixtures in specific locations. For example, it covers fixtures near combustible material and fixtures in closets. However, the *NEC®*

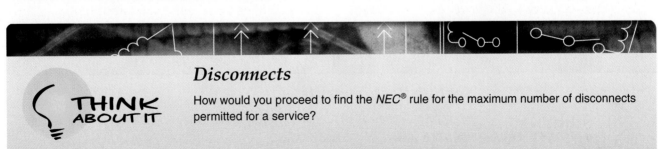

THINK ABOUT IT

Disconnects

How would you proceed to find the *NEC®* rule for the maximum number of disconnects permitted for a service?

does not describe how many fixtures will be needed in a given area to provide a certain amount of illumination.

NEC Article 422 covers the use of electric appliances in any occupancy. It includes kitchen appliances, various heating appliances, cord-and-plug connected equipment, and other appliances.

NEC Article 424 covers fixed electric space-heating equipment. It includes heating cable, unit heaters, boilers, central systems, and other approved equipment.

NEC Article 430 covers electric motors, including electrical connections, motor controls, and overload protection.

NEC Articles 440 through 460 cover air conditioning and heating equipment, generators, transformers, phase converters, and capacitors.

NEC Article 480 provides requirements related to battery-operated electrical systems. Storage batteries are seldom thought of as part of a conventional electrical system, but they often provide standby emergency lighting service. They may also supply power to security systems that are separate from the main AC electrical system.

4.5.0 NEC Chapter 5, Special Occupancies

NEC Chapter 5 covers special occupancy areas. These are areas where the sparks generated by electrical equipment may cause an explosion or fire. The hazard may be due to the atmosphere of the area or the presence of a volatile material in the area. Commercial garages, aircraft hangars, and service stations are typical special occupancy locations.

NEC Article 500 covers the different types of special occupancy atmospheres where an explosion is possible. The atmospheric groups were established to make it easy to test and approve equipment for various types of uses.

NEC Articles 501.10, 502.10, and 503.10 cover the installation of explosion-proof wiring. An explosion-proof system is designed to prevent the ignition of a surrounding explosive atmosphere when arcing occurs within the electrical system.

There are three main classes of special occupancy location:

- *Class I* – Areas containing flammable gases or vapors in the air. Class I areas include paint spray booths, dyeing plants where hazardous liquids are used, and gas generator rooms *(NEC Article 501)*.
- *Class II* – Areas where combustible dust is present, such as grain-handling and storage plants, dust and stock collector areas, and sugar-pulverizing plants *(NEC Article 502)*. These are areas where, under normal operating conditions, there may be enough combustible dust in the air to produce explosive or ignitable mixtures.
- *Class III* – Areas that are hazardous because of the presence of easily ignitable fibers or other particles in the air, although not in large enough quantities to produce ignitable mixtures *(NEC Article 503)*. Class III locations include cotton mills, rayon mills, and clothing manufacturing plants.

NEC Articles 511 and 514 regulate garages and fuel dispensing locations where volatile or flammable liquids are used. While these areas are not always considered critically hazardous locations, there may be enough danger to require special precautions in the electrical installation. In these areas, the *NEC®* requires that volatile gases be confined to an area not more than four feet above the floor. So in most cases, conventional raceway systems are permitted above this level. If the area is judged to be critically hazardous, explosion-proof wiring (including seal-offs) may be required.

NEC Article 520 regulates theaters and similar occupancies where fire and panic can cause hazards to life and property. Projection rooms and adjacent areas must be properly ventilated and wired for the protection of operating personnel and others using the area.

NEC Chapter 5 also covers floating buildings, marinas, agricultural buildings, service stations, bulk storage plants, health care facilities, mobile homes and parks, and temporary installations.

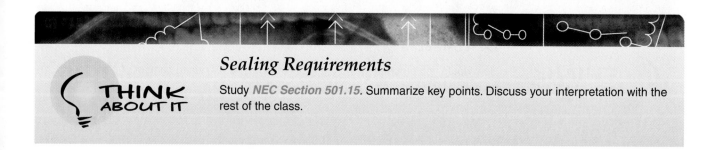

THINK ABOUT IT

Sealing Requirements

Study *NEC Section 501.15*. Summarize key points. Discuss your interpretation with the rest of the class.

4.6.0 NEC Chapter 6, Special Equipment

NEC Article 600 covers electric signs and outline lighting. *NEC Article 610* applies to cranes and hoists. *NEC Article 620* covers the majority of the electrical work involved in the installation and operation of elevators, dumbwaiters, escalators, and moving walks. The manufacturer is responsible for most of this work. The electrician usually just furnishes a feeder terminating in a disconnect means in the bottom of the elevator shaft. The electrician may also be responsible for a lighting circuit to a junction box midway in the elevator shaft for connecting the elevator cage lighting cable and exhaust fans. The articles in this chapter list most of the requirements for these installations.

NEC Article 630 regulates electric welding equipment. It is normally treated as a piece of industrial power equipment requiring a special power outlet, but there are special conditions that apply to the circuits supplying welding equipment. These are outlined in detail in this chapter.

NEC Article 640 covers wiring for sound recording and similar equipment. This type of equipment normally requires low-voltage wiring. Special outlet boxes or cabinets are usually provided with the equipment, but some items may be mounted in or on standard outlet boxes. Some sound recording systems require direct current. It is supplied from rectifying equipment, batteries, or motor generators. Low-voltage alternating current comes from relatively small transformers connected on the primary side to a 120V circuit within the building.

Other items covered in *NEC Chapter 6* include X-ray equipment *(NEC Article 660)*, induction and dielectric heat-generating equipment *(NEC Article 665)*, and industrial machinery *(NEC Article 670)*.

4.7.0 NEC Chapter 7, Special Conditions

In most commercial buildings, the *NEC®* and local ordinances require a means of lighting public rooms, halls, stairways, and entrances. There must be enough light to allow the occupants to exit from the building if the general building lighting is interrupted. Exit doors must be clearly indicated by illuminated exit signs.

NEC Chapter 7 covers the installation of emergency systems. These circuits should be arranged so that they can automatically transfer to an alternate source of current, usually storage batteries or gasoline-driven generators. As an alternative in some types of occupancies, you can connect them to the supply side of the main service, so disconnecting the main service switch would not disconnect the emergency circuits. This chapter also covers fire alarms and a variety of other equipment, systems, and conditions that are not easily categorized elsewhere in the *NEC®*.

4.8.0 NEC Chapter 8, Communications Systems

NEC Chapter 8 is a special category for wiring associated with electronic communications systems including telephone, radio and TV, satellite dish, network powered broadband systems, and community antenna systems.

4.9.0 Examples of Navigating the NEC®

Using your copy of the 2008 *NEC®*, follow along with these sample scenarios to familiarize yourself with the layout of the *NEC®*.

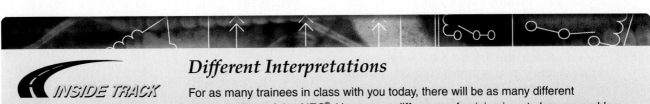

INSIDE TRACK

Different Interpretations

For as many trainees in class with you today, there will be as many different interpretations of the *NEC®*. However, a difference of opinion is not always a problem—discussing the *NEC®* with your peers will allow you to expand your own understanding of it. In your previous discussion about *NEC Section 501.15*, what differences of interpretation did you have to resolve?

4.9.1 Installing Type SE Cable

Suppose you are installing Type SE (service-entrance) cable on the side of a home. You know that this cable must be secured, but you are not sure of the spacing between cable clamps. To find out this information, use the following procedure:

Step 1 Look in the *NEC® Table of Contents* and follow down the list until you find an appropriate category. (Or you can use the index at the end of the book.)

Step 2 *NEC Article 230* will probably catch your eye first, so turn to the page where it begins.

Step 3 Scan down through the section numbers until you come to *NEC Section 230.51, Mounting Supports*. Upon reading this section, you will find in paragraph *(a) Service Cables* that "service cables shall be supported by straps or other approved means within 300 mm (12 in) of every service head, gooseneck, or connection to a raceway or other enclosure and at intervals not exceeding 750 mm (30 in)."

After reading this section, you will know that a cable strap is required within 12 inches of the service head and within 12 inches of the meter base. Furthermore, the cable must be secured in between these two termination points at intervals not exceeding 30 inches.

4.9.2 Installing Track Lighting

Assume that you are installing track lighting in a residential occupancy. The owners want the track located behind the curtain of their sliding glass patio doors. To determine if this is an *NEC®* violation, follow these steps:

Step 1 Look in the *NEC® Table of Contents* and find the chapter that contains information about the general application you are working on. *NEC Chapter 4, Equipment for General Use*, covers track lighting.

Step 2 Now look for the article that fits the specific category you are working on. In this case, *NEC Article 410* covers lighting fixtures, lampholders, and lamps.

Step 3 Next locate the section within *NEC Article 410* that deals with the specific application. For this example, refer to *Part XV, Lighting Track*.

Step 4 Turn to the page listed.

Step 5 Read *NEC Section 410.2, Definitions*, to become familiar with track lighting. Then go to *NEC Section 410.151* and read the information contained therein. Note that paragraph *(c) Locations Not Permitted* under *NEC Section 410.151* states the following: "Lighting track shall not be installed in the following locations: (1) where likely to be subjected to physical damage; (2) in wet or damp locations; (3) where subject to corrosive vapors; (4) in storage battery rooms; (5) in hazardous (classified) locations; (6) where concealed; (7) where extended through walls or partitions; (8) less than 1.5 m (5 ft) above the finished floor except where protected from physical damage or track operating at less than 30 volts rms open-circuit voltage; (9) where prohibited by *NEC Section 410.10(D)*."

Step 6 Read *NEC Section 410.151(C)* carefully. Do you see any conditions that would violate any *NEC®* requirements if the track lighting is installed in the area specified? In checking these items, you will probably note condition (6), "where concealed." Since the track lighting is to be installed behind a curtain, this sounds like an *NEC®* violation. We need to check further.

Step 7 You need the *NEC®* definition of concealed. Therefore, turn to *NEC Article 100, Definitions* and find the main term concealed. It reads: "Concealed. Rendered inaccessible by the structure or finish of the building . . ."

Step 8 Although the track lighting may be out of sight if the curtain is drawn, it will still be readily accessible for maintenance. Consequently, the track lighting is really not concealed according to the *NEC®* definition.

When using the *NEC®* to determine electrical installation requirements, keep in mind that you will nearly always have to refer to more than one section. Sometimes the *NEC®* itself refers the reader to other articles and sections. In some cases, the user will have to be familiar enough with the *NEC®* to know what other sections pertain to the installation at hand. It can be a confusing situation, but time and experience in using the *NEC®* will make it much easier. A pictorial road map of some *NEC®* topics is shown in *Figure 3*.

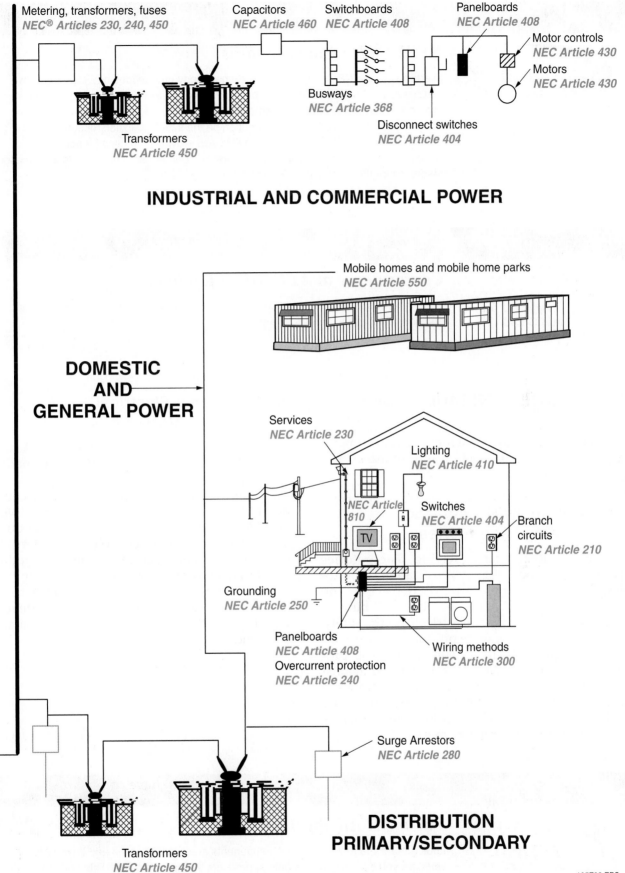

Figure 3 ◆ *NEC*® references for industrial, commercial, and residential power.

5.0.0 ◆ OTHER ORGANIZATIONS

In support of the requirements of the *NEC®* there are other organizations that identify conformance standards for the manufacture and/or use of electrical products.

5.1.0 Nationally Recognized Testing Laboratories

Nationally Recognized Testing Laboratories (NRTLs) are product safety certification laboratories. These laboratories perform extensive testing of new products to make sure they are built to established standards for electrical and fire safety. They establish and operate product safety certification programs to make sure that items produced under the service are safeguarded against reasonably foreseeable risks. NRTLs maintain a worldwide network of field representatives who make unannounced visits to factories to check products bearing their safety marks.

5.2.0 National Electrical Manufacturers Association

The **National Electrical Manufacturers Association (NEMA)** was founded in 1926. It is made up of companies that manufacture equipment used for generation, transmission, distribution, control, and utilization of electric power. The objectives of NEMA are to maintain and improve the quality and reliability of products; to ensure safety standards in the manufacture and use of products; and to develop product standards covering such matters as naming, ratings, performance, testing, and dimensions. NEMA participates in developing the *NEC®* and advocates its acceptance by state and local authorities.

1. The *NEC*® provides the _____ requirements for the installation of electrical systems.
   a. minimum
   b. most stringent
   c. design specification
   d. complete

2. All of the following groups are usually represented on the Code Making Panels *except* _____.
   a. trade associations
   b. electrical inspectors
   c. insurance organizations
   d. government lobbyists

3. Mandatory and permissive rules are defined in _____.
   a. *NEC Article 90*
   b. *NEC Article 100*
   c. *NEC Article 110*
   d. *NEC Article 200*

4. The general design and installation of electrical systems is covered in _____.
   a. *NEC Chapters 1, 2, and 7*
   b. *NEC Chapters 1, 2, 3, and 4*
   c. *NEC Chapters 6, 7, and 8*
   d. *NEC Chapters 5, 6, 7, and 9*

5. Devices such as radios, televisions, and telephones are covered in _____.
   a. *NEC Chapter 8*
   b. *NEC Chapter 7*
   c. *NEC Chapter 6*
   d. *NEC Chapter 5*

6. *NEC Annex* _____ provides examples of branch circuit calculations.
   a. A
   b. C
   c. D
   d. G

7. *NEC Article 110* covers _____.
   a. branch circuits
   b. definitions
   c. general requirements for electrical installations
   d. wiring design and protection

8. Cable trays are covered in _____.
   a. *NEC Article 330*
   b. *NEC Article 342*
   c. *NEC Article 368*
   d. *NEC Article 392*

Questions 9 through 12 are two-part questions that ask you to first locate an article in the *NEC*®, and then answer a question using the article.

9. Where in the *NEC*® would you find information on conductors for general wiring?
   a. *NEC Article 280*
   b. *NEC Article 310*
   c. *NEC Article 404*
   d. *NEC Article 340*

10. Within the article found in Question 9, locate information on stranded conductors. According to this section, conductors of size No. _____ or larger must be stranded when installed in raceways.
    a. 14 AWG
    b. 12 AWG
    c. 10 AWG
    d. 8 AWG

11. Where in the *NEC*® would you find information on switchboards and panelboards?
    a. *NEC Article 285*
    b. *NEC Article 300*
    c. *NEC Article 408*
    d. *NEC Article 540*

12. Within the article found in Question 11, find the section that covers switchboard clearances. Assuming the switchboard does not have a noncombustible shield, what is the minimum clearance between the top of the switchboard and a combustible ceiling?
    a. 12"
    b. 18"
    c. 24"
    d. 36"

13. Installation procedures for lighting fixtures are provided in _____.
    a. *NEC Article 410*
    b. *NEC Article 408*
    c. *NEC Article 366*
    d. *NEC Article 460*

14. Theaters are covered in _____.
    a. *NEC Article 338*
    b. *NEC Article 110*
    c. *NEC Article 430*
    d. *NEC Article 520*

15. *NEC Article 600* covers _____.
    a. track lighting
    b. electric signs and outline lighting
    c. X-ray equipment
    d. emergency lighting systems

Summary

The *NEC*® specifies the minimum provisions necessary for protecting people and property from hazards arising from the use of electricity and electrical equipment. As an electrician, you must be aware of how to use and apply the *NEC*® on the job site. Using the *NEC*® will help you to safely install and maintain the electrical equipment and systems you come into contact with.

Notes

Trade Terms Quiz

1. _____ are the main topics of the *NEC®*.

2. Nine _____ form the broad structure of the *NEC®*.

3. Certain articles in the *NEC®* are subdivided into lettered _____.

4. Parts and articles are subdivided into numbered _____.

5. Although they follow the applicable sections of the *NEC®*, _____ allow alternative methods to be used under specific conditions.

6. A(n) _____ is explanatory material that follows specific *NEC®* sections.

7. The _____ is an organization that maintains and improves the quality and reliability of electrical products.

8. The _____ publishes the *NEC®*; it also develops standards to minimize the possibility and effects of fire.

9. _____ are organizations that are responsible for testing and certifying electrical equipment.

Trade Terms

Articles
Chapters
Exceptions
Fine print note (FPN)

National Electrical
Manufacturers
Association (NEMA)

National Fire Protection
Association (NFPA)
Nationally Recognized
Testing Laboratories
(NRTLs)

Parts
Sections

Christine Thorstensen Porter

Intertek Testing Services

How did you choose a career in the electrical field?
It was "in the blood," literally. My father had his own electrical contracting company and was an instructor for our apprenticeship/training program. One day, my husband, who also worked for my dad, gave me a tool belt and sent me through the Construction Industry Training Council's (CITC) apprenticeship program in Seattle, Washington. Since my father taught many of the classes, the pressure was on. I felt that I had to perform well. My father was a major influence on me both as an electrician, and as a student leader. Eventually, I went into the estimating side of things in my dad's business.

In the fourth year of my apprenticeship program I was taking a lot of estimating and project management courses and applying the skills I gained in the field. My dad's business was high-end residential and some small commercial work. We'd have a crew on site from six months to a year until the projects were complete. A lot of the installations we did were data systems and computer-controlled security systems. When I took the Administrator's exam, that gave me my qualifications to supervise.

What positions have you held in the industry?
In 1981 I was a teaching assistant. That teacher started a new review class for electricians needing to pass journeyman and administrative exams; I took over his second-year apprenticeship/trainee class. To better prepare myself for my new teaching role, I took various classes offered by the International Association of Electrical Inspectors (IAEI), but I was dismayed and disappointed with the instructor because he merely read the *NEC*® to us, instead of giving us any insights as to how the *NEC*® could impact our installations, or why any particular *NEC*®

rule was in place. So I also ended up getting involved with teaching classes for the IAEI.

My experience with IAEI gave me a solid grounding in the Code. It also led to more local involvement. In fact, I'm now the chairman of the city's Electrical Code Advisory Committee. I have also served on the Technical Advisory Committee for our state amendments to the *NEC*®.

What does your current job involve?
I teach the second-year electrical training program at CITC. I teach courses for IAEI. I perform inspections for Intertek Testing Services, and I serve on a variety of committees. In addition to chairing the Electrical Code Advisory Committee, I'm a Senior Associate member of the Puget Sound Chapter of IAEI; I sit on the city of Seattle's Construction Codes Advisory Board and am a subject matter expert for NCCER's *Contren*® Electrical Curriculum.

Do you have any advice for someone just entering the trade?
I'd tell them that the electrical trade is extremely rich and diverse. There is so much that you can do. From installing all different types of systems, to project management, and to estimating, the field is wide open. The job never stays dull. The field changes with all the changes in technology. Because of that, there are so many new and emerging jobs within the trade. As far as becoming a success, I'd say to keep abreast of changes—changes in the Code and in technology— and stay connected. A good way to do that is to join associations such as IAEI. Network and continue your training. It never ends in this field. Also, spend time mentoring others. You learn more that way, by giving yourself.

Trade Terms
Introduced in This Module

Articles: The articles are the main topics of the *NEC*®, beginning with *NEC Article 90, Introduction,* and ending with *NEC Article 830, Network-Powered Broadband Communications Systems.*

Chapters: Nine chapters form the broad structure of the *NEC*®.

Exceptions: Exceptions follow the applicable sections of the *NEC*® and allow alternative methods to be used under specific conditions.

Fine print note (FPN): Explanatory material that follows specific *NEC*® sections.

National Electrical Manufacturers Association (NEMA): The association that maintains and improves the quality and reliability of electrical products.

National Fire Protection Association (NFPA): The publishers of the *NEC*®. The NFPA develops standards to minimize the possibility and effects of fire.

Nationally Recognized Testing Laboratories (NRTLs): Product safety certification laboratories that are responsible for testing and certifying electrical equipment.

Parts: Certain articles in the *NEC*® are subdivided into parts. Parts have letter designations (e.g., Part A).

Sections: Parts and articles are subdivided into sections. Sections have numeric designations that follow the article number and are preceded by a period (e.g., 501.4).

This module is intended to present thorough resources for task training. The following reference work is suggested for further study. This is optional material for continued education rather than for task training.

National Electrical Code® Handbook, Latest Edition. Quincy, MA: National Fire Protection Association.

NCCER makes every effort to keep these textbooks up-to-date and free of technical errors. We appreciate your help in this process. If you have an idea for improving this textbook, or if you find an error, a typographical mistake, or an inaccuracy in NCCER's Contren® textbooks, please write us, using this form or a photocopy. Be sure to include the exact module number, page number, a detailed description, and the correction, if applicable. Your input will be brought to the attention of the Technical Review Committee. Thank you for your assistance.

Instructors – If you found that additional materials were necessary in order to teach this module effectively, please let us know so that we may include them in the Equipment/Materials list in the Annotated Instructor's Guide.

Write: Product Development and Revision
National Center for Construction Education and Research
3600 NW 43rd St., Bldg. G, Gainesville, FL 32606

Fax: 352-334-0932

E-mail: curriculum@nccer.org

Craft _____ Module Name _____

Copyright Date _____ Module Number _____ Page Number(s) _____

Description _____

_____

_____

_____

(Optional) Correction _____

_____

_____

(Optional) Your Name and Address _____

_____

_____

Device Boxes

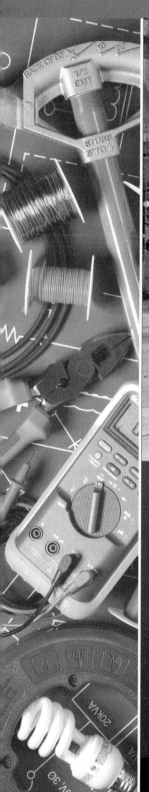

Shell Chemical Norco Subsystems Project

The Automation Group (TAG) and Yokogawa Corporation of America Industrial Specialty Contractors performed a complete reinstrumentation, replacing old-fashioned pneumatic instruments and controls with modern electronics and fiber optics at Shell Chemical's Norco facility.

26106-08

26106-08
Device Boxes

Topics to be presented in this module include:

Overview

The outlet and pull boxes used in an electrical system are selected according to their volume capacity. This volume capacity, called box fill, is measured in cubic inches or centimeters and is regulated by the *National Electrical Code®*.

There are many types of outlet and pull boxes to choose from. It is essential to select the right box for the job in order to make conductor installation easier, and to comply with *NEC®* regulations that govern the types and sizes of boxes. Some of the factors affecting box selection include environmental conditions, mounting surfaces or materials, and raceway design, as well as the size, type, and number of conductors.

This module introduces the factors that must be considered when sizing and installing boxes. It explains how *NEC®* requirements affect box selection and installation. You will learn about the various types of outlet and pull/junction boxes, as well as how to size each type of box using the information in *NEC Article 314*.

Note: *National Electrical Code®* and *NEC®* are registered trademarks of the National Fire Protection Association, Inc., Quincy, MA 02269.
All *National Electrical Code®* and *NEC®* references in this module refer to the 2008 edition of the *National Electrical Code®*.

Objectives

When you have completed this module, you will be able to do the following:

1. Describe the different types of nonmetallic and metallic boxes.
2. Calculate the *NEC*® fill requirements for boxes under 100 cubic inches.
3. Identify the appropriate box type and size for a given application.
4. Select and demonstrate the appropriate method for mounting a given box.

Trade Terms

Connector
Explosion-proof
Handy box
Junction box
Outlet box

Pull box
Raintight
Waterproof
Watertight
Weatherproof

Required Trainee Materials

1. Pencil and paper
2. Appropriate personal protective equipment
3. Copy of the latest edition of the *National Electrical Code*®

Prerequisites

Before you begin this module, it is recommended that you successfully complete *Core Curriculum; Electrical Level One*, Modules 26101-08 through 26105-08.

This course map shows all of the modules in *Electrical Level One*. The suggested training order begins at the bottom and proceeds up. Skill levels increase as you advance on the course map. The local Training Program Sponsor may adjust the training order.

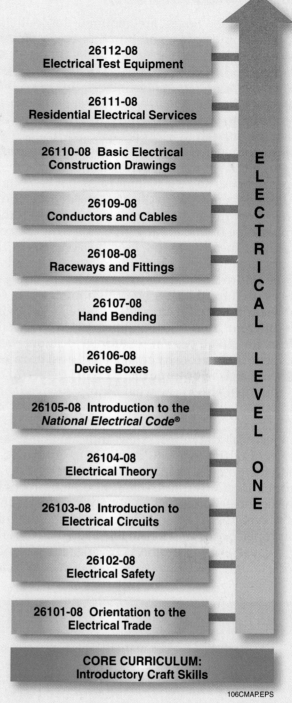

26112-08
Electrical Test Equipment

26111-08
Residential Electrical Services

26110-08 Basic Electrical
Construction Drawings

26109-08
Conductors and Cables

26108-08
Raceways and Fittings

26107-08
Hand Bending

26106-08
Device Boxes

26105-08 Introduction to the
National Electrical Code®

26104-08
Electrical Theory

26103-08 Introduction to
Electrical Circuits

26102-08
Electrical Safety

26101-08 Orientation to the
Electrical Trade

CORE CURRICULUM:
Introductory Craft Skills

ELECTRICAL LEVEL ONE

106CMAP.EPS

1.0.0 ◆ INTRODUCTION

On every job, boxes are required for outlets, switches, pull boxes, and junction boxes. All of these must be sized, installed, and supported to meet current *NEC*® requirements. Since the *NEC*® limits the number of conductors, fittings, and devices allowed in each outlet or switch box according to its size, you must install boxes that are large enough to accommodate the number of conductors that must be spliced in the box or fed through it. Therefore, a knowledge of the various types of boxes and the volume of each is essential.

Besides being able to calculate the required box sizes, you must also know how to select the proper type of box for any given application. For example, metallic boxes used in concrete deck pours are different from those used as device boxes in residential or commercial buildings. Boxes used for the support of lighting fixtures or for securing devices in outdoor installations will be different from the two types just mentioned. Boxes for use in certain hazardous locations will further differ in construction; many must be rated as being explosion-proof.

You must also know what fittings are available for terminating the various wiring methods in these boxes.

Electrical drawings rarely indicate the exact types of outlet boxes to be used in a given area, with the possible exception of boxes used in hazardous locations. Electricians who lack practical on-the-job experience may not always choose the best box for a given application. The use of improper boxes and other ill-adapted materials will cause excessive time to be taken on the job. For example, outlet boxes for use with only Type AC cable should contain built-in clamps; many times boxes will be ordered with knockouts only. This latter case requires the use of additional connectors, which may only take a few additional seconds per connection, but when these are added up over the period of a large project, much additional time is wasted. Because the cost of labor is an expensive item, any excess labor required will more than offset any savings gained from the use of inadequate materials.

Labor is often wasted due to obstructions that enter raceways during the general construction work. Most of these obstructions can be avoided by plugging all raceway openings with capped bushings or similar means of protection (*Figure 1*).

106F01.EPS

Figure 1 ◆ Conduit caps.

Outlet Boxes in Poured Concrete

When installing outlet boxes in poured concrete structures, stuff the inside of the boxes with newspaper and/or cover the openings with duct tape and use capped bushings to keep concrete out of the raceway system.

Special Considerations

Before you can effectively plan and route conductors to their termination points and then select and install the correct types and sizes of boxes, you will need to have the following information:

• Length and number of conductors
• Conductor ampacity
• Allowances for voltage drops
• Environment

Care should also be taken to thoroughly tighten all fittings, couplings, and so on.

2.0.0 ◆ TYPES OF BOXES

Outlet boxes normally fall into three categories:

- Pressed steel boxes with knockouts of various sizes for raceway or cable entrances
- Cast iron, aluminum, or brass boxes with threaded hubs of various sizes and locations for raceway entrances
- Nonmetallic boxes

Pressed steel boxes also fall into two categories:

- Boxes with conduit, electric metallic tubing, and cable
- Boxes designed for use with specific types of surface metal raceways

Outlet boxes vary in size and shape depending upon their use, the size of the raceway, the number of conductors entering the box, the type of building construction, the atmospheric conditions of the area, and special requirements.

Outlet box covers are usually required to adapt the box to the particular use it is to serve. For example, a 4" square box is adapted to one-gang or two-gang switches or receptacles by the use of either one-gang or two-gang flush device covers. A one-gang cast hub box can be adapted to provide a vapor-proof switch or a vapor-proof receptacle cover. Special outlet box hangers are available to facilitate their installation, particularly in frame building construction.

The types of enclosures used as outlet and device boxes for the support of fixtures or for securing devices, such as switches, receptacles, or other equipment, on the same yoke or strap are available in various sizes and shapes. These enclosures may be used in the one-gang, two-gang, three-gang, or four-gang types. Ceiling outlet boxes are available in various shapes. Device boxes are those that are usually installed to support receptacles and switches.

Boxes installed for the support of a lighting fixture are required to be listed for the purpose. Most device boxes are typically not designed or listed for use to support lighting fixtures. The use of a device box to support fixtures is addressed in *NEC Section 314.27(A), Exception.*

A floor box that is listed specifically for installation in a floor is required where receptacles or junction boxes are installed in a floor. Listed floor boxes are provided with covers and gaskets to exclude surface water and cleaning compounds.

A box used at fan outlets is not permitted to be used as the sole support for ceiling (paddle) fans, unless it is listed for the application as the sole means of support. Where a ceiling fan does not exceed 70 pounds in weight, it is permitted to be supported by outlet boxes listed and identified for such use. Boxes designed to support more than 35 pounds must be marked with the maximum weight to be supported. See *NEC Section 314.27(D).* These boxes must be rigidly supported from a structural member of the building. A paddle fan box and its related accessories are shown in *Figure 2.*

Per *NEC Section 314.27(E),* boxes used for support of utilization equipment other than ceiling-suspended (paddle) fans shall meet the requirements of *NEC Sections 314.27(A) and (B)* for support of a luminaire that is the same size and weight.

2.1.0 Octagon and Round Boxes

Octagon boxes are available with knockouts for use with either conduit (using locknuts and bushings) or cable box connectors. They are also available with both Type AC and NM cable clamps. The standard width of octagon boxes is four inches, with depths available in 1¼", 1½", or 2⅛". Extension rings are also available for increasing the depth.

Round boxes are available in the same dimensions, but *NEC Section 314.2* prevents using such boxes where conduit or connectors—requiring locknuts and bushings—are connected to the side

Identifying Boxes

INSIDE TRACK

After installing all the boxes in a complex installation, you can save yourself much time and trouble later on if you label each box location according to type. Using a can of spray paint, paint small dots on the floor below each box. Use one dot for a plug receptacle, two dots for a switch box, and three dots for a wall light. After the drywall installers are finished, you'll be able to easily identify any wall box.

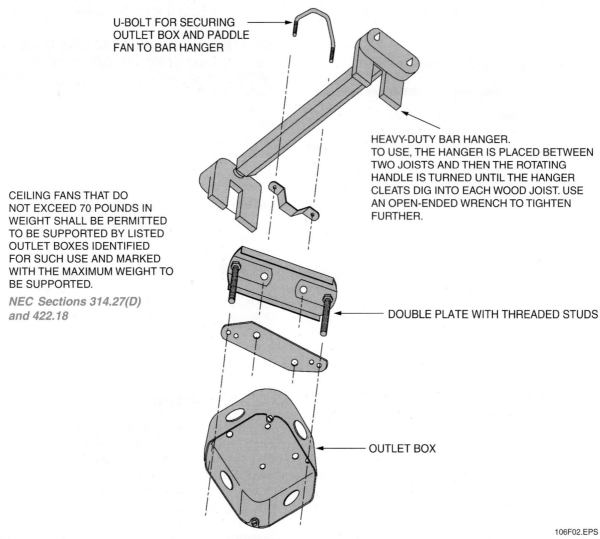

U-BOLT FOR SECURING
OUTLET BOX AND PADDLE
FAN TO BAR HANGER

HEAVY-DUTY BAR HANGER.
TO USE, THE HANGER IS PLACED BETWEEN
TWO JOISTS AND THEN THE ROTATING
HANDLE IS TURNED UNTIL THE HANGER
CLEATS DIG INTO EACH WOOD JOIST. USE
AN OPEN-ENDED WRENCH TO TIGHTEN
FURTHER.

CEILING FANS THAT DO
NOT EXCEED 70 POUNDS IN
WEIGHT SHALL BE PERMITTED
TO BE SUPPORTED BY LISTED
OUTLET BOXES IDENTIFIED
FOR SUCH USE AND MARKED
WITH THE MAXIMUM WEIGHT TO
BE SUPPORTED.
*NEC Sections 314.27(D)
and 422.18*

DOUBLE PLATE WITH THREADED STUDS

OUTLET BOX

106F02.EPS

Figure 2 ◆ Typical fan box for installation in an existing ceiling.

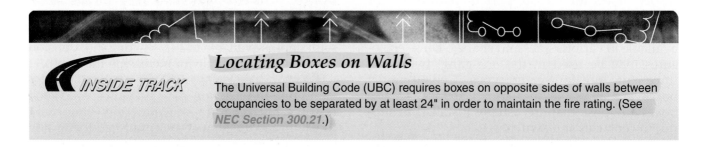

Locating Boxes on Walls

INSIDE TRACK

The Universal Building Code (UBC) requires boxes on opposite sides of walls between occupancies to be separated by at least 24" in order to maintain the fire rating. (See *NEC Section 300.21*.)

of the box. The conduit or box connector must terminate in the top of such boxes. Round boxes with cable clamps, however, are permitted for use with Type AC and NM cables that may terminate in either the side or top of the box. Nonmetallic round boxes are permitted only with open wiring on insulators, concealed knob-and-tube wiring, Type NM cable, and nonmetallic raceways.

Figure 3 shows typical metallic octagon boxes and an octagon extension ring. *Figure 3(A)* shows a box with concentric knockouts for conduit or

box connectors, while *Figure 3(B)* shows a box that utilizes cable clamps. *Figure 3(C)* shows the extension ring. *Figure 4* shows a nonmetallic round box and fixture ring with a bar hanger for mounting between studs or joists.

Octagon and round boxes are used for wall-mounted lighting fixtures (luminaires). However, covers are available for octagon boxes that will support receptacles and switches. Blank covers are also available when the box is used as a junction box.

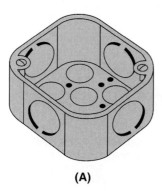

(A)

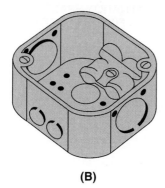

(B)

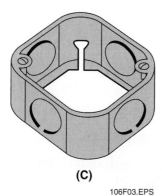

(C)

106F03.EPS

Figure 3 ◆ Typical octagon boxes and octagon extension ring.

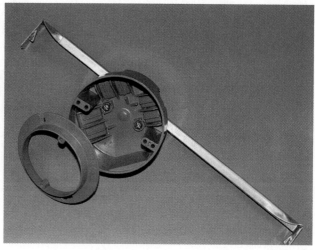

106F04.EPS

Figure 4 ◆ Nonmetallic round box and fixture ring with bar hanger.

Figure 5 shows a cross section of a 3½" round shallow box used for supporting lighting fixtures that have integral wire termination space. *NEC Section 314.24* requires that this and all boxes have a minimum depth of ½". Boxes intended to enclose flush devices must not have a depth of less than ¹⁵⁄₁₆".

2.2.0 Square Boxes

Square boxes are available in 4" and 4¹¹⁄₁₆" square sizes. Both are available in depths of 1¼", 1½", and 2⅛". Extension rings are also available to further increase the depth. These boxes are available with or without mounting brackets (for fastening to structural members). Boxes designed for use with cable may have either Type AC or NM clamps for securing the cable at the box entrance points. Square boxes may be used with a single or two-gang device ring (e.g., plaster ring or tile ring) for mounting receptacles or switches. A ring with a round opening is also available for mounting lighting fixtures. Blank covers are available when the boxes are used as junction boxes.

Figure 6 shows a square box with a rectangular extension ring that is used to bring the box to the finished surface. Square boxes can be used as junction boxes, and when the number of conductors warrants more capacity than is available in other types of boxes.

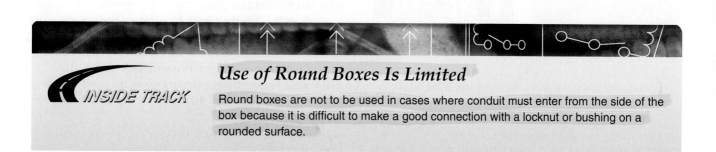

INSIDE TRACK

Use of Round Boxes Is Limited

Round boxes are not to be used in cases where conduit must enter from the side of the box because it is difficult to make a good connection with a locknut or bushing on a rounded surface.

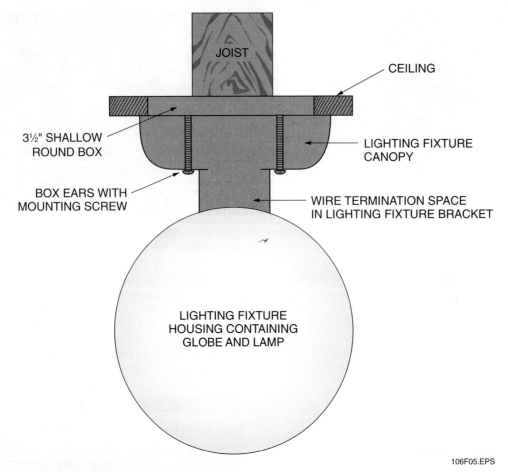

JOIST

CEILING

3½" SHALLOW
ROUND BOX

BOX EARS WITH
MOUNTING SCREW

LIGHTING FIXTURE
CANOPY

WIRE TERMINATION SPACE
IN LIGHTING FIXTURE BRACKET

LIGHTING FIXTURE
HOUSING CONTAINING
GLOBE AND LAMP

106F05.EPS

Figure 5 ◆ Shallow round box used for mounting a lighting fixture.

106F06.EPS

Figure 6 ◆ Square box with extension ring.

2.3.0 Device Boxes

Device boxes are designed for flush mounting mainly in residential and some commercial applications. They are available with or without cable clamps and brackets for mounting to wooden structural members. This type of box is also available with plaster ears for installation in finished wall partitions. Several types of device boxes are shown in *Figure 7*. As you can see, some boxes include integral nails for direct installation into wall studs. Boxes used in metal stud environments are installed using self-tapping sheet metal screws, while those installed in concrete may require the use of special powder-actuated fasteners. These materials are discussed in more detail in a later module.

A special single-gang box with the mounting ears on the inside of the box is called a handy box (also known as a utility box). Such boxes are available in depths of 1½", 1⅞", and 2⅛". Care must be exercised when using these boxes as their limited volume restricts the number of conductors permitted in the box.

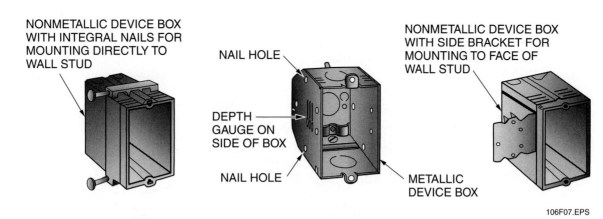

Figure 7 ◆ Typical device boxes.

106F07.EPS

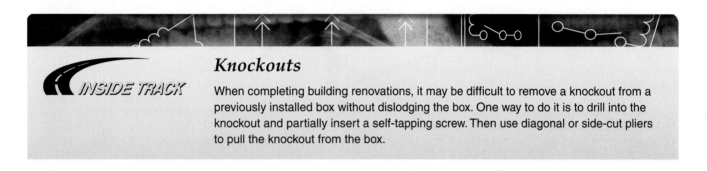

Knockouts

INSIDE TRACK

When completing building renovations, it may be difficult to remove a knockout from a previously installed box without dislodging the box. One way to do it is to drill into the knockout and partially insert a self-tapping screw. Then use diagonal or side-cut pliers to pull the knockout from the box.

2.4.0 Masonry Boxes

Special boxes known as masonry boxes are used in flat-slab construction jobs. These boxes consist of a sleeve with external ears and a plate that is attached after the sleeve is nailed to the deck.

Masonry boxes are manufactured in different heights and care should be taken to use boxes of sufficient height to allow the knockouts to come well above the reinforcing rods. This eliminates the need for offsets in the conduit where it enters a box. *Figure 8* shows a practical application of a masonry box.

2.5.0 Boxes for Damp and Wet Locations

In damp or wet locations, boxes and fittings must be placed or equipped to prevent moisture or water from entering and accumulating within the box or fitting. It is recommended that approved boxes of nonconductive material be used with nonmetallic sheathed cable or approved nonmetallic conduit when the cable or conduit is used in locations where there is likely to be occasional moisture present. Boxes installed in wet locations must be approved for the purpose per *NEC Section 314.15*.

A wet location is any location subject to saturation with water or other liquids, such as locations exposed to weather or water, washrooms, garages, and interiors that might be hosed down. Underground installations or those in concrete slabs or masonry in direct contact with the earth must be considered to be wet locations. Raintight, waterproof, or watertight equipment (including fittings) may satisfy the requirements for weatherproof equipment. Boxes with threaded conduit hubs and gasketed covers will normally prevent water from entering the box except for condensation within the box.

A damp location is a location subject to some degree of moisture. Such locations include partially protected outdoor locations—such as under canopies, marquees, and roofed open porches. It also includes interior locations subject to moderate degrees of moisture—such as some basements, some barns, and cold storage warehouses.

Weatherproof covers for outdoor receptacles must be chosen with care. All 15A and 20A 125 – 250V receptacles in wet locations require a cover that is weatherproof regardless of whether or not the receptacle is in use. Other receptacles may or may not require these covers, depending on the rating. *NEC Section 406.8* covers installation of receptacles in damp or wet locations.

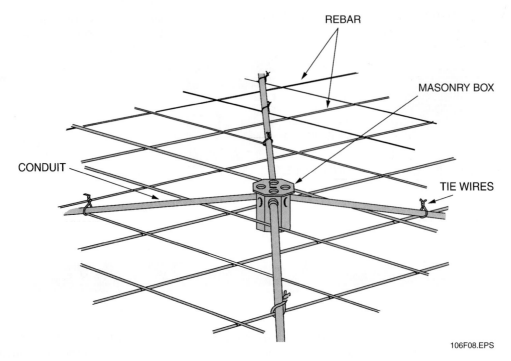

REBAR

MASONRY BOX

CONDUIT

TIE WIRES

106F08.EPS

Figure 8 ◆ Practical application of a masonry box.

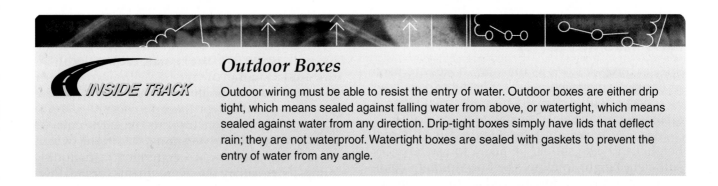

Outdoor Boxes

INSIDE TRACK

Outdoor wiring must be able to resist the entry of water. Outdoor boxes are either drip tight, which means sealed against falling water from above, or watertight, which means sealed against water from any direction. Drip-tight boxes simply have lids that deflect rain; they are not waterproof. Watertight boxes are sealed with gaskets to prevent the entry of water from any angle.

3.0.0 ◆ SIZING OUTLET BOXES

In general, the maximum number of conductors permitted in standard outlet boxes is listed in *NEC Table 314.16(A)*. These figures apply where no fittings or devices such as fixture studs, cable clamps, switches, or receptacles are contained in the box and where no grounding conductors are part of the wiring within the box. Obviously, in all modern residential wiring systems there will be one or more of these items contained in the outlet box. Therefore, where one or more of the above-mentioned items are present, the number of conductors is reduced by one less than that shown in the table for each type of fitting and by two for each device strap. For example, a deduction of two conductors must be made for each strap containing a device such as a switch or duplex recep-

tacle; a further deduction of one conductor shall be made for one or more grounding conductors entering the box. For example, a 3" × 2" × 3½" box is listed in the table as containing a maximum number of eight No. 12 wires. If the box contains a cable clamp and a duplex receptacle, three wires will have to be deducted from the total of eight—providing for only five No. 12 wires. If a ground wire is used, only four No. 12 wires may be used, which might be the case when a three-wire cable with ground is used to feed a three-way wall switch. Also, each looped conductor over 12" used in the box counts as one conductor.

A pictorial definition of stipulated conditions as they apply to *NEC Section 314.16* is shown in the following illustrations. *Figure 9* illustrates an assortment of raised covers and outlet box exten-

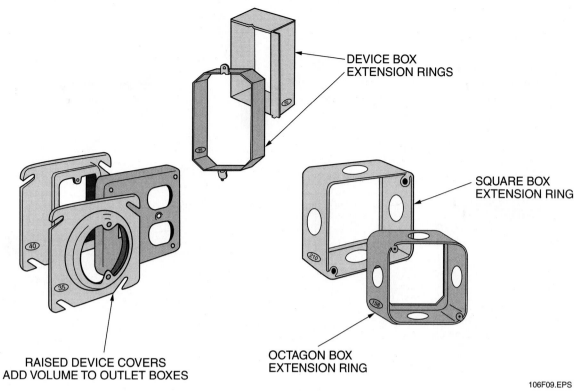

DEVICE BOX
EXTENSION RINGS

SQUARE BOX
EXTENSION RING

OCTAGON BOX
EXTENSION RING

RAISED DEVICE COVERS
ADD VOLUME TO OUTLET BOXES

106F09.EPS

Figure 9 ◆ Devices or components that add to outlet box capacity.

sions. These components, when combined with the appropriate outlet boxes, serve to increase the usable space. Each type is marked with its cubic inch capacity, which may be added to the figures in *NEC Table 314.16(A)* to calculate the increased number of conductors allowed.

Figure 10 shows typical wiring configurations, which must be counted as conductors when calculating the total capacity of outlet boxes. A wire passing through the box without a splice or tap is counted as one conductor. Therefore, a cable containing two wires that passes in and out of an outlet box with a splice or tap is counted as two conductors. However, wires that enter a box and are either spliced or connected to a terminal, and then exit again, are counted as two conductors. In the case of two cables that each have two wires, the total conductors counted will be four. Wires that enter and terminate in the same box are counted as individual conductors and in this case, the total count would be two conductors. Remember, when one or more grounding wires enter the box and are joined, a deduction on only one is required, regardless of their number.

Further components that require deduction adjustments from those specified in *NEC Table*

314.16(A) include fixture studs, hickeys, and fixture stud extensions *[NEC Section 314.16(B)(3)]*. One conductor must be deducted from the total for each type of fitting used. Two conductors must be deducted for each strap-mounted device, such as duplex receptacles and wall switches; a deduction of one conductor is made when one or more internally mounted cable clamps are used *[NEC Section 314.16(B)(2) and (4)]*. When mixed conductor sizes are installed in the box, these items must be counted as the largest wire size in the box.

Figure 11 shows components that may be used in outlet boxes without affecting the total number of conductors. Such items include grounding clips and screws, wire nuts, and cable connectors when the latter are inserted through knockout holes in the outlet box and secured with locknuts. Prewired fixture wires are not counted against the total number of allowable conductors in an outlet box; neither are conductors originating and ending in the box, such as pigtails.

To better understand how outlet boxes are sized, we will take two No. 12 AWG conductors installed in ½" EMT and terminating into a metallic outlet box containing one duplex receptacle. What size outlet box will meet *NEC®* requirements?

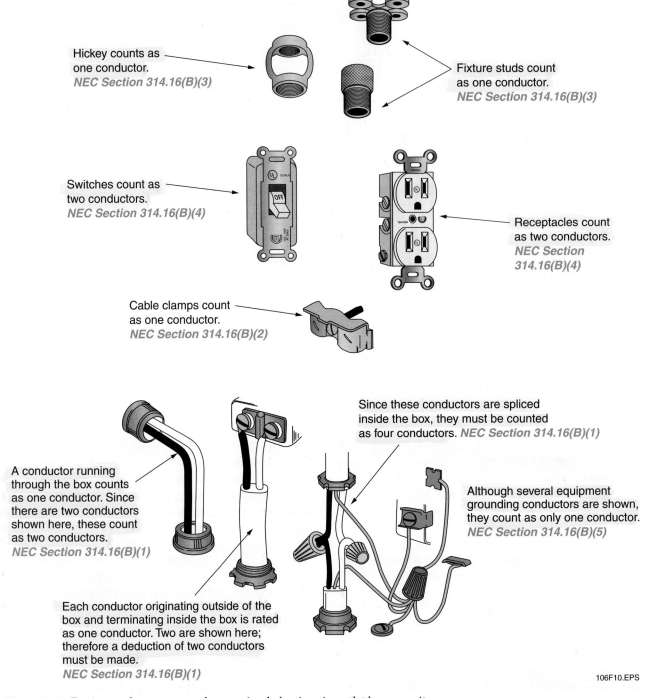

Hickey counts as one conductor. *NEC Section 314.16(B)(3)*

Fixture studs count as one conductor. *NEC Section 314.16(B)(3)*

Switches count as two conductors. *NEC Section 314.16(B)(4)*

Receptacles count as two conductors. *NEC Section 314.16(B)(4)*

Cable clamps count as one conductor. *NEC Section 314.16(B)(2)*

Since these conductors are spliced inside the box, they must be counted as four conductors. *NEC Section 314.16(B)(1)*

A conductor running through the box counts as one conductor. Since there are two conductors shown here, these count as two conductors. *NEC Section 314.16(B)(1)*

Although several equipment grounding conductors are shown, they count as only one conductor. *NEC Section 314.16(B)(5)*

Each conductor originating outside of the box and terminating inside the box is rated as one conductor. Two are shown here; therefore a deduction of two conductors must be made. *NEC Section 314.16(B)(1)*

106F10.EPS

Figure 10 ◆ Devices and components that require deductions in outlet box capacity.

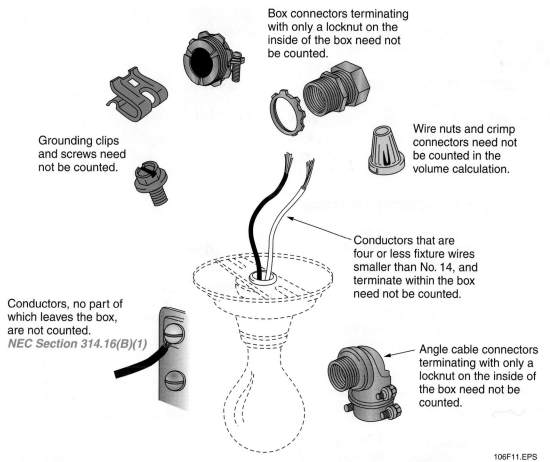

Box connectors terminating with only a locknut on the inside of the box need not be counted.

Grounding clips and screws need not be counted.

Wire nuts and crimp connectors need not be counted in the volume calculation.

Conductors that are four or less fixture wires smaller than No. 14, and terminate within the box need not be counted.

Conductors, no part of which leaves the box, are not counted.
NEC Section 314.16(B)(1)

Angle cable connectors terminating with only a locknut on the inside of the box need not be counted.

106F11.EPS

Figure 11 ◆ Items that may be disregarded when calculating outlet box capacity.

The first step is to count the total number of conductors and equivalents that will be used in the box *(NEC Section 314.16)*.

Step 1 Calculate the total number of conductors and their equivalents:

One receptacle = 2
Two #12 conductors = 2
Total #12 conductors = 4

Step 2 Determine the amount of space required for each conductor. *NEC Table 314.16(B)* gives the box volume required for each conductor:

No. 12 AWG = 2.25 cubic inches

Step 3 Calculate the outlet box space required by multiplying the number of cubic inches required for each conductor by the number of conductors found in Step 1 above.

4 × 2.25 = 9.00 cubic inches

Step 4 Once you have determined the required box capacity, again refer to *NEC Table 314.16(A)* and note that a 3" × 2" × 2" box comes closest to our requirements. This box size is rated for 10.0 cubic inches.

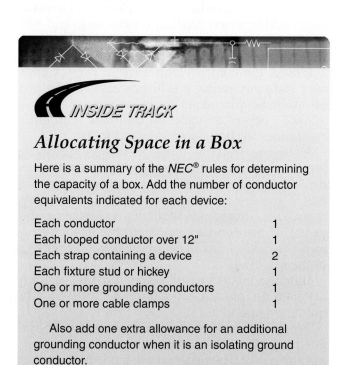

INSIDE TRACK

Allocating Space in a Box

Here is a summary of the *NEC*® rules for determining the capacity of a box. Add the number of conductor equivalents indicated for each device:

Each conductor	1
Each looped conductor over 12"	1
Each strap containing a device	2
Each fixture stud or hickey	1
One or more grounding conductors	1
One or more cable clamps	1

Also add one extra allowance for an additional grounding conductor when it is an isolating ground conductor.

For another example, if four No. 12 conductors enter the box, two additional No. 12 conductors must be added to our previous count for a total of six conductors.

$$6 \times 2.25 = 13.5 \text{ cubic inches}$$

Again, refer to *NEC Table 314.16(A)* and note that a 3" × 2" × 2¾" device box with a rated capacity of 14.0 cubic inches is the closest device box that meets *NEC*® requirements. Of course, any box with a larger capacity is permitted.

4.0.0 ◆ PULL AND JUNCTION BOXES

Pull and junction boxes are provided in an electrical installation to facilitate the installation of conductors, or to provide a junction point for the connection of conductors, or both. In some instances, the location and size of pull boxes are designated on the drawings. In most cases, however, the electricians on the job will have to determine the proper number, location, and sizes of pull or junction boxes to facilitate conductor installation.

Pull boxes should be as large as possible. Workers need space within the box for both hands and in the case of the larger wire sizes, workers will need room for their arms to feed the wire. *NEC Section 314.28* specifies that pull and junction boxes must provide adequate space and dimensions for the installation of conductors. For raceways containing conductors of No. 4 or larger, and for cables containing conductors of No. 4 or larger, the minimum dimensions of pull or junction boxes installed in a raceway or cable run shall comply with the following:

- In straight pulls, the length of the box shall not be less than eight times the trade diameter of the largest raceway.
- Where angle or U pulls are made, the distance between each raceway entry inside the box and the opposite wall of the box shall not be less than six times the trade diameter of the largest raceway in a row. This distance shall be increased for additional entries by the amount of the sum of the trade diameters of all other raceway entries in the same row on the same wall of the box. Each row shall be calculated individually, and the single row that provides the maximum distance shall be used.
- Also where angle or U pulls are made, the distance between raceway entries enclosing the same conductor shall not be less than six times the trade diameter of the larger raceway.
- When transposing cable size into raceway size, the minimum trade size raceway required for the number and size of conductors in the cable shall be used.

INSIDE TRACK

Metal Boxes Must Be Grounded

Metal boxes are good conductors. Therefore, when metal boxes are used, they must be grounded to the circuit grounding system.

Long runs of conductors should not be made in one pull. Pull boxes, installed at convenient intervals, will relieve much of the strain on the conductors. The length of the pull, in many cases, is left to the judgment of the workers or their supervisor, and the condition under which the work is installed.

The installation of pull boxes may seem to cause a great deal of extra work and trouble, but they save a considerable amount of time and hard work when pulling conductors. Properly placed, they eliminate bends and elbows and do away with the necessity of fishing from both ends of a conduit run.

If possible, pull boxes should be installed in a location that allows electricians to work easily and conveniently. For example, in an installation where the conduit comes up a corner of a wall and changes direction at the ceiling, a pull box that is installed too high will force the electrician to stand on a ladder when feeding conductors, and will allow no room for supporting the weight of the wire loop or for the cable-pulling tools.

Unless the contract drawings or project engineer state otherwise, it is just as easy for the pull boxes to be placed at a convenient height that allows workers to stand on the floor with sufficient room for both wire loop and tools.

In some electrical installations, a number of junction boxes must be installed to route the conduit in the shortest, most economical way. The *NEC*® requires all junction boxes to be readily accessible. This means that a person must be able to get to the conductors inside the box without removing plaster, wall covering, or any other part of the building.

Junction boxes or pull boxes must be securely fastened in place on walls or ceilings or adequately suspended.

While certain sizes of factory-constructed boxes are available with concentric knockouts, in many instances it will be necessary to have them custom built to meet the job requirements. When it is not possible to accurately anticipate the raceway entrance requirements, it will be necessary to cut the required knockouts on the job.

In the case of large pull boxes and troughs, shop drawings should be prepared prior to the construction of these items with all required knockouts accurately indicated in relation to the conduit run requirements.

4.1.0 Sizing Pull and Junction Boxes

Figure 12 shows a junction box with several runs of conduit. Since this is a straight pull, and 4" conduit is the largest size in the group, the minimum length required for the box can be determined by the following calculation:

Trade size of conduit × 8 [per *NEC Section 314.28(A)(1)*] = minimum length of box

or:

$$4" \times 8 = 32"$$

Therefore, this particular pull box must be at least 32" in length. The width of the box, however, need only be of sufficient size to enable locknuts and bushings to be installed on all the conduits or connectors entering the enclosure.

Junction or pull boxes in which the conductors are pulled at an angle (*Figure 13*) must have a distance of not less than six times the trade diameter of the largest conduit [*NEC Section 314.28(A)(2)*]. The distance must be increased for additional con-

duit entries by the amount of the sum of the diameter of all other conduits entering the box on the same side. The distance between raceway entries enclosing the same conductors must not be less than six times the trade diameter of the largest conduit.

Since the 4" conduit is the largest size in this case:

$$L_1 = 6 \times 4" + (3 + 2) = 29"$$

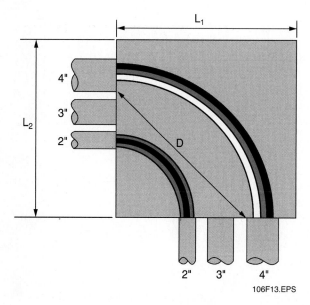

Figure 13 ◆ Pull box with conduit runs entering at right angles.

Using Pull Boxes

INSIDE TRACK

Pull boxes make it easier to install conductors. They may also be installed to avoid having more than 360° worth of bends in a single run. (Remember, if a pull box is used, it is considered the end of the run for the purposes of the *NEC®* 360° rule.)

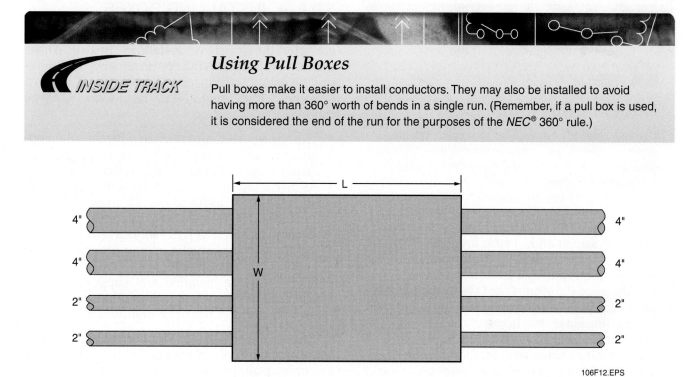

Figure 12 ◆ Pull box with two 4" and two 2" conduit runs.

Knockout Punch Kit

Often, conduit must enter boxes, cabinets, or panels that do not have precut knockouts. In these cases, a knockout punch can be used to make a hole for the conduit connection.

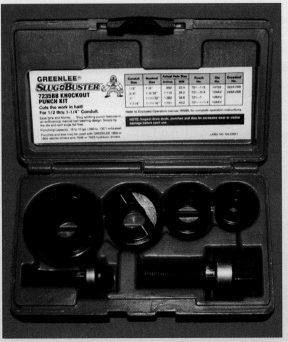

106SA01.EPS

Label Junction Boxes

It's a good idea to label every junction cover with the circuit number, the panel it came from, and its destination. The next person to service the installation will be grateful for this extra help.

Since the same conduit runs are located on the adjacent wall of the box, L_2 is calculated in the same way; therefore, $L_2 = 29"$.

The distance (D) = $6 \times 4"$ or 24". This is the minimum distance permitted between conduit entries enclosing the same conductor.

The depth of the box need only be of sufficient size to permit locknuts and bushings to be properly installed. In this case, a 6" deep box would suffice.

5.0.0 ◆ INSTALLING BOXES

In addition to box fill, there are a number of other considerations when installing boxes. These include additional *NEC®* requirements and making the actual wiring connections.

Some of the general *NEC®* requirements are as follows:

- The box selected must be listed for the given application (for example, a box used in a wet location must be listed for use in that location).
- As discussed previously, the box must have sufficient volume, as listed in *NEC Table 314.16(A)*, and must allow sufficient free space for conductors, as listed in *NEC Table 314.16(B)*.
- Conductors entering boxes and fittings must be protected from abrasion.
- Boxes must be installed and supported properly, and the finished installation must be accessible for later repair or maintenance.

You must also consider the type of box cover or canopy to be used, as well as the type of box (for example, pull/junction or outlet), and in some cases, the direction of the pull. Refer to *NEC Sec-*

tions 314.15 through 314.30 for additional details. Some of the *NEC®* requirements for receptacle locations are discussed in the following section.

5.1.0 *NEC®* Requirements for Receptacle Locations

NEC Section 210.52 states the minimum requirements for the location of receptacles in dwelling units. It specifies that in each kitchen, family room, and dining room, receptacle outlets shall be installed so that no point along the floor line in any wall space is more than six feet, measured horizontally, from an outlet in that space, including any wall space two feet or more in width and the wall space occupied by fixed panels in exterior walls, but excluding sliding panels (*Figure 14*). When spaced in this manner, a six-foot extension cord will reach a receptacle from any point along the wall line. Receptacle outlets shall, insofar as practicable, be spaced equal distances apart. Receptacle outlets in floors shall not be counted as part of the required number of receptacle outlets unless located within 18" of the wall.

The *NEC®* defines wall space as a wall that is unbroken along the floor line by doorways, fireplaces, or similar openings. Each wall space that is two feet or more in width must be treated individually and separately from other wall spaces within the room. This minimizes the use of cords across doorways, fireplaces, and similar openings.

At least one receptacle is required in each laundry area, within three feet of bathroom sinks, on the outside of the building at the front and back (GFCI protected), on any deck larger than 20 square feet and accessible from the inside of the house, in each basement, in each attached and detached garage, in each hallway ten feet or more in length, and at an accessible location for servicing any HVAC equipment.

Although no actual *NEC®* requirements exist for mounting heights and positioning of receptacles, other than the prohibition against mounting receptacles face up on countertops and similar work surfaces, there are certain *NEC®* requirements regarding receptacle placement. For example, *NEC Section 210.52(C)(5)* states that receptacles shall be not more than 20" above a kitchen countertop. Also, where allowed, receptacles may not be located more than 12" below the countertop surface. See *NEC Section 210.52(C)(5) Exception.* In addition to these *NEC®* guidelines, certain installation methods have become standard in the electrical industry. *Figure 15* shows common mounting heights of duplex receptacles used on conventional residential and small

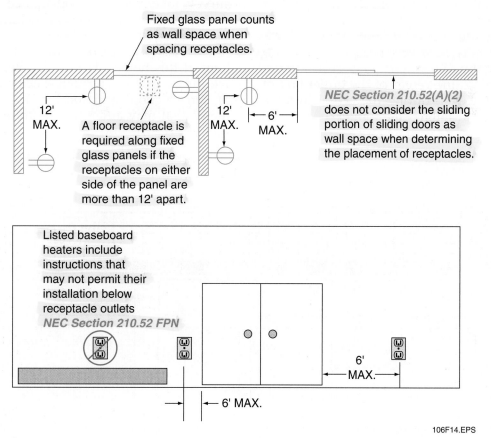

Fixed glass panel counts as wall space when spacing receptacles.

12' MAX.

A floor receptacle is required along fixed glass panels if the receptacles on either side of the panel are more than 12' apart.

12' MAX.

6' MAX.

NEC Section 210.52(A)(2) does not consider the sliding portion of sliding doors as wall space when determining the placement of receptacles.

Listed baseboard heaters include instructions that may not permit their installation below receptacle outlets *NEC Section 210.52 FPN*

6' MAX.

6' MAX.

106F14.EPS

Figure 14 ◆ *NEC®* requirements for receptacle locations in dwelling units.

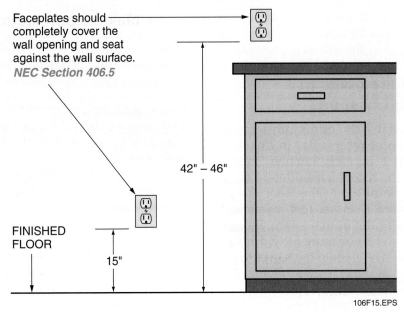

Faceplates should completely cover the wall opening and seat against the wall surface. *NEC Section 406.5*

42" – 46"

FINISHED FLOOR

15"

106F15.EPS

Figure 15 ◆ Mounting heights of duplex receptacles.

commercial installations. However, these dimensions are frequently varied to suit the building structure. For example, ceramic tile might be placed above a kitchen or bathroom countertop. If the dimensions in *Figure 15* put the receptacle part of the way out of the tile, the mounting height should be adjusted to place the receptacle either completely in the tile or completely out of the tile, as shown in *Figure 16*.

Refer again to *Figure 15* and note that the mounting heights are given to the bottom of the outlet box. Many dimensions on electrical drawings are given to the center of the outlet box or receptacle. However, during the actual installation, workers installing the outlet boxes can

mount them more accurately (and in less time) by lining up the bottom of the box with a chalk mark rather than trying to eyeball this mark to the center of the box.

A decade or so ago, most electricians mounted receptacle outlets 12" from the finished floor to the center of the outlet box. However, a recent survey taken of over 500 homeowners shows that they prefer a mounting height of 15" from the finished floor to the bottom of the outlet box. It is easier to plug and unplug the cord assemblies at this height. However, always check the working drawings, written specifications, details of construction, and local codes for measurements that may affect the mounting height of a

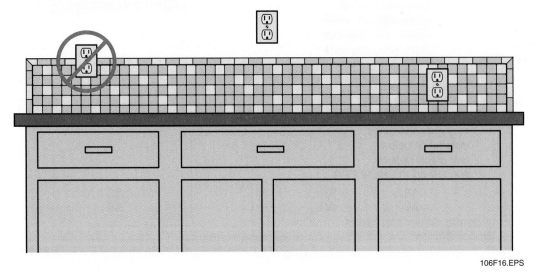

106F16.EPS

Figure 16 ◆ Adjusting mounting heights.

particular receptacle outlet. For example, those confined to wheelchairs may require more specific receptacle height locations to fit their individual needs.

NEC Section 314.20 requires all outlet boxes installed in walls or ceilings of concrete, tile, or other noncombustible material, such as plaster or drywall, to be installed in such a manner that the front edge of the box or fitting is not set back from the finished surface by more than ¼". Where walls and ceilings are constructed of wood or other combustible materials, outlet boxes and fittings must be flush with the finished surface of the wall.

Wall surfaces such as drywall or plaster that contain wide gaps or are broken, jagged, or otherwise damaged, must be repaired so there will be no gaps or open spaces greater than ⅛" between the outlet box and the wall material (*NEC Section 314.21*). These repairs should be made prior to installing the faceplate. Such repairs are best made using a noncombustible caulking or spackling compound. See *Figure 17*.

Many industrial/commercial outlet boxes are surface-mounted and connect directly to the electrical raceway. These boxes are often mounted to the building steel or to a support channel or strut that is welded or clamped to the steel. The installation of outlet boxes or raceways should be coordinated so as not to interfere with piping, ductwork, fireproofing, and other items that may or may not already be in place.

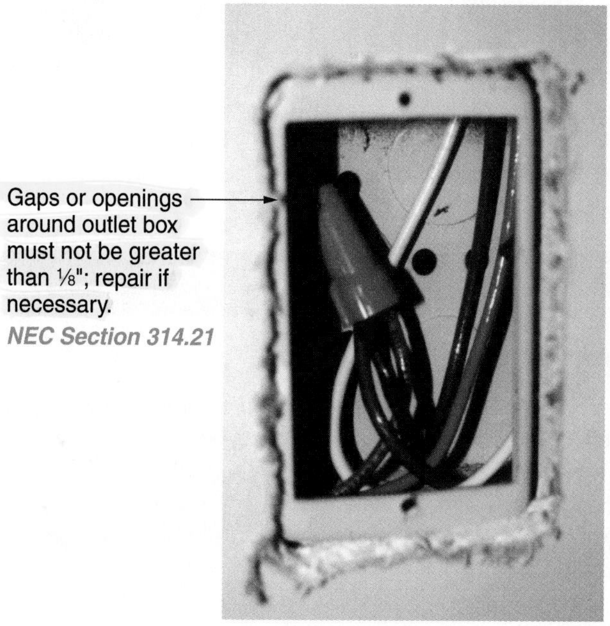

5.2.0 Making Connections

Before any actual installation begins, study the electrical floor plan and consult with the builder or architect to ensure that no last-minute changes have been made in the electrical system. Install all boxes in accordance with the electrical drawings. The spacing should be as even as possible. The box center is the midpoint on the vertical dimension of the box. Make sure that you check the door swing direction so that the switches are not installed behind a door. Measure the height of the switch boxes from the floor so that they will be at the proper height when installed. After the boxes are installed, the wires must be spliced.

Insulated spring connectors, commonly called wire nuts or Wirenuts®, are solderless connectors made in various color-coded sizes that allow for splicing the hundreds of different solid or stranded wire combinations typically encountered in branch-circuit and fixture splicing applications. See *Figure 18*. Several varieties of wire nuts are available, but the following are the ones used most often:

- Those for use on wiring systems 300V and under
- Those for use on wiring systems 600V and under (1,000V in lighting fixtures and signs)

Some types of wire nuts have thin wings on each side of the connector to facilitate their installation. Wire nuts are normally made in sizes to accommodate conductors as small as No. 22 AWG

Gaps or openings around outlet box must not be greater than ⅛"; repair if necessary.
NEC Section 314.21

106F17.EPS

Figure 17 ◆ Gaps or openings around outlet boxes must be repaired.

106F18.EPS

Figure 18 ◆ Wire nuts.

up to as large as No. 10 AWG, with practically any combination of those sizes in between.

Most brands are UL listed for aluminum to copper in dry locations only; aluminum to aluminum only; and copper to copper only. The maximum temperature rating is 105°C (221°F).

The general procedure for splicing wires with wire nuts is as follows:

Step 1 Select the proper size wire nut to accommodate the wires being spliced. Wire nut packages contain charts that list the allowable combinations of wires by size. Refer to the label on the wire nut box or container for this information.

Step 2 Select the appropriate tool (*Figure 19*), and then strip the insulation from the ends of the wires to be spliced. The length of insulation stripped off is typically about ½" (*Figure 20*), but may vary depending on the wire size and the wire nut being used. Follow the manufacturer's directions given on the wire nut package.

Step 3 Stick the ends of the wires into the wire nut and turn clockwise until tight. The wire nut draws the conductors and insulation into the body of the connector. See *Figure 21*.

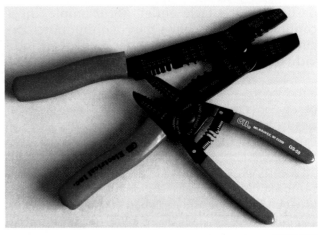

106F19.EPS

Figure 19 ◆ Stripping tools.

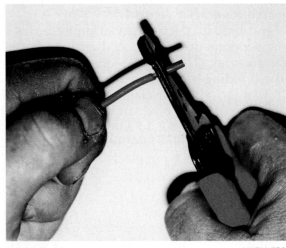

106F20.EPS

Figure 20 ◆ Stripping the insulation.

NOTE

Some manufacturers of wire nuts require that the wires be pre-twisted before screwing on the nut. Also, some manufacturers recommend using a nut driver to tighten the wire nut. Always follow the manufacturer's instructions.

Wire Nuts for Wet Locations

Specially designed wire nuts are made for use in wet locations and/or direct burial applications. These wire nuts have a water repellent, non-hardening sealant inside the body that completely seals out moisture to protect the conductors against moisture, fungus, and corrosion. The sealant remains in a gel state and will not melt or run out of the wire nut body throughout the life of the connection. Unlike other types of wire nuts, this type can be used one time only. The wire nut can be backed off, eliminating the need to cut the wires for future or retrofit applications, but once removed, it must be discarded.

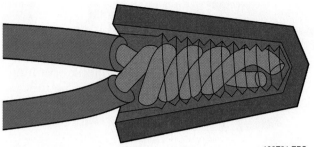

Figure 21 ◆ Wires installed in wire nut.

106F21.EPS

Step 4 After making the connections within a box, tuck the wires neatly into the back of the box. That way, when the painters and plasterers come along to finish the walls, the wires will not be covered in drywall compound or paint.

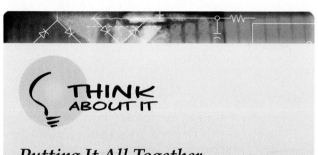

THINK ABOUT IT

Putting It All Together

Turn off the power in one area of your home and then remove some of the switch and receptacle plates. Examine the wiring inside each box. Is the box adequately sized for the number of wires and devices?

INSIDE TRACK

Making Connections

Conductor connections can be made using either push-in connectors or twist-on connectors, as shown here. Get in the habit of making your connections in the following order: first the grounding conductors, then the grounded conductors, and finally, the ungrounded conductors. If you approach each job in a systematic manner, it will soon become second nature to you and you'll be less likely to make mistakes.

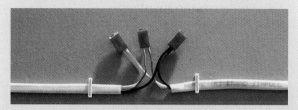

(A) NO. 14/2 NM-B CABLE WITH
PUSH-IN WIRE CONNECTORS

(B) NO. 12/2 NM-B CABLE WITH
TWIST-ON WIRE NUTS

106SA02.EPS

1. The maximum weight allowed by the *NEC®* when ceiling fans are mounted directly to an approved, unmarked outlet box is _____ pounds.
   a. 25
   b. 35
   c. 45
   d. 55

2. Square outlet boxes are available in 4" and _____ sizes.
   a. 4¹¹⁄₁₆"
   b. 5"
   c. 5¼"
   d. 6"

3. A box installed under a roofed open porch is considered a _____ location.
   a. dry
   b. wet
   c. damp
   d. weatherproof

4. Deduct _____ conductor(s) for each strap-mounted device in a device box.
   a. one
   b. two
   c. three
   d. four

5. Using _____ adds to the capacity of an outlet box.
   a. fixture studs
   b. wire nuts
   c. strap-mounted devices
   d. raised device covers

6. A conductor under 12" long running through a box counts as two conductors.
   a. True
   b. False

7. If there are four equipment grounding conductors in a box, you must deduct _____ conductor(s) from the total allowed.
   a. 1
   b. 2
   c. 3
   d. 4

8. If the largest trade diameter of a raceway entering a pull box is 3", and it is a straight pull, the minimum size box allowed is _____.
   a. 20"
   b. 24"
   c. 30"
   d. 36"

9. The depth of a pull box with conduit runs entering at right angles need be only of sufficient depth to permit locknuts and bushings to be properly installed.
   a. True
   b. False

10. Factors to consider when installing boxes are covered in _____.
    a. *NEC Sections 314.1 through 314.5*
    b. *NEC Sections 314.15 through 314.30*
    c. *NEC Sections 314.40 through 314.44*
    d. *NEC Sections 314.70 through 314.72*

Summary

Electricians work with device boxes almost every day on every project. Consequently, you must have a thorough knowledge of the types of boxes available and their applications. *NEC Article 314* covers the installation of boxes.

One of the best ways to learn about boxes and fittings is to study manufacturers' catalogs. You will find a wealth of information in each of these, many of which include detailed instructions on installation techniques. Some of these catalogs also provide simplified *NEC®* explanations of the use of the manufacturer's products.

In this module, you learned that outlet boxes must be sized, installed, and supported to meet current *NEC®* requirements. Since the *NEC®* limits the number of conductors in a given box size, boxes with adequate volume must be selected.

The boxes must also be the proper type for the application.

Outlet boxes fall into three basic categories: pressed steel; cast iron, aluminum, or brass with threaded hubs; and nonmetallic. Each type of box was discussed, along with references to applicable portions of *NEC Article 314*.

The maximum number of conductors permitted in standard outlet boxes is listed in *NEC Table 314.16(A)*. Certain devices and components require deductions in the number of conductors allowed in the box.

In addition to box sizing, there are various other factors to consider when installing boxes. These requirements are discussed in *NEC Sections 314.15 through 314.30*.

Notes

Trade Terms Quiz

1. _RAINTIGHT_ ____ means constructed or protected so that exposure to beating rain will not result in the entrance of water under specified test conditions.

2. A box designed and constructed to withstand an internal explosion without creating an external explosion or fire is called _EXPLOSION PROOF_

3. A box constructed so that water will not enter the enclosure under specified test conditions is ____. _WATERTIGHT_

4. A(n) _PULL BOX_ ____ is an enclosure used to facilitate the installation of cables from point to point in long runs.

5. A box constructed or protected so that exposure to the weather will not interfere with successful operation is referred to as ____. _WEATHERPROOF_

6. A(n) _OUTLET BOX_ ____ is a metallic or nonmetallic box installed in an electrical wiring system from which current is taken to supply some apparatus or device.

7. A single-gang outlet box used for surface mounting to enclose receptacles or wall switches on concrete or concrete block construction is called a(n) _HANDY BOX_

8. A box constructed so that moisture will not interfere with successful operation is said to be _WATERPROOF_

9. A(n) _JUNCTION BOX_ ____ is an enclosure where one or more raceways or cables enter, and in which electrical conductors may be spliced.

10. A(n) _CONNECTOR_ ____ is a device used to physically connect conduit or cable to an outlet box, cabinet, or other enclosure.

Trade Terms

Connector
Explosion-proof
Handy box

Junction box
Outlet box
Pull box

Raintight
Waterproof

Watertight
Weatherproof

Gary Edgington
Baker Electric

What made you decide to become an electrician?

I had an early fascination with electricity and used to work on motorized cars, lights or about anything else that would plug into the wall or run from a battery. My interest continued through high school. After watching an electrician wire a house, I made up my mind to be in the electrical field somewhere.

How did you learn the trade?

After graduating high school in 1972, I called an electrical contractor, Kinsey Electric, on a regular basis until he hired me. Harry was a seasoned electrical contractor, who did residential, commercial, and agricultural wiring. Because it was a one-man operation, I had an advantage in working with the owner and getting some good hands-on training along with acquiring his good work ethics. Harry was a well-respected person in the community, with a reputation of being the person to call if you wanted it done right. He was my first role model. Along with the electrical training, Harry taught me that if you're going to do a job, be the best at it and take pride in what you do.

What factor or factors have contributed most to your success?

After working with Harry for a few years, I went to work at a manufacturing facility as an industrial electrician. I was fortunate to meet two more role models who greatly influenced me in my career. One

was a retired Air Force electrician, who taught me the importance of knowledge and the need to be able to find answers to what you don't know. The second was an electrician from Manchester, England. He helped me recognize my abilities and inspired me to acquire more. Above everything, he gave me confidence in myself which has helped me throughout my life.

What kind of jobs did you hold on the way to your current position?

In addition to my job as an electrician helper/electrician for Kinsey and as an industrial electrician, I became a licensed electrical contractor in 1978. After being in the electrical contracting industry for 22 years, and employing up to fifty people at times, I sold my business and became an electrical consultant. As the owner of the business I have found myself doing everything from electrical work in the field to estimating to project and business management. During this time and since then I have taught electrical apprentice classes and *National Electrical Code®* classes for thirteen years.

What does an electrical consultant do?

As a consultant I have taught electrical classes and code updates, been involved in the electrical design process, done code review on projects, and have been an expert in court cases.

What advice would you give to someone entering the electrical field?
My advice to someone entering the field is to embrace every opportunity to learn something new, to develop good work practices and ethics, and to think safety all of the time.

What advice do you have for trainees?
As a trainee in this field you need to practice at your profession. Be accurate in your work and pursue knowledge. You will find the ability and the confidence you need to succeed in this field.

What would you say was the single greatest factor that contributed to your success?
The single greatest factor for me has been getting close to a professional in the field and trying to follow in their footsteps.

Trade Terms
Introduced in This Module

Connector: Device used to physically connect conduit or cable to an outlet box, cabinet, or other enclosure.

Explosion-proof: Designed and constructed to withstand an internal explosion without creating an external explosion or fire.

Handy box: Single-gang outlet box used for surface mounting to enclose receptacles or wall switches on concrete or concrete block construction of industrial and commercial buildings; nongangable; also made for recessed mounting; also known as a utility box.

Junction box: An enclosure where one or more raceways or cables enter, and in which electrical conductors can be, or are, spliced.

Outlet box: A metallic or nonmetallic box installed in an electrical wiring system from which current is taken to supply some apparatus or device.

Pull box: A sheet metal box-like enclosure used in conduit runs to facilitate the pulling of cables from point to point in long runs, or to provide for the installation of conduit support bushings needed to support the weight of long riser cables, or to provide for turns in multiple-conduit runs.

Raintight: Constructed or protected so that exposure to a beating rain will not result in the entrance of water under specified test conditions.

Waterproof: Constructed so that moisture will not interfere with successful operation.

Watertight: Constructed so that water will not enter the enclosure under specified test conditions.

Weatherproof: Constructed or protected so that exposure to the weather will not interfere with successful operation.

This module is intended to present thorough resources for task training. The following reference works are suggested for further study. These are optional materials for continued education rather than for task training.

American Electrician's Handbook, Latest Edition. New York: Croft and Summers, McGraw-Hill.

National Electrical Code® Handbook, Latest Edition. Quincy, MA: National Fire Protection Association.

NCCER makes every effort to keep these textbooks up-to-date and free of technical errors. We appreciate your help in this process. If you have an idea for improving this textbook, or if you find an error, a typographical mistake, or an inaccuracy in NCCER's Contren® textbooks, please write us, using this form or a photocopy. Be sure to include the exact module number, page number, a detailed description, and the correction, if applicable. Your input will be brought to the attention of the Technical Review Committee. Thank you for your assistance.

Instructors – If you found that additional materials were necessary in order to teach this module effectively, please let us know so that we may include them in the Equipment/Materials list in the Annotated Instructor's Guide.

Write: Product Development and Revision
National Center for Construction Education and Research
3600 NW 43rd St., Bldg. G, Gainesville, FL 32606

Fax: 352-334-0932

E-mail: curriculum@nccer.org

Craft _____ Module Name _____

Copyright Date _____ Module Number _____ Page Number(s) _____

Description _____

_____

_____

_____

(Optional) Correction _____

_____

_____

(Optional) Your Name and Address _____

_____

_____

Hand Bending

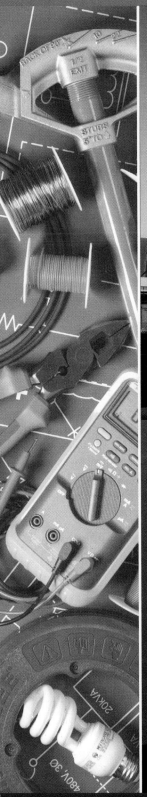

National Air and Space Museum

The Steven F. Udvar-Hazy Center near Washington Dulles International Airport is the companion facility to the Museum on the National Mall. The building opened in December, 2003, and provides enough space for the Smithsonian to display the thousands of aviation and space artifacts that cannot be exhibited on the National Mall.

26107-08

26107-08
Hand Bending

Topics to be presented in this module include:

Overview

An electrical conduit is a pipe or tube that protects electrical wires from accidental damage and exposure to the elements. Electricians working on commercial and industrial jobs must know how to bend and install conduit to go over and around obstacles. The *National Electrical Code®* limits the number and degree of bends allowed in a single run of conduit. The primary purpose of these limits is to permit easy installation and provide physical protection of conductors once the conduit is installed. It is up to the electrician to comply with these regulations. To bend conduit correctly, the electrician must make precise measurements of lengths and angles, refer to tables of predetermined values, and apply knowledge of basic geometry.

Scrap piles of bent conduit are found on many job sites because some electricians rely on a trial-and-error approach instead of learning how to bend conduit properly. When a construction job involves extensive conduit installation, knowing how to cut, bend, and ream conduit correctly can save time and money.

Objectives

When you have completed this module, you will be able to do the following:

1. Identify the methods for hand bending and installing conduit.
2. Determine conduit bends.
3. Make 90-degree bends, back-to-back bends, offsets, kicks, and saddle bends using a hand bender.
4. Cut, ream, and thread conduit.

Trade Terms

90° bend
Back-to-back bend
Concentric bends
Developed length
Gain

Offsets
Rise
Segment bend
Stub-up

Required Trainee Materials

1. Paper and pencil
2. Copy of the latest edition of the *National Electrical Code®*
3. Appropriate personal protective equipment

Prerequisites

Before you begin this module, it is recommended that you successfully complete *Core Curriculum; Electrical Level One*, Modules 26101-08 through 26106-08.

This course map shows all of the modules in *Electrical Level One*. The suggested training order begins at the bottom and proceeds up. Skill levels increase as you advance on the course map. The local Training Program Sponsor may adjust the training order.

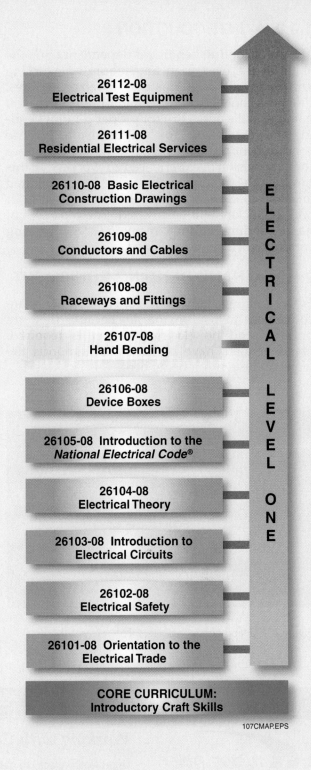

26112-08
Electrical Test Equipment

26111-08
Residential Electrical Services

26110-08 Basic Electrical
Construction Drawings

26109-08
Conductors and Cables

26108-08
Raceways and Fittings

26107-08
Hand Bending

26106-08
Device Boxes

26105-08 Introduction to the
National Electrical Code®

26104-08
Electrical Theory

26103-08 Introduction to
Electrical Circuits

26102-08
Electrical Safety

26101-08 Orientation to the
Electrical Trade

CORE CURRICULUM:
Introductory Craft Skills

ELECTRICAL LEVEL ONE

107CMAP.EPS

1.0.0 ◆ INTRODUCTION

The art of conduit bending is dependent upon the skills of the electrician and requires a working knowledge of basic terms and proven procedures. Practice, knowledge, and training will help you gain the skills necessary for proper conduit bending and installation. You will be able to practice conduit bending in the lab and in the field under the supervision of experienced co-workers. In this module, the techniques for using hand-operated and step conduit benders such as the hand bender and the hickey will be covered. The processes of hand bending, cutting, reaming, and threading conduit will also be explained.

2.0.0 ◆ HAND BENDING EQUIPMENT

Figure 1 shows hand benders. Hand benders are convenient to use on the job because they are portable and no electrical power is required. Hand benders have a shape that supports the walls of the conduit being bent.

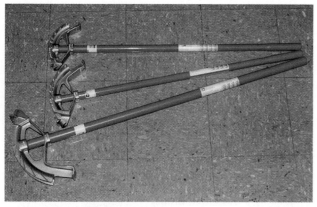

107F01.EPS

Figure 1 ◆ Hand benders.

These benders are used to make various bends in smaller-size conduit (½" to 1¼"). Most hand benders are sized to bend rigid conduit and electrical metallic tubing (EMT) of corresponding sizes. For example, a single hand bender can bend either ¾" EMT or ½" rigid conduit. The next larger size of hand bender will bend either 1" EMT or ¾" rigid conduit. This is because the corresponding sizes of conduit have nearly equal outside diameters.

The first step in making a good bend is familiarizing yourself with the bender. The manufacturer of the bender will typically provide documentation indicating starting points, distance between **offsets**, **gains**, and other important values associated with that particular bender. There is no substitute for taking the time to review this information. It will make the job go faster and result in better bends.

 CAUTION

When making bends, be sure you have a firm grip on the handle to avoid slippage and possible injury.

When performing a bend, it is important to keep the conduit on a stable, firm, flat surface for the entire duration of the bend. Hand benders are designed to have force applied using one foot and the hands. See *Figure 2*. It is important to use constant foot pressure as well as force on the handle to achieve uniform bends. Allowing the conduit to rise up or performing the bend on soft ground can result in distorting the conduit outside the bender.

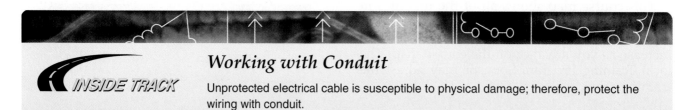

INSIDE TRACK

Working with Conduit

Unprotected electrical cable is susceptible to physical damage; therefore, protect the wiring with conduit.

Bending Conduit

A good way to practice bending conduit is to use a piece of No. 10 or No. 12 solid wire and bend it to resemble the bends you need. This gives you some perspective on how to bend the conduit and it will also help you to anticipate any problems with the bends.

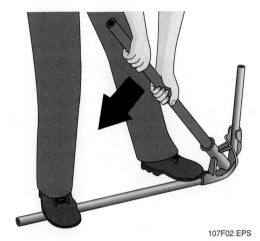

Figure 2 ◆ Pushing down on the bender to complete the bend.

 NOTE

Bends should be made in accordance with the guidelines of *NEC Article 342* (intermediate metal conduit or IMC), *Article 344* (rigid metal conduit or RMC), *Article 352* (rigid polyvinyl chloride conduit or PVC), or *Article 358* (electrical metallic tubing or EMT).

A hickey should not be confused with a hand bender. The hickey, which is used for RMC and IMC only, functions quite differently. See *Figure 3*.

When you use a hickey to bend conduit, you are forming the bend as well as the radius. When using a hickey, be careful not to flatten or kink the conduit. Hickeys should only be used with RMC and IMC because very little support is given to the walls of the conduit being bent. A hickey is a segment bending device. First, a small bend of about 10° is made. Then, the hickey is moved to a new

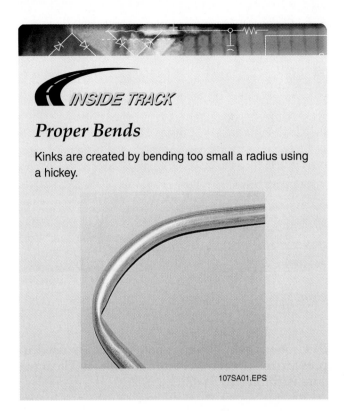
position and another small bend is made. This process is continued until the bend is completed. A hickey can be used for conduit **stub-ups** in slabs and decks.

Polyvinyl chloride (PVC) conduit is bent using a heating unit (*Figure 4*). The PVC must be rotated regularly while it is in the heater so that it heats evenly. Once heated, the PVC is removed, and the bending is performed by hand. Some units use an electric heating element, while others use liquid propane (LP). After bending, a damp sponge or cloth is often used so that the PVC sets up faster.

When bending PVC that is 2" or larger in diameter, there is a risk of wrinkling or flattening the bend. A plug set eliminates this problem (*Figure 5*). A plug is inserted into each end of the piece of

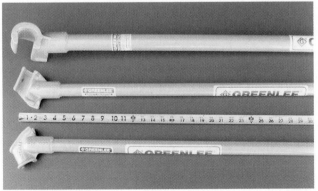

107F03.EPS

Figure 3 ◆ Hickeys.

107F04.EPS

Figure 4 ◆ Typical PVC heating units.

Figure 5 ◆ Typical plug set.

107F05.EPS

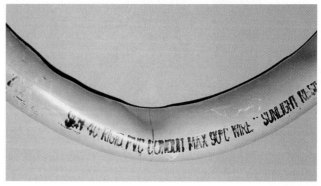

107SA02.EPS

PVC being bent. Then, a hand pump is used to pressurize the conduit before bending it. The pressure is about 3 to 5 psi.

CAUTION

Avoid contact with the case of the heating unit; it can become very hot and cause burns. Also, to avoid a fire hazard, ensure that the unit is cool before storage. If using an LP unit, keep a fire extinguisher nearby.

NOTE

The plugs must remain in place until the pipe is cooled and set.

2.1.0 Geometry Required to Make a Bend

Bending conduit requires that you use some basic geometry. You may already be familiar with most of the concepts needed; however, here is a review of the concepts directly related to this task. A right triangle is defined as any triangle with a 90° angle. The side directly opposite the 90° angle is called the hypotenuse, and the side on which the triangle sits is the base. The vertical side is called the height. On the job, you will apply the relationships in a right triangle when making an offset bend. The offset forms the hypotenuse of a right triangle (*Figure 6*).

NOTE

There are reference tables for sizing offset bends based on these relationships (see *Appendix A*).

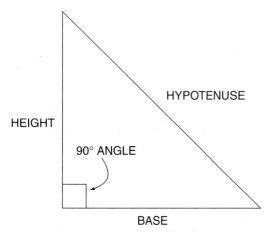

Figure 6 ◆ Right triangle and offset bend.

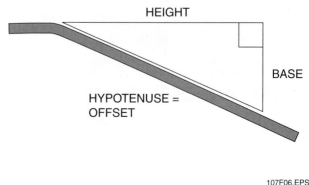

107F06.EPS

A circle is defined as a closed curved line whose points are all the same distance from its center. The distance from the center point to the edge of the circle is called the radius. The length from one edge of the circle to the other edge through the center point is the diameter. The distance around the circle is called the circumference. A circle can be divided into four equal quadrants. Each quadrant accounts for 90°, making a total of 360°. When you make a **90° bend**, you will use ¼ of a circle, or one quadrant. Concentric circles are circles that have a common center but different radii. The concept of concentric circles can be applied to **concentric bends** in conduit. The angle of each bend is 90°. Such bends have the same center point, but the radius of each is different. See *Figure 7*

To calculate the circumference of a circle, use the following formula:

$$C = \pi \times D \text{ or } C = \pi D$$

In this formula, C = circumference, D = diameter, and π = 3.14. Another way of stating the formula for circumference is $C = 2\pi R$, where R equals the radius or ½ the diameter. To figure the arc of a quadrant use:

$$\text{Length of arc} = (0.25)\ 2\pi R = 1.57R$$

For this formula, the arc of a quadrant equals ¼ the circumference of the circle or 1.57 times the radius.

A bending radius table is included in *Appendix B*.

2.2.0 Making a 90° Bend

The 90° stub bend is probably the most basic bend of all. The stub bend is used much of the time, regardless of the type of conduit being installed. Before beginning to make the bend, you need to know two measurements:

- Desired **rise** or stub-up
- Take-up distance of the bender

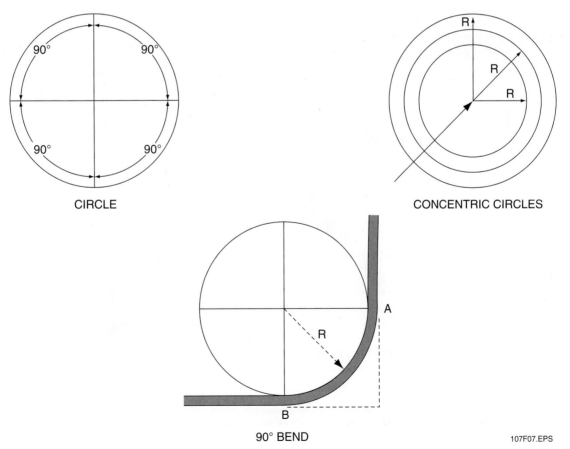

CIRCLE

CONCENTRIC CIRCLES

90° BEND

107F07.EPS

Figure 7 ◆ Circles and 90° bends.

The desired rise is the height of the stub-up. The take-up is the amount of conduit the bender will use to form the bend. Take-up distances are usually listed in the manufacturer's instruction manual. Typical bender take-up distances are shown in *Table 1*.

Once you have determined the take-up, subtract it from the stub-up height. Mark that distance on the conduit (all the way around) at that distance from the end. The mark will indicate the point at which you will begin to bend the conduit. Line up the starting point on the conduit with the starting point on the bender. Most benders have a mark, like an arrow, to indicate this point. *Figure 8* shows the take-up required to achieve an 18" stub-up on a piece of ½" EMT.

Once you have lined up the bender, use one foot to hold the conduit steady. Keep your heel on the floor for balance. Apply constant pressure on the bender foot pedal with your other foot. Make sure you hold the bender handle perpendicular to the floor, and as far up as possible, to get maximum leverage. Then, bend the conduit in one smooth motion, pulling as steadily as possible. Avoid overstretching.

NOTE

When bending conduit using the take-up method, always place the bender on the conduit and make the bend facing the hook of the conduit from which the measurements were taken.

After finishing the bend, check to make sure you have the correct angle and measurement. Use the following steps to check a 90° bend:

Step 1 With the back of the bend on the floor, measure to the end of the conduit stub-up to make sure it is the right length.

Step 2 Check the 90° angle of the bend with a square or at the angle formed by the floor and a wall. A torpedo level may also be used.

NOTE

If you overbend a conduit slightly past the desired angle, you can use the bender to bend the conduit back to the correct angle.

The above procedure will produce a 90° one-shot bend. That means that it took a single bend to form the conduit bend. A **segment bend** is any bend that is formed by a series of bends of a few degrees each, rather than a single one-shot bend. A shot is actually one bend in a segment bend. Segment or sweep bends must conform to the provisions of the *NEC®*.

Table 1 Typical Bender Take-Up Distances

EMT	Rigid/IMC	Take-Up
½"	—	5"
¾"	½"	6"
1"	¾"	8"
1¼"	1"	11"

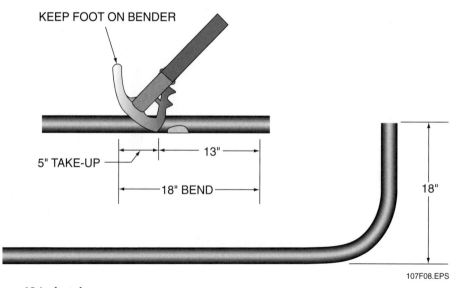

Figure 8 ◆ Bending an 18-inch stub-up.

Matching Bends in Parallel Runs

Suppose you are running 1" rigid conduit along with a 2" rigid conduit in a rack and you come to a 90° bend. If you used a 1" shoe, the radius would not match that of the 2" conduit bend. To match the 2" 90° bend, take your 1" conduit and put it in the 2" shoe of your bender, then bend as usual. This 1" 90° bend will now have the same radius as the 2" 90° bend. This trick will only work on rigid conduit. If done with EMT, it will flatten the pipe.

Take-Up Method

When bending conduit using the take-up method, always place the bender on the conduit and make the bend facing the end of the conduit from which the measurements were taken. It helps to make a narrow mark with a soft lead pencil or marker completely around the conduit. This is called girdling.

2.3.0 Gain

The gain is the distance saved by the arc of a 90° bend. Knowing the gain can help you to precut, ream, and prethread both ends of the conduit before you bend it. This will make your work go more quickly because it is easier to work with conduit while it is straight. *Figure 9* shows that the overall **developed length** of a piece of conduit with a 90° bend is less than the sum of the horizontal and vertical distances when measured square to the corner. This is shown by the following equation:

Developed length = (A + B) − gain

An example of a manufacturer's gain table is also shown in *Figure 9*. These tables are used to determine the gain for a certain size conduit.

THINK ABOUT IT

Gain

What is the difference between the gain and the take-up of a bend?

Smooth Bends

Why are smooth bends so important?

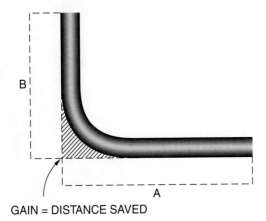

GAIN = DISTANCE SAVED

Conduit Size	*NEC*® Radius	90° Gain
½"	4"	2⅝"
¾"	5"	3¼"
1"	6"	4"
1¼"	8"	5⅝"

TYPICAL GAIN TABLE

107F09.EPS

Figure 9 ◆ Gain.

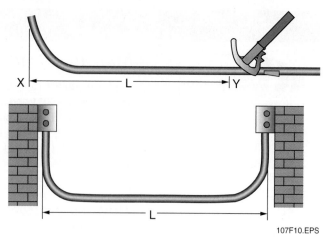
Checking Vertical Rise

Use a torpedo level to check for plumb on a vertical rise.

107SA03.EPS

2.4.0 Back-to-Back 90° Bends

A back-to-back bend consists of two 90° bends made on the same piece of conduit and placed back-to-back (*Figure 10*).

107F10.EPS

Figure 10 ◆ Back-to-back bends.

To make a back-to-back bend, make the first bend (labeled X in *Figure 10*) in the usual manner. To make the second bend, measure the required distance between the bends from the back of the first bend. This distance is labeled L in the figure. Reverse the bender on the conduit, as shown in *Figure 10*. Place the bender's back-to-back indicating mark at point Y on the conduit. Note that outside measurements from point X to point Y are used. Holding the bender in the reverse position and properly aligned, apply foot pressure and complete the second bend.

2.5.0 Making an Offset

Many situations require that the conduit be bent so that it can pass over objects such as beams and other conduits, or enter meter cabinets and junction boxes. Bends used for this purpose are called offsets (kicks). To produce an offset, two equal bends of less than 90° are required, a specified distance apart, as shown in *Figure 11*.

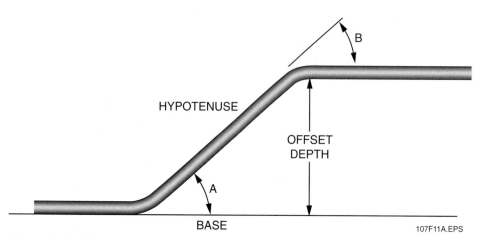

107F11A.EPS

107F11B.EPS

Figure 11 ◆ Offsets.

Offsets are a trade-off between space and the effort it will take to pull the wire. The larger the degree of bend, the harder it will be to pull the wire. The smaller the degree of bend, the easier it will be to pull the wire. Use the shallowest degree of bend that will still allow the conduit to bypass the obstruction and fit in the given space.

When conduit is offset, some of the conduit length is used. If the offset is made into the area, an allowance must be made for this shrinkage. If the offset angle is away from the obstruction, the shrinkage can be ignored.

Table 2 shows the amount of shrinkage per inch of rise for common offset angles.

The formula for figuring the distance between bends is as follows:

Distance between bends =
depth of offset × multiplier

The distance between the offset bends can generally be found in the manufacturer's documentation for the bender. *Table 3* shows the

Table 2	Shrinkage Calculation	
Offset Angle	**Multiplier**	**Shrinkage (per inch of rise)**
10° × 10°	6.0	¹⁄₁₆"
22½° × 22½°	2.6	³⁄₁₆"
30° × 30°	2.0	¼"
45° × 45°	1.4	⅜"
60° × 60°	1.2	½"

distance between bends for the most common offset angles.

Calculations related to offsets are derived from the branch of mathematics known as trigonometry, which deals with triangles. The multipliers shown in *Table 2* represent the cosecant (CSC) of the related offset angle. The multiplier is determined by dividing the hypotenuse of the triangle created by the offset by the depth of the offset (*Figure 11*).

Table 3 Common Offset Factors (in Inches)

Offset Depth	22½° Between Bends	22½° Shrinkage	30° Between Bends	30° Shrinkage	45° Between Bends	45° Shrinkage	60° Between Bends	60° Shrinkage
2	5¼	⅜	—	—	—	—	—	—
3	7¾	⁹⁄₁₆	6	¾	—	—	—	—
4	10½	¾	8	1	—	—	—	—
5	13	¹⁵⁄₁₆	10	1¼	7	1⅞	—	—
6	15½	1⅛	12	1½	8½	2¼	7¼	3
7	18¼	1⁵⁄₁₆	14	1¾	9¾	2⅝	8⅜	3½
8	20¾	1½	16	2	11¼	3	9⅝	4
9	23½	1¾	18	2¼	12½	3⅜	10⅞	4½
10	26	1⅞	20	2½	14	3¾	12	5

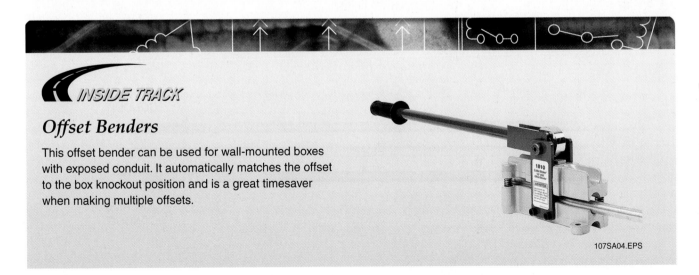

INSIDE TRACK

Offset Benders

This offset bender can be used for wall-mounted boxes with exposed conduit. It automatically matches the offset to the box knockout position and is a great timesaver when making multiple offsets.

107SA04.EPS

Basic trigonometry (trig) functions are briefly covered in *Appendix A*. As you will see in the next section, the tangent (TAN) of the offset angle is also used in calculating parallel offsets. Understanding trig functions will help you understand how offsets are determined. If you have a scientific calculator and understand these functions, you can calculate offset angles when you know the dimensions of the triangle created by the offset and the obstacle.

2.6.0 Parallel Offsets

Often, multiple pieces of conduit must be bent around a common obstruction. In this case, parallel offsets are made. Since the bends are laid out along a common radius, an adjustment must be made to ensure that the ends do not come out uneven, as shown in *Figure 12*.

The center of the first bend of the innermost conduit is found first, as shown in *Figure 13*. Each successive conduit must have its centerline moved farther away from the end of the pipe, as shown in *Figure 14*. The amount to add is calculated as follows:

Amount added = center-to-center spacing
× tangent (TAN) of ½ offset angle

Tangents can be found using the trig tables provided in *Appendix A*.

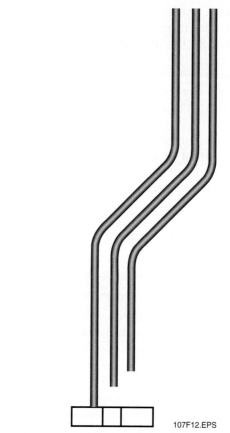

Figure 12 ◆ Incorrect parallel offsets.

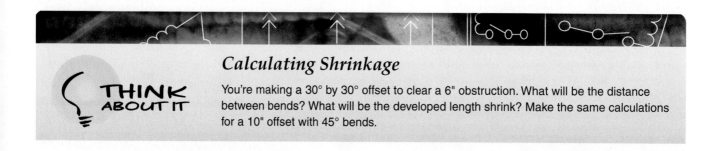

Calculating Shrinkage

You're making a 30° by 30° offset to clear a 6" obstruction. What will be the distance between bends? What will be the developed length shrink? Make the same calculations for a 10" offset with 45° bends.

THINK ABOUT IT

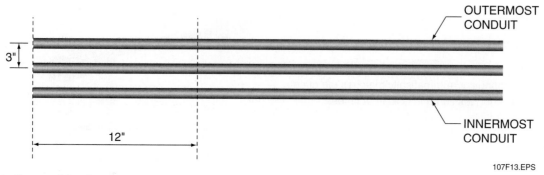

Figure 13 ◆ Center of first bend.

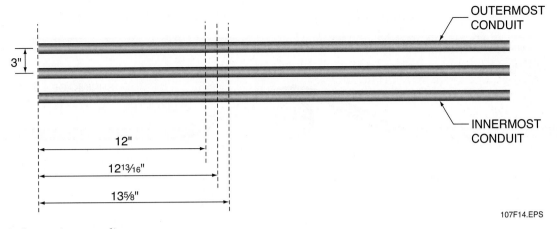

Figure 14 ◆ Successive centerlines.

107F14.EPS

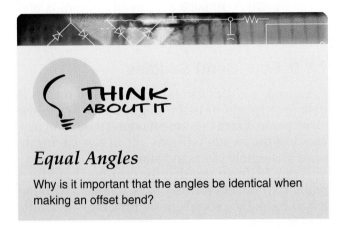

Equal Angles

Why is it important that the angles be identical when making an offset bend?

For example, *Figure 15* shows three pipes laid out as parallel and offset. The angle of the offset is 30°. The center-to-center spacing is 3". The start of the innermost pipe's first bend is 12".

The starting point of the second pipe will be:

$$12" + [\text{center-to-center spacing} \times \text{TAN} (\tfrac{1}{2} \text{ offset angle})]$$
$$12" + (3" \times \text{TAN } 15°) = 12" + (3" \times 0.2679)$$
$$= 12" + 0.8037"$$

This is approximately 12¹³⁄₁₆".
The starting point for the outermost pipe is:

$$12^{13}\!/_{16}" + {}^{13}\!/_{16}" = 13^{5}\!/_{8}"$$

2.7.0 Saddle Bends

A saddle bend is used to go around obstructions. *Figure 16* illustrates an example of a saddle bend that is required to clear a pipe obstruction. Making a saddle bend will cause the center of the saddle to shorten ³⁄₁₆" for every inch of saddle depth (see *Table 4*). For example, if the pipe diameter is 2 inches, this would cause a ⅜" shortening of the conduit on each side of the bend. When making saddle bends, the following steps should apply:

Step 1 Locate the center mark A on the conduit by using the size of the obstruction (i.e., pipe diameter) and calculate the shrink rate of the obstruction (for example, if the pipe diameter is 2 inches, ⅜" of conduit will be lost on each side of the bend for a total shrinkage of ¾"). This figure will be added to the measurement from the end of the conduit to the centerline of the obstruction (for example, if the distance measured from the conduit end to the obstruction centerline was 15", the distance to A would be 15 ⅜").

Step 2 Locate marks B and C on the conduit by measuring 2½" for every 1" of saddle depth from the A mark (i.e., for the saddle depth of 2 inches, the B mark would be 5" before the A mark and the C mark would be 5" after the A mark). See *Figure 17*.

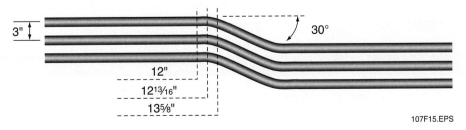

Figure 15 ◆ Parallel offset pipes.

107F15.EPS

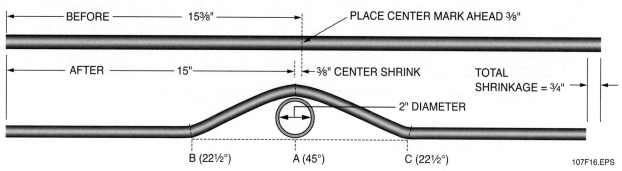

Figure 16 ◆ Saddle measurement.

Table 4	Shrinkage Chart for Saddle Bends with a 45° Center Bend and Two 22½° Bends	
Obstruction Depth	Shrinkage Amount (Move Center Mark Forward)	Make Outside Marks from New Center Mark
1	³⁄₁₆"	2½"
2	⅜"	5"
3	⁹⁄₁₆"	7½"
4	¾"	10"
5	¹⁵⁄₁₆"	12½"
6	1⅛"	15"
For each additional inch, add	³⁄₁₆"	2½"

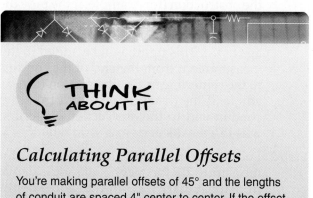

Calculating Parallel Offsets

You're making parallel offsets of 45° and the lengths of conduit are spaced 4" center to center. If the offset starts 12" down the pipe, what is the starting point for the bend on the second pipe?

Step 3 Refer to *Figure 18* and make a 45° bend at point A, make a 22½° bend at point B, and make a 22½° bend at point C. (Be sure to check the manufacturer's specifications.)

2.8.0 Four-Bend Saddles

Four-bend saddles can be difficult. The reason is that four bends must be aligned exactly on the same plane. Extra time spent laying it out and performing the bends will pay off in not having to scrap the whole piece and start over.

Figure 19 illustrates that the four-bend saddle is really two offsets formed back-to-back. Working left to right, the procedure for forming this saddle is as follows:

Step 1 Determine the height of the offset.

Step 2 Determine the correct spacing for the first offset and mark the conduit.

Step 3 Bend the first offset.

Step 4 Mark the start point for the second offset at the trailing edge of the obstruction.

Step 5 Mark the spacing for the second offset.

Step 6 Bend the second offset.

Using *Figure 20* as an example, a four-bend saddle using ½" EMT is laid out as follows:

- Height of the box = 6"
- Width of the box = 8"
- Distance to the obstruction = 36"

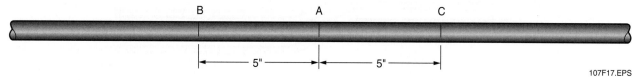

Figure 17 ◆ Measurement locations.

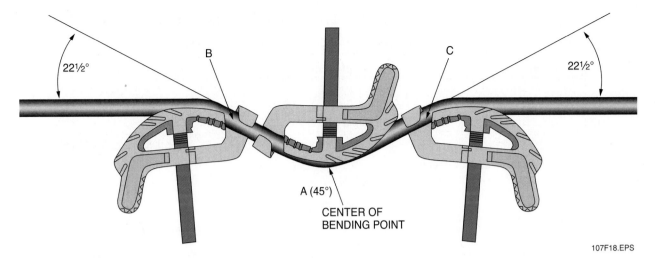

Figure 18 ◆ Location of bends.

Figure 19 ◆ Typical four-bend saddle.

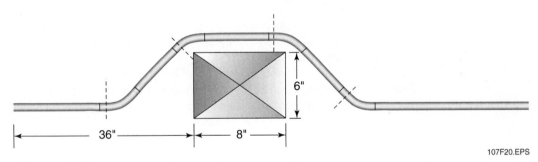

Figure 20 ◆ Four-bend saddle.

Two 30° offsets will be used to form the saddle. It is created as follows:

Step 1 See *Figure 21*. Working from left to right, calculate the start point for the first bend. The distance to the obstruction is 36", the offset is 6", and the 30° multiplier from *Table 2* is 2.0:

Distance to the obstruction – (offset × constant for the angle) + shrinkage = distance to the first bend

36" – (6" × 2.0) + 1½" = 25½"

Step 2 Determine where the second bend will end to ensure the conduit clears the obstruction. See *Figure 22*.

Distance to the first bend + distance to second bend + shrinkage = total length of the first offset

25½" + 12" + 1½" = 39"

Step 3 Determine the start point of the second offset. The width of the box is 8"; therefore, the start point of the second offset should be 8" beyond the end of the first offset:

8" + 39" = 47"

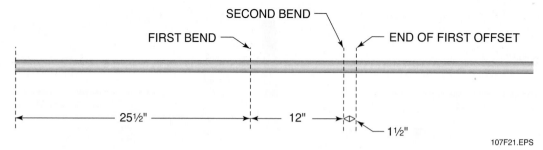

Figure 21 ◆ Four-bend saddle measurements.

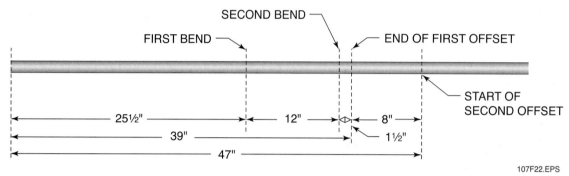

Figure 22 ◆ Bend and offset measurements.

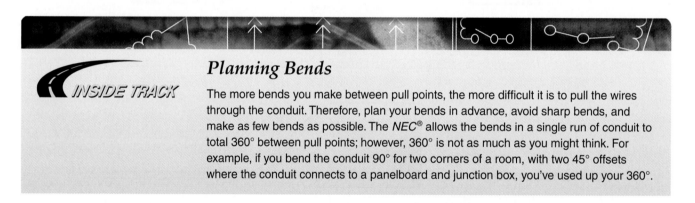

Planning Bends

INSIDE TRACK

The more bends you make between pull points, the more difficult it is to pull the wires through the conduit. Therefore, plan your bends in advance, avoid sharp bends, and make as few bends as possible. The *NEC*® allows the bends in a single run of conduit to total 360° between pull points; however, 360° is not as much as you might think. For example, if you bend the conduit 90° for two corners of a room, with two 45° offsets where the conduit connects to a panelboard and junction box, you've used up your 360°.

Step 4 Determine the spacing for the second off-set. Since the first and second offsets have the same rise and angle, the distance between bends will be the same, or 12".

3.0.0 ◆ CUTTING, REAMING, AND THREADING CONDUIT

RMC, IMC, and EMT are available in standard 10-foot lengths. When installing conduit, it is cut to fit the job requirements.

3.1.0 Hacksaw Method of Cutting Conduit

Conduit is normally cut using a hacksaw. To cut conduit with a hacksaw:

Step 1 Inspect the blade of the hacksaw and re-place it, if needed. A blade with 18, 24, or 32 cutting teeth per inch is recommended for conduit. Use a higher tooth count for EMT and a lower tooth count for rigid conduit and IMC. If the blade needs to be replaced, point the teeth toward the front of the saw when installing the new blade.

Step 2 Secure the conduit in a pipe vise.

Step 3 Rest the middle of the hacksaw blade on the conduit where the cut is to be made. Position the saw so the end of the blade is pointing slightly down and the handle is pointing slightly up. Push forward gently until the cut is started. Make even strokes until the cut is finished.

INSIDE TRACK

Using Your Bender Head to Secure Conduit

To secure conduit while cutting, insert the conduit into the bender head, brace your foot against the bender to secure it, then proceed to cut the conduit.

EMT Reaming Tools

There are specialty tools available for reaming EMT. One example is shown here. This tool slips over the end of a square-shank screwdriver and is secured in place with setscrews. The tool is inserted into the end of the EMT and rotated back and forth to deburr the conduit.

107SA06.EPS

 CAUTION

To avoid bruising your knuckles on the newly cut pipe, use gentle strokes for the final cut.

Step 4 Check the cut. The end of the conduit should be straight and smooth. *Figure 23* shows correct and incorrect cuts. Ream the conduit.

3.2.0 Pipe Cutter Method

A pipe cutter can also be used to cut RMC and IMC. To use a pipe cutter:

Step 1 Secure the conduit in a pipe vise and mark a place for the cut.

Step 2 Open the cutter and place it over the conduit with the cutter wheel on the mark.

Step 3 Tighten the cutter by rotating the screw handle.

CAUTION

Do not overtighten the cutter. Overtightening can break the cutter wheel and distort the wall of the conduit.

Step 4 Rotate the cutter counterclockwise to start the cut. *Figure 24* shows the proper way to rotate the cutter.

Step 5 Tighten the cutter handle ¼ turn for each full turn around the conduit. Again, make sure that you do not overtighten it.

Step 6 Add a few drops of cutting oil to the groove and continue cutting. Avoid skin contact with the oil.

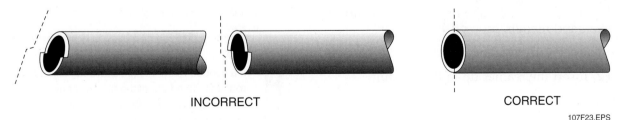

INCORRECT   CORRECT

107F23.EPS

Figure 23 ◆ Conduit ends after cutting.

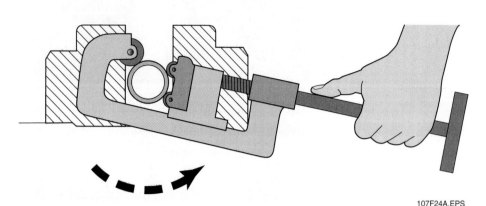

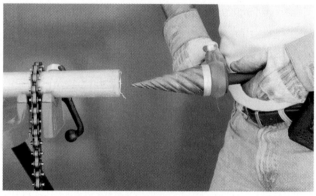

Figure 24 ◆ Cutter rotation.

107F24A.EPS

107F24B.EPS

Step 7 When the cut is almost finished, stop cutting and snap the conduit to finish the cut. This reduces the ridge that can be formed on the inside of the conduit.

Step 8 Clean the conduit and cutter with a shop towel rag.

Step 9 Ream the conduit.

3.3.0 Reaming Conduit

When the conduit is cut, the inside edge is sharp. This edge will damage the insulation of the wire when it is pulled through. To avoid this damage, the inside edge must be smoothed or reamed using a reamer (*Figure 25*).

To ream the inside edge of a piece of conduit using a hand reamer, proceed as follows:

Step 1 Place the conduit in a pipe vise.

Step 2 Insert the reamer tip in the conduit.

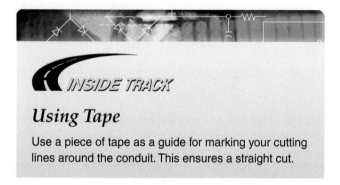

107F25.EPS

Figure 25 ◆ Rigid conduit reamer.

Step 3 Apply light forward pressure and start rotating the reamer. *Figure 26* shows the proper way to rotate the reamer. It should be rotated using a downward motion. The reamer can be damaged if you rotate it in the wrong direction. The reamer should bite as soon as you apply the proper pressure.

Step 4 Remove the reamer by pulling back on it while continuing to rotate it. Check the progress and then reinsert the reamer. Rotate the reamer until the inside edge is smooth. You should stop when all burrs have been removed.

NOTE

If a conduit reamer is not available, use a half-round file (the tang of the file must have a handle attached). EMT may be reamed using the nose of diagonal cutters or small hand reamers.

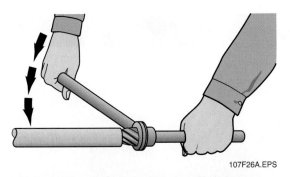

107F26A.EPS

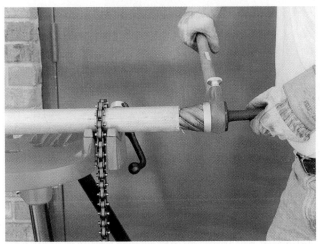

102F26B.EPS

Figure 26 ◆ Reamer rotation.

3.4.0 Threading Conduit

After conduit is cut and reamed, it is usually threaded so it can be properly joined. Only RMC and IMC have walls thick enough for threading.

The tool used to cut threads in conduit is called a die. Conduit dies are made to cut a taper of ¾ inch per foot. The number of threads per inch varies from 8 to 18, depending upon the diameter of the conduit. A thread gauge is used to measure how many threads per inch are cut.

The threading dies are contained in a die head. The die head can be used with a hand-operated ratchet threader (*Figure 27*) or with a portable power drive.

To thread conduit using a hand-operated threader, proceed as follows:

Step 1 Insert the conduit in a pipe vise. Make sure the vise is fastened to a strong surface. Place supports, if necessary, to help secure the conduit.

Step 2 Determine the correct die and head. Inspect the die for damage such as broken teeth. Never use a damaged die.

Step 3 Insert the die securely in the head. Make sure the proper die is in the appropriately numbered slot on the head.

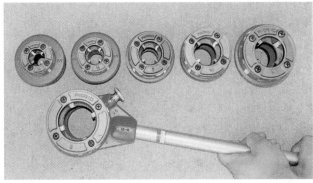

107F27.EPS

Figure 27 ◆ Hand-operated ratchet threader.

Step 4 Determine the correct thread length to cut for the conduit size used (match the manufacturer's thread length).

Step 5 Lubricate the die with cutting oil at the beginning and throughout the threading operation. Avoid skin contact with the oil.

Step 6 Cut threads to the proper length. Make sure that the conduit enters the tapered side of the die. Apply pressure and start turning the head. You should back off the head each quarter-turn to clear away chips.

Step 7 Remove the die when the proper cut is made. Threads should be cut only to the length of the die. Overcutting will leave the threads exposed to corrosion.

Step 8 Inspect the threads to make sure they are clean, sharp, and properly made. Use a thread gauge to measure the threads. The finished end should allow for a wrench-tight fit with one or two threads exposed.

NOTE

The conduit should be reamed again after threading to remove any burrs and edges. Cutting oil must be swabbed from the inside and outside of the conduit. Use a sandbox or drip pan under the threader to collect drips and shavings.

Die heads can also be used with portable power drives. You will follow the same steps when using a portable power drive. Threading machines are often used on larger conduit and where frequent threading is required. Threading machines hold and rotate the conduit while the die is fed onto the conduit for cutting. When using a threading machine, make sure you secure the legs properly and follow the manufacturer's instructions.

Oiling the Threader

For smoother operation, oil the threader often while threading the conduit.

107SA07.EPS

3.5.0 Cutting and Joining PVC Conduit

PVC conduit may be easily cut with a handsaw. To ensure square cuts, a miter box or similar device is recommended for cutting 2" and larger PVC. You can deburr the cut ends using a pocket knife. Smaller diameter PVC conduit, up to 1½", may be cut using a PVC cutter.

Use the following steps to join PVC conduit sections or attachments to plastic boxes:

Step 1 Wipe all the contacting surfaces clean and dry.

Step 2 Apply a coat of cement (a brush or aerosol can is recommended) on both ends of the conduit.

Step 3 Press the conduit and fitting together and rotate about a half-turn to evenly distribute the cement.

Forming PVC in the field requires a special tool called a hot box or other specialized methods. PVC may not be threaded when it is used for electrical applications.

CAUTION

Solvents and cements used with PVC are hazardous. Wear gloves and eye protection, and always follow the product instructions. Ensure that the area is well ventilated.

NOTE

Cementing the PVC must be done quickly. The aerosol spray cans of cement or the cement/brush combination are usually provided by the PVC manufacturer. Make sure you use the recommended cement.

Threading Conduit

The key to threading conduit is to start with a square cut. If you don't get it right, the conduit won't thread properly.

PVC Cutters

A nylon string can be used to cut PVC in place in awkward locations. However, it is best to use a PVC cutter to cut smaller trade sizes of PVC.

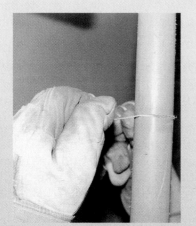

107SA08.EPS

107SA09.EPS

Putting It All Together

This module has stressed the precision necessary for creating accurate and uniform bends. Why is this important? What practical problems can result from sloppy or inaccurate bends?

1. The field bending of PVC requires a _____.
   a. hickey
   b. heating unit
   c. segmented bender
   d. one-shot bender

2. A hickey can be used to bend _____.
   a. RMC
   b. EMT
   c. PVC
   d. HDPE

3. What is the key to accurate bending with a hand bender?
   a. Correct size and length of handle
   b. Constant foot pressure on the back piece
   c. Using only the correct brand of bender
   d. Correct inverting of the conduit bender

4. In a right triangle, the side directly opposite the 90° angle is called the _____.
   a. right side
   b. hypotenuse
   c. altitude
   d. base

5. Prior to making a 90° bend, what two measurements must be known?
   a. Length of conduit and size of conduit
   b. Desired rise and length of conduit
   c. Size of bender and size of conduit
   d. Stub-up distance and take-up distance

6. A back-to-back bend is _____.
   a. a two-shot 90° bend
   b. two 90° bends made back-to-back
   c. an offset with four bends back-to-back
   d. a segmented bend

7. To prevent the ends of the conduit from being staggered, what additional information must be used when making parallel offset bends?
   a. Center-to-center spacing and tangent of ½ the offset angle
   b. Length of conduit and size of conduit
   c. Stub-up distance and take-up distance
   d. Offset angle and length of conduit

8. When making a saddle bend, the center of the saddle will cause the conduit to shrink _____ for every inch of saddle depth.
   a. ⅜"
   b. ³⁄₁₆"
   c. ¾"
   d. ³⁄₃₂"

9. When using a pipe cutter, always rotate the cutter _____ to start the cut.
   a. in a clockwise direction
   b. with the grain
   c. in a counterclockwise direction
   d. against the grain

10. EMT is threaded using a die.
    a. True
    b. False

Summary

You must choose a conduit bender to suit the kind of conduit being installed and the type of bend to be made. Some knowledge of the geometry of right triangles and circles needs to be mastered to make the necessary calculations. You must be able to calculate, lay out, and perform bending opera- tions on a single run of conduit and also on two or more parallel runs of conduit. At times, data tables for the figures may be consulted for the cal- culations. All work must conform to the require- ments of the *NEC®*.

Notes

Trade Terms Quiz

1. A right-angle bend is also called a(n) _____.

2. The rise in a section of conduit is called a(n) _____.

3. The _____ length is the actual length of the conduit that will be bent.

4. Also called a kick, a(n) _____ is two bends placed in a piece of conduit in order to navigate around obstructions.

5. _____ bends are large bends that are formed by multiple short bends or shots.

6. Two 90° bends with a straight section of conduit between them constitute a(n) _____ bend.

7. _____ bends are 90° bends made in two or more parallel sections of conduit, where the radius of each bend in conduit after the inside bend is respectively increased.

8. _____ is the distance that is saved by the arc of a 90° bend.

9. _____ is the length of the bent section of conduit measured from the bottom, centerline, or top of the straight section to the end of the bent section.

Trade Terms

90° bend	Developed length	Offsets	Segment bend
Back-to-back bend	Gain	Rise	Stub-up
Concentric bends			

Timothy Ely
Beacon Electric Company

Tim Ely is a man who believes in giving something back to the industry that nurtured his successful career. Despite working in a demanding executive position, he serves on many industry committees and was instrumental in the development of the NCCER Electrical Program.

What made you decide to become an electrician?
During my last two years of high school, I worked for a do-it-all construction company. We laid concrete, installed roofs, hung drywall, installed plumbing, and did electrical work. I liked the electrical work the best.

How did you learn the trade?
I learned through on-the-job training, hard work, and studying on my own. I had good teachers who were patient with me and took the time to help me succeed.

What kinds of jobs did you hold on the way to your current position?
I started out wiring houses and did that for the first two years. Then I switched over to commercial and industrial work, and worked as an apprentice in that area for two more years before becoming a journeyman. From there, I served as a lead electrician, then foreman, then city superintendent, then finally general superintendent before being promoted to my current job as vice president of construction.

What factor or factors have contributed most to your success?
Hard work helps a lot. I also try to bring a positive attitude to work with me every day. My family and friends have supported me throughout my career.

What does a vice president of construction do in your company?
In my job, I have responsibility for all the job sites, as well as the warehouse and service trucks. I also have responsibility for employee hiring, safety training, job planning and scheduling, quality control, and licensing. I personally hold 28 different state and city licenses, and I firmly believe that getting the training to obtain your licenses and then doing the in-service training to keep your licenses current are important factors in an electrician's success. For example, an electrical contractor can bid on jobs in a wide geographical area. Electricians working for that contractor can work on projects in different cities, even different states. Every place you go will require you to have a valid license.

What advice would you give to someone entering the electrical trade?
Work hard, treat people with respect, and keep an open mind. Be careful how you deal with people. Someone you offend today may wind up being your boss or a potential customer tomorrow.

Trade Terms Introduced in This Module

90° bend: A bend that changes the direction of the conduit by 90°.

Back-to-back bend: Any bend formed by two 90° bends with a straight section of conduit between the bends.

Concentric bends: 90° bends made in two or more parallel runs of conduit with the radius of each bend increasing from the inside of the run toward the outside.

Developed length: The actual length of the conduit that will be bent.

Gain: Because a conduit bends in a radius and not at right angles, the length of conduit needed for a bend will not equal the total determined length. Gain is the distance saved by the arc of a 90° bend.

Offsets: An offset (kick) is two bends placed in a piece of conduit to change elevation to go over or under obstructions or for proper entry into boxes, cabinets, etc.

Rise: The length of the bent section of conduit measured from the bottom, centerline, or top of the straight section to the end of the bent section.

Segment bend: A large bend formed by multiple short bends or shots.

Stub-up: Another name for the rise in a section of conduit. Also, a term used for conduit penetrating a slab or the ground.

Using Trigonometry to Determine
Offset Angles and Multipliers

You do not have to be a mathematician to use trigonometry. Understanding the basic trig functions and how to use them can help you calculate unknown distances or angles. Assume that the right triangle below represents a conduit offset. If you know the length of one side and the angle, you can calculate the length of the other sides, or if you know the length of any two of the sides of the triangle, you can then find the offset angle using one or more of these trig functions. You can use a trig table such as that shown on the following pages or a scientific calculator to determine the offset angle. For example, if the cosecant of angle A is 2.6, the trig table tells you that the offset angle is 22½°.

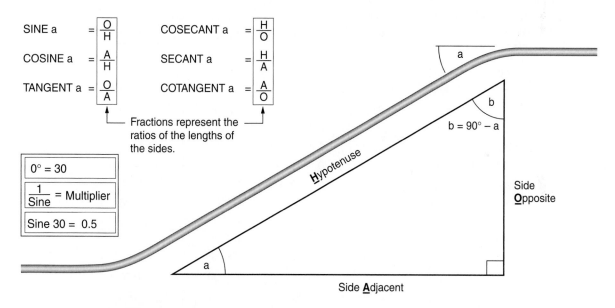

To determine the multiplier for the distance between bends in an offset:

1. Determine the angle of the offset: 30°

2. Find the sine of the angle: 0.5

3. Find the inverse (reciprocal) of the sine: $\frac{1}{0.5}$ = 2. This is also listed in trig tables as the cosecant of the angle.

4. This number multiplied by the height of the offset gives the hypotenuse of the triangle, which is equal to the distance between bends.

107A01.EPS

ANGLE	SINE	COSINE	TANGENT	COTANGENT	COSECANT
1°	0.0175	0.9998	0.0175	57.3000	57.3065
2°	0.0349	0.9994	0.0349	28.6000	28.6532
3°	0.0523	0.9986	0.0524	19.1000	19.1058
4°	0.0698	0.9976	0.0699	14.3000	14.3348
5°	0.0872	0.9962	0.0875	11.4000	11.4731
6°	0.1045	0.9945	0.1051	9.5100	9.5666
7°	0.1219	0.9925	0.1228	8.1400	8.2054
8°	0.1392	0.9903	0.1405	7.1200	7.1854
9°	0.1564	0.9877	0.1584	6.3100	6.3926
10°	0.1736	0.9848	0.1763	5.6700	5.7587
11°	0.1908	0.9816	0.1944	5.1400	5.2408
12°	0.2079	0.9781	0.2126	4.7000	4.8097
13°	0.2250	0.9744	0.2309	4.3300	4.4454
14°	0.2419	0.9703	0.2493	4.0100	4.1335
15°	0.2588	0.9659	0.2679	3.7300	3.8636
16°	0.2756	0.9613	0.2867	3.4900	3.5915
17°	0.2924	0.9563	0.3057	3.2700	3.4203
18°	0.3090	0.9511	0.3249	3.0800	3.2360
19°	0.3256	0.9455	0.3443	2.9000	3.0715
20°	0.3420	0.9397	0.3640	2.7500	2.9238
21°	0.3584	0.9336	0.3839	2.6100	2.7904
22°	0.3746	0.9272	0.4040	2.4800	2.6694
23°	0.3907	0.9205	0.4245	2.3600	2.5593
24°	0.4067	0.9135	0.4452	2.2500	2.4585
25°	0.4226	0.9063	0.4663	2.1400	2.3661
26°	0.4384	0.8988	0.4877	2.0500	2.2811
27°	0.4540	0.8910	0.5095	1.9600	2.2026
28°	0.4695	0.8829	0.5317	1.8800	2.1300
29°	0.4848	0.8746	0.5543	1.8000	2.0626
30°	0.5000	0.8660	0.5774	1.7300	2.0000
31°	0.5150	0.8572	0.6009	1.6600	1.9415
32°	0.5299	0.8480	0.6249	1.6000	1.8870
33°	0.5446	0.8387	0.6494	1.5400	1.8360
34°	0.5592	0.8290	0.6745	1.4800	1.7883
35°	0.5736	0.8192	0.7002	1.4300	1.7434
36°	0.5878	0.8090	0.7265	1.3800	1.7012
37°	0.6018	0.7986	0.7536	1.3300	1.6616
38°	0.6157	0.7880	0.7813	1.2800	1.6242
39°	0.6293	0.7771	0.8098	1.2300	1.5890
40°	0.6428	0.7660	0.8391	1.1900	1.5557
41°	0.6561	0.7547	0.8693	1.1500	1.5242
42°	0.6691	0.7431	0.9004	1.1100	1.4944
43°	0.6820	0.7314	0.9325	1.0700	1.4662
44°	0.6947	0.7193	0.9657	1.0400	1.4395
45°	0.7071	0.7071	1.0000	1.0000	1.4142

107A02.EPS

ANGLE	SINE	COSINE	TANGENT	COTANGENT	COSECANT
46°	0.7193	0.6947	1.0355	0.9660	1.4395
47°	0.7314	0.6820	1.0724	0.9330	1.3673
48°	0.7431	0.6691	1.1106	0.9000	1.3456
49°	0.7547	0.6561	1.1504	0.8690	1.3250
50°	0.7660	0.6428	1.1918	0.8390	1.3054
51°	0.7771	0.6293	1.2349	0.8100	1.2867
52°	0.7880	0.6157	1.2799	0.7810	1.2690
53°	0.7986	0.6018	1.3270	0.7540	1.2521
54°	0.8090	0.5878	1.3764	0.7270	1.2360
55°	0.8192	0.5736	1.4281	0.7000	1.2207
56°	0.8290	0.5592	1.4826	0.6750	1.2062
57°	0.8387	0.5446	1.5399	0.6490	1.1923
58°	0.8480	0.5299	1.6003	0.6250	1.1791
59°	0.8572	0.5150	1.6643	0.6010	1.1666
60°	0.8660	0.5000	1.7321	0.5770	1.1547
61°	0.8746	0.4848	1.8040	0.5540	1.1433
62°	0.8829	0.4695	1.8807	0.5320	1.1325
63°	0.8910	0.4540	1.9626	0.5100	1.1223
64°	0.8988	0.4384	2.0503	0.4880	1.1126
65°	0.9063	0.4226	2.1445	0.4660	1.1033
66°	0.9135	0.4067	2.2460	0.4450	1.0946
67°	0.9205	0.3907	2.3559	0.4240	1.0863
68°	0.9272	0.3746	2.4751	0.4040	1.0785
69°	0.9336	0.3584	2.6051	0.3840	1.0711
70°	0.9397	0.3420	2.7475	0.3640	1.0641
71°	0.9455	0.3256	2.9042	0.3440	1.0576
72°	0.9511	0.3090	3.0777	0.3250	1.0514
73°	0.9563	0.2924	3.2709	0.3060	1.0456
74°	0.9613	0.2756	3.4874	0.2870	1.0402
75°	0.9659	0.2588	3.7321	0.2680	1.0352
76°	0.9703	0.2419	4.0108	0.2490	1.0306
77°	0.9744	0.2250	4.3315	0.2310	1.0263
78°	0.9781	0.2079	4.7046	0.2130	1.0223
79°	0.9816	0.1908	5.1446	0.1940	1.0187
80°	0.9848	0.1736	5.6713	0.1760	1.0154
81°	0.9877	0.1564	6.3138	0.1580	1.0124
82°	0.9903	0.1392	7.1154	0.1410	1.0098
83°	0.9925	0.1219	8.1443	0.1230	1.0075
84°	0.9945	0.1045	9.5144	0.1050	1.0055
85°	0.9962	0.0872	11.4301	0.0880	1.0038
86°	0.9976	0.0698	14.3007	0.0700	1.0024
87°	0.9986	0.0523	19.0811	0.0520	1.0013
88°	0.9994	0.0349	28.6363	0.0350	1.0006
89°	0.9998	0.0175	57.2900	0.0180	1.0001
90°	1.0000	0.0000	—	0.0000	1.0000

107A03.EPS

Bending Radius Table

Radius (Inches)	Radius Increments (Inches)									
	0	1	2	3	4	5	6	7	8	9
0	0.00	1.57	3.14	4.71	6.28	7.85	9.42	10.99	12.56	14.13
10	15.70	17.27	18.84	20.41	21.98	23.85	25.12	26.69	28.26	29.83
20	31.40	32.97	34.54	36.11	37.68	39.25	40.82	42.39	43.96	45.83
30	47.10	48.67	50.24	51.81	53.38	54.95	56.52	58.09	59.66	61.23
40	62.80	64.37	65.94	67.50	69.03	70.65	72.22	73.79	75.36	76.93
50	87.50	80.07	81.64	83.21	84.78	86.35	87.92	89.49	91.06	92.63
60	94.20	95.77	97.34	98.91	100.48	102.05	103.62	105.19	106.76	108.33
70	109.90	111.47	113.04	114.61	116.18	117.75	119.32	120.89	122.46	124.03
80	125.60	127.17	128.74	130.31	131.88	133.45	135.02	136.59	138.16	139.73
90	141.30	142.87	144.44	146.01	147.58	149.15	150.72	–	–	–

To find the developed length for the following angles, use a fraction of the 90° chart.

For	15°	22½°	30°	45°	60°	67½°	75°	90°
Take	⅙	¼	⅓	½	⅔	¾	⅚	See Chart

For any other degrees: Developed length = 0.01744 × radius × degrees.

107A04.EPS

This module is intended to present thorough resources for task training. The following reference works are suggested for further study. These are optional materials for continued education rather than for task training.

Benfield Conduit Bending Manual, 2nd Edition. Overland Park, KS: EC&M Books.

National Electrical Code® Handbook, Latest Edition. Quincy, MA: National Fire Protection Association.

Tom Henry's Conduit Bending Package (includes video, book, and bending chart). Winter Park, FL: Code Electrical Classes, Inc.

NCCER makes every effort to keep these textbooks up-to-date and free of technical errors. We appreciate your help in this process. If you have an idea for improving this textbook, or if you find an error, a typographical mistake, or an inaccuracy in NCCER's Contren® textbooks, please write us, using this form or a photocopy. Be sure to include the exact module number, page number, a detailed description, and the correction, if applicable. Your input will be brought to the attention of the Technical Review Committee. Thank you for your assistance.

Instructors – If you found that additional materials were necessary in order to teach this module effectively, please let us know so that we may include them in the Equipment/Materials list in the Annotated Instructor's Guide.

Write: Product Development and Revision
National Center for Construction Education and Research
3600 NW 43rd St., Bldg. G, Gainesville, FL 32606

Fax: 352-334-0932

E-mail: curriculum@nccer.org

Craft   Module Name
_____

Copyright Date   Module Number   Page Number(s)
_____

Description
_____

_____

_____

_____

(Optional) Correction
_____

_____

_____

(Optional) Your Name and Address
_____

_____

_____

Raceways and Fittings

The Beauvallon

The Beauvallon, a high-end, mixed-use project in Denver's Golden Triangle features two 14-story towers with a total of 210 luxury, for-sale condos. The project's European flavor comes from what designers call its "modified French Baroque style," which features wide balconies, two-story mansard roofs, ornate cornices, urban landscaping, and other Old World architectural elements.

26108-08
Raceways and Fittings

Topics to be presented in this module include:

Overview

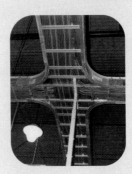

The *National Electrical Code*® defines a raceway as an "enclosed channel of metal or nonmetallic materials designed expressly for holding wires, cables, or busbars." Raceways include conduit, wireways, ducting, and cable trays. Raceways protect the wiring and provide a means of identifying one type of wiring when it is located next to another type.

Each type of raceway is suited to a particular purpose. Different installation methods are required depending on the construction environment in which the raceway is to be installed. Raceways must be supported by securing them to the building structure with fasteners. For that reason, it is important that every electrician be familiar with the various types of fasteners used to attach raceway supports to wood, concrete, and metal.

Every type of raceway is interconnected with a specifically designed series of boxes and fittings. Boxes and fittings provide ease of raceway installation, as well as an access point or outlet for circuit conductors. Boxes are also used as convenient pull points in the raceway system when installing large conductors or a great number of conductors. It is important to always match the boxes and fittings with the type of raceway being installed.

Objectives

When you have completed this module, you will be able to do the following:

1. Identify and select various types and sizes of raceways and fittings for a given application.
2. Identify various methods used to fabricate (join) and install raceway systems.
3. Identify uses permitted for selected raceways.
4. Demonstrate how to install a flexible raceway system.
5. Terminate a selected raceway system.
6. Identify the appropriate conduit body for a given application.

Trade Terms

Accessible
Approved
Bonding wire
Cable trays
Conduit
Exposed location
Kick

Raceways
Splice
Tap
Trough
Underwriters
    Laboratories, Inc. (UL)
Wireways

Required Trainee Materials

1. Paper and pencil
2. Copy of the latest edition of the *National Electrical Code®*
3. Appropriate personal protective equipment

Prerequisites

Before you begin this module, it is recommended that you successfully complete *Core Curriculum* and *Electrical Level One*, Modules 26101-08 through 26107-08.

This course map shows all of the modules in *Electrical Level One*. The suggested training order begins at the bottom and proceeds up. Skill levels increase as you advance on the course map. The local Training Program Sponsor may adjust the training order.

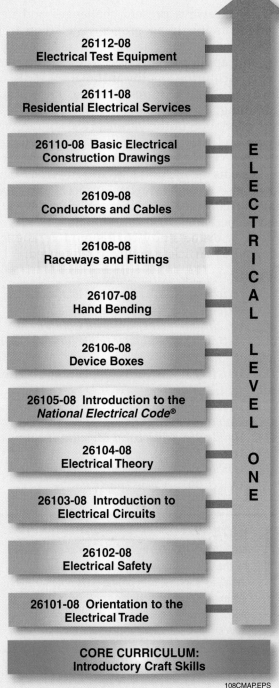

26112-08
Electrical Test Equipment

26111-08
Residential Electrical Services

26110-08 Basic Electrical
Construction Drawings

26109-08
Conductors and Cables

26108-08
Raceways and Fittings

26107-08
Hand Bending

26106-08
Device Boxes

26105-08 Introduction to the
National Electrical Code®

26104-08
Electrical Theory

26103-08 Introduction to
Electrical Circuits

26102-08
Electrical Safety

26101-08 Orientation to the
Electrical Trade

CORE CURRICULUM:
Introductory Craft Skills

ELECTRICAL LEVEL ONE

108CMAP.EPS

1.0.0 ◆ INTRODUCTION

Electrical **raceways** present challenges and new requirements involving proper installation techniques, general understanding of raceway systems, and applications of the *NEC*® to raceway systems. Acquiring quality installation skills for raceway systems requires practice, knowledge, and training.

A presentation of the various types of raceway systems and fittings, basic raceway installation skills, and *NEC*® requirements applicable to raceway systems is included in this module. This module also covers raceway supports and environmental considerations for raceway systems, as well as general raceway information.

Along with the study of this module, the following *NEC*® Articles should be referenced:

- *NEC Article 250 – Grounding*
- *NEC Article 342 – Intermediate Metal Conduit*
- *NEC Article 344 – Rigid Metal Conduit*
- *NEC Article 348 – Flexible Metal Conduit*
- *NEC Article 350 – Liquidtight Flexible Metal Conduit*
- *NEC Article 352 – Rigid Polyvinyl Chloride Conduit*
- *NEC Article 356 – Liquidtight Flexible Nonmetallic Conduit*
- *NEC Article 358 – Electrical Metallic Tubing*
- *NEC Article 376 – Metal Wireways*
- *NEC Article 378 – Nonmetallic Wireways*
- *NEC Article 392 – Cable Trays*

> **NOTE**
> Mandatory rules in the *NEC*® are characterized by the use of the word *shall*. Explanatory material is in the form of fine print notes (FPNs). When referencing specific sections of the *NEC*®, always check to see if any exceptions apply.

2.0.0 ◆ RACEWAYS

Raceway is a general term referring to a wide range of circular and rectangular enclosed channels used to house electrical wiring. Raceways can be metallic or nonmetallic and come in different shapes. Depending on the particular purpose for which they are intended, raceways include enclosures such as underfloor raceways, flexible metal **conduit**, tubing, **wireways**, surface metal raceways, surface nonmetallic raceways, and support systems such as cable trays.

3.0.0 ◆ CONDUIT

Conduit is a raceway with a circular cross section, similar to pipe, that contains wires or cables. Conduit is used to provide protection for conductors and route them from one place to another. In addition, conduit makes it easier to replace or add wires to existing structures. Metal conduit also provides a permanent electrical path to ground. This equipment should be listed per the *NEC*®.

3.1.0 Conduit as a Ground Path

For safety reasons, most equipment that receives electrical power and has a metallic frame is bonded. In order to bond the equipment, an electrical connection must be made to connect the metal frame of the electrically powered equipment to the grounding point at the service-entrance equipment. This is usually done in one or both of the following ways:

- The frame of the equipment is connected to a wire (equipment grounding conductor), which is directly connected to the ground point at the grounding terminal.
- The frame of the equipment is connected (bonded) to a metal conduit or other type of raceway system, which provides an uninterrupted and low-impedance circuit to the ground point at the service-entrance equipment. The metal raceway or conduit acts as the equipment grounding conductor.

> **NOTE**
> According to *NEC Section 250.96*, metal raceways, cable trays, cable armor, cable sheath, enclosures, frames, fittings, and other metal noncurrent-carrying parts that are to serve as grounding conductors with or without the use of supplementary equipment grounding conductors shall be effectively bonded where necessary to ensure electrical continuity and the capacity to safely conduct any fault current likely to be imposed on them. The purpose of the equipment grounding conductor is to provide a low-resistance path to ground for all equipment that receives power. This is done so that if an ungrounded conductor comes in contact with the frame of a piece of equipment, the circuit overcurrent device immediately acts to open the circuit. It also reduces the voltage to ground that would be present on the faulted equipment if a person came in contact with the equipment frame.

3.2.0 Types of Conduit and Tubing

There are many types of conduit used in the construction industry. The size of conduit to be used is determined by engineering specifications, local codes, and the *NEC*®. Refer to *NEC Chapter 9, Tables 1 through 8 and Annex C* for conduit fill with various conductors. There are several common types of conduit to examine.

3.2.1 Electrical Metallic Tubing

Electrical metallic tubing (EMT) is the lightest duty tubing available for enclosing and protecting electrical wiring. EMT is widely used for residential, commercial, and industrial wiring systems. It is lightweight, easily bent and/or cut to shape, and is the least costly type of metallic conduit. Because the wall thickness of EMT is less than that of rigid conduit, it is often referred to as thinwall conduit. A comparison of inside and outside diameters of EMT to rigid metal conduit (RMC) and intermediate metal conduit (IMC) is shown in *Figure 1*.

NEC Section 358.10(A) permits the installation of EMT for either exposed or concealed work where it will not be subject to severe physical damage during installation or after construction. Installation of EMT is permitted in wet locations such as outdoors or indoors in dairies, laundries, and canneries using waterproof fittings.

NOTE
Refer to *NEC Section 358.12* for restrictions that apply to the use of EMT.

EMT shall not be used (1) where, during installation or afterward, it will be subject to severe physical damage; (2) where protected from corrosion solely by enamel; (3) in cinder concrete or cinder fill where subject to permanent moisture unless protected on all sides by a layer of non-cinder concrete at least two inches thick or unless the tubing is at least 18 inches under the fill; (4) in any hazardous (classified) locations except as permitted by *NEC Sections 502.10(B)(2), 503.10, and 504.20;* or (5) for the support of fixtures or other equipment.

In a wet area, EMT and other conduit must be installed to prevent water from entering the conduit system. In locations where walls are subject to regular wash-down [see *NEC Section 300.6(D)*], the entire conduit system must be installed to provide a ¼-inch air space between it and the wall or supporting surface. The entire conduit system is considered to include conduit, boxes, and fittings. To ensure resistance to corrosion caused by wet environments, EMT is galvanized. The term galvanized is used to describe the procedure in which the interior and exterior of the conduit are coated with a corrosion-resistant zinc compound.

EMT, being a good conductor of electricity, may be used as an equipment grounding conductor. In order to qualify as an equipment grounding conductor [see *NEC Section 250.118(4)*], the conduit system must be tightly connected at each joint and provide a continuous grounding path from each electrical load to the service equipment. The connectors used in an EMT system ensure electrical and mechanical continuity throughout the system (see *NEC Sections 250.96, 300.10, and 358.42*).

NOTE
Support requirements for EMT are also covered in *NEC Section 358.30.* The types of supports will be discussed later in this module.

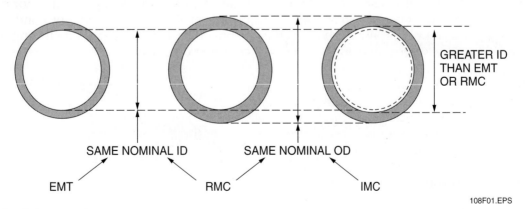

Figure 1 ◆ Conduit comparison.

EMT Use

Because EMT is too thin for threads, fittings listed for EMT must be used. For wet or damp locations, compression fittings such as those shown in *Figure 2* are used. These fittings contain a compression ring made of metal that forms a raintight seal.

When EMT compression couplings are used, they must be securely tightened, and when installed in masonry concrete, they must be of the concrete-tight type. If installed in a wet location, they must be the raintight type. Refer to *NEC Section 358.42*.

EMT fittings for dry locations can be either the setscrew type or the indenting type. To use the setscrew type, the ends of the EMT are inserted into the sleeve and the setscrews are tightened to make the connection. Various types of setscrew fittings are shown in *Figure 3*.

EMT sizes of 2½ inches and larger have the same outside diameter as corresponding sizes of galvanized RMC. RMC threadless connectors may be used to connect EMT.

NOTE

EMT connectors smaller then 2½", although they are the same size as RMC threadless connectors, may not be used to connect RMC.

Both setscrew and compression couplings are available in die-cast or steel construction. Steel couplings are stronger, but may not seal as well.

Support requirements for EMT are presented in *NEC Section 358.30.* As with most other metal

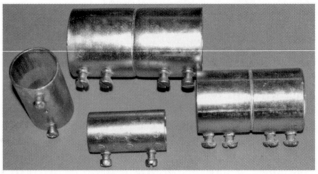

108F03.EPS

Figure 3 ◆ Setscrew fittings.

conduit, EMT must be supported at least every 10 feet and within 3 feet of each outlet box, junction box, cabinet, fitting, or terminating end of the conduit. An exception to *NEC Section 358.30(A), Exception 1* allows the fastening of unbroken lengths of EMT to be increased to a distance of 5 feet where structural members do not readily permit fastening within 3 feet.

Electrical nonmetallic tubing (ENT) is also available. It provides an economical alternative to EMT, but it can only be used in certain applications. See *NEC Article 362.*

3.2.2 Rigid Metal Conduit

Rigid metal conduit (RMC) is conduit that is constructed of metal of sufficient thickness to permit the cutting of pipe threads at each end. RMC provides the best physical protection for conductors of any of the various types of conduit. RMC is supplied in 10-foot lengths including a threaded coupling on one end.

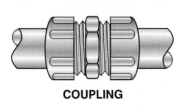

COUPLING

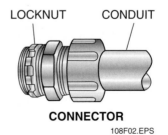

LOCKNUT   CONDUIT

CONNECTOR

108F02.EPS

Figure 2 ◆ Compression fittings.

RMC may be made from steel or aluminum. Rigid metal steel conduit may be stainless, galvanized, or enamel-coated inside and out. Because of its threaded fittings, RMC provides an excellent equipment grounding conductor as defined in *NEC Section 250.118 (2).* A piece of RMC is shown in *Figure 4(A).* The support requirements for RMC are presented in *NEC Section 344.30(A) and (B)* and *NEC Table 344.30(B)(2).*

RMC is mostly used in industrial applications. RMC is heavier than EMT and IMC. It is more difficult to cut and bend, usually requires threading of cut ends, and has a higher purchase price than EMT and IMC. As a result, the cost of installing RMC is generally higher than the cost of installing EMT and IMC.

3.2.3 Plastic-Coated RMC

Plastic-coated RMC has a thin coating of polyvinyl chloride (PVC) over the RMC. See *Figure*

(A) RIGID METAL CONDUIT (RMC)

(B) PLASTIC-COATED RMC

108F04.EPS

Figure 4 ◆ Types of rigid metal conduit (RMC).

4(B). This combination is useful when an environment calls for the ruggedness of RMC along with the corrosion resistance of rigid nonmetallic conduit (PVC). Typical installations where plastic-coated RMC may be required are:

- Chemical plants
- Food plants
- Refineries
- Fertilizer plants
- Paper mills
- Wastewater treatment plants

Plastic-coated RMC requires special threading and bending techniques.

3.2.4 Aluminum Conduit

Aluminum conduit has several characteristics that distinguish it from steel conduit. Because it has better resistance to wet environments and some chemical environments, aluminum conduit generally requires less maintenance in installations such as sewage treatment plants.

NEC Section 300.6(B) states that aluminum conduit used in concrete or in direct contact with soil requires supplementary corrosion protection. According to Underwriters Laboratories *Electrical Construction Equipment Directory* (UL Green Book), examples of supplementary protection are paints approved for the purpose (such as bitumastic paint), tape wraps approved for the purpose, or PVC-coated conduit.

3.2.5 Black Enamel Steel Conduit

Rigid black enamel steel conduit (often called black conduit) is steel conduit that is coated with a black enamel. In the past, this type of conduit was used exclusively for indoor wiring. Black enamel steel conduit is no longer manufactured for sale in the United States. It is mentioned only because it may still be found in existing installations.

EMT Installation

When installing EMT, hook your index finger up through the box to check that the conduit is seated in the connector. If you feel a lip between the conduit and the connector, the conduit is not properly seated.

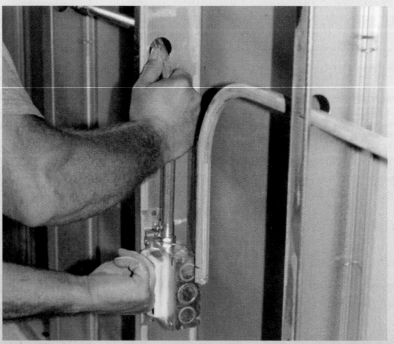

108SA01.EPS

RMC Installations

Use RMC in hazardous environments, such as feed mills, or in areas where there is a chance of physical abuse or extreme moisture, such as outdoor environments. The *NEC®* does allow EMT to be buried in the ground or in concrete, but galvanized RMC is more commonly used.

Use of Aluminum Conduit

Aluminum conduit is used for special purposes such as high-cycle lines (400 cycles or above); around cooling towers, food service areas, and other applications in which corrosion is a factor; or where magnetic induction is a concern, such as near magnetic resonance imaging (MRI) equipment in hospitals.

3.2.6 Intermediate Metal Conduit

Intermediate metal conduit (IMC) has a wall thickness that is less than RMC but greater than that of EMT. The weight of IMC is approximately ⅔ that of RMC. Because of its lower purchase price, lighter weight, and thinner walls, IMC installations are generally less expensive than comparable RMC installations. However, IMC installations still have high strength ratings.

 NOTE
Additional information on IMC may be found in *NEC Article 342.*

The outside diameter of a given size of IMC is the same as that of the comparable size of RMC. Therefore, RMC fittings may be used with IMC. Since the threads on IMC and RMC are the same size, no special threading tools are needed to thread IMC. Some electricians feel that threading

IMC is more difficult than threading RMC because IMC is somewhat harder.

The internal diameter of a given size of IMC is somewhat larger than the internal diameter of the same size of RMC because of the difference in wall thickness. Bending IMC is considered easier than bending RMC because of the reduced wall thickness. However, bending is sometimes complicated by kinking, which may be caused by the increased hardness of IMC.

The *NEC*® requires that IMC be identified along its length at 5-foot intervals with the letters IMC. *NEC Sections 110.21 and 342.120* describe this marking requirement.

Like RMC, IMC is permitted to act as an equipment grounding conductor, as defined in *NEC Section 250.118(3)*. The use of IMC may be restricted in some jurisdictions. It is important to investigate the requirements of each jurisdiction before selecting any materials.

3.2.7 Rigid Polyvinyl Chloride Conduit

The most common type of rigid nonmetallic conduit is manufactured from polyvinyl chloride (PVC). Because PVC is noncorrosive, chemically inert, and non-aging, it is often used for installation in wet or corrosive environments. Corrosion problems found with steel and aluminum RMC do not occur with PVC. However, PVC may deteriorate under some conditions, such as extreme sunlight, unless marked sunlight resistant.

All PVC is marked according to standards established by the National Electrical Manufacturers Association (NEMA) or **Underwriters Laboratories, Inc. (UL).** A section of PVC is shown in *Figure 5.*

Since PVC is lighter than steel or aluminum rigid conduit, IMC, or EMT, it is considered easier to handle. PVC can usually be installed much faster than other types of conduit because the joints are made up with cement and require no threading.

PVC contains no metal. This characteristic reduces the voltage drop of conductors carrying alternating current in PVC compared to identical conductors in steel conduit.

Because PVC is nonconducting, it cannot be used as an equipment grounding conductor. An equipment grounding conductor sized in accordance with *NEC Table 250.122* must be pulled in each PVC conductor run (except for underground service-entrance conductors).

PVC is available in lengths up to 20 feet. However, some jurisdictions require it to be cut to 10-foot lengths prior to installation. PVC is subject to expansion and contraction directly related to the difference in temperature, plus any radiating effects on the conduit. In moderate climates, even a 10-foot installation of PVC would require an expansion joint per the *NEC*®. Each straight section of conduit run must be treated independently from other sections when connected by elbows. To avoid damage to PVC caused by temperature changes, expansion couplings are used. See *Figure 6.* The inside of the coupling is sealed with one or more O-rings. This type of coupling may allow up to six inches of movement. Check the requirements of the local jurisdiction prior to installing PVC.

PVC is manufactured in two types:

- *Type EB* – Thin wall for underground use only when encased in concrete. Also referred to as Type I.
- *Type DB* – Thick wall for underground use without encasement in concrete. Also referred to as Type II.

Type DB is available in two wall thicknesses, Schedule 40 and Schedule 80.

- Schedule 40 is heavy wall for direct burial in the earth and aboveground installations.
- Schedule 80 is extra heavy wall for direct burial in the earth, aboveground installations for general applications, and installations where the conduit is subject to physical damage.

PVC is affected by higher-than-usual ambient temperatures. Support requirements for PVC are found in *NEC Table 352.30(B).* As with other conduit, it must be supported within three feet of each termination, but the maximum spacing between supports depends upon the size of the

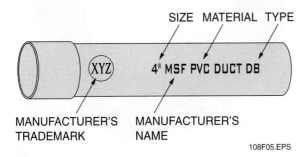

SIZE MATERIAL TYPE

4" MSF PVC DUCT DB

XYZ

MANUFACTURER'S TRADEMARK

MANUFACTURER'S NAME

108F05.EPS

Figure 5 ◆ Rigid nonmetallic conduit.

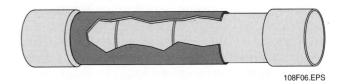

108F06.EPS

Figure 6 ◆ PVC expansion coupling.

Applications of PVC

What installations would be suitable for the use of PVC? For what situations would PVC be a poor choice?

conduit. Some of the regulations for the maximum spacing of supports are:

- ½- to 1-inch conduit: every 3 feet
- 1¼- to 2-inch conduit: every 5 feet
- 2½- to 3-inch conduit: every 6 feet
- 3½- to 5-inch conduit: every 7 feet
- 6-inch conduit: every 8 feet

3.2.8 High-Density Polyethylene Conduit

High-density polyethylene conduit (HDPE) is a rigid nonmetallic conduit listed for underground installations. See *NEC Article 353*. It is suitable for direct burial or where encased in concrete. In many signaling and communications applications, it is provided on reels with conductors pre-installed and may be laid in a trench or plowed into the earth.

NOTE

This raceway is not listed for aboveground use.

3.2.9 Liquidtight Flexible Nonmetallic Conduit

Liquidtight flexible nonmetallic conduit (LFNC) was developed as a raceway for industrial equipment where flexibility was required and protection of conductors from liquids was also necessary. This is covered by *NEC Article 356*. Usage of LFNC has been expanded from industrial applications to outside and direct burial usage where listed.

Several varieties of LFNC have been introduced. The first product (LFNC-A) is commonly referred to as hose. It consists of an inner and outer layer of neoprene with a nylon reinforcing web between the layers. A second-generation product (LFNC-B), and most widely used, consists of a smooth wall, flexible PVC with a rigid PVC integral reinforcement rod. The third product (LFNC-C) is a nylon corrugated shape without any integral reinforcements. These three

permitted LFNC raceway designs must be flame resistant with fittings **approved** for installation of electrical conductors. Nonmetallic connectors are listed for use and some liquidtight metallic flexible conduit connectors are dual-listed for both metallic and nonmetallic liquidtight flexible conduit.

LFNC is sunlight-resistant and suitable for use at conduit temperatures of 80°C dry and 60°C wet. It is available in ⅜"-inch through 4-inch sizes. *NEC Section 356.12* states that LFNC cannot be used where subject to physical damage or LFNC-A in

INSIDE TRACK

Liquidtight Conduit

Liquidtight conduit protects conductors from vapors, liquids, and solids. Liquidtight conduit that includes an inner metal core is widely used in commercial and industrial construction.

108SA02.EPS

lengths longer than 6 feet, except where properly secured, where flexibility is required, or as permitted by *NEC Section 356.10.* Also, it cannot be used to contain conductors in excess of 600 volts nominal except as permitted by *NEC Section 600.32 (A).*

Liquidtight flexible metal conduit is a raceway of circular cross section having an outer liquidtight, nonmetallic, sunlight-resistant jacket over an inner flexible metal core with associated couplings and connectors covered by *NEC Article 350.*

Flex connectors are used to connect flexible conduit to boxes or equipment. They are available in straight, 45°, and 90° configurations (*Figure 7*).

3.2.10 Flexible Metal Conduit

Flexible metal conduit, also called flex, may be used for many kinds of wiring systems. Flexible metal conduit is made from a single strip of steel or aluminum, wound and interlocked. It is typically available in sizes from ⅜ inch to 4 inches in diameter. An illustration of flexible metal conduit is shown in *Figure 8.*

Flexible metal conduit is often used to connect equipment or machines that vibrate or move slightly during operation. Also, final connection to equipment having an electrical connection point that is marginally **accessible** is often accomplished with flexible metal conduit.

Flexible metal conduit is easily bent, but the minimum bending radius is the same as for other types of conduit. It should not be bent more than the equivalent of four quarter bends (360° total) between pull points (e.g., conduit bodies and boxes). It can be connected to boxes with a flexible conduit connector and to rigid conduit or EMT by using a combination coupling.

Two types of combination couplings are shown in *Figure 9.*

Flexible metal conduit is generally available in two types: nonliquidtight and liquidtight. *NEC Articles 348 and 350* cover the uses of flexible metal conduit.

Liquidtight flexible metal conduit has an outer covering of liquidtight, sunlight-resistant flexible material that acts as a moisture seal. It is intended for use in wet locations. It is used primarily for equipment and motor connections when movement of the equipment is likely to occur. The number of bends, size, and support requirements for liquidtight conduit are the same as for all flexible conduit. Fittings used with liquidtight conduit must also be of the liquidtight type.

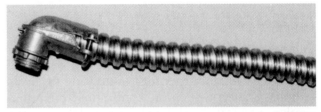

108F08.EPS

Figure 8 ◆ Flexible metal conduit.

FLEXIBLE TO EMT

FLEXIBLE TO RIGID

108F09.EPS

Figure 9 ◆ Combination couplings.

STRAIGHT CONNECTOR

45° CONNECTOR

90° CONNECTOR

108F07.EPS

Figure 7 ◆ Flex connectors.

Support requirements for flexible metal conduit are found in *NEC Sections 348.30 and 350.30.* Straps or other means of securing the flexible metal conduit must be spaced every 4½ feet and within 12 inches of each end. (This spacing is closer together than for rigid conduit.) However, at terminals where flexibility is necessary, lengths of up to 36 inches without support are permitted. Failure to provide proper support for flexible conduit can make pulling conductors difficult.

4.0.0 ◆ METAL CONDUIT FITTINGS

A large variety of conduit fittings are available to do electrical work. Manufacturers design and construct fittings to permit a multitude of applications. The type of conduit fitting used in a particular application depends upon the size and type of conduit, the type of fitting needed for the application, the location of the fitting, and the installation method. The requirements and proper applications of boxes and fittings (conduit bodies) are found in *NEC Section 300.15.* Some of the more common types of fittings are examined in the following sections.

NOTE

When using a combination coupling, be sure the flexible conduit is pushed as far as possible into the coupling. This covers the end and protects the conductors from damage.

4.1.0 Couplings

Couplings are sleeve-like fittings that are typically threaded inside to join two male threaded pieces of rigid conduit or IMC. A piece of conduit with a coupling is shown in *Figure 10.*

Other types of couplings may be used depending upon the location and type of conduit. Several types are shown in *Figure 11.*

4.2.0 Conduit Bodies

Conduit bodies, also called condulets, are a separate portion of a conduit or tubing system that provide access through a removable cover(s) to the interior of the system at a junction of two or more sections of the system, a pull point, or at a terminal point of the system. They are usually cast and are significantly higher in cost than the stamped steel boxes permitted with EMT. However, there are situations in which conduit bodies are preferable, such as in outdoor locations, for appearance's sake in an **exposed location**, or to

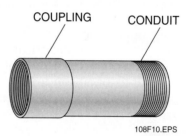

Figure 10 ◆ Conduit and coupling.

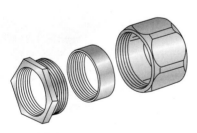

THREE-PIECE COUPLING

HINGED COUPLING

CONCRETE-TIGHT SETSCREW

EMT TO RIGID

Figure 11 ◆ Metal conduit couplings.

---

change types or sizes of raceways. Also, conduit bodies do not have to be supported, as do stamped steel boxes. They are also used when elbows or bends would not be appropriate.

NEC Section 314.16(C)(2) states that conduit bodies cannot contain **splices**, **taps**, or devices unless they are durably and legibly marked by the manufacturer with their cubic inch capacity and wire size. The maximum number of conductors permitted in a conduit body is found using *NEC Table 314.16(B)*. (See *Table 1*.)

4.2.1 Type C Conduit Bodies

Type C conduit bodies may be used to provide a pull point in a long conduit run or a conduit run that has bends totaling more than 360°. A Type C conduit body is shown in *Figure 12*.

4.2.2 Type L Conduit Bodies

When referring to conduit bodies, the letter L represents an elbow. A Type L conduit body is used as a pulling point for conduit that requires a 90° change in direction. The cover is removed, then the wire is pulled out, coiled on the ground or floor, reinserted into the other conduit body's opening, and pulled. The cover and its associated gasket are then replaced. Type L conduit bodies are available with the cover on the back (Type LB), on the sides (Type LL or LR), or on both sides (Type LRL). Several Type L conduit bodies are shown in *Figure 13*.

Table 1	Volume Required per Conductor [Data from *NEC Table 314.16(B)*]	
Size of Conductor	**Free Space Within Box for Each Conductor**
No. 18	1.5 cubic inches
No. 16	1.75 cubic inches
No. 14	2.0 cubic inches
No. 12	2.25 cubic inches
No. 10	2.5 cubic inches
No. 8	3.0 cubic inches
No. 6	5.0 cubic inches

Reprinted with permission from NFPA 70, the *National Electrical Code*®. Copyright © 2007, National Fire Protection Association, Quincy, MA 02269. This reprinted material is not the complete and official position of the National Fire Protection Association on the referenced subject, which is represented only by the standard in its entirety.

108F12.EPS

Figure 12 ◆ Type C conduit body.

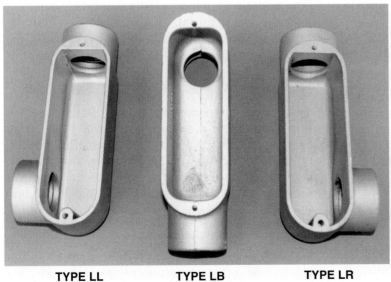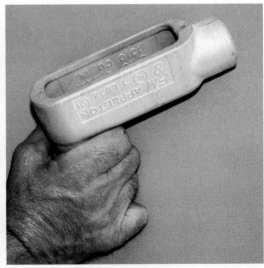

| TYPE LL | TYPE LB | TYPE LR | TYPE LB |

108F13A.EPS

108F13B.EPS

Figure 13 ◆ Type L conduit bodies and how to identify them.

To identify Type L conduit bodies, use the following method:

Step 1 Hold the body like a pistol.

Step 2 Locate the opening on the body:
- If the opening is to the left, it is a Type LL.
- If the opening is to the right, it is a Type LR.
- If the opening is on top (back), it is a Type LB.
- If there are openings on both the left and the right, it is a Type LRL.

4.2.3 Type T Conduit Bodies

Type T conduit bodies are used to provide a junction point for three intersecting conduits and are used extensively in conduit systems. A Type T conduit body is shown in *Figure 14*.

4.2.4 Type X Conduit Bodies

Type X conduit bodies are used to provide a junction point for four intersecting conduits. The removable cover provides access to the interior of the X so that wire pulling and splicing may be performed. A Type X conduit body is shown in *Figure 15*.

4.2.5 Threaded Weatherproof Hub

Threaded weatherproof hubs are used for conduit entering a box in a wet location. *Figure 16* shows typical threaded weatherproof hubs.

Figure 14 ◆ Type T conduit body.

108F15.EPS

Figure 15 ◆ Type X conduit body.

108F16.EPS

Figure 16 ◆ Threaded weatherproof hubs.

4.3.0 Insulating Bushings

An insulating bushing is either nonmetallic or has an insulated throat. Insulating bushings are installed on the threaded end of conduit that enters a sheet metal enclosure.

4.3.1 Nongrounding Insulating Bushings

The purpose of a nongrounding insulating bushing is to protect the conductors from being damaged by the sharp edges of the threaded conduit end. *NEC Section 300.15(C)* states that where a conduit enters a box, fitting, or other enclosure, a bushing must be provided to protect the wire from abrasion unless the design of the box, fitting, or enclosure is such as to afford equivalent protection. *NEC Section 312.6(C)* references *Section 300.4(G),* which states that where ungrounded conductors of No. 4 or larger enter a raceway in a cabinet or box enclosure, the conductors shall be protected by a substantial fitting providing a smoothly rounded insulating surface, unless the conductors are separated from the raceway fitting by substantial insulating material securely fastened in place. An exception is where threaded hubs or bosses that are an integral part of a cabinet, box enclosure, or raceway provide a smoothly rounded or flared entry for conductors. Insulating bushings are shown in *Figure 17*.

4.3.2 Grounding Insulating Bushings

Grounded insulating bushings, usually called grounding bushings, are used to protect conductors and also have provisions for connection of an equipment grounding conductor. The ground wire, once connected to the grounding bushing, may be connected to the enclosure to which the conduit is connected. Grounding insulating bushings are shown in *Figure 18*.

4.4.0 Offset Nipples

Offset nipples are used to connect two pieces of electrical equipment in close proximity where a slight offset is required. They come in sizes ranging from ½" to 2" in diameter. See *Figure 19*.

108F18.EPS

Figure 18 ◆ Grounding insulating bushings.

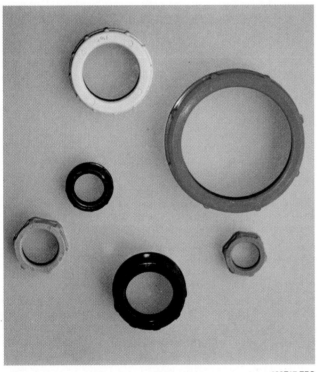

108F17.EPS

Figure 17 ◆ Insulating bushings.

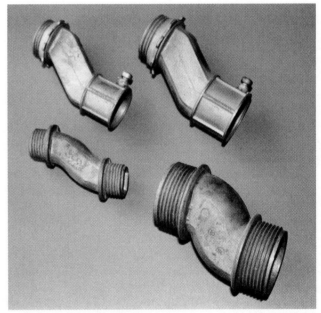

108F19.EPS

Figure 19 ◆ Offset nipples.

Installation of Conduit Bodies

It will be much easier to identify conduit bodies once you begin to see them in use. Here we show liquidtight nonmetallic conduit entering a Type T conduit body (A) and a Type LB conduit body in an outdoor commercial application (B).

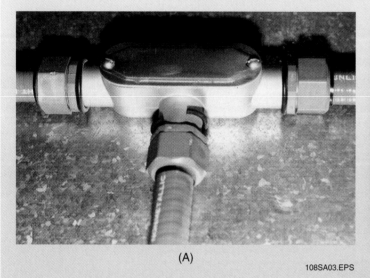

(A)

108SA03.EPS

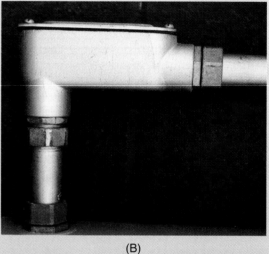

(B)

108SA04.EPS

5.0.0 ◆ MAKING A CONDUIT-TO-BOX CONNECTION

Conduit is joined to boxes by connectors, adapters, threaded hubs, or locknuts.

Bushings protect the wires from the sharp edges of the conduit. As previously discussed, bushings are usually made of plastic or metal. Some metal bushings have a grounding screw to permit a bonding wire to be installed.

Locknuts (*Figure 20*) are used on the inside and outside walls of the box to which the conduit is connected. A grounding locknut may be needed if a bonding wire is to be installed. Special sealing locknuts are also used in wet locations.

When joining metal conduit to metal boxes, a means must be provided in each metal box for the connection of an equipment grounding conductor. The means shall be permitted to be a tapped hole or equivalent per *NEC Section 314.40(D)*.

A proper conduit-to-box connection is shown in *Figure 21*.

In order to make a good connection, use the following procedure:

Step 1 Thread the external locknut onto the conduit. Run the locknut to the bottom of the threads.

Step 2 Insert the conduit into the box opening.

Step 3 If an inside locknut or grounding locknut is required, screw it onto the conduit inside the box opening.

Step 4 Screw the bushing onto the threads projecting into the box opening. Make sure the bushing is tightened as much as possible.

Step 5 Tighten the external locknut to secure the conduit to the box.

SEALING LOCKNUT

STANDARD LOCKNUT

STANDARD LOCKNUT

GROUNDING LOCKNUT

108F20.EPS

Figure 20 ◆ Locknuts.

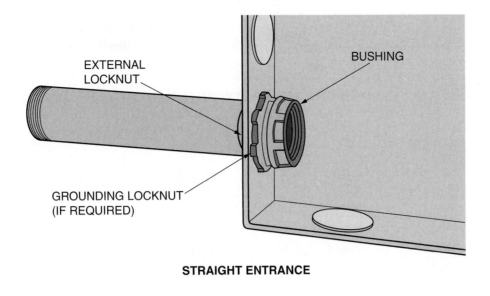

EXTERNAL
LOCKNUT

BUSHING

GROUNDING LOCKNUT
(IF REQUIRED)

STRAIGHT ENTRANCE

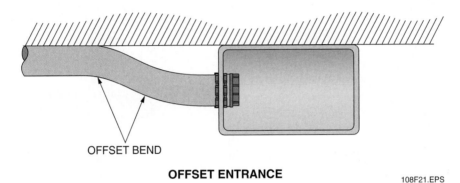

OFFSET BEND

OFFSET ENTRANCE

108F21.EPS

Figure 21 ◆ Conduit-to-box connections.

It is important that the bushings and locknuts fit tightly. For this reason, the conduit must enter straight into the box. This may require that a box offset or kick be made in the conduit.

6.0.0 ◆ SEALING FITTINGS

Hazardous locations in manufacturing plants and other industrial facilities involve a wide variety of flammable gases and vapors and ignitable dusts. These hazardous substances have widely different flash points, ignition temperatures, and flammable limits requiring fittings that can be sealed. Sealing fittings are installed in conduit runs to minimize the passage of gases, vapors, or flames through the conduit and reduce the accumulation of moisture. They are required by *NEC Article 500* in hazardous locations where explosions may occur. They are also required where conduit passes from a hazardous location of one classification to another or to an unclassified location. Several types of sealing fittings are shown in *Figure 22*.

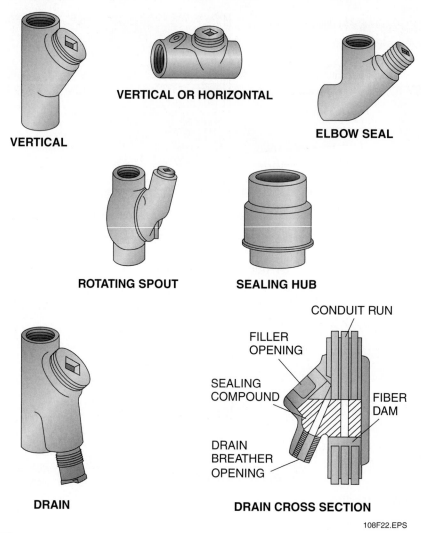

VERTICAL OR HORIZONTAL

VERTICAL

ELBOW SEAL

ROTATING SPOUT

SEALING HUB

DRAIN

DRAIN CROSS SECTION

CONDUIT RUN

FILLER OPENING

SEALING COMPOUND

FIBER DAM

DRAIN BREATHER OPENING

108F22.EPS

Figure 22 ◆ Sealing fittings.

7.0.0 ◆ FASTENERS AND ANCHORS

Conduit and other types of raceways used to carry wiring and cables must be properly supported. This generally means attaching the raceway to the building structure. Depending on the type of construction, the raceways may have to be attached to wood, concrete, or metal. Each of these materials requires the use of fasteners designed for the specific use. Using the wrong fastener, or installing the right fastener incorrectly, can lead to a failure of the raceway support.

The project specifications and manufacturer's installation instructions may specify the type and size of fasteners to use and how to install them. In other instances, the electrician will be expected to select the right type of fastener for a given application. It is therefore important that every electrician be familiar with the different types of fasteners, their uses, and their limitations.

7.1.0 Tie Wraps

A tie wrap is a one-piece, self-locking cable tie, usually made of nylon, that is used to fasten a bundle of wires and cables together. Tie wraps can be quickly installed either manually or using a special installation tool. Black tie wraps resist ultraviolet light and are recommended for outdoor use.

Tie wraps are made in standard, cable strap and clamp, and identification configurations (*Figure 23*). All types function to clamp bundled wires or cables together. In addition, the cable strap and clamp has a molded mounting hole in the head used to secure the tie with a rivet, screw, or bolt after the tie wrap has been installed around the wires or cable. Identification tie wraps have a large flat area provided for imprinting or writing cable identification information. There is also a releasable version available. It is a non-permanent

Installing Sealing Fittings

These fittings must be sealed after the wires are pulled. A fiber dam is first packed into the base of the fitting between and around the conductors, then the liquid sealing compound is poured into the fitting.

108SA05.EPS

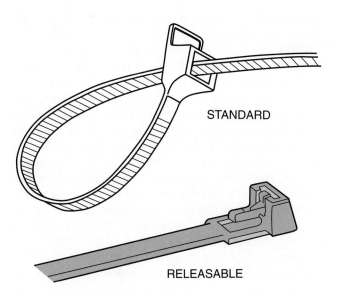

STANDARD

RELEASABLE

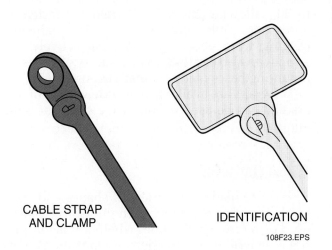

CABLE STRAP
AND CLAMP

IDENTIFICATION

108F23.EPS

Figure 23 ◆ Tie wraps.

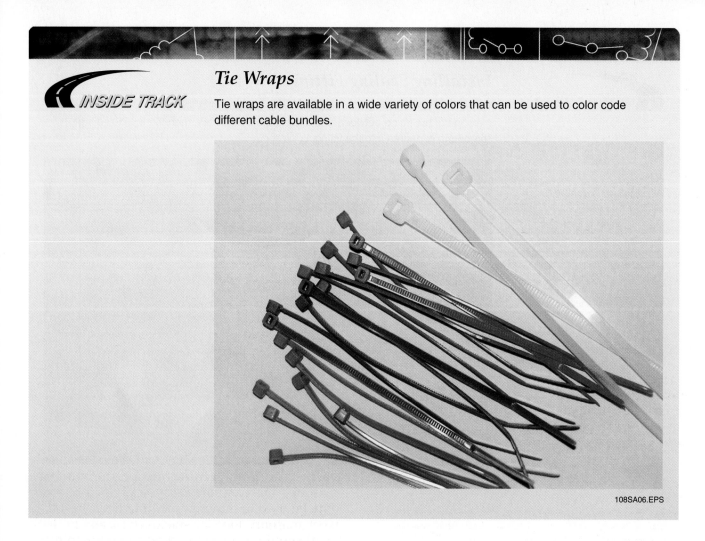

INSIDE TRACK

Tie Wraps

Tie wraps are available in a wide variety of colors that can be used to color code different cable bundles.

108SA06.EPS

tie used for bundling wires or cables that may require frequent additions or deletions. Cable ties are made in various lengths ranging from about 3" to 30", allowing them to be used for fastening wires and cables into bundles with diameters ranging from about ½" to 9", respectively. Tie wraps can also be attached to a variety of adhesive mounting bases made for that purpose.

Tie wraps are available in a wide variety of colors that can be used to color code different cable bundles.

7.2.0 Screws

Screws are made in a variety of shapes and sizes for different fastening jobs. The finish or coating used on a screw determines whether it is for interior or exterior use, corrosion resistant, etc. Screws of all types have heads with different shapes and slots similar to those previously described for machine screws. Some have machine threads and are self-drilling. The size or diameter of a screw body or shank is given in gauge numbers ranging from No. 0 to No. 24, and in fractions of an inch for

screws with diameters larger than ¼". The higher the gauge number, the larger the diameter of the shank. Screw lengths range from ¼" to 6", measured from the tip to the part of the head that is flush to the surface when driven in. When choosing a screw for an application, you must consider the type and thickness of the materials to be fastened, the size of the screw, the material it is made of, the shape of its head, and the type of driver. Because of the wide diversity in the types of screws and their application, always follow the manufacturer's recommendation to select the right screw for the job. To prevent damage to the screw head or the material being fastened, always use a screwdriver or power driver bit with the proper size and shape tip to fit the screw.

Some of the more common types of screws are:

- Wood screws
- Lag screws
- Masonry/concrete screws
- Thread-forming and thread-cutting screws
- Deck screws
- Drywall screws
- Drive screws

7.2.1 Wood Screws

Wood screws (*Figure 24*) are typically used to fasten boxes, panel enclosures, etc. to wood framing or structures where greater holding power is needed than can be provided by nails. They are also used to fasten equipment to wood in applications where it may occasionally need to be unfastened and removed. Wood screws are commonly made in lengths from ¼" to 4", with shank gauge sizes ranging from 0 to 24. The shank size used is normally determined by the size hole provided in the box, panel, etc. to be fastened. When determining the length of a wood screw to use, a good rule of thumb is to select screws long enough to allow about ⅔ of the screw length to enter the piece of wood that is being gripped.

7.2.2 Lag Screws and Shields

Lag screws (*Figure 25*) or lag bolts are heavy-duty wood screws with square- or hex-shaped heads that provide greater holding power. Lag screws with diameters ranging between ¼" and ½" and lengths ranging from 1" to 6" are common. They are typically used to fasten heavy equipment to wood, but can also be used to fasten equipment to concrete when a lag shield is used.

A lag shield is a lead tube that is split lengthwise but remains joined at one end. It is placed in a predrilled hole in the concrete. When a lag screw

is screwed into the lag shield, the shield expands in the hole, firmly securing the lag screw. In hard masonry, short lag shields (typically 1" to 2" long) may be used to minimize drilling time. In soft or weak masonry, long lag shields (typically 1½" to 3" long) should be used to achieve maximum holding strength.

Make sure to use the proper length lag screw to achieve proper expansion. The length of the lag screw used should be equal to the thickness of the component being fastened plus the length of the lag shield. Also, drill the hole in the masonry to a depth approximately ½" longer than the shield being used. If the head of a lag screw rests directly

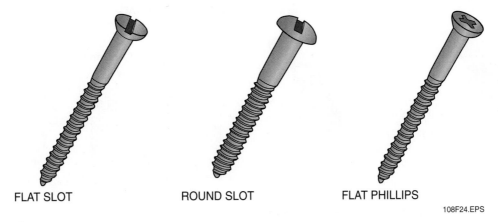

FLAT SLOT   ROUND SLOT   FLAT PHILLIPS

108F24.EPS

Figure 24 ◆ Wood screws.

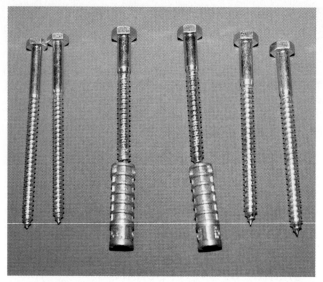

Figure 25 ◆ Lag screws and shields.

108F25.EPS

on wood when installed, a flat washer should be placed under the head to prevent the head from digging into the wood as the lag screw is tightened down. Be sure to take the thickness of any washers used into account when selecting the length of the screw.

7.2.3 Concrete/Masonry Screws

Concrete/masonry screws (*Figure 26*), commonly called self-threading anchors, are used to fasten a device or fixture to concrete, block, or brick. No anchor is needed. To provide a matched tolerance anchoring system, the screws are installed using specially designed carbide drill bits and installation tools made for use with the screws. These tools are typically used with a standard rotary drill hammer. The installation tool, along with an appropriate drive socket or bit, is used to drive the screws directly into predrilled holes that have a diameter and depth specified by the screw manu-

facturer. When being driven into the concrete, the widely spaced threads on the screws cut into the walls of the hole to provide a tight friction fit. Most types of concrete/masonry screws can be removed and reinstalled to allow for shimming and leveling of the fastened device.

7.2.4 Thread-Forming and Thread-Cutting Screws

Thread-forming screws (*Figure 27*), commonly called sheet metal screws, are made of hard metal. They form a thread as they are driven into the work. This thread-forming action eliminates the need to tap a hole before installing the screw. To achieve proper holding, it is important to make sure to use the proper size bit when drilling pilot holes for thread-forming screws. The correct drill bit size used for a specific size screw is usually marked on the box containing the screws. Some types of thread-forming screws also drill their own holes, eliminating drilling, punching, and aligning parts. Thread-forming screws are primarily used to fasten light-gauge metal parts together. They are made in the same diameters and lengths as wood screws.

Hardened steel thread-cutting metal screws with blunt points and fine threads (*Figure 28*) are used to join heavy-gauge metals, metals of different gauges, and nonferrous metals. They are also used to fasten sheet metal to building structural members. These screws are made of hardened steel that is harder than the metal being tapped. They cut threads by removing and cutting a portion of the metal as they are driven into a pilot hole and through the material.

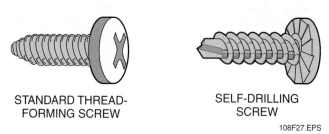

STANDARD THREAD-FORMING SCREW   SELF-DRILLING SCREW

108F27.EPS

Figure 27 ◆ Thread-forming screws.

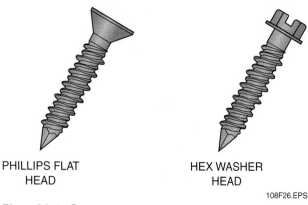

PHILLIPS FLAT HEAD   HEX WASHER HEAD

108F26.EPS

Figure 26 ◆ Concrete screws.

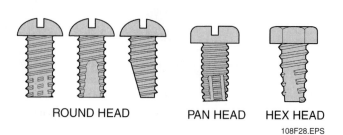

ROUND HEAD   PAN HEAD   HEX HEAD

108F28.EPS

Figure 28 ◆ Thread-cutting screws.

Self-Drilling Screws

Can you name an electrical application for self-drilling screws?

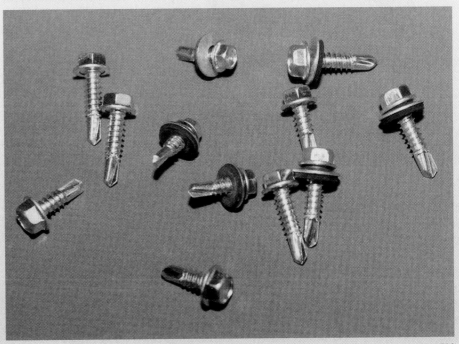

108SA07.EPS

7.2.5 Drywall Screws

Drywall screws (*Figure 29*) are thin, self-drilling screws with bugle-shaped heads. Depending on the type of screw, it cuts through the wallboard and anchors itself into wood and/or metal studs, holding the wallboard tight to the stud. Coarse thread screws are normally used to fasten wallboard to wood studs. Fine thread and high-and-low thread types are generally used for fastening to metal studs. Some screws are made for use in either wood or metal. A Phillips or Robertson drive head allows the drywall screw to be countersunk without tearing the surface of the wallboard.

7.2.6 Drive Screws

Drive screws do not require that the hole be tapped. They are installed by hammering the screw into a drilled or punched hole of the proper size. Drive screws are mostly used to fasten parts that will not be exposed to much pressure. A typical use of drive screws is to attach permanent name plates on electric motors and other types of equipment. *Figure 30* shows typical drive screws.

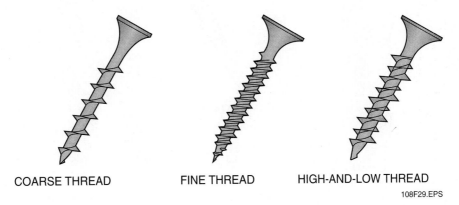

COARSE THREAD   FINE THREAD   HIGH-AND-LOW THREAD

108F29.EPS

Figure 29 ◆ Drywall screws.

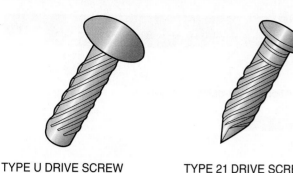

TYPE U DRIVE SCREW   TYPE 21 DRIVE SCREW

108F30.EPS

Figure 30 ◆ Drive screws.

7.3.0 Hammer-Driven Pins and Studs

Hammer-driven pins or threaded studs (*Figure 31*) can be used to fasten wood or steel to concrete or block without the need to predrill holes. The pin or threaded stud is inserted into a hammer-driven tool designed for its use. The pin or stud is inserted in the tool point end out with the washer seated in the recess. The pin or stud is then positioned against the base material where it is to be fastened and the drive rod of the tool tapped lightly until the striker pin contacts the pin or stud. Following this, the tool's drive rod is struck using heavy blows with about a two-pound engineer's hammer. The force of the hammer blows is transmitted through the tool directly to the head of the fastener, causing it to be driven into the concrete or block. For best results, the drive pin or stud should be embedded a minimum of ½" in hard concrete to 1¼" in softer concrete block.

7.4.0 Powder-Actuated Tools and Fasteners

Powder-actuated tools (*Figure 32*) can be used to drive a wide variety of specially designed pin and threaded stud-type fasteners into masonry and steel. These tools look and fire like a gun and use the force of a detonated gunpowder load (typically .22, .25, or .27 caliber) to drive the fastener into the material. The depth to which the pin or stud is driven is controlled by the density of the base material in which the pin or stud is being installed and by the power level or strength of the cased powder load.

Powder loads and their cases are designed for use with specific types and/or models of powder-actuated tools and are not interchangeable. Typically, powder loads are made in 12 increasing power or load levels used to achieve the proper penetration. The different power levels are identified by a color-code system and load case types. Note that different manufacturers may use different color codes to identify load strength. Power level 1 is the lowest power level while 12 is the highest. Higher number power levels are used when driving into hard materials or when a deeper penetration is needed. Powder loads are available as single-shot units for use with single-shot tools. They are also made in multi-shot strips or disks for semiautomatic tools.

 WARNING!

Powder-actuated fastening tools are to be used only by trained and licensed operators and in accordance with the tool operator's manual. You must carry your license with you whenever you are using a powder-actuated tool.

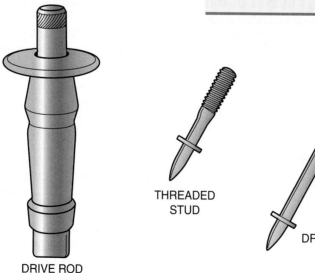

DRIVE ROD

THREADED STUD

DRIVE PINS

108F31.EPS

Figure 31 ◆ Hammer-driven pins and installation tool.

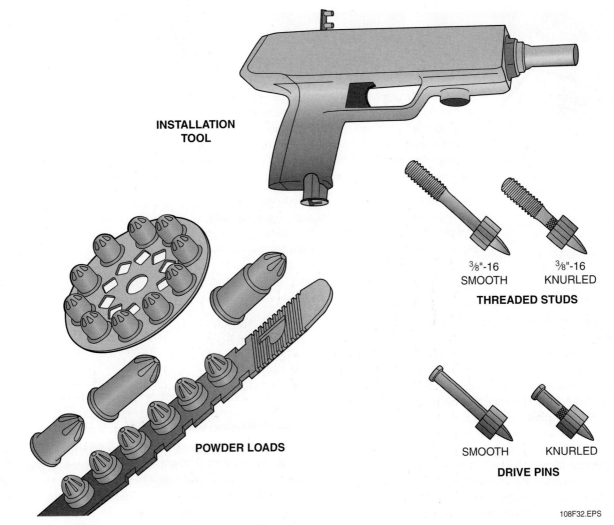

INSTALLATION TOOL

⅜"-16 SMOOTH

⅜"-16 KNURLED

THREADED STUDS

POWDER LOADS

SMOOTH

KNURLED

DRIVE PINS

108F32.EPS

Figure 32 ◆ Powder-actuated installation tools and fasteners.

OSHA Standard 29 CFR 1926.302(e) governs the use of powder-actuated tools and states that only those individuals who have been trained in the operation of the particular powder-actuated tool in use be allowed to operate it. Authorized instructors available from the various powder-actuated tool manufacturers generally provide such training and licensing. Trained operators must take precautions to protect both themselves and others in the area when using a powder-actuated driver tool:

• Always use the tool in accordance with the published tool operation instructions. The instructions should be kept with the tool. Never attempt to override the safety features of the tool.
• Never place your hand or other body parts over the front muzzle end of the tool.
• Use only fasteners, powder loads, and tool parts specifically made for use with the tool. Use of other materials can cause improper and unsafe functioning of the tool.

• Operators and bystanders must wear eye and hearing protection along with hard hats. Other personal safety gear, as required, must also be used.
• Always post warning signs that state *Powder-Actuated Tool in Use* within 50 feet of the area where tools are used.
• Before using a tool, make sure it is unloaded and perform a proper function test. Check the functioning of the unloaded tool as described in the published tool operation instructions.
• Do not guess before fastening into any base material; always perform a center punch test.
• Always make a test firing into a suitable base material with the lowest power level recommended for the tool being used. If this does not set the fastener, try the next higher power level. Continue this procedure until the proper fastener penetration is obtained.
• Always point the tool away from operators or bystanders.

Powder-Actuated Tools

A 22-year-old apprentice was killed when he was struck in the head by a nail fired from a powder-actuated tool in an adjacent room. The tool operator was attempting to anchor plywood to a hollow wall and fired the gun, causing the nail to pass through the wall, where it traveled nearly thirty feet before striking the victim. The tool operator had never received training in the proper use of the tool, and none of the employees in the area were wearing personal protective equipment.

The Bottom Line: Never use a powder-actuated tool to secure fasteners into easily penetrated materials; these tools are designed primarily for installing fasteners into masonry. The use of powder-actuated tools requires special training and certification. In addition, all personnel in the area must be aware that the tool is in use and should be wearing appropriate personal protective equipment.

- Never use the tool in an explosive or flammable area.
- Never leave a loaded tool unattended. Do not load the tool until you are prepared to complete the fastening. Should you decide not to make a fastening after the tool has been loaded, always remove the powder load first, then the fastener. Always unload the tool before cleaning or servicing, when changing parts, prior to work breaks, and when storing the tool.
- Always hold the tool perpendicular to the work surface and use the spall (chip or fragment) guard or stop spall whenever possible.
- Always follow the required spacing, edge distance, and base material thickness requirements.
- Never fire through an existing hole or into a weld area.
- In the event of a misfire, always hold the tool depressed against the work surface for at least 30 seconds. If the tool still does not fire, follow the published tool instructions. Never carelessly discard or throw unfired powder loads into a trash receptacle.
- Always store the powder loads and unloaded tool under lock and key.

7.5.0 Mechanical Anchors

Mechanical anchors are devices used to give fasteners a firm grip in a variety of materials, where the fasteners by themselves would otherwise have a tendency to pull out. Anchors can be classified in many ways by different manufacturers. In this module, anchors have been divided into five broad categories:

- One-step anchors
- Bolt anchors
- Screw anchors
- Self-drilling anchors
- Hollow-wall anchors

7.5.1 One-Step Anchors

One-step anchors are designed so that they can be installed through the mounting holes in the component to be fastened. This is because the anchor and the drilled hole into which it is installed have the same size diameter. They come in various diameters ranging from ¼" to 1¼" with lengths ranging from 1¾" to 12". Wedge, stud, sleeve, one-piece, screw, and nail anchors (*Figure 33*) are common types of one-step anchors.

- *Wedge anchors* – Wedge anchors are heavy-duty anchors supplied with nuts and washers. The drill bit size used to drill the hole is the same diameter as the anchor. The depth of the hole is not critical as long as the minimum length recommended by the manufacturer is drilled. After the hole is blown clean of dust and other material, the anchor is inserted into the hole and driven with a hammer far enough so that at least six threads are below the top surface of the component. Then, the component is fastened by tightening the anchor nut to expand the anchor and tighten it in the hole.
- *Stud bolt anchors* – Stud bolt anchors are heavy-duty threaded anchors. Because this type of anchor is made to bottom in its mounting hole, it is a good choice to use when jacking or leveling of the fastened component is needed. The depth of the hole drilled in the masonry must

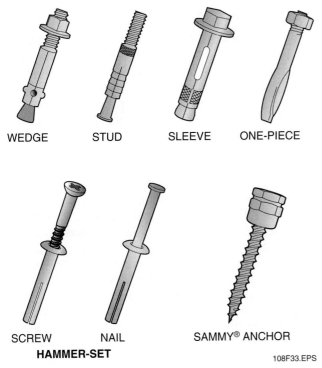

WEDGE   STUD   SLEEVE   ONE-PIECE

SCREW   NAIL   SAMMY® ANCHOR
HAMMER-SET

108F33.EPS

Figure 33 ◆ One-step anchors.

be as specified by the manufacturer in order to achieve proper expansion. After the hole is blown clean of dust and other material, the anchor is inserted in the hole with the expander plug end down. Following this, the anchor is driven into the hole with a hammer (or setting tool) to expand the anchor and tighten it in the hole. The anchor is fully set when it can no longer be driven into the hole. The component is fastened using the correct size and thread bolt for use with the anchor stud.

- *Sleeve anchors* – Sleeve anchors are multipurpose anchors. The depth of the anchor hole is not critical as long as the minimum length recommended by the manufacturer is drilled. After the hole is blown clean of dust and other material, the anchor is inserted into the hole and tapped until flush with the component. Then, the anchor nut or screw is tightened to expand the anchor and tighten it in the hole.
- *One-piece anchors* – One-piece anchors are multi-purpose anchors. They work on the principle that as the anchor is driven into the hole, the spring force of the expansion mechanism is compressed and flexes to fit the size of the hole. Once set, it tries to regain its original shape. The depth of the hole drilled in the masonry must be at least ½" deeper than the required embedment. The proper depth is crucial. Overdrilling is as bad as underdrilling. After the hole is

blown clean of dust and other material, the anchor is inserted through the component and driven with a hammer into the hole until the head is firmly seated against the component. It is important to make sure that the anchor is driven to the proper embedment depth. Note that manufacturers also make specially designed drivers and manual tools that are used instead of a hammer to drive one-piece anchors. These tools allow the anchors to be installed in confined spaces and help prevent damage to the component from stray hammer blows.

- *Hammer-set anchors* – Hammer-set anchors are made for use in concrete and masonry. There are two types: nail and screw. An advantage of the screw-type anchors is that they are removable. Both types have a diameter the same size as the anchoring hole. For both types, the anchor hole must be drilled to the diameter of the anchor and to a depth of at least ¼" deeper than that required for embedment. After the hole is blown clean of dust and other material, the anchor is inserted into the hole through the mounting holes in the component to be fastened; then the screw or nail is driven into the anchor body to expand it. It is important to make sure that the head is seated firmly against the component and is at the proper embedment.
- *Threaded-rod anchors* – Threaded-rod anchors, such as the Sammy® anchor, are available for installation in concrete, steel, or wood. The anchor is designed to support a threaded rod, which is screwed into the head of the anchor after the anchor is installed. A special nut driver is available for installing the screws.

7.5.2 Bolt Anchors

Bolt anchors are designed to be installed flush with the surface of the base material. They are used in conjunction with threaded machine bolts or screws. In some types, they can be used with threaded rod. Drop-in, single and double expansion, and caulk-in anchors (*Figure 34*) are commonly used types of bolt anchors.

- *Drop-in anchors* – Drop-in anchors are typically used as heavy-duty anchors. There are two types of drop-in anchors. The first type, made for use in solid concrete and masonry, has an internally threaded expansion anchor with a preassembled internal expander plug. The anchor hole must be drilled to the specific diameter and depth specified by the manufacturer. After

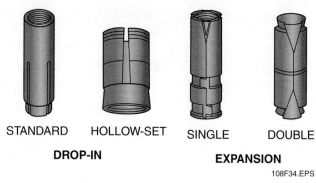

STANDARD   HOLLOW-SET   SINGLE   DOUBLE

DROP-IN   **EXPANSION**

108F34.EPS

Figure 34 ◆ Bolt anchors.

the hole is blown clean of dust and other material, the anchor is inserted into the hole and tapped until it is flush with the surface. Following this, a setting tool supplied with the anchor is driven into the anchor to expand it. The component to be fastened is positioned in place and fastened by threading and tightening the correct size machine bolt or screw into the anchor.

The second type, called a hollow set drop-in anchor, is made for use in hollow concrete and masonry base materials. Hollow set drop-in anchors have a slotted, tapered expansion sleeve and a serrated expansion cone. They come in various lengths compatible with the outer wall thickness of most hollow base materials. They can also be used in solid concrete and masonry. The anchor hole must be drilled to the specific diameter specified by the manufacturer. When installed in hollow base materials, the hole is drilled into the cell or void. After the hole is blown clean of dust and other material, the anchor is inserted into the hole and tapped until it is flush with the surface. Following this, the component to be fastened is positioned in place; then the proper size machine bolt or screw is threaded into the anchor and tightened to expand the anchor in the hole.

- *Single- and double-expansion anchors* – Single- and double-expansion anchors are both made for use in concrete and other masonry. The double-expansion anchor is used mainly when fastening into concrete or masonry of questionable strength. For both types, the anchor hole must be drilled to the specific diameter and depth specified by the manufacturer. After the hole is blown clean of dust and other material, the anchor is inserted into the hole, threaded cone end first. It is then tapped until it is flush with the surface. Following this, the component to be fastened is positioned in place; then the proper size machine bolt or screw is threaded into the anchor and tightened to expand the anchor in the hole.

7.5.3 Screw Anchors

Screw anchors are lighter-duty anchors made to be installed flush with the surface of the base material. They are used in conjunction with sheet metal, wood, or lag screws depending on the anchor type. Fiber, lead, and plastic anchors are common types of screw anchors (*Figure 35*). The lag shield anchor used with lag screws was described earlier in this module.

Fiber, lead, and plastic anchors are typically used in concrete and masonry. Plastic anchors are also commonly used in wallboard and similar base materials. The installation of all types is simple. The anchor hole must be drilled to the diameter specified by the manufacturer. The minimum depth of the hole must equal the anchor length. After the hole is blown clean of dust and other material, the anchor is inserted into the hole and tapped until it is flush with the surface. Following this, the component to be fastened is positioned in place; then the proper type and size screw is driven through the component mounting hole and into the anchor to expand the anchor in the hole.

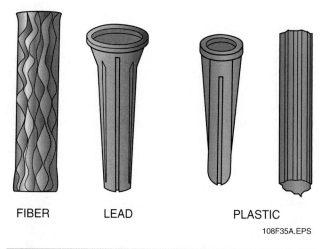

FIBER   LEAD   PLASTIC

108F35A.EPS

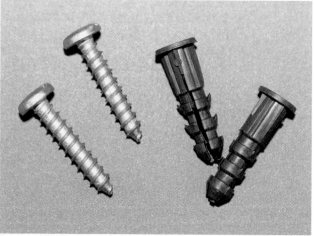

108F35B.EPS

Figure 35 ◆ Screw anchors and screws.

7.5.4 Self-Drilling Anchors

Some anchors made for use in masonry are self-drilling anchors. *Figure 36* is typical of those in common use. This fastener has a cutting sleeve that is first used as a drill bit and later becomes the expandable fastener itself. A rotary hammer is used to drill the hole in the concrete using the anchor sleeve as the drill bit. After the hole is drilled, the anchor is pulled out and the hole cleaned. This is followed by inserting the anchor's expander plug into the cutting end of the sleeve. The anchor sleeve and expander plug are driven back into the hole with the rotary hammer until they are flush with the surface of the concrete. As the fastener is hammered down, it hits the bottom, where the tapered expander causes the fastener to expand and lock into the hole. The anchor is then snapped off at the shear point with a quick lateral movement of the hammer. The component to be fastened can then be attached to the anchor using the proper size bolt.

7.6.0 Guidelines for Drilling Anchor Holes in Hardened Concrete or Masonry

When selecting masonry anchors, regardless of the type, always take into consideration and follow the manufacturer's recommendations pertaining to hole diameter and depth, minimum embedment in concrete, maximum thickness of material to be fastened, and the pullout and shear load capacities.

When installing anchors and/or anchor bolts in hardened concrete, make sure the area where the equipment or component is to be fastened is smooth so that it will have solid footing. Uneven footing might cause the equipment to twist, warp, not tighten properly, or vibrate when in operation. Before starting, carefully inspect the rotary hammer or hammer drill and the drill bit(s) to ensure they are in good operating condition. Be sure to use the type of carbide-tipped

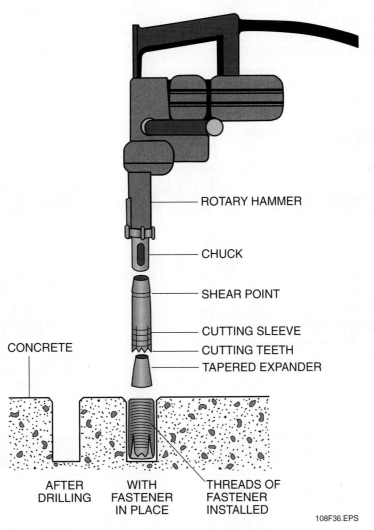

ROTARY HAMMER

CHUCK

SHEAR POINT

CUTTING SLEEVE
CUTTING TEETH
TAPERED EXPANDER

CONCRETE

AFTER DRILLING

WITH FASTENER IN PLACE

THREADS OF FASTENER INSTALLED

108F36.EPS

Figure 36 ◆ Self-drilling anchor.

masonry or percussion drill bits recommended by the drill/hammer or anchor manufacturer because these bits are made to take the higher impact of the masonry materials. Also, it is recommended that the drill or hammer tool depth gauge be set to the depth of the hole needed. The trick to using masonry drill bits is not to force them into the material by pushing down hard on the drill. Use a little pressure and let the drill do the work. For large holes, start with a smaller bit, then change to a larger bit.

The methods for installing the different types of anchors in hardened concrete or masonry were briefly described in the sections above. Always install the selected anchors according to the manufacturer's directions. Here is an example of a typical procedure used to install many types of expansion anchors in hardened concrete or masonry.

Refer to *Figure 37* as you study the procedure.

WARNING!

Drilling in concrete generates noise, dust, and flying particles. Always wear safety goggles, ear protectors, and gloves. Make sure other workers in the area also wear protective equipment.

Step 1 Drill the anchor bolt hole the same size as the anchor bolt. The hole must be deep enough for six threads of the bolt to be below the surface of the concrete (see *Figure 37, Step 1*). Clean out the hole using a squeeze bulb.

Step 2 Drive the anchor bolt into the hole using a hammer (*Figure 37, Step 2*). Protect the threads of the bolt with a nut that does not allow any threads to be exposed.

Step 3 Put a washer and nut on the bolt, and tighten the nut with a wrench until the anchor is secure in the concrete (*Figure 37, Step 3*).

7.7.0 Hollow-Wall Anchors

Hollow-wall anchors are used in hollow materials such as concrete plank, block, structural steel, wallboard, and plaster. Some types can also be used in solid materials. Toggle bolts, sleeve-type wall anchors, wallboard anchors, and metal drive-in anchors are common anchors used when fastening to hollow materials.

When installing anchors in hollow walls or ceilings, regardless of the type, always follow the

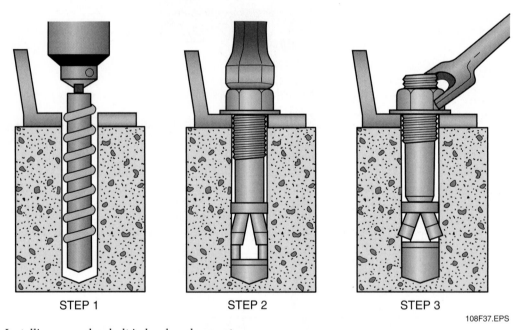

STEP 1   STEP 2   STEP 3

108F37.EPS

Figure 37 ◆ Installing an anchor bolt in hardened concrete.

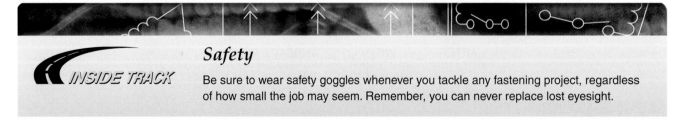

INSIDE TRACK

Safety

Be sure to wear safety goggles whenever you tackle any fastening project, regardless of how small the job may seem. Remember, you can never replace lost eyesight.

manufacturer's recommendations pertaining to use, hole diameter, wall thickness, grip range (thickness of the anchoring material), and the pullout and shear load capacities.

7.7.1 Toggle Bolts

Toggle bolts (*Figure 38*) are used to fasten equipment, hangers, supports, and similar items into hollow surfaces such as walls and ceilings. They consist of a slotted bolt or screw and spring-loaded wings. When inserted through the item to be fastened, then through a predrilled hole in the wall or ceiling, the wings spring apart and provide a firm hold on the inside of the hollow wall or ceiling as the bolt is tightened. Note that the hole drilled in the wall or ceiling should be just large enough for the compressed wing-head to pass through. Once the toggle bolt is installed, be careful not to completely unscrew the bolt because the wings will fall off, making the fastener useless. Screw-actuated plastic toggle bolts are also made. These are similar to metal toggle bolts, but they come with a pointed screw and do not require as large a hole. Unlike the metal version, the plastic wings remain in place if the screw is removed.

Toggle bolts are used to fasten a part to hollow block, wallboard, plaster, panel, or tile. The following general procedure can be used to install toggle bolts.

WARNING!
Follow all safety precautions when using an electric drill.

Step 1 Select the proper size drill bit or punch and toggle bolt for the job.

Step 2 Check the toggle bolt for damaged or dirty threads or a malfunctioning wing mechanism.

Step 3 Drill a hole completely through the surface to which the part is to be fastened.

Step 4 Insert the toggle bolt through the opening in the item to be fastened.

Step 5 Screw the toggle wing onto the end of the toggle bolt, ensuring that the flat side of the toggle wing is facing the bolt head.

Step 6 Fold the wings completely back and push them through the drilled hole until the wings spring open.

Step 7 Pull back on the item to be fastened in order to hold the wings firmly against the inside surface to which the item is being attached.

Step 8 Tighten the toggle bolt with a screwdriver until it is snug.

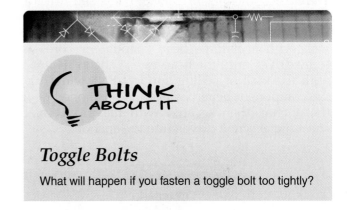

THINK ABOUT IT

Toggle Bolts

What will happen if you fasten a toggle bolt too tightly?

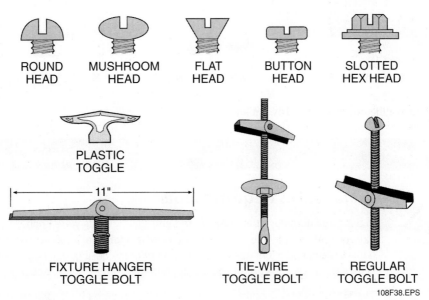

Figure 38 ◆ Toggle bolts.

7.7.2 Sleeve-Type Wall Anchors

Sleeve-type wall anchors (*Figure 39*) are suitable for use in concrete, block, plywood, wallboard, hollow tile, and similar materials. The two types made are standard and drive. The standard type is commonly used in walls and ceilings and is installed by drilling a mounting hole to the required diameter. The anchor is inserted into the hole and tapped until the gripper prongs embed in the base material. Following this, the anchor's screw is tightened to draw the anchor tight against the inside of the wall or ceiling. Note that the drive-type anchor is hammered into the material without the need for drilling a mounting hole. After the anchor is installed, the anchor screw is removed, the component being fastened is positioned in place, then the screw is reinstalled through the mounting hole in the component and into the anchor. The screw is tightened into the anchor to secure the component.

7.7.3 Wallboard Anchors

Wallboard anchors (*Figure 39*) are self-drilling medium- and light-duty anchors used for fastening in wallboard. The anchor is driven into the wall with a Phillips head manual or cordless screwdriver until the head of the anchor is flush with the wall or ceiling surface. Following this, the component being fastened is positioned over the anchor, then secured with the proper size sheet metal screw driven into the anchor.

THINK ABOUT IT

Sleeve-Type Drive Anchors

What happens when you remove the screw when a sleeve-type drive anchor is in place?

108SA08.EPS

STANDARD

DRIVE

SLEEVE-TYPE   **WALLBOARD**   **METAL DRIVE-IN**

108F39.EPS

Figure 39 ◆ Sleeve-type, wallboard, and metal drive-in anchors.

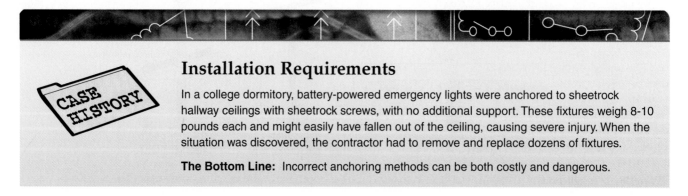

Installation Requirements

In a college dormitory, battery-powered emergency lights were anchored to sheetrock hallway ceilings with sheetrock screws, with no additional support. These fixtures weigh 8-10 pounds each and might easily have fallen out of the ceiling, causing severe injury. When the situation was discovered, the contractor had to remove and replace dozens of fixtures.

The Bottom Line: Incorrect anchoring methods can be both costly and dangerous.

Ceiling Installations

THINK ABOUT IT

In the dormitory problem discussed above, which of the following fasteners could have been used to safely secure the emergency lights?

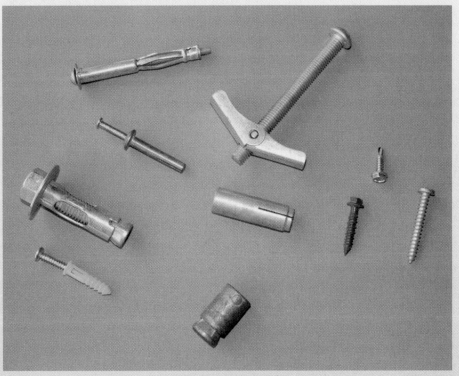

108SA09.EPS

7.7.4 Metal Drive-In Anchors

Metal drive-in anchors (*Figure 39*) are used to fasten light to medium loads to wallboard. They have two pointed legs that stay together when the anchor is hammered into a wall and spread out against the inside of the wall when a No. 6 or 8 sheet metal screw is driven in.

7.8.0 Epoxy Anchoring Systems

Epoxy resin compounds can be used to anchor threaded rods, dowels, and similar fasteners in solid concrete, hollow wall, and brick. For one manufacturer's product, a two-part epoxy is packaged in a two-chamber cartridge that keeps the resin and hardener ingredients separated until use. This cartridge is placed into a special tool similar to a caulking gun. When the gun handle is pumped, the epoxy resin and hardener components are mixed within the gun; then the epoxy is ejected from the gun nozzle.

To use the epoxy to install an anchor in solid concrete (*Figure 40*), a hole of the proper size is drilled in the concrete and cleaned using a nylon (not metal) brush. Following this, a small amount of epoxy is dispensed from the gun to make sure that the resin and hardener have mixed properly. This is indicated by the epoxy being of a uniform color. The gun nozzle is then placed into the hole,

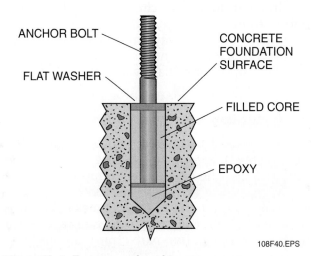

Figure 40 ◆ Fastener anchored in epoxy.

108F40.EPS

and the epoxy is injected into the hole until half the depth of the hole is filled. Following this, the selected fastener is pushed into the hole with a slow twisting motion to make sure that the epoxy fills all voids and crevices, then is set to the required plumb (or level) position. After the recommended cure time for the epoxy has elapsed, the fastener nut can be tightened to secure the component or fixture in place.

The procedure for installing a fastener in a hollow wall or brick using epoxy is basically the same as described above. The difference is that the epoxy is first injected into an anchor screen to fill the screen, then the anchor screen is installed into the drilled hole. Use of the anchor screen is necessary to hold the epoxy intact in the hole until the anchor is inserted into the epoxy.

8.0.0 ◆ RACEWAY SUPPORTS

Raceway supports are available in many types and configurations. This section discusses the most common conduit supports found in electrical installations. *NEC Section 300.11* discusses the requirements for branch circuit wiring that is supported from above suspended ceilings. Electrical equipment and raceways must have their own supporting methods and may not be supported by the supporting hardware of a fire-rated roof/ceiling assembly.

8.1.0 Straps

Straps are used to support conduit to a surface (see *Figure 41*). The spacing of these supports must conform to the minimum support spacing requirements for each type of conduit. One- and two-hole straps are used for all types of conduit: EMT, RMC, IMC, PVC, and flex. The straps can be flexible or rigid. Two-part straps are used to secure conduit to electrical framing channels

INSIDE TRACK

Wall-Mounted Supports

This wall-mounted support has been fabricated to hold the conduit away from the metal building. While it is shown with only one raceway, additional raceways can be added to the framing channel.

108SA10.EPS

(struts). Parallel and right angle beam clamps are also used to support conduit from structural members.

Clamp back straps can also be used with a backplate to maintain the ¼-inch spacing from the surface required for installations in wet locations.

8.2.0 Standoff Supports

The standoff support, often referred to as a Minerallac® (the name of a manufacturer of this type

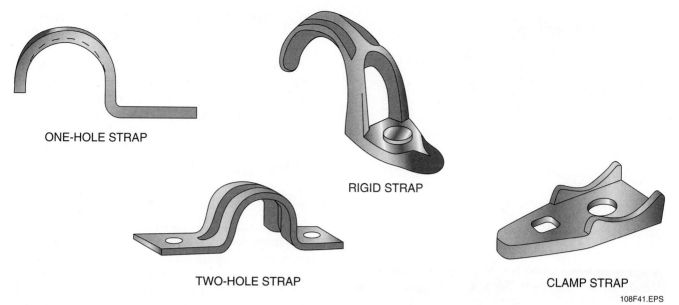

ONE-HOLE STRAP

RIGID STRAP

TWO-HOLE STRAP

CLAMP STRAP

108F41.EPS

Figure 41 ◆ Straps.

of support), is used to support conduit away from the supporting structure. In the case of the one-hole and two-hole straps, the conduit must be off-set wherever a fitting occurs. If standoff supports are used, the conduit is held away from the sup-porting surface, and no offsets are required in the conduit at the fittings. Standoff supports may be used to support all types of conduit including RMC, IMC, EMT, PVC, and flex, as well as tubing installations. A standoff support is shown in *Figure 42*.

8.3.0 Electrical Framing Channels

Electrical framing channels or other similar framing materials are used together with Unistrut®-type conduit clamps to support conduit (see *Figure 43*). They may be attached to a ceiling, wall, or other sur-face or be supported from a trapeze hanger.

8.4.0 Beam Clamps

Beam clamps are used with suspended hangers. The raceway is attached to or laid in the hanger. The hanger is suspended by a threaded rod. One end of the threaded rod is attached to the hanger and the other end is attached to a beam clamp. The beam clamp is then attached to a beam. A beam clamp with wireway support assembly is shown in *Figure 44*.

108F42.EPS

Figure 42 ◆ Standoff support.

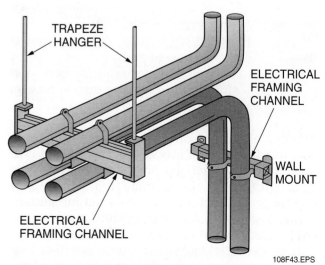

TRAPEZE HANGER

ELECTRICAL FRAMING CHANNEL

WALL MOUNT

ELECTRICAL FRAMING CHANNEL

108F43.EPS

Figure 43 ◆ Electrical framing channels.

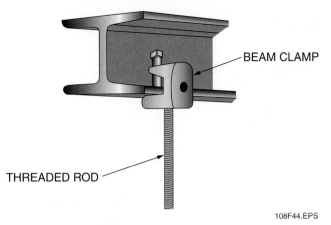

THREADED ROD

BEAM CLAMP

108F44.EPS

Figure 44 ◆ Beam clamp.

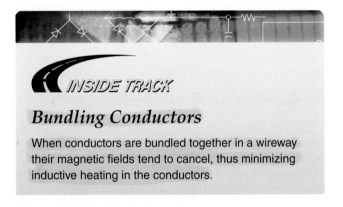

INSIDE TRACK

Bundling Conductors

When conductors are bundled together in a wireway their magnetic fields tend to cancel, thus minimizing inductive heating in the conductors.

9.0.0 ◆ WIREWAYS

Wireways are sheet metal **troughs** provided with hinged or screw-on removable covers. Like other types of raceways, wireways are used for housing electric wires and cables. Wireways are available in various lengths, including 1, 2, 3, 4, 5, and 10 feet. The availability of various lengths allows runs of any exact number of feet to be made without cutting the wireway ducts. Wireways are dealt with specifically in *NEC Article 376.*

As listed in *NEC Section 376.22(A),* the sum of the cross-sectional areas of all contained conductors at any cross section of a wireway shall not exceed 20% of the interior cross-sectional area of the wireway. The derating factors in *NEC Table 310.15(B)(2)(a)* shall be applied only where the number of current-carrying conductors exceeds 30, including neutral conductors classified as current-carrying under the provisions of *NEC Section 310.15(B)(4).* Conductors for signaling or controller conductors between a motor and its starter used only for starting duty shall not be considered current-carrying conductors.

It is also noted in *NEC Section 376.56* that conductors, together with splices and taps, must not fill the wireway to more than 75% of its cross-

sectional area. No conductor larger than that for which the wireway is designed shall be installed in any wireway. Be sure to check *NEC Article 378* for the requirements of nonmetallic wireways.

NEC Section 376.23(A) requires that the dimensions of *NEC Table 312.6(A)* be applied where insulated conductors are deflected in a wireway. *NEC Section 376.23(B)* requires that the provisions of *NEC Section 314.28* apply where wireways are used as pull boxes.

9.1.0 Auxiliary Gutters

Strictly speaking, an auxiliary gutter is a wireway that is intended to add to wiring space at switchboards, meters, and other distribution locations. Auxiliary gutters are dealt with specifically in *NEC Article 366.* Even though the component parts of wireways and auxiliary gutters are identical, you should be familiar with the differences in their use. Auxiliary gutters are used as parts of complete assemblies of apparatus such as switchboards, distribution centers, and control equipment. However, an auxiliary gutter may only contain conductors or busbars, even though it looks like a surface metal raceway that may contain devices and equipment. Unlike auxiliary gutters, wireways represent a type of wiring because they are used to carry conductors between points located considerable distances apart.

The allowable ampacities for insulated conductors in wireways and gutters are given in *NEC Tables 310.16 and 310.18.* It should be noted that these tables are used for raceways in general. These *NEC*® tables and the notes are often used to determine if the correct materials are on hand for an installation. They are also used to determine if it is possible to add conductors in an existing wireway or gutter.

In many situations, it is necessary to make extensions from the wireways to wall receptacles and control devices. In these cases, *NEC Section 376.70* specifies that these extensions be made using any wiring method presented in *NEC Chapter 3* that includes a means for equipment grounding. Finally, as required in *NEC Section 376.120,* wireways must be marked in such a way that their manufacturer's name or trademark will be visible.

As you can see in *Figure 45,* a wide range of fittings is required for connecting wireways to one another and to fixtures such as switchboards, power panels, and conduit.

9.2.0 Types of Wireways

Rectangular duct-type wireways come as either hinged-cover or screw-cover troughs. Typical

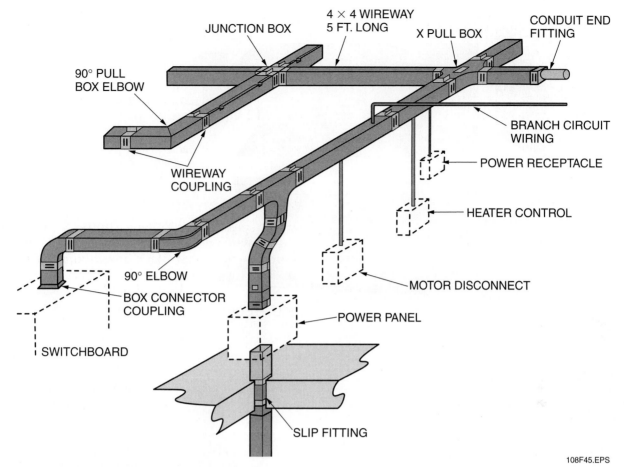

JUNCTION BOX

4 × 4 WIREWAY 5 FT. LONG

X PULL BOX

CONDUIT END FITTING

90° PULL BOX ELBOW

BRANCH CIRCUIT WIRING

POWER RECEPTACLE

WIREWAY COUPLING

HEATER CONTROL

90° ELBOW

BOX CONNECTOR COUPLING

MOTOR DISCONNECT

POWER PANEL

SWITCHBOARD

SLIP FITTING

108F45.EPS

Figure 45 ◆ Wireway system layout.

lengths are 1, 2, 3, 4, 5, and 10 feet. Shorter lengths are also available. Raintight troughs are permitted to be used in environments where moisture is not permitted within the raceway. However, the raintight trough should not be confused with the raintight lay-in wireway, which has a hinged cover. *Figure 46* shows a raintight trough with a removable side cover.

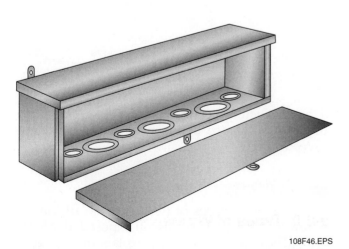

108F46.EPS

Figure 46 ◆ Raintight trough.

Wireway troughs are exposed when first installed. Whenever possible, they are mounted on the ceilings or walls, although they may sometimes be suspended from the ceiling. Note that in *Figure 47*, the trough has knockouts similar to those found on junction boxes. After the wireway system has been installed, branch circuits are brought from the distribution panels using conduit. The conduit is joined to the wireway at the most convenient knockout possible.

Wireway components such as trough crosses, 90° internal elbows, and tee connectors serve the same function as fittings on other types of raceways. The fittings are attached to the duct using slip-on connectors. All attachments are made with nuts and bolts or screws. When assembling wireways, always place the head of the bolt on the inside and the nut on the outside so that the conductors will not be resting against a sharp edge. It is usually best to assemble sections of the wireway system on the floor, and then raise the sections into position. An exploded view of a section of wireway is shown in *Figure 48*. Both the wireway fittings and the duct come with screw-on, hinged, or snap-on covers to permit conductors to be laid in or pulled through.

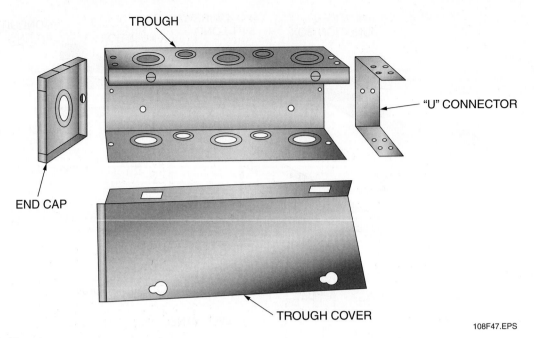

Figure 47 ◆ Trough.

108F47.EPS

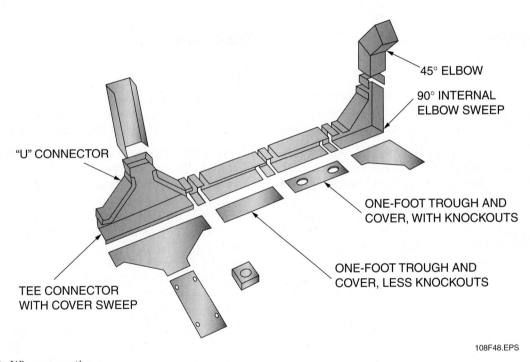

Figure 48 ◆ Wireway sections.

108F48.EPS

The *NEC*® specifies that wireways may be used only for exposed work. Therefore, they cannot be used in underfloor installations. If they are used for outdoor work, they must be of an approved raintight construction. It is important to note that wireways must not be installed where they are subject to severe physical damage, corrosive vapors, or hazardous locations.

Wireway troughs must be installed so that they are supported at distances not exceeding 5 feet. When specially approved supports are used, the distance between supports must not exceed 10 feet.

9.2.1 Wireway Fittings

Many different types of fittings are available for wireways, especially for use in exposed, dry locations. The following sections explain fittings commonly used in the electrical craft.

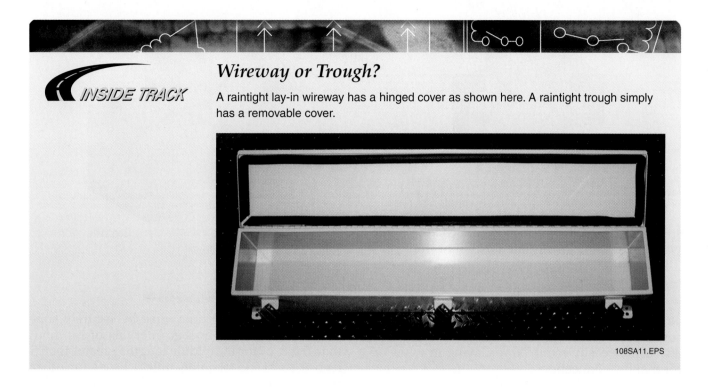

Wireway or Trough?

A raintight lay-in wireway has a hinged cover as shown here. A raintight trough simply has a removable cover.

108SA11.EPS

9.2.2 Connectors

Connectors (*Figure 49*) are used to join wireway sections and fittings. Connectors are slipped inside the end of a wireway section and are held in place by small bolts and nuts. Alignment slots allow the connector to be moved until it is flush with the inside surface of the wireway. After the connector is in position, it can be bolted to the wireway. This helps to ensure a strong rigid connection. Connectors have a friction hinge that helps hold the wireway cover open when needed.

9.2.3 End Plates

End plates, or closing plates (*Figure 50*), are used to seal the ends of wireways. They are inserted into the end of the wireway and fastened by screws and bolts. End plates contain knockouts so that conduit or cable may be extended from the wireway.

9.2.4 Tees

Tee fittings (*Figure 51*) are used when a tee connection is needed in a wireway system. A tee connection is used where circuit conductors may branch in different directions. The tee fitting's covers and sides can be removed for access to

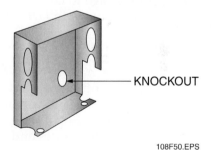

KNOCKOUT

108F50.EPS

Figure 50 ◆ End plate.

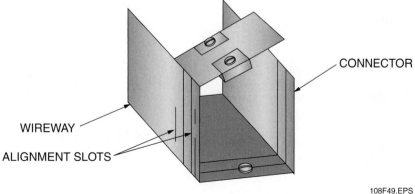

CONNECTOR

WIREWAY

ALIGNMENT SLOTS

108F49.EPS

Figure 49 ◆ Connector.

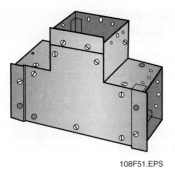

Figure 51 ◆ Tee.

splices and taps. Tee fittings are attached to other wireway sections using standard connectors.

9.2.5 Crosses

Crosses (*Figure 52*) have four openings and are attached to other wireway sections with standard connectors. The cover is held in place by screws and can be easily removed for laying in wires or for making connections.

9.2.6 Elbows

Elbows are used to make a bend in the wireway. They are available in angles of 22½°, 45°, or 90°, and are either internal or external. They are attached to wireway sections with standard connectors. Covers and sides can be removed for wire installation. The inside corners of elbows are rounded to prevent damage to conductor insulation. An inside elbow is shown in *Figure 53*.

9.2.7 Telescopic Fittings

Telescopic or slip fittings may be used between lengths of wireway. Slip fittings are attached to standard lengths by setscrews and usually adjust from ½ inch to 11½ inches. Slip fittings have a removable cover for installing wires and are similar in appearance to a nipple.

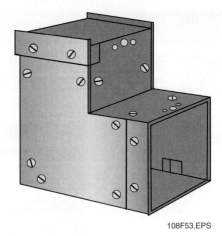

Figure 53 ◆ 90° inside elbow.

9.3.0 Wireway Supports

Horizontal wireway runs must be securely supported at each end and at intervals of no more than 5 feet or for individual lengths greater than 5 feet at each end or joint, unless listed for other support intervals. In no case shall the support distance be greater than 10 feet, in accordance with *NEC Section 376.30.* If possible, wireways can be mounted directly to a surface. Otherwise, wireways are supported by hangers or brackets.

9.3.1 Suspended Hangers

In many cases, the wireway is supported from a ceiling, beam, or other structural member. In such installations, a suspended hanger (*Figure 54*) may be used to support the wireway.

The wireway is attached to or laid in the hanger. The hanger is suspended by a threaded rod. One end of the rod is attached to the hanger with hex nuts. The other end of the rod is attached to a beam clamp or anchor.

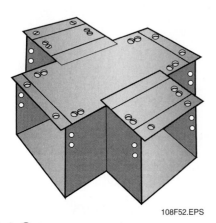

Figure 52 ◆ Cross.

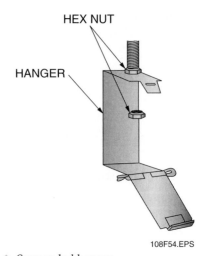

HEX NUT

HANGER

Figure 54 ◆ Suspended hanger.

9.3.2 Gusset Brackets

Another type of support used to mount wireways is a gusset bracket (*Figure 55*). This is an L-type bracket that is mounted to a wall. The wireway rests on the bracket and is attached by screws or bolts.

9.3.3 Standard Hangers

Standard hangers (*Figure 56*) are made in two pieces. The two pieces are combined in different ways for different installation requirements. The wireway is attached to the hanger by bolts and nuts.

9.3.4 Wireway Hangers

When a larger wireway must be suspended, a wireway hanger may be used. A wireway hanger is made by suspending a piece of strut from a ceiling, beam, or other structural member. The strut is suspended by threaded rods attached to beam clamps or other ceiling anchors, as shown in *Figure 57*.

9.4.0 Other Types of Raceways

In this section, other types of raceways will be discussed. Depending on the particular purpose for which they are intended, raceways include enclosures such as surface metal and nonmetallic raceways, and underfloor raceways.

9.4.1 Surface Metal and Nonmetallic Raceways

Surface metal raceways consist of a wide variety of special raceways designed primarily to carry power and communications wiring to locations on the surface of ceilings or walls of building interiors.

Installation specifications of both surface metal raceways and surface nonmetallic raceways are listed in detail in *NEC Articles 386 and 388*, respectively. All these raceways must be installed in dry, interior locations. The number of conductors, their amperage, and the allowable cross-sectional area of the conductors, as well as regulations for combination raceways, are specified in *NEC Tables 310.16 and 310.18* and *NEC Articles 386 and 388*.

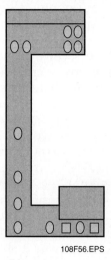

108F56.EPS

Figure 56 ◆ Standard hanger.

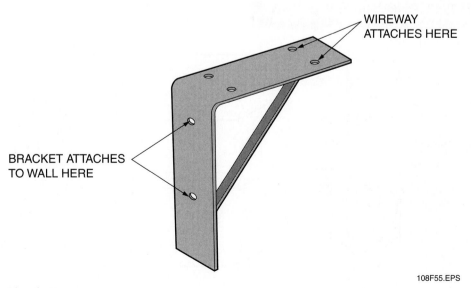

108F55.EPS

Figure 55 ◆ Gusset bracket.

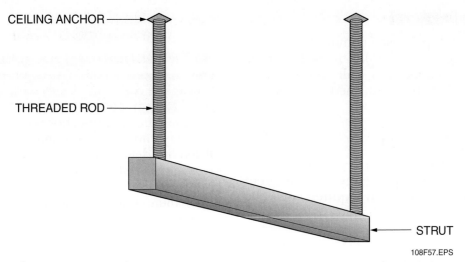

CEILING ANCHOR

THREADED ROD

STRUT

108F57.EPS

Figure 57 ◆ Wireway hanger.

One use of surface metal raceways is to protect conductors that run to non-accessible outlets.

Surface metal and nonmetallic raceways have been divided into subgroups based on the specific purpose for which they are intended. There are three small surface raceways that are primarily used for extending power circuits from one point to another. In addition, there are six larger surface raceways that have a much wider range of applications. Typical cross sections of the first three smaller raceways are shown in *Figure 58*.

Additional surface metal raceway designs are referred to as pancake raceways, because their flat cross sections resemble pancakes. Their primary use is to extend power, lighting, telephone, or signal wire across a floor to locations away from the walls of a room without embedding them under the floor. A pancake raceway is shown in *Figure 59*.

There are also surface metal raceways available that house two or three different conductor raceways. These are referred to as twinduct or tripleduct. These raceways permit different circuits, such as power and signal, to be placed within the same raceway.

Nonmetallic raceways come in a variety of styles. The perimeter raceways shown in *Figure 60* are available in sizes ranging from ¾" to more than

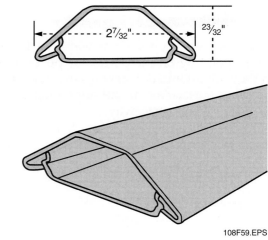

108F59.EPS

Figure 59 ◆ Pancake raceway.

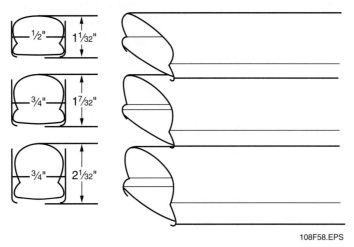

108F58.EPS

Figure 58 ◆ Smaller surface raceways.

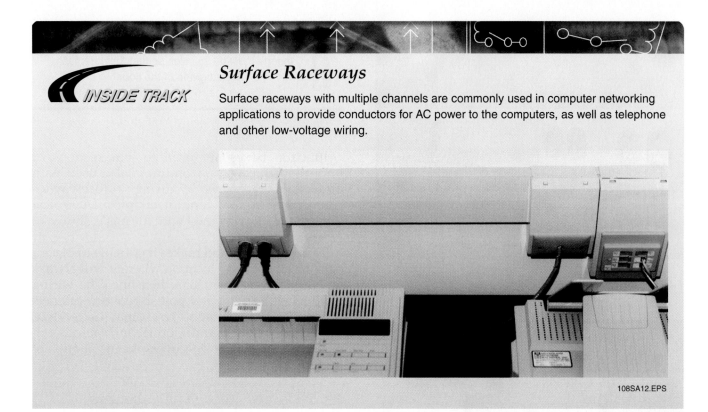

INSIDE TRACK

Surface Raceways

Surface raceways with multiple channels are commonly used in computer networking applications to provide conductors for AC power to the computers, as well as telephone and other low-voltage wiring.

108SA12.EPS

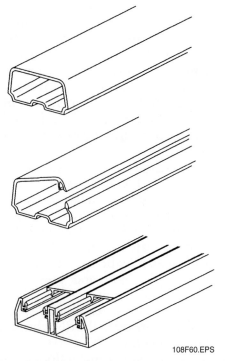

108F60.EPS

Figure 60 ◆ Examples of surface raceway.

7" wide. Many of these raceways contain barriers that allow them to carry both low voltage and power wiring.

The number and types of conductors permitted to be installed and the capacity of a particular surface raceway must be calculated and matched with *NEC®* requirements, as discussed previously. *NEC Tables 310.16 through 310.19* are used for surface raceways in the same manner in which they are used for wireways. For surface raceway installations with more than three conductors in each raceway, particular reference must be made to *NEC Table 310.15(B)(2)(a).*

9.4.2 Multi-Outlet Assemblies

Manufacturers offer a wide variety of multi-outlet surface raceways. Their function is to hold receptacles and other devices within the raceway. When surface raceways are used in this manner, the assembly is referred to as a multi-outlet assembly. Multi-outlet assemblies (*Figure 61*) are covered in *NEC Article 380.* Multi-outlet systems are either wired in the field or come pre-wired from the factory.

9.4.3 Pole Systems

There are many situations in which power and other electric circuits have to be carried from overhead wiring systems to devices that are not located near existing wall outlets or control circuits. This type of wiring is typically used in open office spaces where cubicles are provided by temporary dividers. Poles are used to accomplish this. The poles usually come in lengths suitable for 10-, 12-, or 15-foot ceilings. *Figure 62* shows a typical pole base.

Figure 61 ◆ Multi-outlet assembly.

108F61.EPS

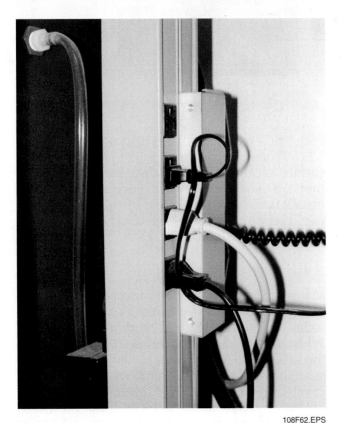

108F62.EPS

Figure 62 ◆ Power pole.

9.4.4 Underfloor Systems

Underfloor raceway systems were developed to provide a practical means of bringing conductors for lighting, power, and signaling to cabinets and consoles. Underfloor raceways are available in 10-foot lengths and widths of 4 and 8 inches. The sections are made with inserts spaced every 24 inches. The inserts can be removed for outlet installation. These are explained in *NEC Article 390.*

Junction boxes are used to join sections of underfloor raceways. Conduit is also used with underfloor raceways by using a raceway-to-conduit connector (conduit adapter). A typical underfloor raceway duct with fittings is shown in *Figure 63.*

This wiring method makes it possible to place a desk or table in any location where it will always be over, or very near to, a duct line. The wiring method for lighting and power between cabinets and the raceway junction boxes may be conduit, underfloor raceway, wall elbows, and cabinet connectors. *NEC Article 390* covers the installation of underfloor raceways.

9.4.5 Cellular Metal Floor Raceways

A cellular metal floor raceway is a type of floor construction designed for use in steel-frame buildings. In these buildings, the members supporting the floor between the beams consist of sheet steel rolled into shapes. These shapes are combined to form cells, or closed passageways, which extend across the building. The cells are of various shapes and sizes, depending upon the structural strength required. The cells of this type of floor construction form the raceways, as shown in *Figure 64.*

Connections to the cells are made using headers that extend across the cells. A header connects only to those cells to be used as raceways for conductors. A junction box or access fitting is necessary at each joint where a header connects to a cell. Two or three separate headers, connecting to different sets of cells, may be used for different systems. For example, light and power, signaling systems, and public telephones would each have a separate header. A special elbow fitting is used to extend the headers up to the distribution equipment on a wall or column. *NEC Article 374* covers the installation of cellular metal floor raceways.

9.4.6 Cellular Concrete Floor Raceways

The term precast cellular concrete floor refers to a type of floor used in steel-frame, concrete-frame, and wall-bearing construction. In this type of system, the floor members are precast with hollow voids that form smooth, round cells. The cells

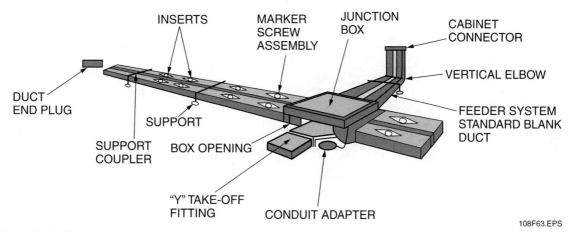

Figure 63 ◆ Underfloor raceway duct.

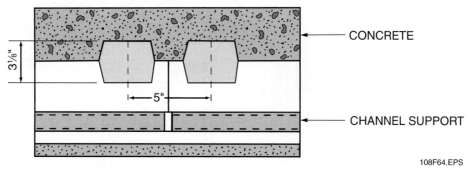

Figure 64 ◆ Cross section of a cellular floor.

form raceways, which can be adapted, using fittings, for use as underfloor raceways. A precast cellular concrete floor is fire-resistant and requires no further fireproofing. The precast reinforced concrete floor members form the structural floor and are supported by beams or bearing walls. Connections to the cells are made with headers that are secured to the precast concrete floor. *NEC Article 372* covers the installation of cellular concrete floor raceways.

10.0.0 ◆ CABLE TRAYS

Cable trays function as a support for conductors and tubing (see *NEC Article 392*). A cable tray has the advantage of easy access to conductors, and thus lends itself to installations where the addition or removal of conductors is a common practice. Cable trays are fabricated from aluminum, steel, and fiberglass. Cable trays are available in two basic forms: ladder and trough. Ladder tray, as the name implies, consists of two parallel channels connected by rungs. Trough consists of two parallel channels (side rails) having a corrugated, ventilated bottom, or a corrugated, solid bottom. (There is also a special center rail cable tray available for use in light-duty applications such as tele-

phone and sound wiring. We will discuss this type of cable tray in more detail in Level 2.)

Cable trays are commonly available in 12- and 24-foot lengths. They are usually available in widths of 6, 9, 12, 18, 24, 30, and 36 inches, and load depths of 4, 6, and 8 inches.

Cable trays may be used in most electrical installations. Cable trays may be used in air handling ceiling space, but only to support the wiring methods permitted in such spaces by *NEC Section 300.22(C)(1)*. Also, cable trays may be used in Class 1, Division 2 locations according to *NEC Section 501.10(B)*. Cable trays may also be used above a suspended ceiling that is not used as an air handling space. Some manufacturers offer an aluminum cable tray that is coated with PVC for installation in caustic environments. A typical cable tray system with fittings is shown in *Figure 65*.

Wire and cable installation in cable trays is defined by the *NEC®*. Read *NEC Article 392* to become familiar with the requirements and restrictions made by the *NEC®* for safe installation of wire and cable in a cable tray.

Metallic cable trays that support electrical conductors must be grounded as required by *NEC Article 250.* Where steel and aluminum cable tray

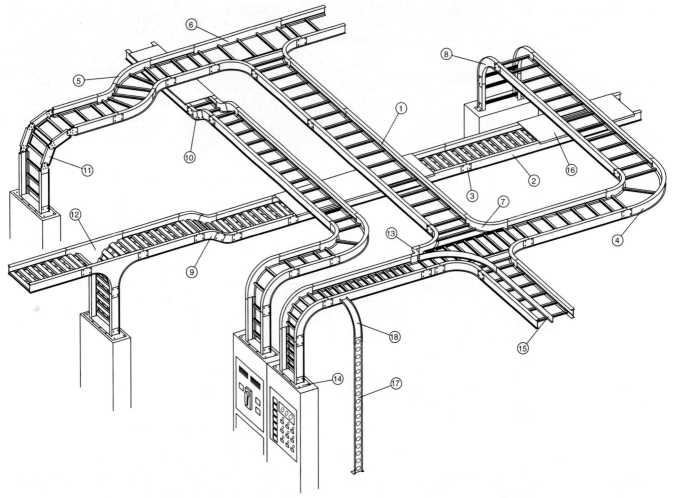

Legend

1. LADDER TYPE CABLE TRAY
2. VENTILATED TROUGH TYPE CABLE TRAY
3. STRAIGHT SPLICE PLATE
4. 90° HORIZONTAL BEND, LADDER TYPE CABLE TRAY
5. 45° HORIZONTAL BEND, LADDER TYPE CABLE TRAY
6. HORIZONTAL TEE, LADDER TYPE CABLE TRAY
7. HORIZONTAL CROSS, LADDER TYPE CABLE TRAY
8. 90° VERTICAL OUTSIDE BEND, LADDER TYPE CABLE TRAY
9. 45° VERTICAL OUTSIDE BEND, VENTILATED TYPE CABLE TRAY

10. 30° VERTICAL INSIDE BEND, LADDER TYPE CABLE TRAY
11. VERTICAL BEND SEGMENT (VBS)
12. VERTICAL TEE DOWN, VENTILATED TROUGH TYPE CABLE TRAY
13. LEFT HAND REDUCER, LADDER TYPE CABLE TRAY
14. FRAME TYPE BOX CONNECTOR
15. BARRIER STRIP STRAIGHT SECTION
16. SOLID FLANGED TRAY COVER
17. VENTILATED CHANNEL STRAIGHT SECTION
18. CHANNEL CABLE TRAY, 90° VERTICAL OUTSIDE BEND

108F65.EPS

Figure 65 ◆ Cable tray system.

systems are used as an equipment grounding conductor, all of the provisions of *NEC Section 392.7* must be complied with.

WARNING!
Do not stand on, climb in, or walk on a cable tray.

10.1.0 Cable Tray Fittings

Cable tray fittings are part of the cable tray system and provide a means of changing the direction or dimension of the different trays. Some of the uses of horizontal and vertical tees, horizontal and vertical bends, horizontal crosses, reducers, barrier strips, covers, and box connectors are shown in *Figure 65*.

Cable Trays and Wireways

THINK ABOUT IT

What is the difference between a wireway and a cable tray? What kinds of conductors would you expect to find in a cable tray as compared to a wireway?

Cable Tray Systems

INSIDE TRACK

Cable tray systems must be continuous and grounded. One of the advantages of using a cable tray system is that it makes it easy to expand or modify the wiring system following installation. Unlike conduit systems, wires can be added or changed by simply laying them into (or lifting them out of) the tray.

108SA13.EPS

10.2.0 Cable Tray Supports

Cable trays are usually supported in one of five ways: direct rod suspension, trapeze mounting, center hung, wall mounting, and pipe rack mounting.

10.2.1 Direct Rod Suspension

The direct rod suspension method of supporting cable tray uses threaded rods and hanger clamps. One end of the threaded rod is connected to the ceiling or other overhead structure. The other end is connected to hanger clamps that are attached to the cable tray side rails. A direct rod suspension assembly is shown in *Figure 66*.

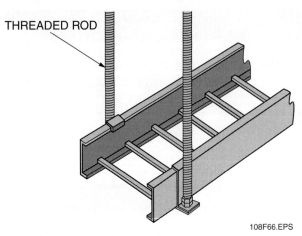

THREADED ROD

108F66.EPS

Figure 66 ◆ Direct rod suspension.

10.2.2 Trapeze Mounting and Center Hung Support

Trapeze mounting of cable tray is similar to direct rod suspension mounting. The difference is in the method of attaching the cable tray to the threaded rods. A structural member, usually a steel channel or strut, is connected to the vertical supports to provide an appearance similar to a swing or trapeze. The cable tray is mounted to the structural member. Often, the underside of the channel or strut is used to support conduit. A trapeze mounting assembly is shown in *Figure 67*.

A method that is similar to trapeze mounting is a center hung tray support. In this case, only one rod is used and it is centered between the cable tray side rails.

10.2.3 Wall Mounting

Wall mounting is accomplished by supporting the cable tray with structural members attached to the wall (*Figure 68*). This method of support is often used in tunnels and other underground or sheltered installations where large numbers of conductors interconnect equipment that is separated by long distances.

10.2.4 Pipe Rack Mounting

Pipe racks are structural frames used to support piping that interconnects equipment in outdoor industrial facilities. Usually, some space on the rack is reserved for conduit and cable tray. Pipe rack mounting of cable tray is often used when power distribution and electrical wiring is routed over a large area.

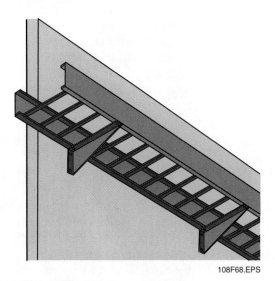

108F68.EPS

Figure 68 ◆ Wall mounting.

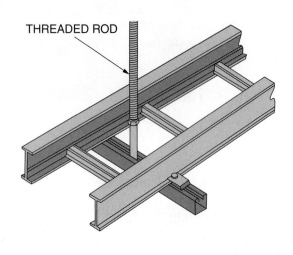

THREADED ROD

THREADED ROD

TRAPEZE

CENTER HUNG

108F67.EPS

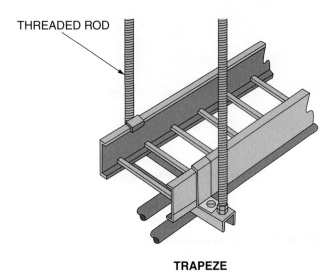

Figure 67 ◆ Trapeze mounting and center hung support.

11.0.0 ◆ STORING RACEWAYS

Proper and safe methods of storing conduit, wireways, raceways, and cable trays may sound like a simple task, but improper storage techniques can result in wasted time and damage to the raceways, as well as personal injury. There are correct ways to store raceways that will help avoid costly damage, save time in identifying stored raceways, and reduce the chance of personal injury.

Pipe racks are commonly used for storing conduit. The racks provide support to prevent bending, sagging, distorting, scratching, or marring of conduit surfaces. Most racks have compartments where different types and sizes of conduit can be separated for ease of identification and selection. The storage compartments in racks are usually elevated to help avoid damage that might occur at floor level. Conduit that is stored at floor level is easily damaged by people and other materials or equipment in the area.

The ends of stored conduit should be sealed to help prevent contamination and damage. Conduit ends can be capped, taped, or plugged.

Always inspect raceway before storing it to make sure that it is clean and not damaged. It is discouraging to get raceway for a job and find that it is dirty or damaged. Also, make sure that the raceway is stored securely so that when someone comes to get it for a job, it will not fall in any way that could cause injury.

To prevent contamination and corrosion of stored raceway, it should be covered with a tarpaulin or other suitable covering. It should also be separated from noncompatible materials such as hazardous chemicals.

Wireways, surface metal raceways, and cable trays should always be stored off the ground on boards in an area where people will not step on it and equipment will not run over it. Stepping or running over raceway bends the metal and makes it unusable.

12.0.0 ◆ HANDLING RACEWAYS

Raceway is made to strict specifications. It can be easily damaged by careless handling. From the time raceway is delivered to a job site until the installation is complete, use proper and safe handling techniques. These are a few basic guidelines for handling raceway that will help avoid damaging or contaminating it:

- Never drag raceway off a delivery truck or off other lengths of raceway.
- Never drag raceway on the ground or floor. Dragging raceway can cause damage to the ends.

- Keep the thread protection caps on when handling or transporting conduit raceway.
- Keep raceway away from any material that might contaminate it during handling.
- Flag the ends of long lengths of raceway when transporting it to the job site.
- Never drop or throw raceway when handling it.
- Never hit raceway against other objects when transporting it.
- Always use two people when carrying long pieces of raceway. Make sure that you both stay on the same side and that the load is balanced. Each person should be about ¼ of the length of the raceway from the end. Lift and put down the raceway at the same time.

13.0.0 ◆ DUCTING

In the common vocabulary of the electrical trade, a duct is a single enclosed raceway, or runway, through which conductors or cables can be led. Basically, ducting is a system of ducts. However, underground duct systems include manholes, transformer vaults, and risers.

There are several reasons for running power lines underground rather than overhead. In some situations, an overhead high-voltage line would be dangerous, or the space may not be adequate. For aesthetic reasons, architectural plans may require buried lines throughout a subdivision or a planned community. Tunnels may already exist, or be planned, for carrying steam or water lines. In any of these situations, underground installations are appropriate. Underground cables may be buried directly in the ground or run through tunnels or raceways, including conduit and recognized ducts.

In underground construction, a duct system provides a safe passageway for power lines, communication cables, or both. In buildings, underfloor raceways and cellular floor raceways are built to provide ducting so that electricity will be available throughout a large area. As an electrician, you need to know the approved methods of constructing underground ducting. You also need to know how to avoid potential electrical hazards in both original construction and maintenance. It is essential to understand the requirements and limitations imposed on running wires through underfloor and cellular floor raceways and ducts.

13.1.0 Underground Ducts

A duct consists of conduit or an approved duct system (such as HDPE) placed in a trench and covered with earth or concrete. The minimum depth at which the duct will be placed is determined using *NEC Table 300.5.* Encasing the duct in concrete or other materials provides mechanical strength and helps dissipate heat. *Figure 69* shows a duct bank in place and ready for backfill. In this case, it will be covered in concrete.

Manholes are set at intervals in an underground duct run. *Figure 70* shows a manhole with pull strings installed and tied off in preparation for the conductor installation. Manholes provide access through throats (sometimes called chimneys). At ground level, or street surface level, a manhole cover closes off the manhole area tightly. A duct line may consist of a single conduit or several, each carrying a cable length from one manhole to the next.

Manholes provide room for conductor installation and maintenance. Workers enter a manhole from above. In a two-way manhole, cables enter

and leave in only two directions. There are also three-way and four-way manholes. Often manholes are located at the intersection of two streets so that they can be used for cables leaving in four directions. Manholes are usually constructed of brick or concrete. Their design must provide room for drainage and for workers to move around inside them. A similar opening known as a handhole is sometimes provided for splicing on lateral two-way duct lines.

Transformer vaults house power transformers, voltage regulators, network protectors, meters, and circuit breakers. A cable may end at a transformer vault. Other cables end at a customer's substation or terminate as risers that connect with overhead lines.

13.2.0 Duct Materials

Underground duct lines can be made of fiber, vitrified tile, rigid metal or nonmetallic conduit, or poured concrete. The inside diameter of the ducting for a specific job is determined by the size of the cable that will be drawn into the duct. Sizes from two to six inches (inside diameter) are available for most types of ducting.

Figure 69 ◆ Duct bank.

 WARNING!

Be careful when working with unfamiliar duct materials. In older installations, asbestos/cement duct may have been used. You must be certified to remove or disturb asbestos.

Rigid nonmetallic conduit may be made of PVC (polyvinyl chloride), PE (polyethylene), or styrene. Since this type of conduit is available in lengths up to 20 feet, fewer couplings are needed than with other types of ducting. PVC is popular because it is easy to install, requires less labor than other types of conduit, and is low in cost.

13.3.0 Monolithic Concrete Duct

Monolithic concrete duct is poured at the job site. Multiple duct lines can be formed using rubber tubing cores on spacers. The cores may be removed after the concrete has set. A die containing steel tubes, known as a boat, can also be used to form ducts. It is pulled slowly through the trench on a track as concrete is poured from the top. Poured concrete ducting made by either method is relatively expensive, but offers the advantage of creating a very clean duct interior with no residue that can decay. The rubber core method is especially useful for curving or turning part of a duct system.

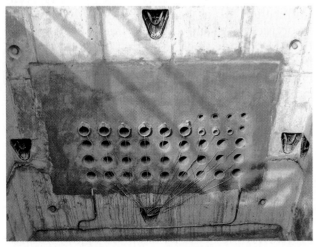

108F69.EPS

108F70.EPS

Figure 70 ◆ Manhole.

13.4.0 Cable-in-Duct

One of the most popular duct types is the cable-in-duct. This type of duct comes from the manufacturer with cables already installed. The duct comes in a reel and can be laid in the trench with ease. The installed cables can be withdrawn in the future, if necessary. This type of duct, because of the form in which it comes, reduces the need for fittings and couplings. It is most frequently used for street lighting systems.

14.0.0 ◆ CONSTRUCTION METHODS

Conduit and box installation varies with the type of construction. This section discusses some special requirements for masonry and concrete, metal framing, wood, and structural steel construction.

14.1.0 Masonry and Concrete Flush-Mount Construction

In a reinforced concrete construction environment, the conduit and boxes must be embedded in the concrete to achieve a flush surface. Ordinary boxes may be used, but special concrete boxes are preferred and are available in depths up to six inches. These boxes have special ears by which they are nailed to the wooden forms for the concrete. When installing them, stuff the boxes tightly with paper to prevent concrete from seeping in. *Figure 71* shows an installed box.

Flush construction can also be done on existing concrete walls, but this requires chiseling a channel and box opening, anchoring the box and conduit, and then resealing the wall.

To achieve flush construction with masonry walls, the most acceptable method is for the electrician to work closely with the mason laying the blocks. When the construction blocks reach the convenience outlet elevation, boxes are made up as shown in *Figure 72*. The figure shows a raised tile ring or box device cover.

Figure 73 shows a masonry box that needs no extension or deep plaster ring to bring it to the surface.

NOTE
The electrician must work with the mason to ensure the box is properly grouted and sealed.

Sections of conduit are then coupled in short (4- or 5-foot) lengths. This is done because it is impractical for the mason to maneuver blocks over 10-foot sections of conduit.

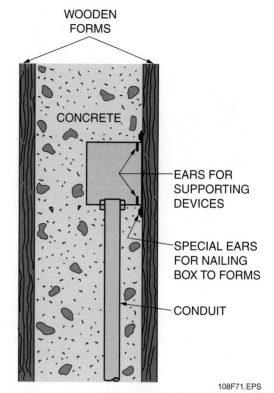

Figure 71 ◆ Concrete flush-mount installation.

Labels in figure: WOODEN FORMS, CONCRETE, EARS FOR SUPPORTING DEVICES, SPECIAL EARS FOR NAILING BOX TO FORMS, CONDUIT, 108F71.EPS

Figure 72 ◆ Box with raised ring.

108F72.EPS

14.2.0 Metal Stud Environment

Metal stud walls are a popular method of construction for the interior walls of commercial buildings. Metal stud framing consists of relatively thin metal channel studs, usually constructed of galvanized steel and with an overall dimension the same as standard 2 × 4 wooden studs. Wiring in this type of construction is relatively easy when compared to masonry.

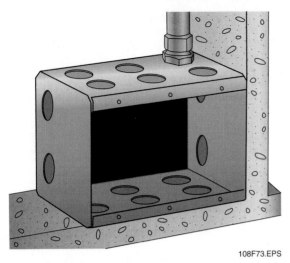

Figure 73 ◆ Three-gang concrete box.

EMT conduit and MC cable are the most common type of wiring methods for metal stud environments. Metal studs usually have some number of pre-punched holes that can be used to route the conduit. If a pre-punched hole is not located where it needs to be, holes can be easily punched in the metal stud with a hole cutter or knockout punch (*Figure 74*).

WARNING!

Cutting or punching metal studs can create sharp edges. Avoid contact that can result in cuts.

Boxes can be secured to the metal stud using self-tapping screws or one of the many types of box supports available. EMT conduit is supported by the metal studs using conduit straps or other

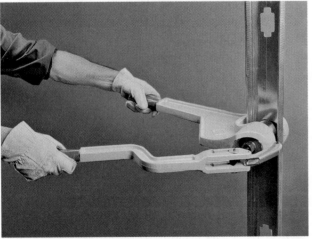

108F74.EPS

Figure 74 ◆ Metal stud punch.

approved methods. It is important that the conduit be properly supported to facilitate pulling the conductors through the tubing. Boxes are mounted on the metal studs so that the box will be flush with the finished walls. You must know what the finished wall thickness is going to be to properly secure the boxes to the metal studs. For example, if the finished wall will be ⅝-inch drywall, then the box must be fastened so that it protrudes ⅝ of an inch from the metal stud.

WARNING!

When using a screw gun or cordless drill to mount boxes to studs, keep the hand holding the box away from the gun/drill to avoid injury.

Per *NEC Section 300.4(B)(1)*, NM cable run through metal studs must be protected by listed bushings or listed grommets (*Figure 75*). This protects the cables from the friction of pulling during installation and from the weight of the cable and vibrations following the installation.

14.3.0 Wood Frame Environment

At one time, the use of rigid conduit in partitions and ceilings was a time-consuming operation. Thinwall conduit makes an easier and quicker job, largely because of the types of fittings that are specially adapted to it.

Figure 76 shows two methods of running thinwall conduit in these locations: boring timbers

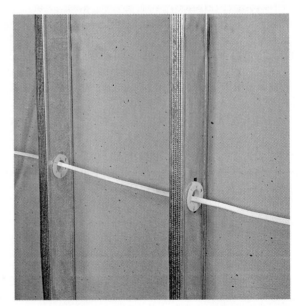

108F75.EPS

Figure 75 ◆ NM cable protected by grommets.

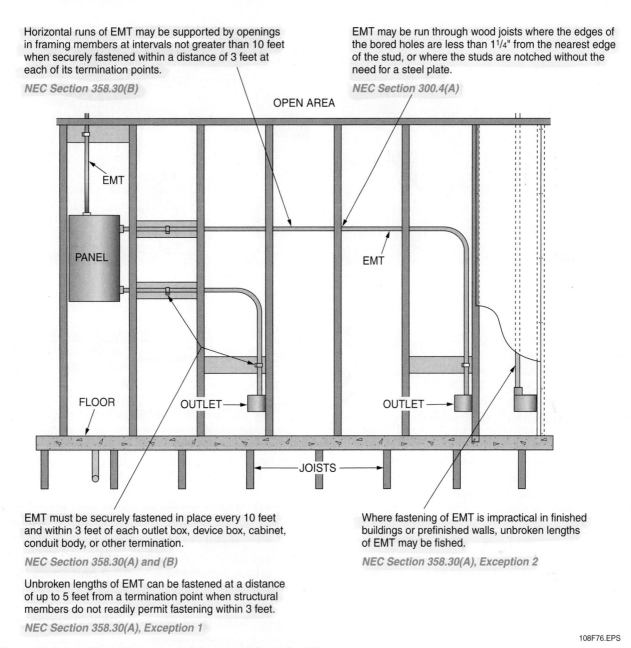

Horizontal runs of EMT may be supported by openings in framing members at intervals not greater than 10 feet when securely fastened within a distance of 3 feet at each of its termination points.

NEC Section 358.30(B)

EMT may be run through wood joists where the edges of the bored holes are less than 1¹/₄" from the nearest edge of the stud, or where the studs are notched without the need for a steel plate.

NEC Section 300.4(A)

OPEN AREA

EMT

PANEL

EMT

FLOOR

OUTLET

OUTLET

JOISTS

EMT must be securely fastened in place every 10 feet and within 3 feet of each outlet box, device box, cabinet, conduit body, or other termination.

NEC Section 358.30(A) and (B)

Unbroken lengths of EMT can be fastened at a distance of up to 5 feet from a termination point when structural members do not readily permit fastening within 3 feet.

NEC Section 358.30(A), Exception 1

Where fastening of EMT is impractical in finished buildings or prefinished walls, unbroken lengths of EMT may be fished.

NEC Section 358.30(A), Exception 2

108F76.EPS

Figure 76 ◆ Installing wire or conduit in a wood-frame building.

and notching them. When boring, holes must be drilled large enough for the tubing to be inserted between the studs. The tubing is cut rather short, calling for multiple couplings. EMT can be bowed quite a bit while threading through holes in studs. Boring is the preferred method.

 WARNING!
Always wear safety goggles when boring wood.

NEC Section 300.4 addresses the requirements to prevent physical damage to conductors and cabling in wood members. By keeping the edge of the drilled hole 1¼" from the closest edge of the stud, nails are not likely to penetrate the stud far enough to damage the cables. The building codes provide maximum requirements for bored or notched holes in studs.

NEC Section 300.4(A)(1) requires the use of a steel plate or bushing at least ¹⁄₁₆" thick or a listed steel nail plate where wiring is installed through bored wooden members less than 1¼" from the nearest edge. See *Figure 77*. Nail plates are also

Figure 77 ◆ Steel nail plate.

108F77.EPS

Figure 78 ◆ Metal building.

108F78.EPS

required to protect the conductors in all notched wooden members per *NEC Section 300.4(A)(2)*.

The exception in the *NEC®* permits IMC, RMC, PVC, and EMT to be installed through bored holes or laid in notches less than 1¼" from the nearest edge without a steel plate or bushing.

Because of its weakening effect upon the structure, notching should be resorted to only where absolutely necessary. Notches should be as narrow as possible and in no case deeper than ⅙ the stock of a bearing timber. A bearing timber supports floor joists or other weight.

 NOTE
Always check with the architect before notching or drilling.

Some wood I-beams are manufactured with perforated knockouts in their web, approximately 12" apart. Never notch or drill through the beam flange or cut other openings in the web without checking the manufacturer's specification sheet.

Also, do not drill or notch other types of engineered lumber without first checking the specification sheets.

14.4.0 Metal Buildings

Many commercial and industrial buildings are prefabricated structures with steel structural supports, and roofing and siding made of light-gauge metal sheets (*Figure 78*). Conduit can be routed across the structural members that support the roof. *NEC Section 300.4(E)* states that cable or raceway-type wiring in exposed or concealed locations under metal-corrugated sheet roof decking must be installed and supported so the nearest outside surface of the cable or raceway is not less than 1½" from the nearest surface of the roof decking. The roof structure can consist of beams and purlins (*Figure 79*) or open-web steel joists (*Figure 80*).

Beams and purlins should not be drilled through; consequently, the conduit is supported from the metal beams by anchoring devices designed especially for that purpose. The supports attach to the beams or supports and have clamps to secure the conduit to the structure. All conduit runs should be plumb since they are exposed. Bends should be correct and have a neat and orderly appearance.

Rigid metal conduit is often required in metal buildings. If a large number of conduits are run along the same path, strut-type systems are used. These systems are sometimes referred to as Unistrut® systems (Unistrut® is a manufacturer of these systems). Another manufacturer of strut systems is B-Line systems. Both are very similar. These systems use a channel-type member that can support conduits from the ceiling by using threaded rod supports for the channel, as shown in *Figure 81*. Strut channel can also be secured to masonry walls to support vertical runs of conduit, wireways, and various types of boxes.

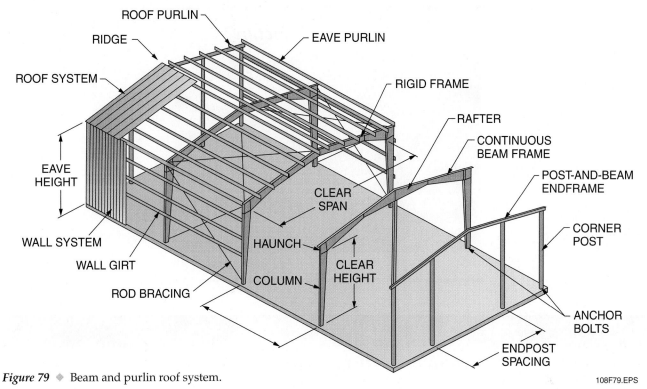

Figure 79 ◆ Beam and purlin roof system.

108F79.EPS

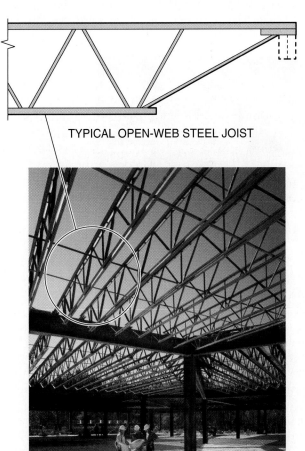

TYPICAL OPEN-WEB STEEL JOIST

Figure 80 ◆ Open-web steel joist roof supports.

108F80.EPS

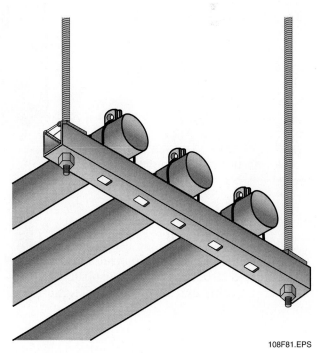

Figure 81 ◆ Steel strut system.

108F81.EPS

What's wrong with this picture?

108SA14.EPS

Putting It All Together

Think about the effort that goes into the design of a large industrial installation. If you were to design a large complex, such as the one shown here, where would you start and why?

108SA15.EPS

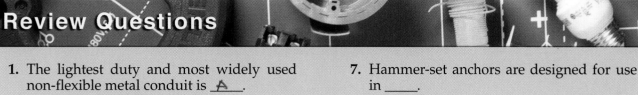

1. The lightest duty and most widely used non-flexible metal conduit is __A__.
   a. electrical metallic tubing
   b. rigid metal conduit
   c. aluminum conduit
   d. plastic-coated RMC

2. EMT and RMC connectors are interchangeable.
   a. True
   b. False

3. Because the wall thickness of _____ is less than that of rigid conduit, it is often referred to as thinwall conduit.
   a. IMC
   b. EMT
   c. RMC
   d. galvanized rigid steel conduit

4. RMC is made of _____.
   a. cast iron
   b. steel or aluminum
   c. copper or aluminum
   d. PVC

5. A Type LB conduit body has a cover on _____.
   a. the left
   b. the right
   c. the back
   d. both sides

6. The device used to protect conductors from the sharp edges of conduit where it enters a box is called a _____.
   a. bushing
   b. locknut
   c. coupling
   d. nipple

7. Hammer-set anchors are designed for use in _____.
   a. wood
   b. metal studs
   c. concrete
   d. structural steel

8. *NEC Section 376.56* limits wireway fill to no more than _____ of the cross-sectional area of the wireway.
   a. 40%
   b. 60%
   c. 75%
   d. 90%

9. Raceways designed to extend conductors across a floor without embedding it in the floor are called _____.
   a. cellular raceways
   b. raceway ducts
   c. cellular ductways
   d. pancake raceways

10. Which of the following regulations applies to drilling of wood joists and girders?
    a. A joist cannot be drilled unless approved by the manufacturer.
    b. They can only be drilled in the center third.
    c. The hole must be at least 1 inch from an edge.
    d. The hole diameter must not exceed one-half of the depth of the girder or joist.

Summary

This module discussed the various types of raceways, boxes, and fittings, including their uses and procedures for installation. The primary purpose of raceways is to house electric wire used for power distribution, communication, or electronic signal transmission. Raceways provide protection to the wiring and even a means of identifying one type of wire from another when run adjacent to each other. This process requires proper planning to allow for current needs, future expansion, and a neat and orderly appearance.

Notes

Trade Terms Quiz

1. A(n) _ACCESSIBLE_ area is one that can be reached for service or repair.

2. When something is in a(n) _EXPOSED LOCAT_, it is not permanently closed in by the structure or finish of a building.

3. When materials meet a regulatory agency's requirements, the material is then said to be _APPROVED_

4. _(UL)_ is a regulatory agency that evaluates and approves electrical components and equipment.

5. A(n) _BONDING WIRE_ is used to make a continuous grounding path between equipment and ground.

6. _CABLE TRAYS_ are rigid structures, either suspended or mounted, that are used to support electrical conductors.

7. Similar to pipe, _CONDUIT_ is a round raceway that houses conductors.

8. A(n) _KICK_ is a bend made in a piece of conduit to alter its course.

9. _RACEWAYS_ are enclosed channels that are used to house wires and cables.

10. _WIREWAYS_ are steel troughs designed to carry electrical wire and cable.

11. A(n) _SPLICE_ is the connection of two or more conductors.

12. An intermediate point on a main circuit where another wire is connected to supply electrical current to another circuit is called a(n) _TAP_.

13. Electrical connectors that could be exposed to the environment are housed in long, narrow boxes, or _TROUGH_.

Trade Terms

Accessible
Approved
Bonding wire
Cable trays

Conduit
Exposed location
Kick
Raceways

Splice
Tap
Trough

Underwriters
 Laboratories, Inc. (UL)
Wireways

Leonard "Skip" Layne

Rust Constructors Inc.

How did you choose a career in the electrical field?
I think the electrical field chose me. My father was a contractor for several years before closing shop and accepting a job as an electrical superintendent with the Rust Engineering Company. That happened when I was nine years old. After being moved around the country for the next several years and working as an apprentice on Dad's projects during my college summers, I couldn't think of anything that I would rather do.

Tell us about your apprenticeship experience.
I've never attended a formal apprenticeship school. There are probably several in our group who might say that they suspected this. My electrical education came from field work exposure and several electrical and engineering courses and seminars I've attended over the years.

I'm happy to say that I'm still learning and I've learned a great deal while working on the NCCER Electrical Committee and from my association with the other subject matter experts.

What positions have you held in the industry?
I started as a field apprentice on a tire plant in Madison, Tennessee, in 1959. I've held field positions as an apprentice, journeyman, field engineer, start-up manager, and superintendent. I spent a number of years estimating work, and I established the material control department for another major open-shop contractor several years ago. I managed the project controls group on a nuclear project for another open-shop contractor. I even spent a few years as vice

president with an underground utility/treatment plant contractor.

What would you say is the primary factor in achieving success?
Keep learning. Work hard. I've had to work sixteen-hour days on the job site and in the office in order to meet the schedule and incorporate changes. Do what is asked of you and do it well.

What does your current job involve?
My job title says that I'm the Construction Engineering Manager for Rust, but the lack of a definitive job title means that I do whatever the company needs me to do at the time. I qualify the company's electrical licenses in seventeen states where we work.

Recently, Rust volunteered my services to the Gulf Coast Workforce Initiative, a business roundtable initiative to train 20,000 new construction workers for the Gulf Coast area devastated by hurricanes Katrina and Rita.

Do you have any advice for someone just entering the trade?
Get all of the classroom learning you can. Go through all four levels of the *Contren®* Electrical Program while working in the field. Ask questions and try to get assigned to as many new and different tasks as you can. All of our larger ABC contractors have excellent supervisory training programs and you need to get into those after your craft training. Be adaptable and keep learning.

John Autrey
Trident Technical College

What positions have you held in the industry?
I made a career of the U.S. Navy and retired after 28 years of service. While in the Navy, I maintained complex electromechanical and electronic systems, taught electrical/electronics courses in Navy vocational schools, got progressively more responsibilities, and was ultimately commissioned a Chief Warrant Officer. In this capacity, I managed the repair and troubleshooting efforts of 125 technicians in thirteen separate shops at an ashore repair facility and was the electrical/electronics repair officer on surface ships. After retiring from the U.S. Navy in 1992, I obtained a master electrician and an electrical contractor's license in South Carolina and did various contracting work. In 1996, I was hired by Trident Technical College as the Program Coordinator for the Industrial Electricity and Electronics Program. Additionally, I'm a certified NCCER Master Trainer, Electrical craft instructor, and serve as an SME on NCCER's Electrical Committee.

What would you say is the primary factor in achieving success?
Truly enjoying your work and having confidence in your abilities and competence in your actions.

What does your current job involve?
I'm responsible for all aspects of the program curriculum, scheduling, and sequencing of classes, student advisement, and liaison with industry for both generalized and specialized electrical/electronic training for their maintenance technicians.

Do you have any advice for someone just entering the trade?
Do the very best job you're capable of instead of the bare minimum. Don't be afraid to ask questions and take every opportunity to enhance or advance your knowledge. In short, be professional in your work and your outlook.

Trade Terms
Introduced in This Module

Accessible: Able to be reached, as for service or repair.

Approved: Meeting the requirements of an appropriate regulatory agency.

Bonding wire: A wire used to make a continuous grounding path between equipment and ground.

Cable trays: Rigid structures used to support electrical conductors.

Conduit: A round raceway, similar to pipe, that houses conductors.

Exposed location: Not permanently closed in by the structure or finish of a building; able to be installed or removed without damage to the structure.

Kick: A bend in a piece of conduit, usually less than 45°, made to change the direction of the conduit.

Raceways: Enclosed channels designed expressly for holding wires, cables, or busbars, with additional functions as permitted in the *NEC®*.

Splice: Connection of two or more conductors.

Tap: Intermediate point on a main circuit where another wire is connected to supply electrical current to another circuit.

Trough: A long, narrow box used to house electrical connections that could be exposed to the environment.

Underwriters Laboratories, Inc. (UL): An agency that evaluates and approves electrical components and equipment.

Wireways: Steel troughs designed to carry electrical wire and cable.

This module is intended to present thorough resources for task training. The following reference works are suggested for further study. These are optional materials for continued education rather than for task training.

Benfield Conduit Bending Manual, 2nd Edition. Overland Park, KS: EC&M Books.

National Electrical Code® Handbook, Latest Edition. Quincy, MA: National Fire Protection Association.

NCCER makes every effort to keep these textbooks up-to-date and free of technical errors. We appreciate your help in this process. If you have an idea for improving this textbook, or if you find an error, a typographical mistake, or an inaccuracy in NCCER's Contren® textbooks, please write us, using this form or a photocopy. Be sure to include the exact module number, page number, a detailed description, and the correction, if applicable. Your input will be brought to the attention of the Technical Review Committee. Thank you for your assistance.

Instructors – If you found that additional materials were necessary in order to teach this module effectively, please let us know so that we may include them in the Equipment/Materials list in the Annotated Instructor's Guide.

Write: Product Development and Revision
National Center for Construction Education and Research
3600 NW 43rd St., Bldg. G, Gainesville, FL 32606

Fax: 352-334-0932

E-mail: curriculum@nccer.org

Craft _____ Module Name _____

Copyright Date _____ Module Number _____ Page Number(s) _____

Description _____

_____

_____

_____

(Optional) Correction _____

_____

_____

(Optional) Your Name and Address _____

_____

_____

Conductors and Cables

Forest Park-DeBaliviere Station

This MetroLink expansion project in downtown St. Louis used concrete in prolific ways, using more than 500 feet of concrete sewer bridges, installing three cut-and-cover tunnels totaling 1,400 linear feet, and constructing a pedestrian tunnel under a major thoroughfare. The expansion, which opened in August 2006, provides MetroLink customers with a new, eight-mile light rail line.

26109-08

26109-08
Conductors and Cables

Topics to be presented in this module include:

Overview

A conductor is the current-carrying portion of a wire along with its insulation. Electricians spend much of their time installing conductors. Conductors must be selected and installed very carefully.

When selecting the right conductor for the job, electricians are faced with many options, including conductor material, ampacity, type of insulation, and color coding. The *National Electrical Code®* regulates the color of conductors that may be used as grounding or grounded conductors.

Quality installation of conductors depends on using the right tool for the situation. The installation of large conductors in enclosed raceways sometimes requires power-driven equipment. Smaller conductors can be installed manually using a common tool called a fish tape. Conductor installation can be dangerous, and electricians must take necessary safety precautions. Conductor installations must be planned carefully to complete the job properly and without injury to the workers or the conductors.

Note: *National Electrical Code®* and *NEC®* are registered trademarks of the National Fire Protection Association, Inc., Quincy, MA 02269.
All *National Electrical Code®* and *NEC®* references in this module refer to the 2008 edition of the *National Electrical Code®*.

Objectives

When you have completed this module, you will be able to do the following:

1. From the cable markings, describe the insulation and jacket material, conductor size and type, number of conductors, temperature rating, voltage rating, and permitted uses.
2. Determine the allowable ampacity of a conductor for a given application.
3. Identify the *NEC®* requirements for color coding of conductors.
4. Install conductors in a raceway system.

Trade Terms

Ampacity
Capstan
Fish tape

Mouse
Wire grip

Required Trainee Materials

1. Paper and pencil
2. Copy of the latest edition of the *National Electrical Code®*
3. Appropriate personal protective equipment

Prerequisites

Before you begin this module, it is recommended that you successfully complete *Core Curriculum* and *Electrical Level One*, Modules 26101-08 through 26108-08.

This course map shows all of the modules in *Electrical Level One*. The suggested training order begins at the bottom and proceeds up. Skill levels increase as you advance on the course map. The local Training Program Sponsor may adjust the training order.

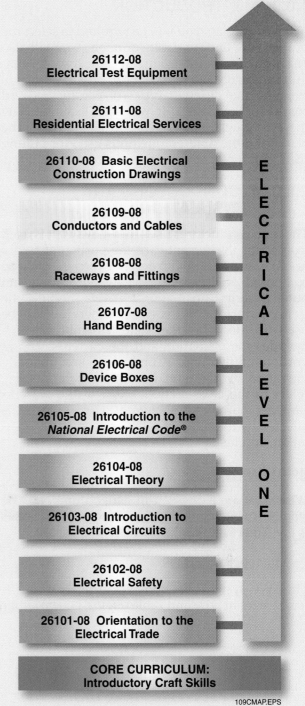

109CMAP.EPS

1.0.0 ◆ INTRODUCTION

As an electrician, you will be required to select the proper wire and/or cable for a job. You will also be required to pull this wire or cable through conduit runs in order to terminate it. This module will examine the different types of conductors and conductor insulation. It will also examine how these conductors are rated and classified by the *NEC*® and the different methods used for pulling these conductors through conduit runs.

2.0.0 ◆ CONDUCTORS AND INSULATION

The term conductor is used in two ways. It is used to describe the current-carrying portion of a wire or cable, and it is used to describe the wire or cable composed of a current-carrying portion and an outer covering (insulation). In this module, the term conductor, if not specified otherwise, will be used to describe the wire assembly, which includes the insulation and the current-carrying portion of the wire. Conductors are uniquely identified by size and insulation material. Size refers to the cross-sectional area of the current-carrying portion of the wire. The ampacity is affected by the conductor material and size, insulation, and installation location.

2.1.0 Wire Size

Wire sizes are expressed in gauge numbers. The standard system of wire sizes in the United States is the American Wire Gauge (AWG) system.

2.1.1 AWG System

The AWG system uses numbers to identify the different sizes of wire and cable (*Figure 1*). The larger the number, the smaller the cross-sectional area of the wire. The larger the cross-sectional area of the current-carrying portion of a conductor, the higher the amount of current the wire can conduct. The AWG numbers range from 50 to 1; then 0, 00, 000, and 0000 (one aught [1/0], two aught [2/0], three aught [3/0], and four aught [4/0]). Any wire larger than 0000 is identified by its area in circular mils. Wire sizes smaller than No. 18 AWG are usually solid, but may be stranded in some cases. Wire sizes of No. 6 AWG or larger are stranded.

For wire sizes larger than No. 16 AWG, the wire size is marked on the insulation (*Figure 2*).

NEC Chapter 9, Table 8 has descriptive information on wire sizes. Again, note that all wires smaller than No. 6 are available as solid or stranded. Wire sizes of No. 6 or larger are shown only as stranded. Solid wire larger than No. 6 is manufactured; however, the *NEC*® only permits the use of solid wire in a raceway for sizes smaller than No. 8 (*NEC Section 310.3*).

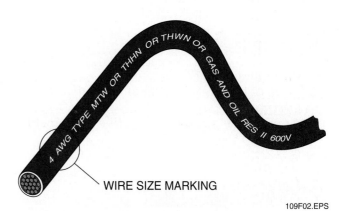

WIRE SIZE MARKING

109F02.EPS

Figure 2 ◆ Wire size marking.

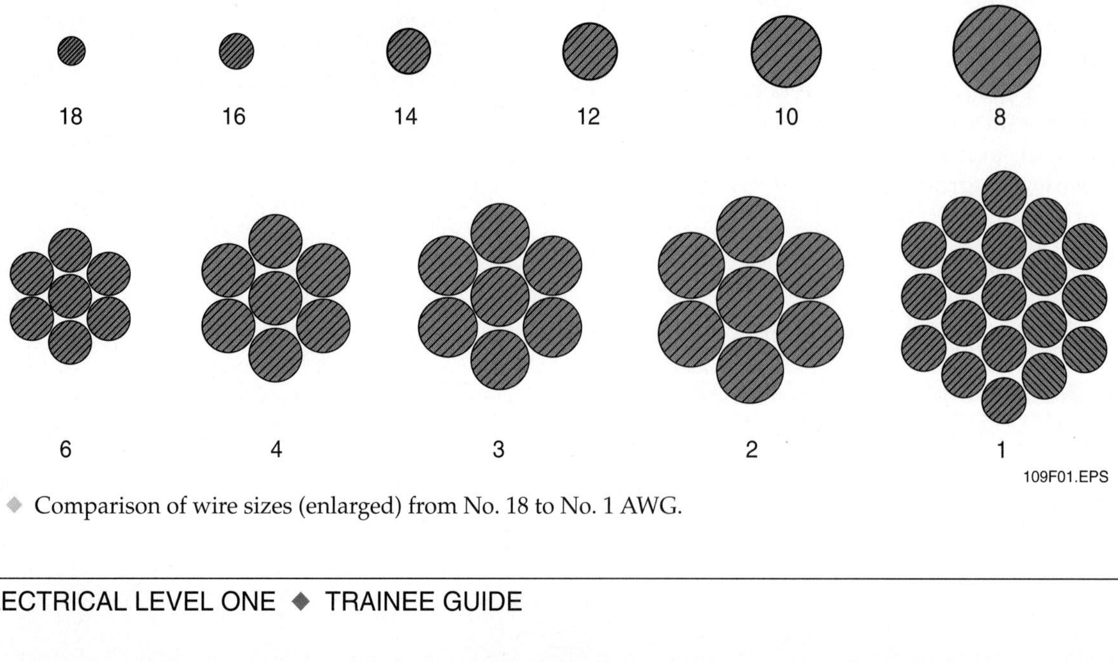

18   16   14   12   10   8

6   4   3   2   1

109F01.EPS

Figure 1 ◆ Comparison of wire sizes (enlarged) from No. 18 to No. 1 AWG.

Wire Size

2.1.2 Stranding

According to *NEC Chapter 9, Table 8,* wire sizes No. 18 to No. 2 have seven strands; wire sizes No. 1 to No. 4/0 have 19 strands; and wire sizes between 250 kcmil and 500 kcmil have 37 strands. The purpose of stranding is to increase the flexibility of the wire. Terminating solid wire sizes larger than No. 8 in pull boxes, disconnect switches, and panels would not only be very difficult, but might also result in damage to equipment and wire insulation.

Pulling solid wire in conduit around bends could pose a major problem and cause damage to equipment for wire sizes larger than No. 8. The reason for choosing 7, 19, and 37 strands for stranded conductors is that it is necessary to provide a flexible, almost round conductor. In order for a conductor to be flexible, individual strands must not be too large. *Figure 3* shows how these conductors are configured.

Aluminum conductors (*Figure 4*) are designed with compact stranding.

2.1.3 Circular Mils

A circular mil is a circle that has a diameter of 1 mil. A mil is 0.001 inch. When a wire size is 250 kcmil, the cross-sectional area of the current-carrying portion of the wire is the same as 250,000 circles having a diameter of 0.001 inch. This may seem to be a rather clumsy way of sizing wire at first; however, the alternative would be to size the wire as a function of its cross-sectional area expressed in square inches.

According to *NEC Chapter 9, Table 8,* the cross-sectional area of a 250 kcmil conductor is 0.260 square inch.

If a conductor is to be sized by cross-sectional area, it is much easier to express the wire size in circular mils (or thousands of circular mils) than in square inches.

2.2.0 Ampacity

Ampacity is the current in amperes a conductor can carry continuously under the conditions of use without exceeding its temperature rating. The

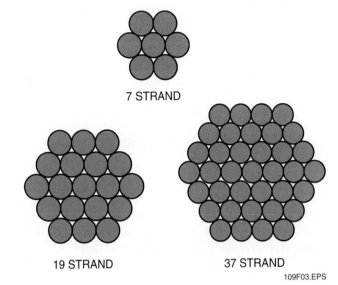

Figure 3 ◆ Strand configurations.

Figure 4 ◆ Aluminum conductors.

ampacity of conductors for given conditions of use are listed in *NEC Tables 310.16 through 310.19. NEC Table 310.16* covers conductors rated up to 2,000 volts where not more than three conductors are installed in a raceway or cable or are directly buried in the earth, based on an ambient temperature of 30°C (86°F).

NEC Table 310.17 covers both copper conductors and aluminum or copper-clad aluminum conductors up to 2,000 volts where conductors are used as single conductors in free air, based on an ambient temperature of 30°C (86°F).

NEC Tables 310.18 and 19 apply to conductors rated at 150°C to 250°C (302°F to 482°F), used either in raceway or cable or as single conductors in free air, based on an ambient temperature of 40°C (104°F).

Example:

Determine the ampacity of a No. 12 Cu (copper) THW conductor.

Solution:

25 amps (from *NEC Table 310.16*).

2.3.0 Conductor Material

The most common conductor material is copper. Copper is used because of its excellent conductivity (low resistance), ease of use, and value. The value of a material as an ingredient for wire is determined by several factors, including conductivity, cost, availability, and workability.

2.3.1 Conductivity

Conductivity is a word that describes the ease (or difficulty) of travel presented to an electric current by a conductor. If a conductor has a low resistance, it has a high conductivity. Silver is one of the best conductors since it has very low resistance and high conductivity. Copper has high conductivity and a lower price than silver. Aluminum, another material with good conductivity, is also a good choice for conductor material. The conductivity of aluminum is approximately two-thirds that of copper.

2.3.2 Cost

Cost is always an issue that contributes to the selection of a material to be used for a given application. Often, a material that has low cost may be selected as a conductor material even though it

has physical properties that are inferior to the more expensive material. Such is the case in the selection of copper over platinum. Here, the cost of platinum is very high, and very little thinking is required to determine that copper is a better choice. The choice between copper and aluminum is often more difficult to make.

2.3.3 Availability

The availability of some material is often a concern when selecting components for a job. As applied to wire, the mining industry often controls the availability of raw materials, which could produce shortages of some material. The availability of a substance such as copper or aluminum affects the price of the finished product (copper or aluminum wire).

2.3.4 Workability

It is a good idea to select a material that requires less expense for tools and is easier to work with. Aluminum conductors are lighter than copper conductors of the same size. They are also much more flexible than copper conductors and, in general, are easier to work with. However, terminating aluminum conductors often requires special tools and treatment of termination surfaces with an anti-oxidation material. Splicing and terminating aluminum conductors often requires a higher degree of training on the part of the electrician than do similar efforts with copper wire. This is partly due to the fact that aluminum expands and contracts with heat more than copper.

2.4.0 Conductor Insulation

The first attempt to insulate wire was made in the early 1800s during the development of the telegraph. This insulation was designed to provide physical protection rather than electrical protection. Electrical insulation was not an important issue because the telegraph operated at low-voltage DC. This early form of insulation was a substance composed of tarred hemp or cotton

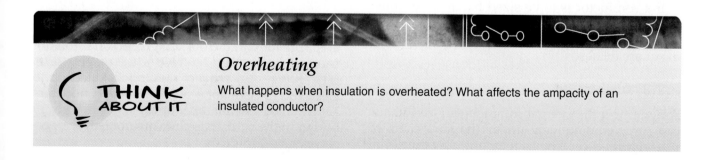

Overheating

What happens when insulation is overheated? What affects the ampacity of an insulated conductor?

fiber and shellac and was used primarily for weatherproofing long-distance distribution lines to mines, industrial sites, and railroads.

Some early electrical distribution systems utilized the knob-and-tube technique of installing wire. The wire was often bare and was pulled between and wrapped around ceramic knobs that were affixed to the building structure. When it was necessary to pull wire through structural members, it was pulled through ceramic tubes. The structural member (usually wood) was drilled, the tube was pressed into the hole, and the wire was pulled through the hole in the tube. As dangerous as this may appear, older homes still exist that have knob-and-tube wiring that was installed in the early 1900s and is still operational. Knob-and-tube wiring was revised to use insulated conductors and was in use up to 1957 in some areas.

The grounded or neutral conductor in overhead services may be bare. Furthermore, the concentric grounded conductor in Type SE cable may be bare when used as a service-entrance cable. However, all current-carrying conductors (including the grounded conductor) must be insulated when used on the inside of buildings, or after the first overcurrent protection device.

NEC Table 310.13(A) presents application and construction data on the wide range of 600-volt insulated, individual conductors recognized by the *NEC®*, with the appropriate letter designation used to identify each type of insulated conductor.

2.4.1 Thermoplastic

Thermoplastic is a popular and effective insulation material. The following thermoplastics are widely used:

- *Polyvinyl chloride (PVC)* – The base material used for the manufacture of TW and THW insulation.
- *Polyethylene (PE)* – An excellent weatherproofing material used primarily for insulation of control and communications wiring. It is not used for high-voltage conductors (those exceeding 5,000 volts).
- *Cross-linked polyethylene (XLP)* – An improved PE with superior heat- and moisture-resistant qualities. Used for THHN, THWN, and THHW wiring as well as many high-voltage cables.
- *Nylon* – Primarily used as jacketing material. THHN building wire has an outer coating of nylon.
- *Teflon®* – A high-temperature insulation. Widely used for telephone wiring in a plenum (where other insulated conductors require conduit routing).

2.4.2 Thermoset

Many thermoplastic materials deform when heated. Thermoset materials maintain their form when heated. Thermoset insulations include RHH, RHW, XHH, XHHW, and SIS.

2.4.3 Letter Coding

Conductor insulation as applied to building wire is coded by letters. The letters generally, but not always, indicate the type of insulation or its environmental rating. The types of conductor insulation described in this module will be those indicated at the top of *NEC Table 310.16*. The various insulation designations are shown in *Table 1*.

NOTE

Any conductor used in a wet location (see definition under Location, Wet, in *NEC Article 100*) must be listed for use in wet locations. Any conduit run underground is assumed to be subject to water infiltration and is, therefore, in a wet location.

2.4.4 Color Coding

A color code is used to help identify wires by the color of the insulation. This makes it easier to install and properly connect the wires. A typical color code is as follows:

- *Two-conductor cable* – One white or gray wire, one black wire, and a grounding wire (usually bare)
- *Three-conductor cable* – One white or gray, one black, one red, and a grounding wire
- *Four-conductor cable* – Same as three-conductor cable plus fourth wire (blue)
- *Five-conductor cable* – Same as four-conductor cable plus fifth wire (yellow)

Table 1 Insulation Coding

Letter	Description
B	Braid
E	Ethylene or Entrance
F	Fluorinated or Feeder
H	Heat-Rated or Flame-Retardant
N	Nylon
P	Propylene
R	Rubber
S	Silicon or Synthetic
T	Thermoplastic
U	Underground
W	Weather-Rated
X	Cross-Linked Polyethylene
Z	Modified Ethylene Tetrafluoroethylene
TW	Weather-Rated Thermoplastic (60°C/140°F)
FEP	Fluorinated Ethylene Propylene
FEPB	Fluorinated Ethylene Propylene with Glass Braid
MI	Mineral Insulation
MTW	Moisture, Heat, and Oil-Resistant Thermoplastic
PFA	Perfluoroalkoxy
RH	Heat-Rated Rubber (75°C/167°F)
RHH	Flame-Retardant Heat-Rated Rubber
RHW	Weather-Rated, Heat-Rated Rubber (75°C/167°F)
SA	Silicon
SIS	Synthetic Heat-Resistant
TBS	Thermoplastic Braided Silicon
TFE	Extended Polytetrafluoroethylene
THHN	Heat-Resistant Thermoplastic
THHW	Moisture and Heat-Resistant Thermoplastic
THW	Moisture and Heat-Resistant Thermoplastic
THWN	Weather-Rated, Heat-Rated Thermoplastic with Nylon Cover
UF	Underground Feeder
USE	Underground Service Entrance
XHH	Thermoset
XHHW	Heat-Rated, Flame-Retardant, Weather-Rated Thermoset
ZW	Weather-Rated Modified Ethylene Tetrafluoroethylene

109T01.EPS

The grounding conductor may be bare, green, or green with a yellow stripe. Power cable color codes are shown in *Figure 5*.

The *NEC*® does not require color coding of ungrounded conductors except where more than one nominal voltage system is present [*NEC Section 210.5(C)*]. The ungrounded conductors may be any color with the exception of white, gray, or green; however, it is a good practice to color code conductors as described here. In fact, many construction specifications require color coding. Furthermore, on a four-wire, delta-connected secondary where the midpoint of one phase is grounded to supply lighting and similar loads, the phase conductor having the higher voltage to ground must be identified by an outer finish that is orange in color, by tagging, or by other effective means. Such identification must be placed at each point where a connection is made if the grounded conductor is also present. In most cases, orange tape is used at all termination points when such a condition exists.

2.4.5 Wire Ratings

A critical factor in selecting conductors is the conductor's maximum operating temperature. Consider how and where a conductor will be used so that the conductor's limiting (maximum) temperature rating will not be exceeded. A conductor's operating temperature is determined by the ambient temperature, current flow in the conductor (including harmonic current), current flow in bundled conductors (which raises the ambient temperature), and how fast or slow heat is dissipated into the surrounding medium (which is affected by the conductor insulation).

Another significant factor to consider when selecting and installing conductors is where the conductor will be terminated. The temperature rating of the termination may limit the allowable ampacity of the conductor.

The amount of current a conductor can safely carry, and thus the maximum safe temperature the conductor can reach, is determined in general by conductor size (diameter in circular mils), the ambient (surrounding) temperature, the number of conductors in a bundle, and where the conductors are installed (raceways, conduit, ducts, underground, etc.).

THINK ABOUT IT

Conductor Insulation

What are the functions of conductor insulation? Under what conditions does the *NEC*® allow uninsulated conductors?

Insulation Types

Use *NEC Table 310.13(A)* to identify two types of insulation that are suitable for use in wet locations.

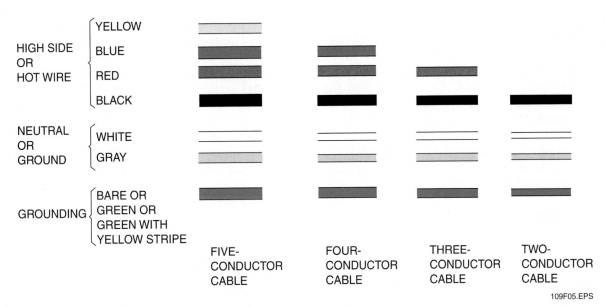

HIGH SIDE OR HOT WIRE
- YELLOW
- BLUE
- RED
- BLACK

NEUTRAL OR GROUND
- WHITE
- GRAY

GROUNDING
- BARE OR GREEN OR GREEN WITH YELLOW STRIPE

FIVE-CONDUCTOR CABLE FOUR-CONDUCTOR CABLE THREE-CONDUCTOR CABLE TWO-CONDUCTOR CABLE

109F05.EPS

Figure 5 ◆ Typical power cable insulation color codes.

Conductor selection is based largely on the temperature rating of the wire. This requirement is extremely important and is the basis of safe operation for insulated conductors. As shown in *NEC Table 310.13(A)*, conductors have various ratings (60°C, 75°C, 90°C, etc.). Since *NEC Tables 310.16 through 310.19* are based on an assumed ambient temperature of 30°C (86°F), conductor ampacities are based on the ambient temperature plus the heat (I²R) produced by the conductor while carrying current. Therefore, the type of insulation used on the conductor is the first consideration in determining the maximum permitted conductor ampacity.

For example, a No. 3/0 THW copper conductor for use in a raceway has an ampacity of 200 according to *NEC Table 310.16.* In a 30°C ambient temperature, the conductor is subjected to this temperature when it carries no current. Since a THW-insulated conductor is rated at 75°C, this leaves 45°C (75 − 30) for increased temperature due to current flow. If the ambient temperature exceeds 30°C, the conductor maximum load-current rating must be reduced proportionally (see Correction Factors at the bottom of *NEC Table 310.16*) so that the total temperature (ambient plus conductor temperature rise due to current flow) will not exceed the temperature rating of the conductor insulation (60°C, 75°C, etc.). For the same reason, the allowable ampacity must be reduced when more than three conductors are contained in a raceway or cable. See *NEC Section 310.15(B)(2).*

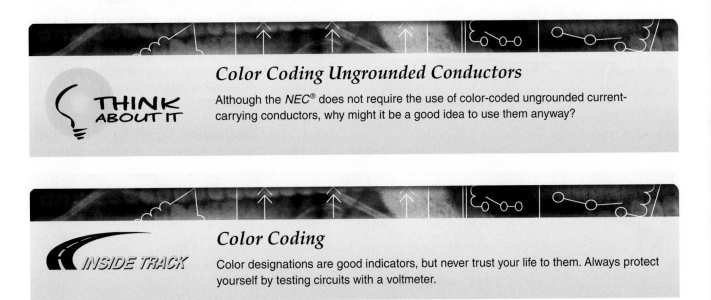

Color Coding Ungrounded Conductors

THINK ABOUT IT

Although the *NEC*® does not require the use of color-coded ungrounded current-carrying conductors, why might it be a good idea to use them anyway?

Color Coding

INSIDE TRACK

Color designations are good indicators, but never trust your life to them. Always protect yourself by testing circuits with a voltmeter.

Using the ampacity tables—An important step in circuit design is the selection of the type of conductor to be used (TW, THW, THWN, RHH, THHN, XHHW, etc.). The various types of conductors are covered in *NEC Article 310,* and the ampacities of conductors are given in *NEC Tables 310.16 through 310.19* for the varying conditions of use (e.g., in a raceway, in open air, at normal or higher-than-normal ambient temperatures). Conductors must be used in accordance with the data in these tables and notes.

2.5.0 Fixture Wires

Fixture wire is used for the interior wiring of fixtures and for wiring fixtures to a power source. Guidelines concerning fixture wire are given in *NEC Article 402.* The list of approved types of fixture wire is given in *NEC Table 402.3. Figure 6* shows one example of fixture wire. The wires are composed of insulated conductors with or without an outer jacket. The conductors range in size from No. 18 to No. 10 AWG.

The decision of which fixture wire to use depends primarily upon the operating temperature that is expected within the fixture. Therefore, it is the character of the insulation that will determine the wire selected. For instance, fixture wires insulated with perfluoroalkoxy (PFA) or extruded polytetrafluoroethylene (PTF) would be selected if the operating temperature of the fixture is expected to reach a maximum of 482°F. This is the highest operating temperature allowed for any fixture wire.

As indicated by *NEC Section 402.3,* fixture wires are suitable for service at 600 volts unless otherwise specified in *NEC Table 402.3.* The allowed ampacities of fixture wire are given in *NEC Table 402.5.*

Although the primary use for fixture wire is the internal wiring of fixtures, several of the wires listed in *NEC Table 402.3* may be used for wiring remote-control, signaling, or power-limited circuits in accordance with *NEC Section 725.49.* Fixture wires may never be used as substitutes for branch circuit conductors.

2.6.0 Cables

Cables are two or more insulated wires and may contain a grounding wire covered by an outer jacket or sheath. Cable is usually classified by the type of covering it has, either nonmetallic (plastic) or metallic, also called armored cable.

Cable may also be classified according to where it can be used (see *NEC Table 400.4*). Because water is such a good conductor of electricity, moisture on conductors can cause power loss or short circuits. For this reason, cables are classified for either dry, damp, or wet locations. Cables can also be classified regarding exposure to sunlight and rough use.

2.6.1 Cable Markings

All cables are marked to show important properties and uses. Cable markings show the wire size, number of conductors, cable type, and voltage rating. In addition, a marking may be included to signify approved service or applications. This information is printed on nonmetallic cable (*Figure 7*). On metallic cable, marking information is usually included on a tag.

2.6.2 Nonmetallic-Sheathed Cable

Nonmetallic-sheathed cable (Type NM and Type NMC) is widely used for branch circuits and feeders in residential and commercial systems. See *Figure 8.* Both types are commonly called Romex®, even though the cable manufacturer only calls Type NM cable Romex®. Guidelines for the use of nonmetallic-sheathed cable are given in *NEC Article 334.* This cable consists of two or three

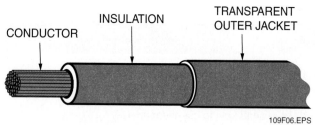

CONDUCTOR   INSULATION   TRANSPARENT OUTER JACKET

109F06.EPS

Figure 6 ◆ Fixture wire.

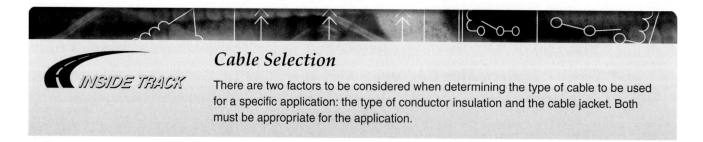

Cable Selection

INSIDE TRACK

There are two factors to be considered when determining the type of cable to be used for a specific application: the type of conductor insulation and the cable jacket. Both must be appropriate for the application.

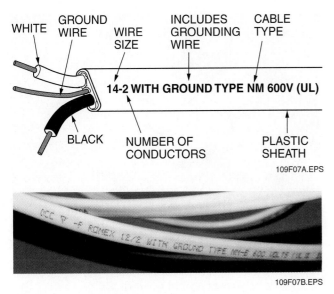

Figure 7 ◆ Nonmetallic cable markings.

insulated conductors and one bare conductor enclosed in a nonmetallic sheath. The conductors may be wrapped individually with paper, and the spaces between the conductors may be filled with jute, paper, or other material to protect the conductors and help the cable keep its shape. The sheath covering both Type NM cable and Type NMC cable is flame-retardant and moisture-resistant. The sheath covering Type NMC cable has the additional characteristics of being fungus- and corrosion-resistant.

NEC Article 334 lists the allowed and prohibited uses for Type NM cable and Type NMC cable. Both are allowed to be installed in either exposed or concealed work. The primary difference in their applied uses is that Type NM cable is suitable for dry locations only, whereas Type NMC is permitted for dry, moist, damp, or corrosive locations.

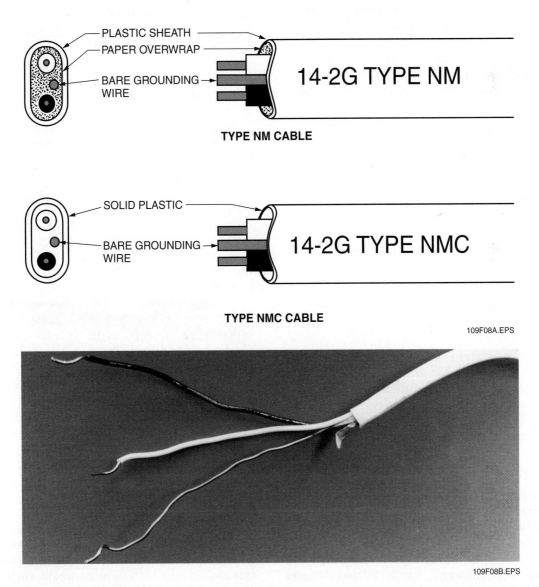

Figure 8 ◆ Nonmetallic-sheathed cable.

Types NM and NMC cables may be used in one- and two-family dwellings, and in certain multifamily dwellings, depending on the type of construction. See *NEC Section 334.10*.

In general, NM and NMC cable cannot be used in ducts or plenums because toxic gases from burning cable insulation would be spread throughout the structure. NM and NMC cables cannot be installed exposed in the space above suspended ceilings. They cannot be used as service-entrance cable, embedded in concrete, or in hazardous locations. There are many other requirements for NM and NMC specified in *NEC Article 334*. Before installing this type of cable, be sure you read and understand the applicable sections of the *NEC®*.

2.6.3 Type UF Cable

Guidelines for the use of Type UF (underground feeder and branch circuit) cable are given in *NEC Article 340*. Type UF cable is very similar in appearance, construction, and use to Type NMC cable. The main difference between these two cables is that Type UF cable is suitable for direct burial, whereas Type NMC cable is not.

Some of the permitted uses of Type UF cable are: underground and direct burial, as a single-conductor cable, in wet, dry, or corrosive conditions, as a nonmetallic-sheathed cable, in solar photovoltaic systems, and in cable trays.

Typically, Type UF cable may not be used as service-entrance cable, in commercial garages, in theaters, in hoistways or elevators, or in hazardous locations. Type UF cable cannot generally be embedded in poured cement, concrete, or aggregate, exposed to sunlight (unless designed for that use), or used as an overhead cable. Refer to *NEC Article 340* for specifics on where and when to use Type UF cable.

2.6.4 Type NMS Cable

Refer to *NEC Sections 334.10 and 334.12* for the applications of Type NMS cable. Type NMS cable is a form of nonmetallic-sheathed cable that contains a factory assembly of power, communications, and signaling conductors enclosed within a moisture-resistant, flame-retardant sheath.

2.6.5 Type MV Cable

Type MV (medium-voltage) cable is covered in *NEC Article 328*. It consists of one or more insulated conductors encased in an outer jacket. This cable is suitable for use with voltages ranging from 2,001 to 35,000 volts. It may be installed in wet and dry locations and may be buried directly in the earth. See *Figure 9*.

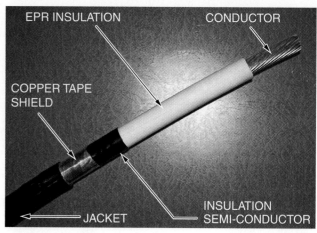

Figure 9 ◆ Type MV cable.

2.6.6 Type MC Cable

Type MC (metal clad) cable consists of one or more insulated conductors encased in a metal tape or a metallic sheath. *NEC Article 330* covers Type MC cable. Further information can be found in *UL 1569, Standard for Metal Clad Cables*.

MC cable is used in a wide variety of applications, from small instrumentation cable up to medium voltage feeders. The conductors are coated with a thermoset or thermoplastic insulation. Type MC cable can also be a composite of electrical conductors and optical fiber conductors.

Typical markings on the cable include the maximum rated voltage, AWG size (or circular mil area), and insulation type. If the outer covering will not accept markings, the markings will be on a tape inside the cable along the entire length of the cable. If on the outside, the markings typically have a 24-inch spacing.

The three types of MC cable are: interlocked metal tape, corrugated metal tube, and smooth metal tube. Cables with special uses will be marked accordingly. The outer covering may be a nonmetallic jacket over the metal sheath. One type of MC cable is shown in *Figure 10*.

Some of the typical uses for the three types of MC cable are for services, feeders, and branch circuits; for power, lighting, control, and signal circuits; indoors or outdoors; exposed or concealed; direct burial (if identified for that use); in any raceway; and other uses specified in *NEC Article 330*.

Type MC cable may not be used in corrosive or damaging conditions unless the metal cladding protects the conductors, or some other protective material is used. Uses typically not permitted are in areas where the cable is subject to physical damage, direct burial, in concrete, or where subject to caustic materials.

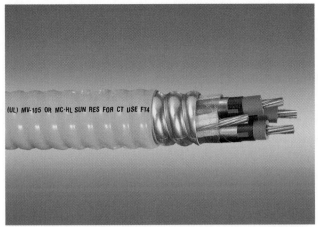

Figure 10 ◆ Type of MC cable.

109F10.EPS

Typically, UL does not recognize the interlocking metal cladding of Type MC cable as the only means of grounding equipment. For this reason, Type MC cable is not allowed in certain applications, such as patient care areas in hospitals.

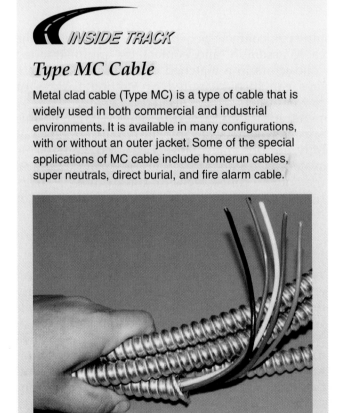
Both armored (AC) and MC cables provide advantages during installation. The flexible metal sheath protects the conductors and allows them to bend around corners without kinking or damage to the conductor. Also, since the conductors are already protected by the sheathing, there is no need to pull conductors into a raceway, nor is there concern about conductor contact with pipes or other hard surfaces. Other advantages of metal clad cables are their relatively easy installation without the need for wire pullers, fish tapes, or lubricants.

There are some fundamental differences between Types AC and MC cables. The significant differences are:

• AC cable has a maximum of four conductors, plus a grounding conductor, and comes in sizes from 14 AWG to 1 AWG. Conversely, MC cable has no limitations on the number of conductors, and is sized from 18 AWG to 2,000 kcmil.
• AC cable has a bonding strip (16 AWG). This strip is in constant contact with the armor and, with the armor, forms an equipment ground. MC cable has no bonding strip. The MC cladding is not a ground, although it can supplement the ground.
• AC cable uses moisture-resistant and fire-retardant paper wraps on individual conductors. MC cables have no such paper wrap, but do incorporate a polyester tape used on the assembly.

2.6.7 High-Voltage Shielded Cable

Shielding of high-voltage cables protects the conductor assembly against surface discharge or burning due to corona discharge in ionized air, which can be destructive to the insulation and jacketing.

Electrostatic shielding of cables makes use of both nonmetallic and metallic materials (*Figures 11* and *12*).

2.6.8 Channel Wire Assemblies

Channel wire assemblies (Type FC) comprise an entire wiring system, which includes the cable, cable supports, splicers, circuit taps, fixture hangers, and fittings (*Figure 13*). Guidelines for the use of this system are given in *NEC Article 322*. Type FC cable is a flat cable assembly with three or four parallel No. 10 special stranded copper conductors. The assembly is installed in an approved U-channel surface metal raceway with one side open. Tap devices can be inserted anywhere along the run. Connections from the tap devices to the flat cable assembly are made by pin-type contacts when the tap devices are fastened in place. The

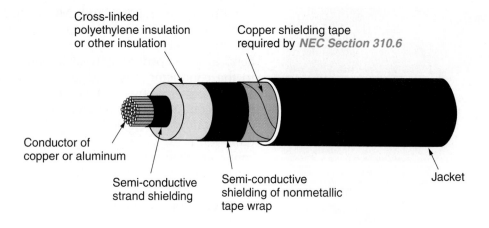

Cross-linked
polyethylene insulation
or other insulation

Copper shielding tape
required by *NEC Section 310.6*

Conductor of
copper or aluminum

Semi-conductive
strand shielding

Semi-conductive
shielding of nonmetallic
tape wrap

Jacket

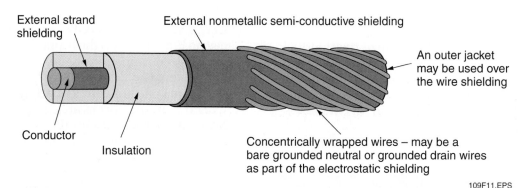

External strand
shielding

External nonmetallic semi-conductive shielding

An outer jacket
may be used over
the wire shielding

Conductor

Insulation

Concentrically wrapped wires – may be a
bare grounded neutral or grounded drain wires
as part of the electrostatic shielding

109F11.EPS

Figure 11 ◆ Various types of shielding.

Six corrugated copper drain wires embedded in
semi-conductive jacket provide shielding instead of tape shield,
and can be pulled out of the way (ripped out of the jacket) to
allow stress cone assembly at the correct point.

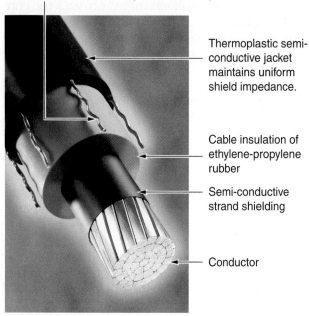

Thermoplastic semi-
conductive jacket
maintains uniform
shield impedance.

Cable insulation of
ethylene-propylene
rubber

Semi-conductive
strand shielding

Conductor

109F12.EPS

Figure 12 ◆ Corrugated drain wire shielding.

pin-type contacts penetrate the insulation of the cable assembly and contact the multi-stranded conductors in a matched phase sequence. These taps can then be wired to lighting fixtures or power outlets (*Figure 14*).

As indicated in *NEC Section 322.10,* this wiring system is suitable for branch circuits that only supply small appliances and lights. This system is suitable for exposed wiring only and may not be concealed within the building structure. It is ideal for quick branch circuit wiring at field installations.

2.6.9 Flat Conductor Cable

Type FCC (flat conductor) cable comprises an entire branch wiring system similar in many respects to Type FC flat conductor assemblies. Guidelines for the use of this system are given in *NEC Article 324.* Type FCC cable consists of three to five flat conductors placed edge-to-edge, separated, and enclosed in a moisture-resistant and flame-retardant insulating assembly. Accessories include cable connectors, terminators, power source adapters, and receptacles.

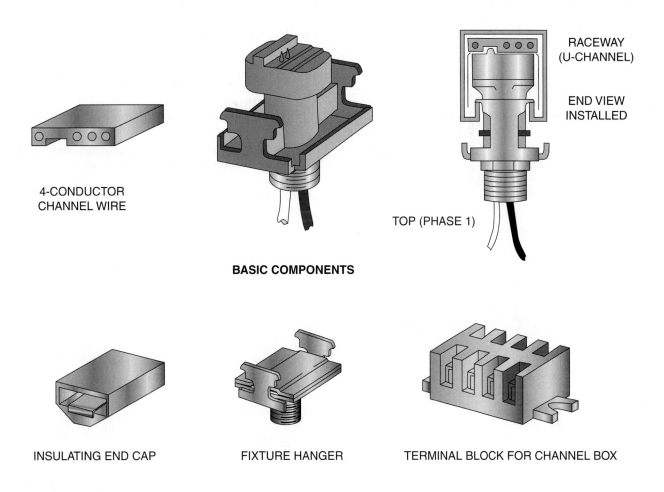

RACEWAY
(U-CHANNEL)

END VIEW
INSTALLED

4-CONDUCTOR
CHANNEL WIRE

TOP (PHASE 1)

BASIC COMPONENTS

INSULATING END CAP   FIXTURE HANGER   TERMINAL BLOCK FOR CHANNEL BOX

ACCESSORIES

109F13.EPS

Figure 13 ◆ Channel wire components and accessories.

This wiring system has been designed to supply floor outlets in office areas and other commercial and institutional interiors. It is meant to be run under carpets so that no floor drilling is required. This system is also suitable for wall mounting. As indicated in *NEC Article 324*, telephone and other communications circuits may share the same enclosure as Type FCC flat cable. The main advantage of the system is its ease of installation. It is the ideal wiring system for use when remodeling or expanding existing office facilities.

2.6.10 Type TC Cable

Guidelines for the use of Type TC (power and control tray) cable are given in *NEC Article 336*. Type TC cable consists of two or more insulated conductors twisted together, with or without associated bare or fully insulated grounding conductors, and covered with a nonmetallic jacket. The cables are rated at 600 volts. The cable is listed in conductor sizes No. 18 AWG to 2,000 kcmil copper or No. 12 AWG to 2,000 kcmil aluminum or copper-clad aluminum (*Figure 15*).

As the T in the letter designator indicates, this cable is tray cable. It can be used in cable trays and raceways. It may also be buried directly if the sheathing material is suitable for this use. Type TC cable is also good for use in sunlight when indicated by the cable markings.

2.6.11 SE and USE Cable

Guidelines for the use of Types SE (service-entrance) and USE (underground service-entrance) cable are given in *NEC Article 338*. The *NEC*® contains no specifications for the construction of this cable; it is left to UL to determine what types of cable should be approved for this purpose. Currently, service-entrance cable is labeled in sizes No. 12 AWG and larger for copper, and No. 10 AWG and larger for aluminum or copper-clad aluminum, with Types RH, RHW, RHH, or

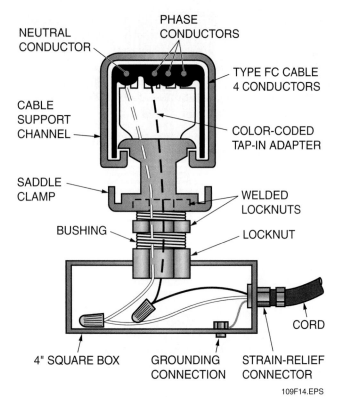

NEUTRAL CONDUCTOR

PHASE CONDUCTORS

TYPE FC CABLE 4 CONDUCTORS

CABLE SUPPORT CHANNEL

COLOR-CODED TAP-IN ADAPTER

SADDLE CLAMP

WELDED LOCKNUTS

BUSHING

LOCKNUT

CORD

4" SQUARE BOX

GROUNDING CONNECTION

STRAIN-RELIEF CONNECTOR

109F14.EPS

Figure 14 ◆ Type FC connection.

109F15.EPS

Figure 15 ◆ Type TC cable.

XHHW conductors. If the type designation for the conductor is marked on the outside surface of the cable, the temperature rating of the cable corresponds to the rating of the individual conductor. When this marking does not appear, the temperature rating of the cable is 75°C (167°F). Type SE cable is for aboveground installation only.

When used as a service-entrance cable, Type SE must be installed as specified in *NEC Article 230.* Service-entrance cable may also be used as feeder and branch circuit cable. Guidelines for the use of service-entrance cable are given in *NEC Section 338.10. Figure 16* shows SE cable with a bare aluminum conductor.

Type USE cable is for underground installation including burial directly in the earth. Type USE cable in sizes No. 4/0 AWG and smaller with all conductors insulated is suitable for all of the underground uses for which Type UF cable is permitted by the *NEC*®.

Type USE cable may consist of either single conductors or a multi-conductor assembly provided with a moisture-resistant covering, but it is not required to have a flame-retardant covering. This type of cable may have a bare copper conductor cabled with the assembly. Furthermore, Type USE single, parallel, or cabled conductor assemblies recognized for underground use may

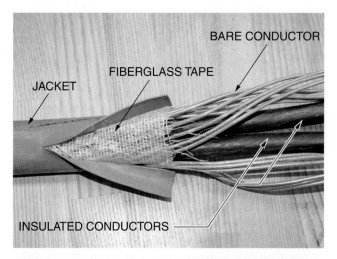

BARE CONDUCTOR

FIBERGLASS TAPE

JACKET

INSULATED CONDUCTORS

END VIEW

109F16.EPS

Figure 16 ◆ SE cable.

have a bare copper concentric conductor applied. These constructions do not require an outer overall covering. Guidelines for the use of Type USE cable are specified in *NEC Article 338.*

When used as a service-entrance cable, Type USE cable must be installed as specified in *NEC Article 230.* Take the time to read *NEC Article 230* to ensure proper installation. Type USE service-entrance cable may also be used as feeder and branch circuit cable. Guidelines for this use of service-entrance cable are given in *NEC Section 338.10.*

2.7.0 Instrumentation Control Wiring

Instrumentation control wiring links the field-sensing, controlling, printout, and operating devices that form an electronic instrumentation control system. The style and size of instrumentation control wiring must be matched to a specific job.

Instrumentation control wiring usually has two or more insulated conductor wires. These wires may also have a shield and a ground wire. An outer layer called the jacket protects the wiring (*Figure 17*). Instrumentation conductor wires come in pairs. The number of pairs in a multi-conductor cable depends on the size of the wire used. A multi-pair cable typically has 12, 24, or 36 pairs of conductors.

2.7.1 Shields

Shields are provided on instrumentation control wiring to protect the electrical signals traveling through the conductors from electrical interference or noise. Shields are usually constructed of aluminum foil bonded to a plastic film (*Figure 18*). If the wiring is not properly shielded, electrical noise may cause erratic or erroneous control signals, false indications, and improper operation of control devices.

2.7.2 Shield Drain

A shield drain is a bare copper wire used in continuous contact with a specified grounding terminal. A shield drain allows connection of all the instruments within a loop to a common grounding

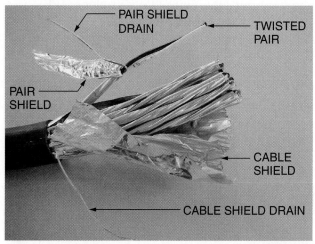

109F18.EPS

Figure 18 ◆ Multi-conductor instrumentation control cable with overall cable shield and individually shielded pairs.

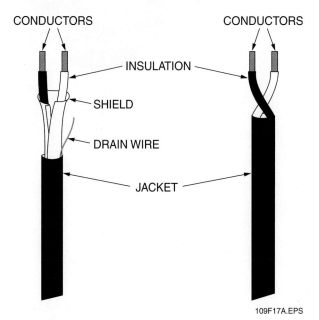

109F17A.EPS

Figure 17 ◆ Instrumentation control cable.

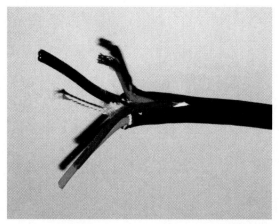

109F17B.EPS

point. Always refer to the loop diagram to determine whether or not the shield is to be terminated.

Typically, the shielding in instrumentation circuits is grounded at one end of the conductor only. The purpose of this is to drain induced charges to ground but not allow a circulating path for the flow of induced current. If the ground is not to be connected at the end of the wire you are installing, do not remove the ground wire. Fold it back and tape it to the cable. This is called floating the ground.

2.7.3 Jackets

A plastic jacket covers and protects the components within the wire. Polyethylene (PE) and polyvinyl chloride (PVC) jackets are the most commonly used (*Figure 19*). Some jackets have a nylon rip cord that allows the jacket to be peeled back without the use of a knife or cable cutter. This eliminates nicking of the conductor insulation when preparing for termination.

3.0.0 ◆ INSTALLING CONDUCTORS IN CONDUIT SYSTEMS

Conductors are installed in all types of conduit by pulling them through the conduit. This is done by using **fish tape**, pull lines, and pulling equipment.

3.1.0 Fish Tape

Fish tape can be made of flexible steel or nylon and is available in coils of 25 to 200 feet. It should be kept on a reel to avoid twisting. Fish tape has a hook or loop on one end to attach to the conductors to be pulled (*Figure 20*). Broken or damaged fish tape should not be used. To prevent electrical shock, fish tape should not be used near or in live circuits.

Fish tape is fed through the conduit from its reel. The tape usually enters at one outlet or junction box and is fed through to another outlet or junction box (*Figure 21*).

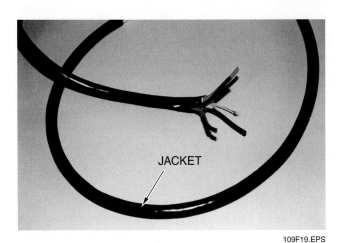

109F19.EPS

Figure 19 ◆ Wire jacket.

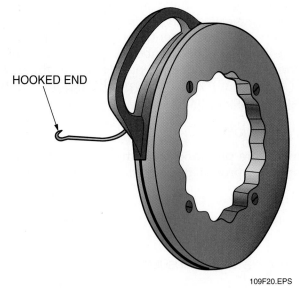

HOOKED END

109F20.EPS

Figure 20 ◆ Fish tape.

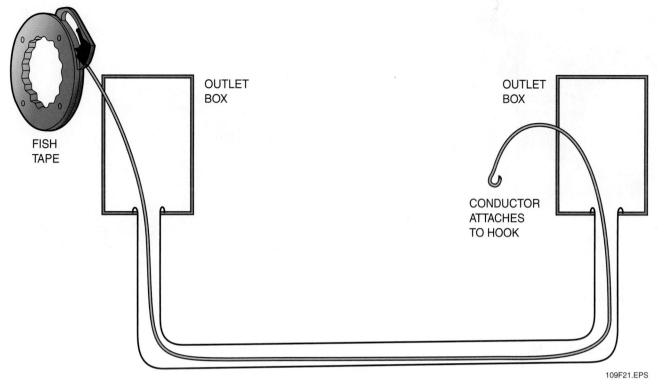

OUTLET
BOX

OUTLET
BOX

FISH
TAPE

CONDUCTOR
ATTACHES
TO HOOK

109F21.EPS

Figure 21 ◆ Fish tape installation.

Sometimes fish tape can get hung up in very long conduit runs. These situations call for a rigid fishing tool known as a rodder. Rodders are available in various sizes and in lengths up to 1,000 feet. A typical rodder is shown in *Figure 22*.

109F22.EPS

Figure 22 ◆ Rodder.

3.1.1 Power Conduit Fishing Systems

String lines can be installed by using different types of power systems. The power system is similar to an industrial vacuum cleaner and pulls a string or rope attached to a piston-like plug (sometimes called a mouse) through the conduit. Once the string emerges at the opposite end, either the conductor or a pull rope is then attached and pulled through the conduit, either manually or with power tools. See *Figure 23*.

The hose connection on these vacuum systems can also be reversed to push the mouse through the conduit. In other words, the system can either suck or blow the mouse through the conduit, depending on which method is best in a given situation. In either case, a fish tape is then attached to the string for retrieving through the conduit.

3.1.2 Connecting Wire to a String Line

Once the string is installed in the conduit run, a fish tape is connected to it and pulled back through the conduit. Conductors are then attached to the hooked end of the fish tape or else connected to a basket grip. In most cases, all required conductors are pulled at one time.

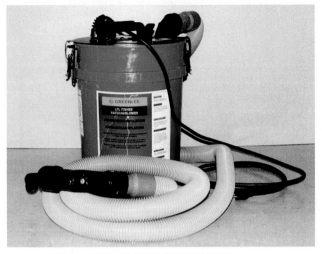

VACUUM BLOWER UNIT

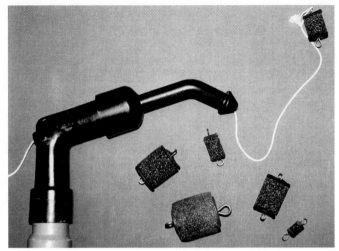

FOAM PLUGS

109F23.EPS

Figure 23 ◆ Power fishing system.

3.2.0 Wire Grips

Wire grips are used to attach the cable to the pull tape. One type of wire grip used is a basket grip (sometimes called Chinese Fingers). A basket grip is a steel mesh basket that slips over the end of a large wire or cable (*Figure 24*). The fish tape hooks onto the end and the pull on the fish tape tightens the basket over the conductor.

3.3.0 Pull Lines

If a pull is going to be difficult because of bends in the conduit, the size of the conductors, or the length of the pull, a pull line should be used.

 WARNING!

When using pull lines, exercise extreme caution and never stand in a direct line with the pulling rope. If the rope breaks, the line will whip back with great force. This can result in serious injury or death.

109F24.EPS

Figure 24 ◆ Basket grip.

A pull line is usually made of nylon or some other synthetic fiber. It is made with a factory-spliced eye for easy connection to fish tape or conductors.

3.4.0 Safety Precautions

The following are several important safety precautions that will help to reduce the chance of being injured while pulling cable.

- To avoid electrical shock, never use fish tape near or in live circuits.
- Read and understand both the operating and safety instructions for the pull system before pulling cable.
- When moving reels of cable, use mechanical lifts for longer spools. For smaller spools, avoid back strain by using your legs to lift (rather than your back) and asking for help with heavy loads. Also, when manually pulling wire, spread your legs to maintain your balance and do not stretch.
- Be careful to avoid any pinchpoints in the capstans and sheaves.
- Select a rope that has a pulling load rating greater than the estimated forces required for the pull.
- Use only low-stretch rope such as multiplex and double-braided polyester for cable pulling. High-stretch ropes store energy much like a stretched rubber band. If there is a failure of the rope, pulling grip, conductors, or any other component in the pulling system, this potential energy will suddenly be unleashed. The whipping action of a rope can cause considerable damage, serious injury, or death.

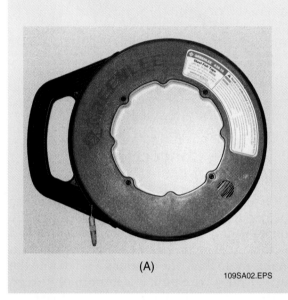

Fish Tape Selection

Metal fish tape (A) generally comes in longer lengths and is the type used most often. Nylon fish tape (B) generally comes in shorter lengths and is more flexible than metal fish tape.

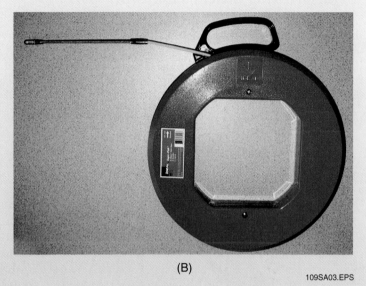

(A)

109SA02.EPS

(B)

109SA03.EPS

- Inspect the rope thoroughly before use. Make sure there are no cuts or frays in the rope. Remember, the rope is only as strong as its weakest point.
- When designing the pull, keep the rope confined in conduit wherever possible. Should the rope break or any other part of the pulling system fail, releasing the stored energy in the rope, the confinement in the conduit will work against the whipping action of the rope by playing out much of this energy within the conduit.
- Do not stand in a direct line with the pulling rope. Some equipment is designed so that you may stand to one side for safety.
- Wrap up the pulling rope after use to prevent others from tripping over it.

3.5.0 Pulling Equipment

Many types of pulling equipment are available to help pull conductors through conduit. Pulling equipment can be operated both manually and electrically (*Figure 25*). A manually operated puller is used mainly for smaller pulling jobs where hand pulling is not possible or practical. It is also used in many locations where hand pulling would put an unnecessary strain on the conductors because of the angle of the pull involved.

Electrically driven power pullers are used where long runs, several bends, or large conductors are involved.

The main parts of a power puller are the electric motor, the chain or sprocket drive, the capstan, the sheave, and the pull line.

The pull line is routed over the sheave to ensure a straight pull. The pull line is wrapped around the capstan two or three times to provide a good grip on the capstan. The capstan is driven by the electric motor and does the actual pulling. The pull line is unwound by hand at the same speed at which the capstan is pulling. This eliminates the need for a large spool on the puller to wind the pull line.

Attachments to power pullers, such as special application sheaves and extensions, are available for most pulling jobs. Follow the manufacturer's instructions for setup and operation of the puller.

CAUTION

Before using power pullers, a qualified person must verify the amount of pull or tension that will be exerted on the conductors being pulled.

Straightening a Bent Fish Tape

To straighten a bent fish tape, drive five 16-penny (16d) nails into a 2 × 4 about 1" apart in a straight line. Then wind the fish tape through the nails in a slalom fashion. This will straighten the tape.

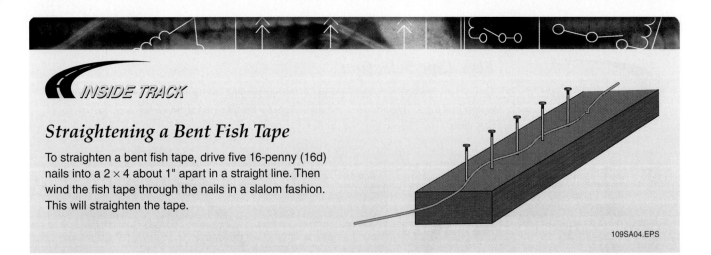

109SA04.EPS

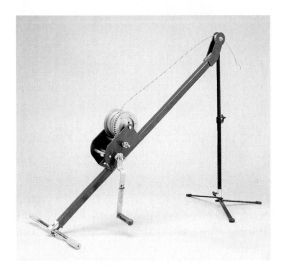

(A) MANUAL WIRE PULLER

(B) POWER PULLER

109F25.EPS

Figure 25 ◆ Pulling equipment.

3.6.0 Feeding Conductors into Conduit

After the fish tape or pull line is attached to the conductors, they must be pulled back through the conduit. As the fish tape is pulled, the attached conductors must be properly fed into the conduit.

Usually, more than one conductor is fed into the conduit during a wire pull. It is important to keep the conductors straight and parallel, and free from kinks, bends, and crossovers. Conductors that are allowed to cross each other will form a bulge and make pulling difficult. This could also damage the conductors.

Spools and rolls of conductors must be set up so that they unwind easily, without kinks and bends.

When several conductors must be fed into the conduit at the same time, a reel cart is used (*Figure 26*). The reel cart will allow the spools to turn freely and help prevent the wires from tangling.

3.7.0 Conductor Lubrication

When conductors are fed into long runs of conduit or conduit with several bends, both the conduit and the wires are lubricated with a compound designed for wire lubrication.

Several types of formulated compounds designed for wire lubrication are available in either dry powder, paste, or gel form. These compounds must be noncorrosive to the insulation material of the conductor and to the conduit itself. The compounds are applied by hand to the conductors as they are fed into the conduit. Battery-operated pumps are also available to lubricate the conduit prior to installing the conductors.

3.8.0 Conductor Termination

The amount of free conductor at each junction or outlet box must meet certain *NEC*® specifications. For example, there must be sufficient free conductor so that bends or terminations inside the box, cabinet, or enclosure may be made to a radius as specified in the *NEC*®. The *NEC*® specifies a minimum of six inches for connections made to wiring devices or for splices. Where conductors pass through junction or pull boxes, enough slack should be provided for splices at a later date.

When a box is used as a pull box, the conductors are not necessarily spliced. They may merely enter the pull box via one conduit run and exit via another conduit run. The purpose of a pull box, as the name suggests, is to facilitate pulling conductors on long runs. A junction box, however, is not only used to facilitate pulling conductors through the raceway system, but it also provides an enclosure for splices in the conductors.

109F26.EPS

Figure 26 ◆ Reel cart.

Cleanup

A professional always takes the time to clean up the work area, removing excess lubricant from the boxes, bushings, and conductors. Completing a job in a professional manner is not only good business, it is also an *NEC®* requirement *(NEC Section 110.12).*

Putting It All Together

Think about the design of conductor installations. How does the location of pull points affect the ease of the pull?

1. A conductor's size refers to the cross-sectional area of both the current-carrying wire and the insulation.
   a. True
   b. False ✓

2. The maximum size solid wire that should be terminated in pull boxes and disconnects is __B__.
   a. No. 6
   b. No. 8 ✓
   c. No. 10
   d. No. 12

3. Compact stranding is used __C__.
   a. when installing larger conduit sizes
   b. when decreasing the ampacity of an existing service
   c. with aluminum conductors ✓
   d. in corrosive environments

4. The ampacity ratings of conductors can be found in __C__.
   a. *NEC Chapter 1*
   b. *NEC Articles 348 through 352*
   c. *NEC Tables 310.16 through 310.19* ✓
   d. *NEC Chapter 9*

5. To determine the ampacity of copper conductors where conductors are used as a single conductor in free air at ambient temperatures of 30°C, use __A__.
   a. *NEC Table 310.16* ✓
   b. *NEC Table 310.17*
   c. *NEC Table 310.18*
   d. *NEC Table 310.19*

6. All of the following help to determine how good a material will be for the construction of wire *except* __D__.
   a. availability
   b. cost
   c. conductivity
   d. molecular weight ✓

7. A type of thermoset insulation is __A__.
   a. RHW ✓
   b. THW
   c. THHW
   d. PVC

8. Polyethylene (PE) is used primarily for __B__.
   a. the manufacture of TW and THW insulation
   b. insulation of control and communication wiring ✓
   c. high-voltage cables
   d. high-temperature insulation

9. What letter must be included in the marking of a conductor if it is to be used in a wet, outdoor application?
   a. D
   b. O
   c. I
   d. W ✓

10. What service conductor may be bare, green, or green with a yellow stripe?
    a. The grounding conductor of a multi-conductor cable ✓
    b. The neutral conductor of a multi-conductor cable
    c. The ungrounded conductor of a multi-conductor cable
    d. The high leg of a four-wire, delta-connected secondary

11. What are the colors of insulation on the conductors for a three-conductor NM cable?
    a. One white or gray, one red, and one black
    b. Two white or gray and one red
    c. One white or gray, one red, one black, and a grounding conductor ✓
    d. One green, one white or gray, one blue, and a grounding conductor

12. A conductor's operating temperature is *not* determined by __D__.
    a. ambient temperature
    b. current flow
    c. current flow in bundled conductors
    d. the number of conductor strands

13. In some cases, fixture wires may be used as a substitute for branch circuit conductors.
    a. True
    b. False

14. Type NMC cable is suitable for all of the following *except* __D__.
    a. dry locations
    b. damp locations
    c. corrosive locations
    d. embedding in concrete

15. The difference between AC cable and MC cable is __C__.
    a. AC cable has a maximum of four conductors; MC has a maximum of six
    b. AC cable has a bonding strip; the MC ground is its metal cladding
    c. AC cable has fire-retardant wraps on individual conductors; MC cable does not
    d. AC cable is sized from 10 AWG to 4/0; MC is sized from 18 AWG to 2,000 kcmil

16. Type USE cable can be used for __C__.
    a. aboveground installation only
    b. underground installation within a special PVC pipeline
    c. underground installation including direct burial
    d. indoor applications only

17. To prevent unwanted ground loops, instrument wiring is __C__.
    a. not grounded at both ends of the wire
    b. grounded at both ends of the wire
    c. floated on both ends
    d. always ungrounded

18. A pulling line is usually made of __D__.
    a. stranded wire
    b. wire rope
    c. steel tape
    d. nylon or other synthetic material

19. When feeding conductors into long runs of conduit, it is important to apply lubricant to the conductors only, and *not* to the conduit.
    a. True
    b. False

20. The *NEC*® specifies a minimum of __6"__ for connections made to wiring devices or for splices.
    a. 4"
    b. 6"
    c. 8"
    d. 10"

Summary

This module introduced key concepts about conductors and cables.

Before selecting a conductor, you must understand wire ratings and why operating temperature is important. The conductor selection process requires that you also consider how and where the conductor will be terminated. The *NEC®* ampacity tables are based on operating temperature. The operating temperature of a conductor is affected by the conductor size (in circular mils), ambient temperature, number of conductors in the bundle, and where the conductors are installed.

Pulling wires and cables through conduit systems is an important part of your job as an electrician. The more you learn about the concepts involved in pulling cable, the safer and more efficient you will be.

Notes

1. A conductor's ___AMPACITY___ is the current the conductor can carry continuously without exceeding its temperature rating.

2. On a cable puller, the ___CAPSTAN___ is the part on which the pulling rope is wrapped and pulled.

3. During a pull, cable may be attached to the pulling rope using a(n) ___WIRE GRIP___.

4. A(n) ___FISH TAPE___ is a manually operated device that is used to pull a wire through conduit.

5. Composed of foam rubber, a(n) ___MOUSE___ fits inside a piece of conduit and is propelled by compressed air or vacuumed through the conduit run, pulling a line or tape.

Trade Terms

Ampacity
Capstan
Fish tape
Mouse
Wire grip

L.J. LeBlanc
Pumba Electric, LLC.

What made you decide to become an electrician?
My dad was an electrician. He had me wiring houses at 9 years old. He was very knowledgeable in the field, having been educated at the Coyne American Institute in Chicago, Illinois. The challenge of the electrical field has always interested me.

How did you learn the trade?
My education came from vo-tech training in Louisiana. It was a self-taught program that was supplemented by on-the-job training. Having the benefit of working side by side with some great craftsmen from all over the U.S. and Canada gave me a broad base of electrical experience. Working maintenance really improved my troubleshooting skills.

What kinds of jobs did you hold on your way to your current position?
I began as a helper and then started trade school in an apprenticeship program in 1965. Work was so plentiful that I was stepped up to a foreman within a year. I took a 2-week job that turned out to last 14 years. I began as a journeyman and ended up as a shop superintendent.

Today I work for my son's company. I do just about all phases of electrical work from estimating, bidding, and installation to troubleshooting.

What factors have contributed most to your success?
I believe in giving eight hours work for eight hours pay. I thirst for knowledge, and I love a challenge. If a man made it, I can fix it. I am never late for work. If I can't be there on time, I show up early.

What advice would you give to someone entering the electrical trade?
Get a good education and complement that education with on-the-job training. Listen to the advice of your peers. What's most important is what you learn after you think you know it all.

Trade Terms
Introduced in This Module

Ampacity: The current in amperes a conductor can carry continuously under the conditions of use without exceeding its temperature rating.

Capstan: The turning drum of the cable puller on which the rope is wrapped and pulled.

Fish tape: A hand device used to pull a wire through a conduit run.

Mouse: A cylinder of foam rubber that fits inside the conduit and is then propelled by compressed air or vacuumed through the conduit run, pulling a line or tape.

Wire grip: A device used to link pulling rope to cable during a pull.

This module is intended to present thorough resources for task training. The following reference work is suggested for further study. This is optional material for continued education rather than for task training.

National Electrical Code® Handbook, Latest Edition. Quincy, MA: National Fire Protection Association.

NCCER makes every effort to keep these textbooks up-to-date and free of technical errors. We appreciate your help in this process. If you have an idea for improving this textbook, or if you find an error, a typographical mistake, or an inaccuracy in NCCER's Contren® textbooks, please write us, using this form or a photocopy. Be sure to include the exact module number, page number, a detailed description, and the correction, if applicable. Your input will be brought to the attention of the Technical Review Committee. Thank you for your assistance.

Instructors – If you found that additional materials were necessary in order to teach this module effectively, please let us know so that we may include them in the Equipment/Materials list in the Annotated Instructor's Guide.

Write: Product Development and Revision
National Center for Construction Education and Research
3600 NW 43rd St., Bldg. G, Gainesville, FL 32606

Fax: 352-334-0932

E-mail: curriculum@nccer.org

Craft _____ Module Name _____

Copyright Date _____ Module Number _____ Page Number(s) _____

Description _____

_____

_____

_____

(Optional) Correction _____

_____

_____

(Optional) Your Name and Address _____

_____

_____

Basic Electrical
Construction Drawings

Fenway Park Pavilion Seat Expansion and EMC/State Street Club Project

This expansion illustrates the accomplishment of an ambitious concept in a sensitive historical environment. The $45 million project was completed within the six-month winter off-season and used precise phasing and logistics to ensure complete protection of Fenway Field. The contractor structurally lifted and shored the historic facility to accommodate the installation of a new ring of columns that would ultimately support the significant addition.

26110-08

26110-08
Basic Electrical Construction Drawings

Topics to be presented in this module include:

Overview

The electrician on a job is usually handed a set of drawings and a book or list of specifications for the job before conduit is installed, wire is pulled, or equipment is mounted. The drawings explain how the customer wants the building wired. Specifications spell out the type and quality of material, components, and equipment that the customer wants, and provide specific instructions for installing them. The drawings and specifications create the roadmap to a successful installation and a satisfied customer.

An electrician must be able to read any set of drawings, even though the style may vary from designer to designer. Standardized symbols are used throughout the industry to represent types of material, raceways, conductors, equipment, and circuit connections. It is the electrician's responsibility to accurately interpret a set of drawings, and to be familiar with the standardized numbering system used in specifications to identify electrical components and their installation.

Note: *National Electrical Code*® and *NEC*® are registered trademarks of the National Fire Protection Association, Inc., Quincy, MA 02269. All *National Electrical Code*® and *NEC*® references in this module refer to the 2008 edition of the *National Electrical Code*®.

Objectives

When you have completed this module, you will be able to do the following:

1. Explain the basic layout of a set of construction drawings.
2. Describe the information included in the title block of a construction drawing.
3. Identify the types of lines used on construction drawings.
4. Using an architect's scale, state the actual dimensions of a given drawing component.
5. Interpret electrical drawings, including site plans, floor plans, and detail drawings.
6. Interpret equipment schedules found on electrical drawings.
7. Describe the type of information included in electrical specifications.

Trade Terms

Architectural drawings
Block diagram
Blueprint
Detail drawing
Dimensions
Electrical drawing
Elevation drawing
Floor plan
One-line diagram

Plan view
Power-riser diagram
Scale
Schedule
Schematic diagram
Sectional view
Shop drawing
Site plan
Written specifications

Required Trainee Materials

1. Paper and pencil
2. Copy of the latest edition of the *National Electrical Code®*
3. Appropriate personal protective equipment

Prerequisites

Before you begin this module, it is recommended that you successfully complete *Core Curriculum* and *Electrical Level One*, Modules 26101-08 through 26109-08.

This course map shows all of the modules in *Electrical Level One.* The suggested training order begins at the bottom and proceeds up. Skill levels increase as you advance on the course map. The local Training Program Sponsor may adjust the training order.

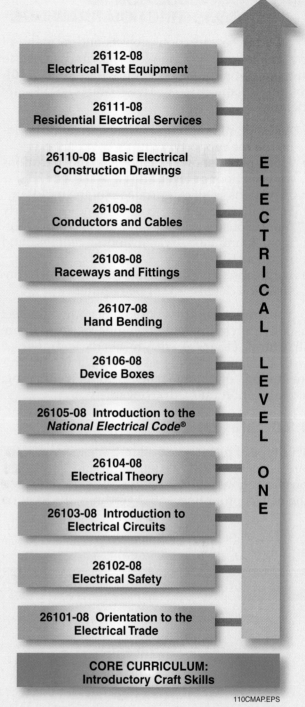

110CMAP.EPS

1.0.0 ◆ INTRODUCTION TO CONSTRUCTION DRAWINGS

In all large construction projects and in many of the smaller ones, an architect is commissioned to prepare complete working drawings and specifications for the project. These drawings usually include:

- A **site plan** indicating the location of the building on the property.
- **Floor plans** showing the walls and partitions for each floor or level.
- Elevations of all exterior faces of the building.
- Several vertical cross sections to indicate clearly the various floor levels and details of the footings, foundation, walls, floors, ceilings, and roof construction.
- Large-scale **detail drawings** showing such construction details as may be required.

For projects of any consequence, the architect usually hires consulting engineers to prepare structural, electrical, and mechanical drawings, with the latter encompassing pipefitting, instrumentation, plumbing, and heating, ventilating, and air conditioning drawings.

1.1.0 Site Plan

This type of plan of the building site looks as if the site is viewed from an airplane and shows the property boundaries, the existing contour lines, the new contour lines (after grading), the location of the building on the property, new and existing roadways, all utility lines, and other pertinent details. The drawing **scale** is also shown. Descriptive notes may also be found on the site (plot) plan listing names of adjacent property owners, the land surveyor, and the date of the survey. A legend or symbol list is also included so that anyone who must work with the site plan can readily read the information. See *Figure 1*.

1.2.0 Floor Plans

The **plan view** of any object is a drawing showing the outline and all details as seen when looking directly down on the object. It shows only two **dimensions**, length and width. The floor plan of a building is drawn as if a horizontal cut were made through the building—at about window height—and then the top portion removed to reveal the bottom part. See *Figure 2*.

If a plan view of a home's basement is needed, the part of the house above the middle of the basement windows is imagined to be cut away. By looking down on the uncovered portion, every detail and partition can be seen. Likewise, imagine the part above the middle of the first floor windows being cut away. A drawing that looks straight down at the remaining part would be called the first floor plan or lower level. A cut through the second floor windows would be called the second floor plan or upper level. See *Figure 3*.

INSIDE TRACK

Then and Now

Years ago, blueprints were created by placing a hand drawing against light-sensitive paper and then exposing it to ultraviolet light. The light would turn the paper blue except where lines were drawn on the original. The light-sensitive paper was then developed, and the resulting print had white lines against a blue background. Modern blueprints usually have blue or black lines against a white background and are generated using computer-aided design programs. Newer programs offer three-dimensional modeling and other enhanced features.

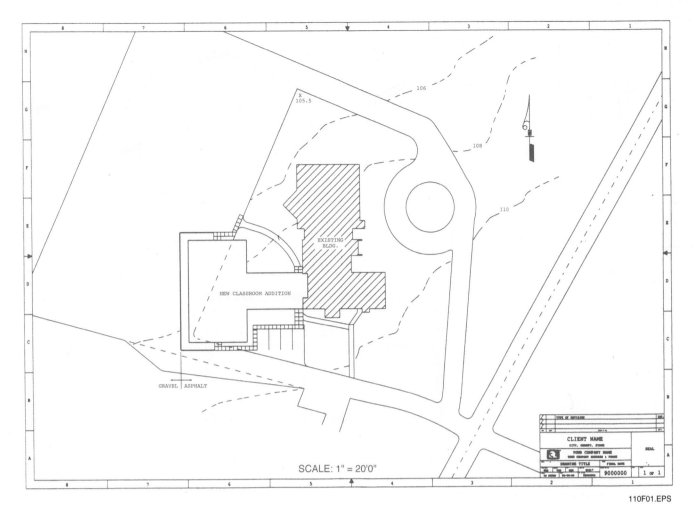

Figure 1 ◆ Typical site plan.

110F01.EPS

Using a Drawing Set

INSIDE TRACK

Always treat a drawing set with care. It is best to keep two sets, one for the office and one for field use. Be sure to use the most current revision. After you use a sheet from a set of drawings, refold the sheet with the title block facing up.

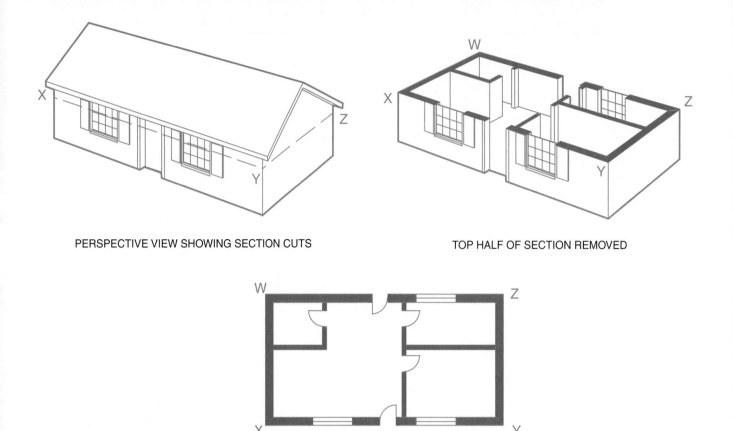

PERSPECTIVE VIEW SHOWING SECTION CUTS

TOP HALF OF SECTION REMOVED

RESULTING FLOOR PLAN IS WHAT THE REMAINING
STRUCTURE LOOKS LIKE WHEN VIEWED FROM ABOVE

110F02.EPS

Figure 2 ◆ Principles of floor plan layout.

1.3.0 Elevations

The elevation is an outline of an object that shows heights and may show the length or width of a particular side, but not depth. *Figures 4* and *5* show **elevation drawings** for a building.

NOTE

These elevation drawings show the heights of windows, doors, and porches, the pitch of roofs, etc., because all of these measurements cannot be shown conveniently on floor plans.

1.4.0 Sections

A section or **sectional view** (*Figure 6*) is a cutaway view that allows the viewer to see the inside of a structure. The point on the plan or elevation showing where the imaginary cut has been made is indicated by the section line, which is usually a dashed line. The section line shows the location of the section on the plan or elevation. It is necessary to know which of the cutaway parts is represented in the sectional drawing. To show this, arrow points are placed at the ends of the section lines.

In **architectural drawings,** it is often necessary to show more than one section on the same drawing. The different section lines must be distinguished by letters, numbers, or other designations placed at the ends of the lines. These section letters are generally large so as to stand out on the drawings. To further avoid confusion, the same letter is usually placed at each end of the section line. The section is named according to these letters (e.g., Section A-A, Section B-B, and so forth).

A longitudinal section is taken lengthwise while a cross section is usually taken straight across the width of an object. Sometimes, however, a section is not taken along one straight line. It is often taken along a zigzag line to show important parts of the object.

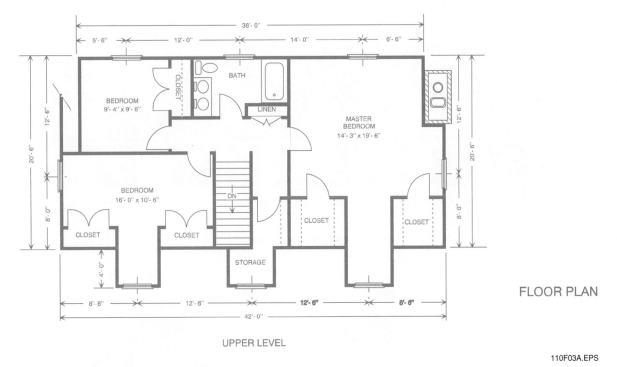

FLOOR PLAN

UPPER LEVEL

110F03A.EPS

Figure 3 ◆ Floor plans of a building (1 of 2).

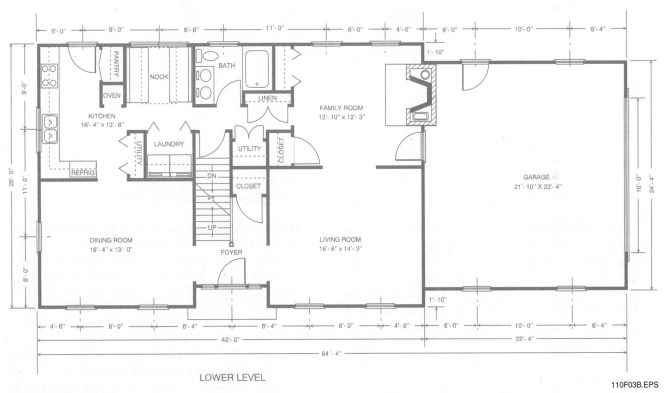

LOWER LEVEL

110F03B.EPS

Figure 3 ◆ Floor plans of a building (2 of 2).

FRONT ELEVATION

REAR ELEVATION

110F04.EPS

Figure 4 ◆ Front and rear elevations.

A sectional view, as applied to architectural drawings, is a drawing showing the building, or portion of a building, as though it were cut through on some imaginary line. This line may be either vertical (straight up and down) or horizontal. Wall sections are nearly always made vertically so that the cut edge is exposed from top to bottom. In some ways, the wall section is one of the most important of all the drawings to construction workers, because it answers the questions as to how a structure should be built. The floor plans of a building show how each floor is arranged, but the wall sections tell how each part is constructed and usually indicate the material to be used. The electrician needs to know this information when determining wiring methods that comply with the *NEC*®.

1.5.0 Electrical Drawings

Electrical drawings show in a clear, concise manner exactly what is required of the electricians. The amount of data shown on such drawings should be sufficient, but not overdone. This means that a complete set of electrical drawings could consist of only one 8½" × 11" sheet, or it could consist of several dozen 24" × 36" (or larger) sheets, depending on the size and complexity of a given project. A **shop drawing,** for example, may contain details of only one piece of equipment, while a set of working drawings for an industrial installation may contain dozens of drawing sheets detailing the electrical system for lighting and power, along with equipment, motor controls, wiring diagrams, **schematic diagrams,** equipment **schedules,** and a host of other pertinent data.

LEFT ELEVATION

RIGHT ELEVATION

110F05.EPS

Figure 5 ◆ Left and right elevations.

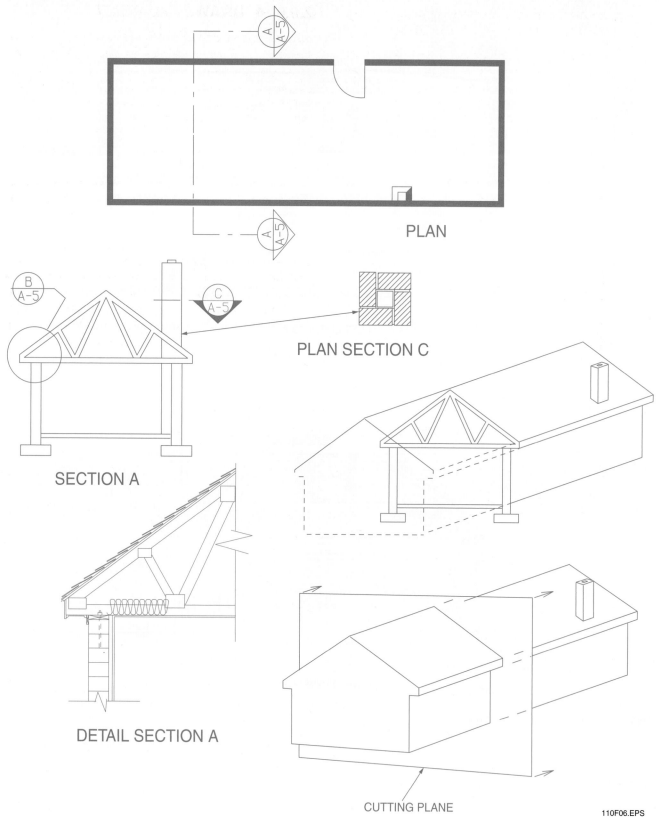

PLAN

PLAN SECTION C

SECTION A

DETAIL SECTION A

CUTTING PLANE

110F06.EPS

Figure 6 ◆ Sectional drawing.

In general, the electrical working drawings for a given project serve three distinct functions:

- They provide electrical contractors with an exact description of the project so that materials and labor may be estimated to calculate a total cost of the project for bidding purposes.
- They provide workers on the project with instructions as to how the electrical system is to be installed.
- They provide a map of the electrical system once the job is completed to aid in maintenance and troubleshooting for years to come.

Electrical drawings from consulting engineering firms will vary in quality from sketchy, incomplete drawings to neat, precise drawings that are easy to understand. Few, however, will cover every detail of the electrical system. Therefore, a good knowledge of installation practices must go hand-in-hand with interpreting electrical working drawings.

Sometimes electrical contractors will have electrical drafters prepare special supplemental drawings for use by the contractors' employees. On certain projects, these supplemental drawings can save supervision time in the field once the project has begun.

2.0.0 ◆ DRAWING LAYOUT

Although a strong effort has been made to standardize drawing practices in the building construction industry, the drawings or **blueprints** prepared by different architectural or engineering firms will rarely be identical. Similarities, however, will exist between most sets of drawings, and with a little experience, you should have no trouble interpreting any set of drawings that might be encountered.

Most drawings used for building construction projects will be drawn on sheets in various sizes. Each drawing sheet has border lines framing the overall drawing and one or more title blocks, as shown in *Figure 7*. The type and size of title blocks varies with each firm preparing the drawings. In addition, some drawing sheets will also contain a revision block near the title block, and perhaps an approval block. This information is normally found on each drawing sheet, regardless of the type of project or the information contained on the sheet.

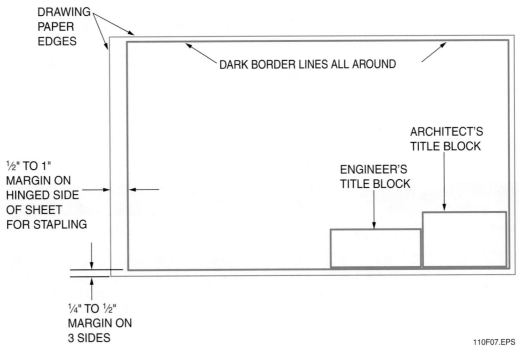

Figure 7 ◆ Typical drawing layout.

2.1.0 Title Block

The architect's title block for a drawing is usually boxed in the lower right-hand corner of the drawing sheet; the size of the block varies with the size of the drawing and with the information required. See *Figure 8*.

In general, the title block of an electrical drawing should contain the following information:

- Name of the project
- Address of the project
- Name of the owner or client
- Name of the architectural firm
- Date of completion
- Scale(s)
- Initials of the drafter, checker, and designer, with dates under each
- Job number
- Sheet number
- General description of the drawing

Interpreting Electrical Drawings

A good example of when an electrician must interpret the drawings is when wiring a log cabin. The drawings will show the receptacle and switch locations in branch circuits as usual, but the electrician must figure out how to route wires and install boxes where there is no hollow wall and sometimes no ceiling space.

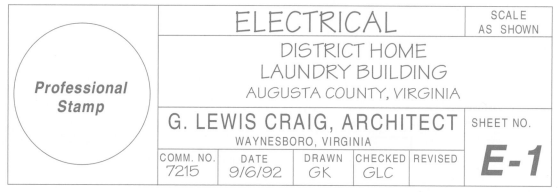

110F08.EPS

Figure 8 ◆ Typical architect's title block.

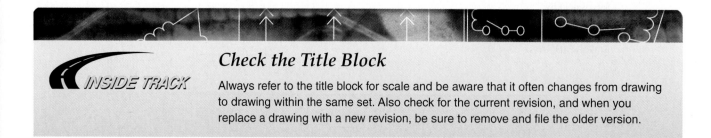

Check the Title Block

Always refer to the title block for scale and be aware that it often changes from drawing to drawing within the same set. Also check for the current revision, and when you replace a drawing with a new revision, be sure to remove and file the older version.

Applying Your Skills

Once you learn how to interpret blueprints, you can apply that knowledge to any type of construction, from simple residential applications to large industrial complexes.

Often, the consulting engineering firm will also be listed, which means that an additional title block will be applied to the drawing, usually next to the architect's title block. *Figure 9* shows completed architectural and engineering title blocks as they appear on an actual drawing.

2.2.0 Approval Block

The approval block, in most cases, will appear on the drawing sheet as shown in *Figure 10*. The various types of approval blocks (drawn, checked, etc.) will be initialed by the appropriate personnel. This type of approval block is usually part of the title block and appears on each drawing sheet.

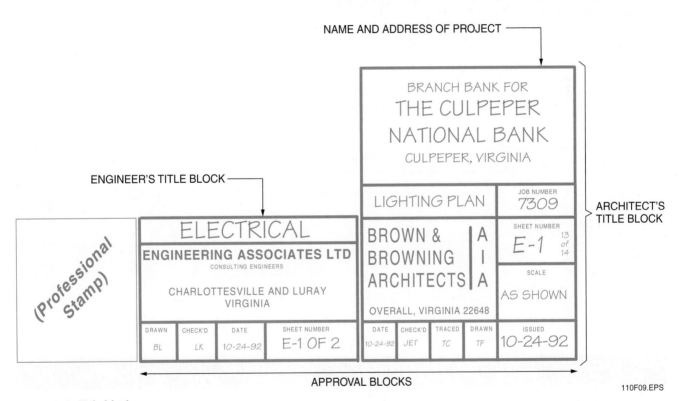

Figure 9 ◆ Title blocks.

Figure 10 ◆ Typical approval block.

Orient Yourself

When reading a drawing, find the north arrow to orient yourself to the structure. Knowing where north is enables you to accurately describe the locations of walls and other parts of the building.

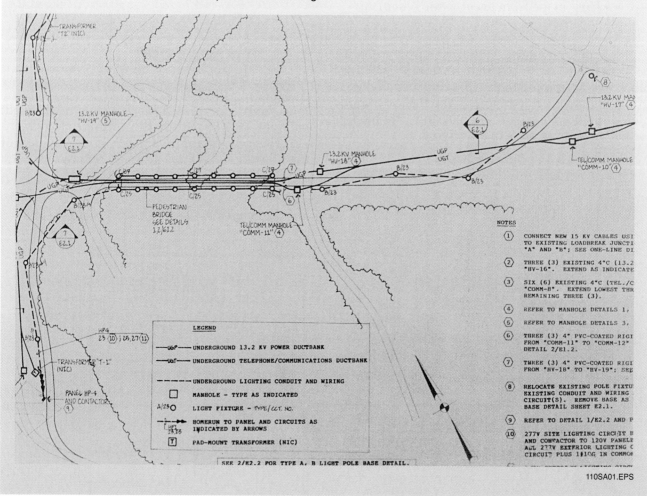

110SA01.EPS

Using All of the Drawings

Look back over the information on floor plans, elevations, and sections. What kinds of information would an electrician get from each of these drawings? What could a sectional drawing show that a floor plan could not?

On some projects, authorized signatures are required before certain systems may be installed, or even before the project begins. An approval block such as the one shown in *Figure 11* indicates that all required personnel have checked the drawings for accuracy, and that the set meets with everyone's approval. Such an approval block usually appears on the front sheet of the blueprint set and may include:

- *Professional stamp* – Registered seal of approval by the licensed architect or consulting engineer.
- *Design supervisor* – Signature of the person who is overseeing the design.
- *Drawn (by)* – Signature or initials of the person who drafted the drawing and the date it was completed.
- *Checked (by)* – Signature or initials of the person who reviewed the drawing and the date of approval.
- *Approved* – Signature or initials of the architect/engineer and the date of the approval.
- *Owner's approval* – Signature of the project owner or the owner's representative along with the date signed.

2.3.0 Revision Block

Sometimes electrical drawings will have to be partially redrawn or modified during the construction of a project. It is extremely important that such modifications are noted and dated on the drawings to ensure that the workers have an up-to-date set of drawings to work from. In some situations, sufficient space is left near the title block for dates and descriptions of revisions, as shown in *Figure 12*. In other cases, a revision block is provided (again, near the title block), as shown in *Figure 13*. The area on the drawing where the revision has been made will often be circled with a cloud shape.

NOTE

Architects, engineers, designers, and drafters have their own methods of showing revisions, so expect to find deviations from those shown here.

CAUTION

When a set of electrical working drawings has been revised, always make certain that the most up-to-date set is used for all future layout work. Either destroy the old, obsolete set of drawings or else clearly mark on the affected sheets, *Obsolete Drawing—Do Not Use*. Also, when working with a set of working drawings and written specifications for the first time, thoroughly check each page to see if any revisions or modifications have been made to the originals. Doing so can save much time and expense to all concerned with the project.

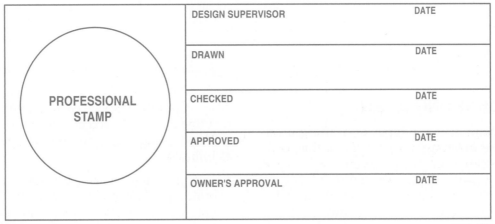

110F11.EPS

Figure 11 ◆ Alternate approval block.

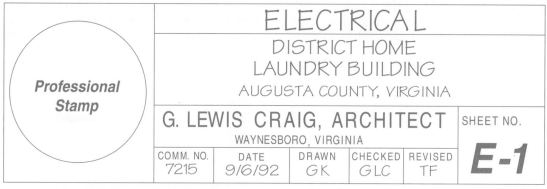

REVISIONS
10/12/92 - REVISED LIGHTING FIXTURE
NO. 3 IN. LIGHTING FIXTURE SCHEDULE

ELECTRICAL
DISTRICT HOME
LAUNDRY BUILDING
AUGUSTA COUNTY, VIRGINIA

G. LEWIS CRAIG, ARCHITECT
WAYNESBORO, VIRGINIA

SHEET NO.

E-1

COMM. NO.	DATE	DRAWN	CHECKED	REVISED
7215	9/6/92	GK	GLC	TF

Professional Stamp

110F12.EPS

Figure 12 ◆ One method of showing revisions on working drawings.

REVISIONS

REV	DESCRIPTION	DR	APPD	DATE
1	FIXTURE NO. 3 IN. LIGHTING-FIXTURE SCHDL	GK	GLC	10/12/92

ELECTRICAL
DISTRICT HOME
LAUNDRY BUILDING
AUGUSTA COUNTY, VIRGINIA

G. LEWIS CRAIG, ARCHITECT
WAYNESBORO, VIRGINIA

SHEET NO.

E-1

COMM. NO.	DATE	DRAWN	CHECKED	REVISED
7215	9/6/92	GK	GLC	TF

Professional Stamp

110F13.EPS

Figure 13 ◆ Alternative method of showing revisions on working drawings.

3.0.0 ◆ DRAFTING LINES

You will encounter many types of drafting lines. To specify the meaning of each type of line, contrasting lines can be made by varying the width of the lines or breaking the lines in a uniform way.

Figure 14 shows common lines used on architectural drawings. However, these lines can vary. Architects and engineers have strived for a common standard for the past century, but unfortunately, their goal has yet to be reached. Therefore, you will find variations in lines and symbols from drawing to drawing, so always consult the legend or symbol list when referring to any drawing.

Also, carefully inspect each drawing to ensure that line types are used consistently.

The drafting lines shown in *Figure 14* are used as follows:

- *Light full line* – This line is used for section lines, building background (outlines), and similar uses where the object to be drawn is secondary to the system being shown (e.g., HVAC or electrical).
- *Medium full line* – This type of line is frequently used for hand lettering on drawings. It is further used for some drawing symbols, circuit lines, etc.

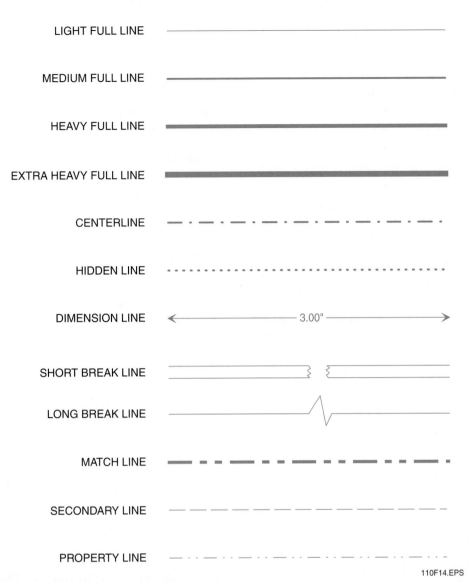

LIGHT FULL LINE

MEDIUM FULL LINE

HEAVY FULL LINE

EXTRA HEAVY FULL LINE

CENTERLINE

HIDDEN LINE

DIMENSION LINE 3.00"

SHORT BREAK LINE

LONG BREAK LINE

MATCH LINE

SECONDARY LINE

PROPERTY LINE

110F14.EPS

Figure 14 ◆ Typical drafting lines.

- *Heavy full line* – This line is used for borders around title blocks, schedules, and for hand lettering drawing titles. Some types of symbols are frequently drawn with a heavy full line.
- *Extra heavy full line* – This line is used for border lines on architectural/engineering drawings.
- *Centerline* – A centerline is a broken line made up of alternately spaced long and short dashes. It indicates the centers of objects such as holes, pillars, or fixtures. Sometimes, the centerline indicates the dimensions of a finished floor.
- *Hidden line* – A hidden line consists of a series of short dashes that are closely and evenly spaced. It shows the edges of objects that are not visible in a particular view. The object outlined by hidden lines in one drawing is often fully pictured in another drawing.

- *Dimension line* – These are thin lines used to show the extent and direction of dimensions. The dimension is usually placed in a break inside the dimension lines. Normal practice is to place the dimension lines outside the object's outline. However, it may sometimes be necessary to draw the dimensions inside the outline.
- *Short break line* – This line is usually drawn free-hand and is used for short breaks.
- *Long break line* – This line, which is drawn partly with a straightedge and partly with freehand zigzags, is used for long breaks.
- *Match line* – This line is used to show the position of the cutting plane. Therefore, it is also called the cutting plane line. A match or cutting plane line is a heavy line with long dashes alternating with two short dashes. It is used on drawings of large structures to show where one drawing stops and the next drawing starts.

- *Secondary line* – This line is frequently used to outline pieces of equipment or to indicate reference points of a drawing that are secondary to the drawing's purpose.
- *Property line* – This is a light line made up of one long and two short dashes that are alternately spaced. It indicates land boundaries on the site plan.

Other uses of the lines just mentioned include the following:

- *Extension lines* – Extension lines are lightweight lines that start about 1/16 inch away from the edge of an object and extend out. A common use of extension lines is to create a boundary for dimension lines. Dimension lines meet extension lines with arrowheads, slashes, or dots. Extension lines that point from a note or other reference to a particular feature on a drawing are called leaders. They usually end in either an arrowhead or a dot and may include an explanatory note at the end.
- *Section lines* – These are often referred to as *cross-hatch lines*. Drawn at a 45° angle, these lines show where an object has been cut away to reveal the inside.
- *Phantom lines* – Phantom lines are solid, light lines that show where an object will be installed. A future door opening or a future piece of equipment can be shown with phantom lines.

3.1.0 Electrical Drafting Lines

Besides the architectural lines shown in *Figure 14* consulting electrical engineers, designers, and drafters use additional lines to represent circuits and their related components. Again, these lines may vary from drawing to drawing, so check the symbol list or legend for the exact meaning of lines on the drawing with which you are working. *Figure 15* shows lines used on some electrical drawings.

EXPOSED WIRING

WIRING CONCEALED IN CEILING OR WALL

WIRING CONCEALED IN FLOOR

WIRING TURNED UP

WIRING TURNED DOWN

BRANCH CIRCUIT HOMERUN TO PANELBOARD*

* Number of arrowheads indicates number of circuits. A number at each arrowhead may be used to identify circuit numbers.

** Half arrowheads are sometimes used for homeruns to avoid confusing them with drawing callouts.

110F15.EPS

Figure 15 ◆ Electrical drafting lines.

4.0.0 ◆ ELECTRICAL SYMBOLS

The electrician must be able to correctly read and understand electrical working drawings. This includes a thorough knowledge of electrical symbols and their applications.

An electrical symbol is a figure or mark that stands for a component used in the electrical system. *Figure 16* shows a list of electrical symbols that are currently recommended by the American National Standards Institute (ANSI). It is evident from this list of symbols that many have the same basic form, but, because of some slight difference, their meaning changes. For example, the

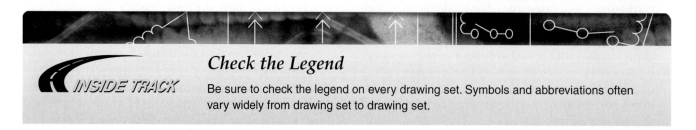

Check the Legend

INSIDE TRACK

Be sure to check the legend on every drawing set. Symbols and abbreviations often vary widely from drawing set to drawing set.

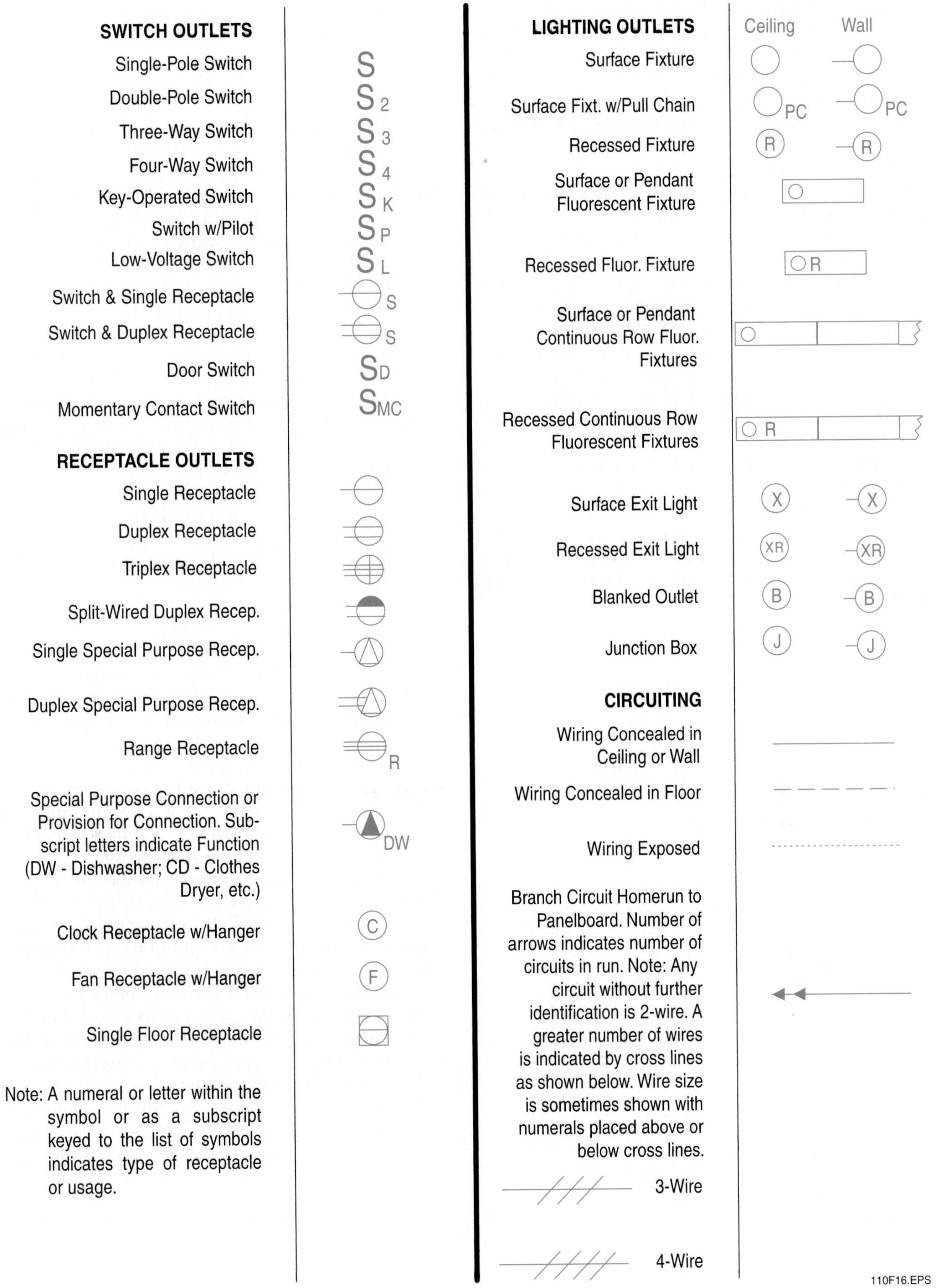

SWITCH OUTLETS

Single-Pole Switch S

Double-Pole Switch S_2

Three-Way Switch S_3

Four-Way Switch S_4

Key-Operated Switch S_K

Switch w/Pilot S_P

Low-Voltage Switch S_L

Switch & Single Receptacle

Switch & Duplex Receptacle

Door Switch S_D

Momentary Contact Switch S_{MC}

RECEPTACLE OUTLETS

Single Receptacle

Duplex Receptacle

Triplex Receptacle

Split-Wired Duplex Recep.

Single Special Purpose Recep.

Duplex Special Purpose Recep.

Range Receptacle R

Special Purpose Connection or Provision for Connection. Subscript letters indicate Function (DW - Dishwasher; CD - Clothes Dryer, etc.) DW

Clock Receptacle w/Hanger C

Fan Receptacle w/Hanger F

Single Floor Receptacle

Note: A numeral or letter within the symbol or as a subscript keyed to the list of symbols indicates type of receptacle or usage.

LIGHTING OUTLETS

Ceiling   Wall

Surface Fixture

Surface Fixt. w/Pull Chain   PC   PC

Recessed Fixture   R   R

Surface or Pendant Fluorescent Fixture

Recessed Fluor. Fixture   R

Surface or Pendant Continuous Row Fluor. Fixtures

Recessed Continuous Row Fluorescent Fixtures   R

Surface Exit Light   X   X

Recessed Exit Light   XR   XR

Blanked Outlet   B   B

Junction Box   J   J

CIRCUITING

Wiring Concealed in Ceiling or Wall

Wiring Concealed in Floor

Wiring Exposed

Branch Circuit Homerun to Panelboard. Number of arrows indicates number of circuits in run. Note: Any circuit without further identification is 2-wire. A greater number of wires is indicated by cross lines as shown below. Wire size is sometimes shown with numerals placed above or below cross lines.

3-Wire

4-Wire

110F16.EPS

Figure 16 ◆ ANSI electrical symbols.

receptacle symbols in *Figure 17* each have the same basic form (a circle), but the addition of a line or an abbreviation gives each an individual meaning. A good procedure to follow in learning symbols is to first learn the basic form and then apply the variations for obtaining different meanings.

It would be much simpler if all architects, engineers, electrical designers, and drafters used the same symbols; however, this is not the case. Although standardization is getting closer to a reality, existing symbols are still modified, and new symbols are created for almost every new project.

The electrical symbols described in the following paragraphs represent those found on actual electrical working drawings throughout the United States and Canada. Many are similar to those recommended by ANSI and the Consulting Engineers Council/US; others are not. Understanding how these symbols were devised will help you to interpret unknown electrical symbols in the future.

Some of the symbols used on electrical drawings are abbreviations, such as WP for weatherproof and AFF for above finished floor. Others are simplified pictographs, such as those shown in *Figure 18*.

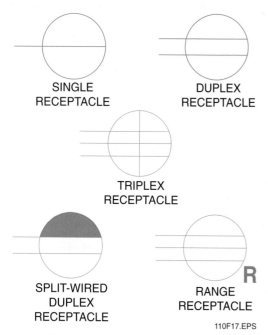

Figure 17 ◆ Various receptacle symbols used on electrical drawings.

In some cases, the symbols are combinations of abbreviations and pictographs, such as in *Figure 18* for a fusible safety switch, a nonfusible safety switch, and a double-throw safety switch. In each example, a pictograph of a switch enclosure has been combined with an abbreviation: F (fusible), DT (double-throw), and NF (nonfusible), respectively.

Lighting outlet symbols have been devised that represent incandescent, fluorescent, and high-intensity discharge lighting; a circle usually represents an incandescent fixture, and a rectangle is used to represent a fluorescent fixture. These symbols are designed to indicate the physical shape of a particular fixture, and while the circles representing incandescent lamps are frequently enlarged somewhat, symbols for fluorescent fixtures are usually drawn as close to scale as possible. The type of mounting used for all lighting fixtures is usually indicated in a lighting fixture schedule, which is shown on the drawings or in the **written specifications**.

The type of lighting fixture is identified by a numeral placed inside a triangle or other symbol, and placed near the fixture to be identified. A complete description of the fixtures identified by the symbols must be given in the lighting fixture schedule and should include the manufacturer, catalog number, number and type of lamps, voltage, finish, mounting, and any other information needed for proper installation of the fixture.

Switches used to control lighting fixtures are also indicated by symbols (usually the letter S followed by numerals or letters to define the exact type of switch). For example, S_3 indicates a three-way switch; S_4 identifies a four-way switch; and S_P indicates a single-pole switch with a pilot light. A subscript letter is often used to identify the fixtures that are controlled by that switch.

Main distribution centers, panelboards, transformers, safety switches, and other similar electrical components are indicated by electrical symbols on floor plans and by a combination of symbols and semipictorial drawings in riser diagrams.

A detailed description of the service equipment is usually given in the panelboard schedule or in the written specifications. However, on small projects, the service equipment is sometimes indicated only by notes on the drawings.

Circuit and feeder wiring symbols are getting closer to being standardized. Most circuits concealed in the ceiling or wall are indicated by a

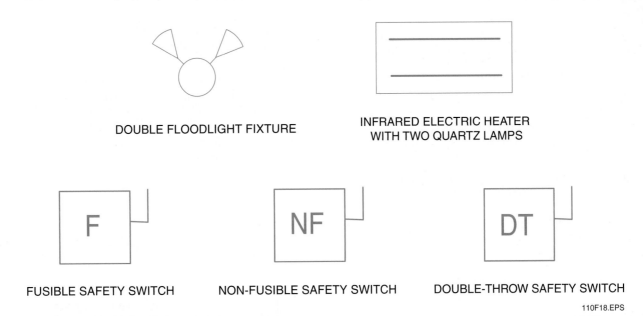

DOUBLE FLOODLIGHT FIXTURE

INFRARED ELECTRIC HEATER
WITH TWO QUARTZ LAMPS

FUSIBLE SAFETY SWITCH

NON-FUSIBLE SAFETY SWITCH

DOUBLE-THROW SAFETY SWITCH

110F18.EPS

Figure 18 ◆ General types of symbols used on electrical drawings.

solid line; a broken line is used for circuits concealed in the floor or ceiling below; and exposed raceways are indicated by short dashes or else the letter *E* placed in the same plane with the circuit line at various intervals. The number of conductors in a conduit or raceway system may be indicated in the panelboard schedule under the appropriate column, or the information may be shown on the floor plan.

Symbols for communication and signal systems, as well as symbols for light and power, are drawn to an appropriate scale and accurately located with respect to the building. This reduces the number of references made to the architectural drawings. Where extreme accuracy is required in locating outlets and equipment, exact dimensions are given on larger-scale drawings and shown on the plans.

Each different category in an electrical system is usually represented by a basic distinguishing symbol. To further identify items of equipment or outlets in the category, a numeral or other identifying mark is placed within the open basic symbol. In addition, all such individual symbols used on the drawings should be included in the symbol list or legend. The electrical symbols shown in *Figure 19* were modified by a consulting engineering firm for use on a small industrial electrical installation. The symbols shown in *Figure 20* are those recommended by the Consulting Engineers Council/US. You should become familiar with these symbols.

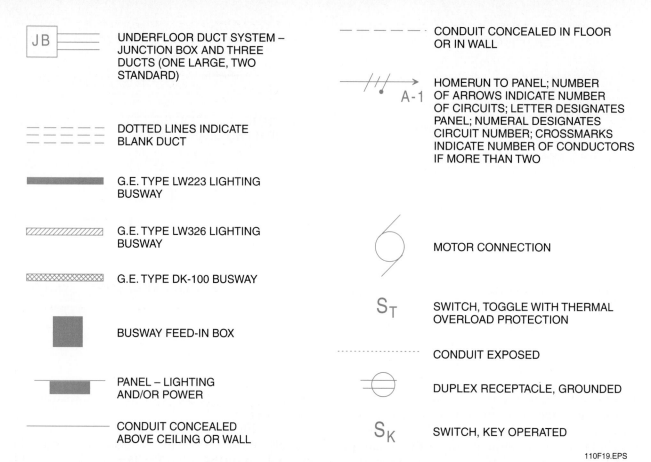

JB — UNDERFLOOR DUCT SYSTEM –
JUNCTION BOX AND THREE
DUCTS (ONE LARGE, TWO
STANDARD)

DOTTED LINES INDICATE
BLANK DUCT

G.E. TYPE LW223 LIGHTING
BUSWAY

G.E. TYPE LW326 LIGHTING
BUSWAY

G.E. TYPE DK-100 BUSWAY

BUSWAY FEED-IN BOX

PANEL – LIGHTING
AND/OR POWER

CONDUIT CONCEALED
ABOVE CEILING OR WALL

CONDUIT CONCEALED IN FLOOR
OR IN WALL

HOMERUN TO PANEL; NUMBER
OF ARROWS INDICATE NUMBER
A-1 OF CIRCUITS; LETTER DESIGNATES
PANEL; NUMERAL DESIGNATES
CIRCUIT NUMBER; CROSSMARKS
INDICATE NUMBER OF CONDUCTORS
IF MORE THAN TWO

MOTOR CONNECTION

S_T SWITCH, TOGGLE WITH THERMAL
OVERLOAD PROTECTION

CONDUIT EXPOSED

DUPLEX RECEPTACLE, GROUNDED

S_K SWITCH, KEY OPERATED

110F19.EPS

Figure 19 ◆ Electrical symbols used by one consulting engineering firm.

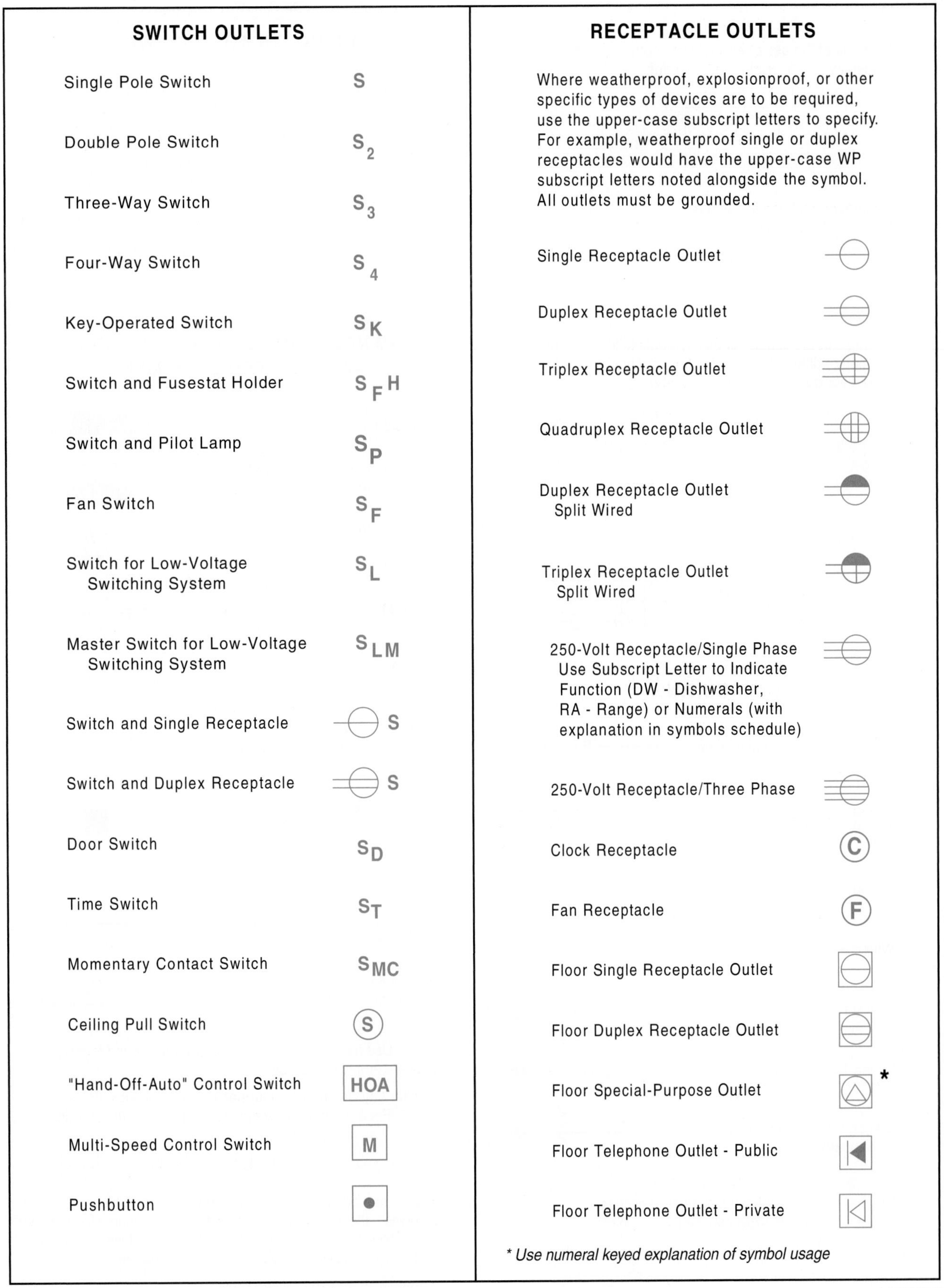

SWITCH OUTLETS

Single Pole Switch	S
Double Pole Switch	S_2
Three-Way Switch	S_3
Four-Way Switch	S_4
Key-Operated Switch	S_K
Switch and Fusestat Holder	S_FH
Switch and Pilot Lamp	S_P
Fan Switch	S_F
Switch for Low-Voltage Switching System	S_L
Master Switch for Low-Voltage Switching System	S_{LM}
Switch and Single Receptacle	⊖ S
Switch and Duplex Receptacle	⊜ S
Door Switch	S_D
Time Switch	S_T
Momentary Contact Switch	S_{MC}
Ceiling Pull Switch	Ⓢ
"Hand-Off-Auto" Control Switch	HOA
Multi-Speed Control Switch	M
Pushbutton	▪

RECEPTACLE OUTLETS

Where weatherproof, explosionproof, or other specific types of devices are to be required, use the upper-case subscript letters to specify. For example, weatherproof single or duplex receptacles would have the upper-case WP subscript letters noted alongside the symbol. All outlets must be grounded.

Single Receptacle Outlet	
Duplex Receptacle Outlet	
Triplex Receptacle Outlet	
Quadruplex Receptacle Outlet	
Duplex Receptacle Outlet Split Wired	
Triplex Receptacle Outlet Split Wired	
250-Volt Receptacle/Single Phase Use Subscript Letter to Indicate Function (DW - Dishwasher, RA - Range) or Numerals (with explanation in symbols schedule)	
250-Volt Receptacle/Three Phase	
Clock Receptacle	Ⓒ
Fan Receptacle	Ⓕ
Floor Single Receptacle Outlet	
Floor Duplex Receptacle Outlet	
Floor Special-Purpose Outlet	*
Floor Telephone Outlet - Public	
Floor Telephone Outlet - Private	

Use numeral keyed explanation of symbol usage

110F20A.EPS

Figure 20 ◆ Recommended electrical symbols (1 of 7).

Example of the use of several floor outlet symbols to identify a 2, 3, or more gang outlet:

Underfloor duct and junction box for triple, double, or single duct system as indicated by the number of parallel lines

Example of the use of various symbols to identify the location of different types of outlets or connections for underfloor duct or cellular floor systems:

Cellular Floor Heater Duct

CIRCUITING

Wiring Exposed (not in conduit)	—— E ——
Wiring Concealed in Ceiling or Wall	
Wiring Concealed in Floor	– – – – –
Wiring Existing*	··············
Wiring Turned Up	——○
Wiring Turned Down	——●
Branch Circuit Homerun to Panelboard	2 1

Number of arrows indicates number of circuits. (A number at each arrow may be used to identify the circuit number.)**

BUS DUCTS AND WIREWAYS

Trolley Duct***	T	T
Busway (Service, Feeder or Plug-in)***	B	B
Cable Trough Ladder or Channel***	C	C
Wireway***	W	W

PANELBOARDS, SWITCHBOARDS AND RELATED EQUIPMENT

Flush Mounted Panelboard and Cabinet***

Surface Mounted Panelboard and Cabinet***

Switchboard, Power Control Center, Unit Substation (Should be drawn to scale)***

Flush Mounted Terminal Cabinet (In small scale drawings the TC may be indicated alongside the symbol)***

Surface Mounted Terminal Cabinet (In small scale drawings the TC may be indicated alongside the symbol)***

Pull Box (Identify in relation to Wiring System Section and Size)

Motor or Other Power Controller May be a starter or contactor***

Externally Operated Disconnection Switch***

Combination Controller and Disconnection Means***

*Note: Use heavy-weight line to identify service and feeders. Indicate empty conduit by notation CO.

**Note: Any circuit without further identification indicates two-wire circuit. For a greater number of wires, indicate with cross lines, e.g.:

3 wires   4 wires, etc.

Neutral and ground wires may be shown longer. Unless indicated otherwise, the wire size of the circuit is the minimum size required by the specification. Identify different functions of wiring system (e.g., signaling system) by notation or other means.

***Identify by Notation or Schedule

Figure 20 ◆ Recommended electrical symbols (2 of 7).

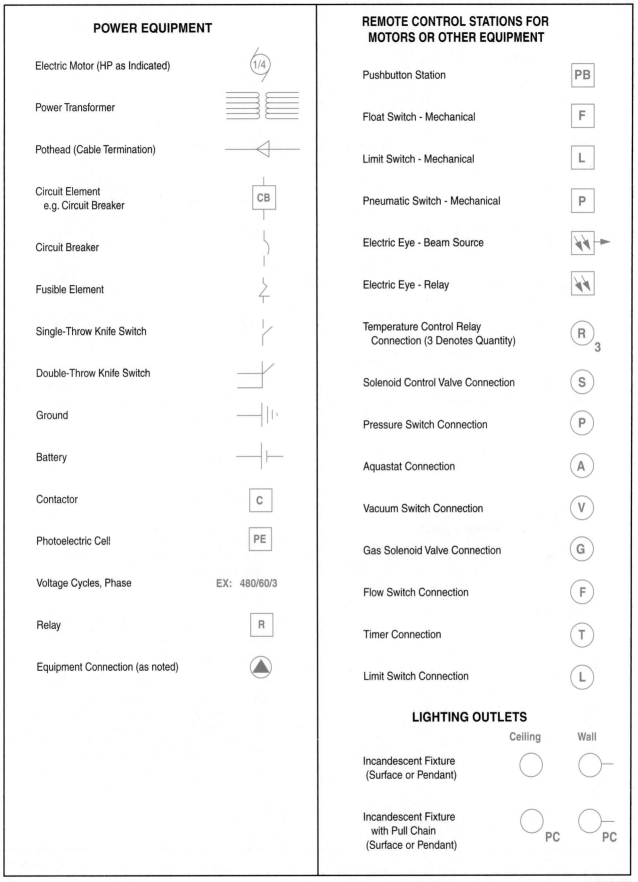

POWER EQUIPMENT

Electric Motor (HP as Indicated)

Power Transformer

Pothead (Cable Termination)

Circuit Element
   e.g. Circuit Breaker

Circuit Breaker

Fusible Element

Single-Throw Knife Switch

Double-Throw Knife Switch

Ground

Battery

Contactor

Photoelectric Cell

Voltage Cycles, Phase   EX: 480/60/3

Relay

Equipment Connection (as noted)

REMOTE CONTROL STATIONS FOR MOTORS OR OTHER EQUIPMENT

Pushbutton Station

Float Switch - Mechanical

Limit Switch - Mechanical

Pneumatic Switch - Mechanical

Electric Eye - Beam Source

Electric Eye - Relay

Temperature Control Relay
   Connection (3 Denotes Quantity)

Solenoid Control Valve Connection

Pressure Switch Connection

Aquastat Connection

Vacuum Switch Connection

Gas Solenoid Valve Connection

Flow Switch Connection

Timer Connection

Limit Switch Connection

LIGHTING OUTLETS

Incandescent Fixture
(Surface or Pendant)

Incandescent Fixture
with Pull Chain
(Surface or Pendant)

110F20C.EPS

Figure 20 ◆ Recommended electrical symbols (3 of 7).

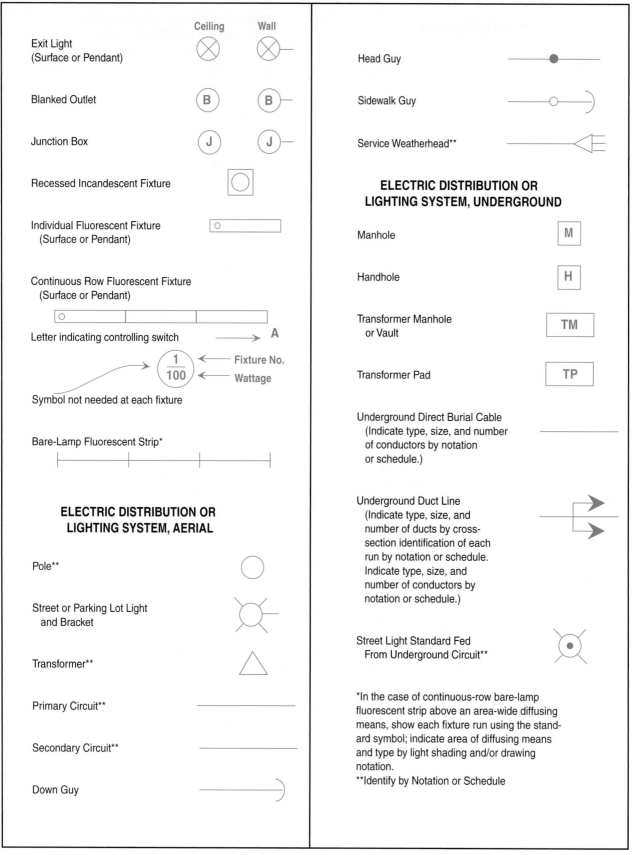

Figure 20 ◆ Recommended electrical symbols (4 of 7).

110F20D.EPS

SIGNALING SYSTEM OUTLETS

INSTITUTIONAL, COMMERCIAL, AND INDUSTRIAL OCCUPANCIES

I NURSE CALL SYSTEM DEVICES (Any Type)

Basic Symbol

(Examples of Individual Item Identification Not a Part of Standard)

Nurses' Annunciator
(Add a number after it as
—① 24 to indicate number
of lamps) ... 1

Call Station, Single Cord,
Pilot Light ... 2

Call Station, Double Cord,
Microphone Speaker ... 3

Corridor Dome Light
1 Lamp ... 4

Transformer ... 5

Any Other Item On Same
System Use Number As
Required ... 6

II PAGING SYSTEM DEVICES

Basic Symbol

(Examples of Individual Item Identification Not a Part of Standard)

Keyboard ... 1

Flush Annunciator ... 2

2-Face Annunciator ... 3

Any Other Item On Same
System Use Numbers As
Required ... 4

III FIRE ALARM SYSTEM DEVICES (Any Type) Including Smoke and Sprinkler Alarm Devices

Basic Symbol

(Examples of Individual Item Identification. Not a Part of Standard)

Control Panel ... 1

Station ... 2

10" Gong ... 3

Pre-Signal Chime ... 4

Any Other Item On Same System
Use Numbers As Required ... 5

IV STAFF REGISTER SYSTEM DEVICES (Any Type)

Basic Symbol

(Examples of Individual Item Identification. Not a Part of Standard)

Phone Operators' Register ... 1

Entrance Register - Flush ... 2

Staff Room Register ... 3

Transformer ... 4

Any Other Item On Same System
Use Numbers As Required ... 5

V ELECTRIC CLOCK SYSTEM DEVICES (Any Type)

Basic Symbol

(Examples of Individual Item Identification. Not a Part of Standard)

110F20E.EPS

Figure 20 ◆ Recommended electrical symbols (5 of 7).

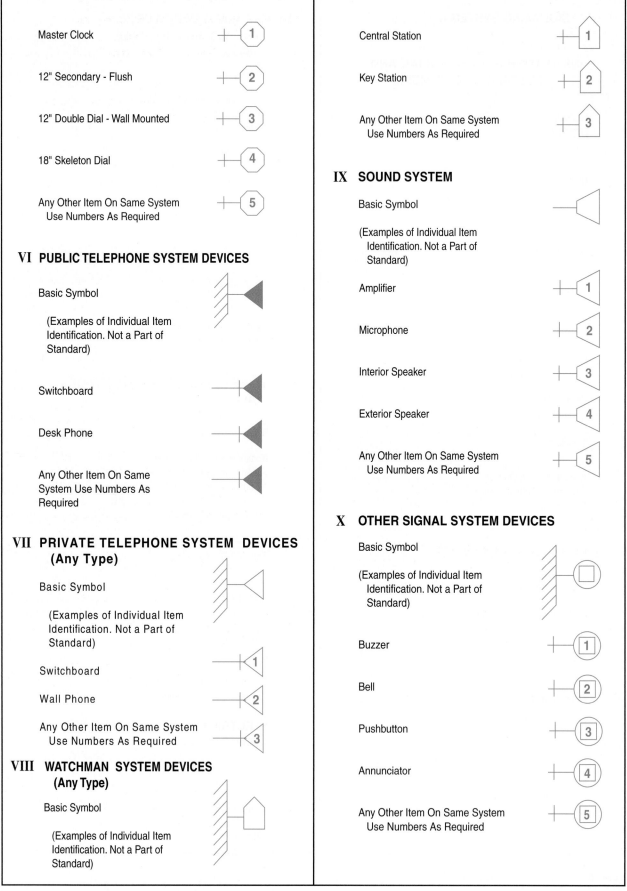

Master Clock   ①

12" Secondary - Flush   ②

12" Double Dial - Wall Mounted   ③

18" Skeleton Dial   ④

Any Other Item On Same System
Use Numbers As Required   ⑤

VI PUBLIC TELEPHONE SYSTEM DEVICES

Basic Symbol

  (Examples of Individual Item
  Identification. Not a Part of
  Standard)

Switchboard

Desk Phone

Any Other Item On Same
System Use Numbers As
Required

VII PRIVATE TELEPHONE SYSTEM DEVICES
(Any Type)

Basic Symbol

  (Examples of Individual Item
  Identification. Not a Part of
  Standard)

Switchboard   1

Wall Phone   2

Any Other Item On Same System
Use Numbers As Required   3

VIII WATCHMAN SYSTEM DEVICES
(Any Type)

Basic Symbol

  (Examples of Individual Item
  Identification. Not a Part of
  Standard)

Central Station   1

Key Station   2

Any Other Item On Same System
Use Numbers As Required   3

IX SOUND SYSTEM

Basic Symbol

(Examples of Individual Item
Identification. Not a Part of
Standard)

Amplifier   1

Microphone   2

Interior Speaker   3

Exterior Speaker   4

Any Other Item On Same System
Use Numbers As Required   5

X OTHER SIGNAL SYSTEM DEVICES

Basic Symbol

(Examples of Individual Item
Identification. Not a Part of
Standard)

Buzzer   1

Bell   2

Pushbutton   3

Annunciator   4

Any Other Item On Same System
Use Numbers As Required   5

110F20F.EPS

Figure 20 ◆ Recommended electrical symbols (6 of 7).

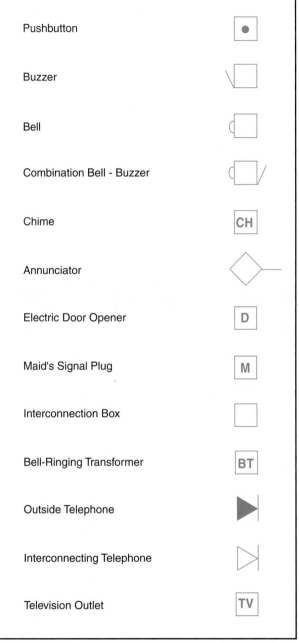

RESIDENTIAL OCCUPANCIES

Signaling system symbols for use in identifying standardized residential-type signal system items on residential drawings where a descriptive symbol list is not included on the drawing. When other signal system items are to be identified, use the above basic symbols for such items together with a descriptive symbol list.

Pushbutton

Buzzer

Bell

Combination Bell - Buzzer

Chime

Annunciator

Electric Door Opener

Maid's Signal Plug

Interconnection Box

Bell-Ringing Transformer

Outside Telephone

Interconnecting Telephone

Television Outlet

110F20G.EPS

Figure 20 ◆ Recommended electrical symbols (7 of 7).

5.0.0 ◆ SCALE DRAWINGS

In most electrical drawings, the components are so large that it would be impossible to draw them actual size. Consequently, drawings are made to some reduced scale; that is, all the distances are drawn smaller than the actual dimensions of the object itself, with all dimensions being reduced in the same proportion. For example, if a floor plan of a building is to be drawn to a scale of ¼" = 1'–0", each ¼" on the drawing would equal 1 foot on the building itself; if the scale is ⅛" = 1'–0", each ⅛" on the drawing equals 1 foot on the building, and so forth.

When architectural and engineering drawings are produced, the selected scale is very important. Where dimensions must be held to extreme accuracy, the scale drawings should be made as large as practical with dimension lines added. Where dimensions require only reasonable accuracy, the object may be drawn to a smaller scale (with dimension lines possibly omitted).

In dimensioning drawings, the dimensions written on the drawing are the actual dimensions of the building, not the distances that are measured on the drawing. To further illustrate this point, look at the floor plan in *Figure 21*; it is drawn to a scale of ½" = 1'–0". One of the walls is drawn to an actual length of 3½" on the drawing paper, but since the scale is ½" = 1'–0" and since 3½" contains 7 halves of an inch (7 × ½ = 3½"), the dimension shown on the drawing will therefore be 7'–0" on the actual building.

As shown in the previous example, the most common method of reducing all the dimensions (in feet and inches) in the same proportion is to choose a certain distance and let that distance represent one foot. This distance can then be divided into 12 parts, each of which represents an inch. If half inches are required, these twelfths are further subdivided into halves, etc. Now the scale represents the common foot rule with its subdivisions into inches and fractions, except that the scaled foot is smaller than the distance known as a foot and, likewise, its subdivisions are proportionately smaller.

When a measurement is made on the drawing, it is made with the reduced foot rule or scale; when a measurement is made on the building, it is made with the standard foot rule. The most common reduced foot rules or scales used in electrical drawings are the architect's scale and the engineer's scale. Drawings may sometimes be encountered that use a metric scale, but using this scale is similar to using the architect's or engineer's scales.

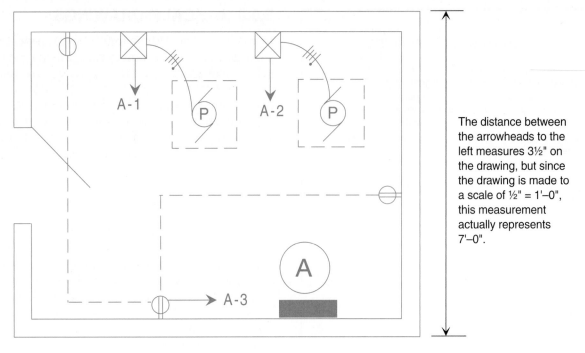

The distance between the arrowheads to the left measures 3½" on the drawing, but since the drawing is made to a scale of ½" = 1'–0", this measurement actually represents 7'–0".

PUMP HOUSE FLOOR PLAN

½" = 1'–0"

110F21.EPS

Figure 21 ◆ Typical floor plan showing drawing scale.

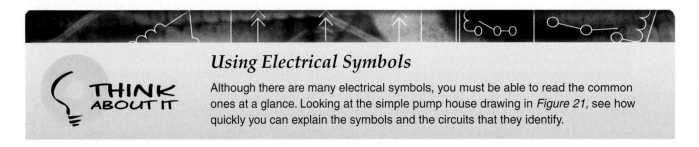

Using Electrical Symbols

Although there are many electrical symbols, you must be able to read the common ones at a glance. Looking at the simple pump house drawing in *Figure 21*, see how quickly you can explain the symbols and the circuits that they identify.

THINK ABOUT IT

5.1.0 Architect's Scale

Figure 22 shows two configurations of architect's scales. The one on the top is designed so that 1" = 1'–0", and the one on the bottom has graduations spaced to represent ⅛" = 1'–0".

Note that on the one-inch scale in *Figure 23*, the longer marks to the right of the zero (with a numeral beneath) represent feet. Therefore, the distance between the zero and the numeral 1 equals one foot. The shorter mark between the zero and 1 represents ½ of a foot, or six inches.

Referring again to *Figure 23*, look at the marks to the left of the zero. The numbered marks are spaced three scaled inches apart and have the numerals 0, 3, 6, and 9 for use as reference points. The other lines of the same length also represent scaled inches, but are not marked with numerals. In use, you can count the number of long marks to the left of the zero to find the number of inches, but after

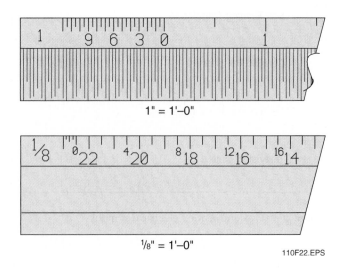

1" = 1'–0"

⅛" = 1'–0"

110F22.EPS

Figure 22 ◆ Two different configurations of architect's scales.

Architect's Scale

Measurements are usually made on architectural drawings using an architect's scale rather than a standard ruler. Architect's scales, like the ones on the left, are divided into feet and inches and usually consist of several scales on one rule. Architect's scales also come in other forms such as tapes or with wheels, like the one shown on the right.

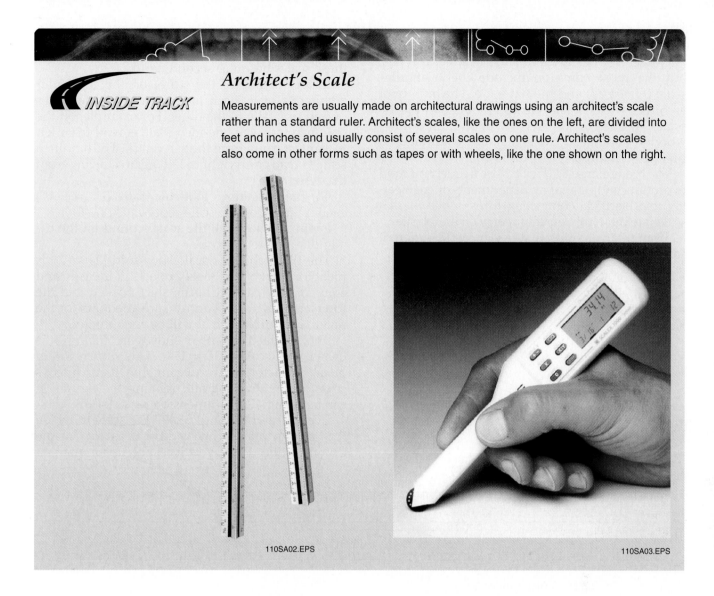

110SA02.EPS

110SA03.EPS

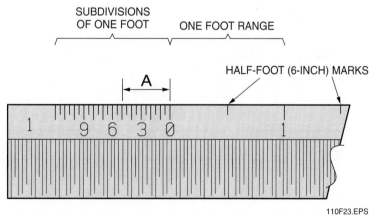

110F23.EPS

Figure 23 ◆ One-inch architect's scale.

some practice, you will be able to tell the exact measurement at a glance. For example, the measurement A represents five inches because it is the fifth inch mark to the left of the zero; it is also one inch mark short of the six-inch line on the scale.

The lines that are shorter than the inch line are the half-inch lines. On smaller scales, the basic unit is not divided into as many divisions. For example, the smallest subdivision on some scales represents two inches.

5.1.1 Types of Architect's Scales

Architect's scales are available in several types, but the most common include the triangular scale (*Figure 24*) and the flat scale. The quality of architect's scales also varies from cheap plastic scales (costing a dollar or two) to high-quality wooden-laminated tools that are calibrated to precise standards.

The triangular scale (*Figure 24*) is frequently found in drafting and estimating departments or engineering and electrical contracting firms, while the flat scales are more convenient to carry on the job site.

Triangular architect's scales have 12 different scales—two on each edge—as follows:

- Common foot rule (12 inches)
- $\frac{1}{16}$" = 1'–0"
- $\frac{3}{32}$" = 1'–0"
- $\frac{3}{16}$" = 1'–0"
- $\frac{1}{8}$" = 1'–0"
- $\frac{1}{4}$" = 1'–0"
- $\frac{3}{8}$" = 1'–0"
- $\frac{3}{4}$" = 1'–0"
- 1" = 1'–0"
- $\frac{1}{2}$" = 1'–0"
- $1\frac{1}{2}$" = 1'–0"
- 3" = 1'–0"

Two separate scales on one face may seem confusing at first, but after some experience, reading these scales becomes second nature.

In all but one of the scales on the triangular architect's scale, each face has one of the scales placed opposite to the other. For example, on the one-inch face, the one-inch scale is read from left to right, starting from the zero mark. The half-inch scale is read from right to left, again starting from the zero mark.

On the remaining foot-rule scale ($\frac{1}{16}$" = 1'–0") each $\frac{1}{16}$" mark on the scale represents one foot.

Figure 25 shows all the scales found on the triangular architect's scale.

The flat architect's scale shown in *Figure 26* is ideal for workers on most projects. It is easily and conveniently carried in the shirt pocket, and the four scales ($\frac{1}{8}$", $\frac{1}{4}$", $\frac{1}{2}$", and 1") are adequate for the majority of projects that will be encountered.

The partial floor plan shown in *Figure 26* is drawn to a scale of $\frac{1}{8}$" = 1'–0". The dimension in question is found by placing the $\frac{1}{8}$" architect's scale on the drawing and reading the figures. It can be seen that the dimension reads 24'–6".

Every drawing should have the scale to which it is drawn plainly marked on it as part of the drawing title. However, it is not uncommon to

110F24.EPS

Figure 24 ◆ Typical triangular architect's scale.

Figure 25 ◆ Various scales on a triangular architect's scale.

110F25.EPS

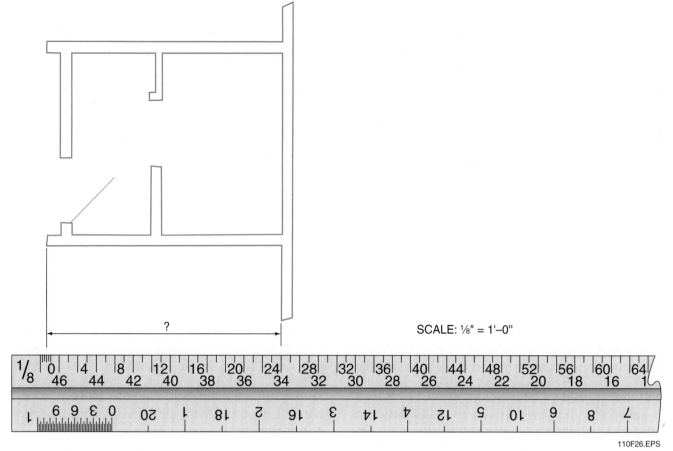

Figure 26 ◆ Using the ⅛" architect's scale to determine the dimensions on a drawing.

110F26.EPS

have several different drawings on one blueprint sheet—all with different scales. Therefore, always check the scale of each different view found on a drawing sheet.

5.2.0 Engineer's Scale

The civil engineer's scale is used in basically the same manner as the architect's scale, with the principal difference being that the graduations on the engineer's scale are decimal units rather than feet, as on the architect's scale.

The engineer's scale is used by placing it on the drawing with the working edge away from the user. The scale is then aligned in the direction of the required measurement. Then, by looking down at the scale, the dimension is read.

Civil engineer's scales commonly show the following graduations:

- 1" = 10 units
- 1" = 20 units
- 1" = 30 units
- 1" = 40 units
- 1" = 60 units
- 1" = 80 units
- 1" = 100 units

The purpose of this scale is to transfer the relative dimensions of an object to the drawing or vice versa. It is used mainly on site plans to determine distances between property lines, manholes, duct runs, direct-burial cable runs, and the like.

Site plans are drawn to scale using the engineer's scale rather than the architect's scale. On small lots, a scale of 1 inch = 10 feet or 1 inch = 20 feet is used. For a 1:10 scale, this means that one inch (the actual measurement on the drawing) is equal to 10 feet on the land itself.

On larger drawings, where a large area must be covered, the scale could be 1 inch = 100 feet or 1 inch = 1,000 feet, or any other integral power of 10. On drawings with the scale in multiples of 10, the engineer's scale marked 10 is used. If the scale is 1 inch = 200 feet, the engineer's scale marked 20 is used, and so on.

Although site plans appear reduced in scale, depending on the size of the object and the size of the drawing sheet to be used, the actual dimensions must be shown on the drawings at all times. When you are reading the drawing plans to scale, think of each dimension in its full size and not in the reduced scale it happens to be on the drawing (*Figure 27*).

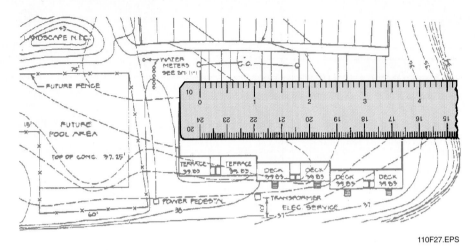

110F27.EPS

Figure 27 ◆ Practical use of the engineer's scale.

5.3.0 Metric Scale

Metric scales are calibrated in units of 10 (*Figure 28*). The two common length measurements used in the metric scale or architectural drawings are the meter and the millimeter, the millimeter being $\frac{1}{1,000}$ of a meter. On drawings drawn to scales between 1:1 and 1:100, the millimeter is typically used. On drawings drawn to scales between 1:200 and 1:2,000, the meter is generally used. Many contracting firms that deal in international trade have adopted a dual-dimensioning system expressed in both metric and English symbols. Drawings prepared for government projects may also require metric dimensions. See *Appendix A*.

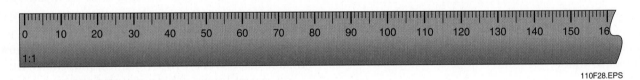

110F28.EPS

Figure 28 ◆ Typical metric scale.

6.0.0 ◆ ANALYZING ELECTRICAL DRAWINGS

The most practical way to learn how to read electrical construction documents is to analyze an existing set of drawings prepared by consulting or industrial engineers.

Engineers or electrical designers are responsible for the complete layout of electrical systems for most projects. Electrical drafters then transform the engineer's designs into working drawings, using either manual drafting instruments or computer-aided design (CAD) systems. The following is a brief outline of what usually takes place in the preparation of electrical design and working drawings:

- The engineer meets with the architect and owner to discuss the electrical needs of the building or project and to discuss various recommendations made by all parties.
- After that, an outline of the architect's floor plan is laid out.
- The engineer then calculates the required power and lighting outlets for the project; these are later transferred to the working drawings.
- All communications and alarm systems are located on the floor plan, along with lighting and power panelboards.
- Circuit calculations are made to determine wire size and overcurrent protection.
- The main electric service and related components are determined and shown on the drawings.
- Schedules are then placed on the drawings to identify various pieces of equipment.
- Wiring diagrams are made to show the workers how various electrical components are to be connected.
- A legend or electrical symbol list is drafted and shown on the drawings to identify all symbols used to indicate electrical outlets or equipment.
- Various large-scale electrical details are included, if necessary, to show exactly what is required of the electricians.
- Written specifications are then made to give a description of the materials and installation methods.

6.1.0 Development of Site Plans

In general practice, it is usually the owner's responsibility to furnish the architect/engineer with property and topographic surveys, which are made by a certified land surveyor or civil engineer. These surveys show:

- All property lines
- Existing public utilities and their location on or near the property (e.g., electrical lines, sanitary sewer lines, gas lines, water-supply lines, storm sewers, manholes, telephone lines, etc.)

A land surveyor does the property survey from information obtained from a deed description of the property. A property survey shows only the property lines and their lengths, as if the property were perfectly flat.

The topographic survey shows both the property lines and the physical characteristics of the land by using contour lines, notes, and symbols. The physical characteristics may include:

- The direction of the land slope
- Whether the land is flat, hilly, wooded, swampy, high, or low, and other features of its physical nature

All of this information is necessary so that the architect can properly design a building to fit the property. The electrical engineer also needs this information to locate existing electrical utilities and to route the new service to the building, provide outdoor lighting and circuits, etc.

Electrical site work is sometimes shown on the architect's plot plan. However, when site work involves many trades and several utilities (e.g., gas, telephone, electric, television, water, and sewage), it can become confusing if all details are shown on one drawing sheet. In cases like these, it is best to have a separate drawing devoted entirely to the electrical work, as shown in *Figure 29*. This project is an office/warehouse building for Virginia Electric, Inc. The electrical drawings consist of four 24" × 36" drawing sheets, along with a set of written specifications, which will be discussed later in this module.

Reading Notes

The notes are crucial elements of the drawing set. Receptacles, for example, are hard to position precisely based on a scaled drawing alone, and yet the designer may call for exact locations. For example, the designer may want receptacles exactly 6" above the kitchen counter backsplash and centered on the sink.

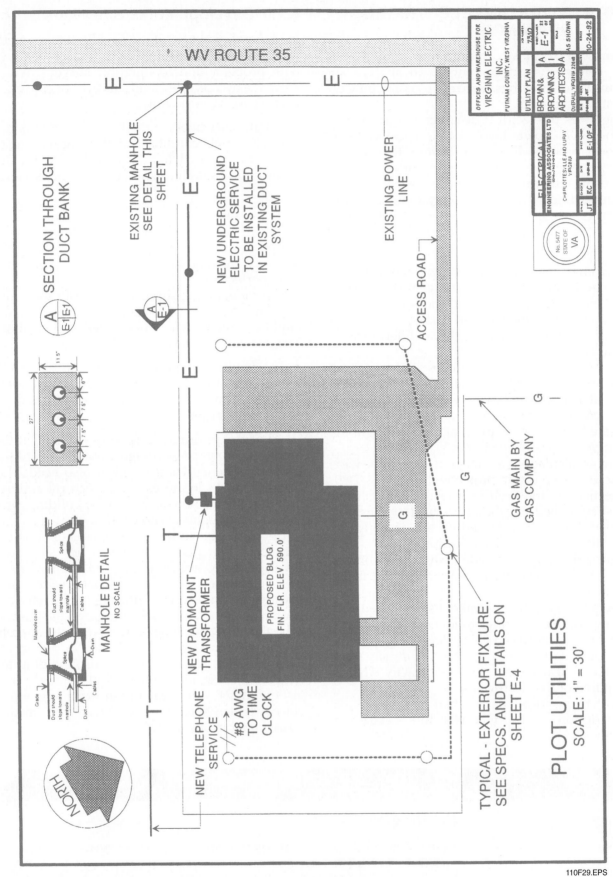

Figure 29 ◆ Typical electrical site plan.

110F29.EPS

The electrical site or plot plan shown in *Figure 29* has the conventional architect's and engineer's title blocks in the lower right-hand corner of the drawing. These blocks identify the project and project owners, the architect, and the engineer. They also show how this drawing sheet relates to the entire set of drawings. Note the engineer's professional stamp of approval to the left of the engineer's title block. Similar blocks appear on all four of the electrical drawing sheets.

When examining a set of electrical drawings for the first time, always look at the area around the title block. This is where most revision blocks or revision notes are placed. If revisions have been made to the drawings, make certain that you have a clear understanding of what has taken place before proceeding with the work.

Refer again to the drawing in *Figure 29* and note the North Arrow in the upper left corner. A North Arrow shows the direction of true north to help you orient the drawing to the site. Look directly down from the North Arrow to the bottom of the page and notice the drawing title, *Plot Utilities*. Directly beneath the drawing title you can see that the drawing scale of 1" = 30' is shown. This means that each inch on the drawing represents 30 feet on the actual job site. This scale holds true for all drawings on the page unless otherwise noted.

An outline of the proposed building is indicated on the drawing along with a callout, *Proposed Bldg. Fin. Flr. Elev. 590.0*. This means that the finished floor level of the building is to be 590 feet above sea level, which in this part of the country will be about two feet above finished grade around the building. This information helps the electrician locate conduit sleeves and stub-ups to the correct height before the finished concrete floor is poured.

The shaded area represents asphalt paving for the access road, drives, and parking lot. Note that the access road leads into a highway, which is designated Route 35. This information further helps workers to orient the drawing to the building site.

Existing manholes are indicated by a solid circle, while an open circle is used to show the position of the five new pole-mounted lighting fixtures that are to be installed around the new building. Existing power lines are shown with a light solid line with the letter E placed at intervals along the line. The new underground electric service is shown in the same way, except the lines are somewhat wider and darker on the drawing. Note that this new high-voltage cable terminates into a padmount transformer near the proposed building. New telephone lines are similar except the letter T is used to identify the telephone lines.

The direct-burial underground cable supplying the exterior lighting fixtures is indicated with dashed lines on the drawing—shown connecting the open circles. A homerun for this circuit is also shown to a time clock.

The manhole detail shown to the right of the North Arrow may seem to serve very little purpose on this drawing since the manholes have already been installed. However, the dimensions and details of their construction will help the electrical contractor or supervisor to better plan the pulling of the high-voltage cable. The same is true of the cross section shown of the duct bank. The electrical contractor knows that three empty ducts are available if it is discovered that one of them is damaged when the work begins.

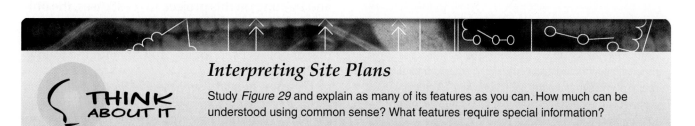

Interpreting Site Plans

THINK ABOUT IT

Study *Figure 29* and explain as many of its features as you can. How much can be understood using common sense? What features require special information?

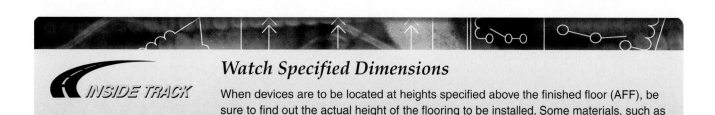

Watch Specified Dimensions

INSIDE TRACK

When devices are to be located at heights specified above the finished floor (AFF), be sure to find out the actual height of the flooring to be installed. Some materials, such as ceramic tile, can add significantly to the height of the finished floor.

Although the electrical work will not involve working with gas, the main gas line is shown on the electrical drawing to let the electrical workers know its approximate location while they are installing the direct-burial conductors for the exterior lighting fixtures.

7.0.0 ◆ POWER PLANS

The electrical power plan (*Figure 30*) shows the complete floor plan of the office/warehouse building with all interior partitions drawn to scale. Sometimes, the physical locations of all wiring and outlets are shown on one drawing; that is, outlets for lighting, power, signal and communications, special electrical systems, and related equipment are shown on the same plan. However, on complex installations, the drawing would become cluttered if both lighting and power were shown on the same floor plan. Therefore, most projects will have a separate drawing for power and another for lighting. Riser diagrams and details may be shown on yet another drawing sheet, or if room permits, they may be shown on the lighting or power floor plan sheets.

A closer look at this drawing reveals the title blocks in the lower right corner of the drawing

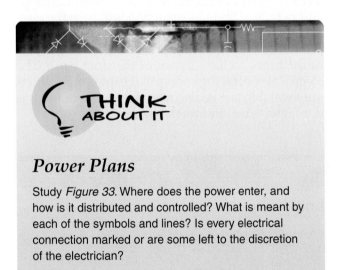

Power Plans

Study *Figure 33*. Where does the power enter, and how is it distributed and controlled? What is meant by each of the symbols and lines? Is every electrical connection marked or are some left to the discretion of the electrician?

sheet. These blocks list both the architectural and engineering firms, along with information to identify the project and drawing sheet. Also note that the floor plan is titled *Floor Plan "B"—Power* and is drawn to a scale of ⅛" = 1'–0". There are no revisions shown on this drawing sheet.

7.1.0 Key Plan

A key plan appears on the drawing sheet immediately above the engineer's title block (*Figure 31*). The purpose of this key plan is to identify that part of the project to which this sheet applies. In this case, the project involves two buildings: Building A and Building B. Since the outline of Building B is cross-hatched in the key plan, this is the building to which this drawing applies. Note that this key plan is not drawn to scale—only its approximate shape.

Although Building A is also shown on this key plan, a note below the key plan title states that there is no electrical work required in Building A.

On some larger installations, the overall project may involve several buildings requiring appropriate key plans on each drawing to help the workers orient the drawings to the appropriate building. In some cases, separate drawing sheets may be used for each room or area in an industrial project—again requiring key plans on each drawing sheet to identify applicable drawings for each room.

7.2.0 Symbol List

A symbol list appears on the electrical power plan (immediately above the architect's title block) to identify the various symbols used for both power and lighting on this project. In most cases, the only symbols listed are those that apply to the particular project. In other cases, however, a standard list of symbols is used for all projects with the following note:

These are standard symbols and may not all appear on the project drawings; however, wherever the symbol on the project drawings occurs, the item shall be provided and installed.

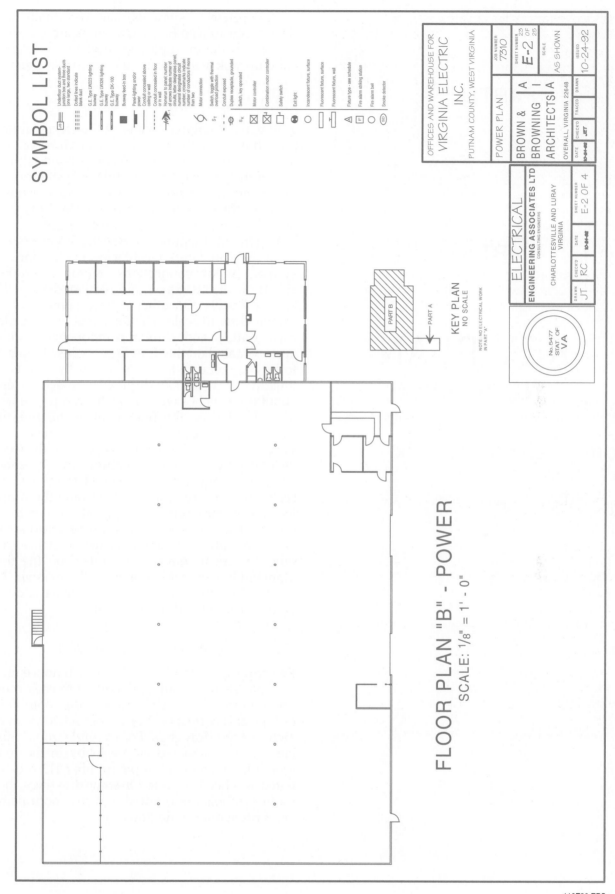

Figure 30 ◆ Electrical power plan.

110F30.EPS

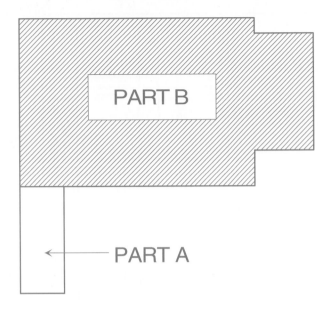

KEY PLAN

NO SCALE

NOTE: NO ELECTRICAL WORK
IN PART "A"

110F31.EPS

Figure 31 ◆ Key plan appearing on electrical power plan.

Only electrical symbols that are actually used for the office/warehouse drawings are shown in the list on the example electrical power plan. A close-up look at these symbols appears in *Figure 32*.

7.3.0 Floor Plan

A somewhat enlarged view of the electrical floor plan drawing is shown in *Figure 33*. However, due to the size of the drawing in comparison with the size of the pages in this module, it is still difficult to see very much detail. This illustration is meant to show the overall layout of the floor plan and how the symbols and notes are arranged.

In general, this plan shows the service equipment (in plan view), receptacles, underfloor duct system, motor connections, motor controllers, electric heat, busways, and similar details. The electric panels and other service equipment are drawn close to scale. The locations of other electrical outlets and similar components are only approximated on the drawings because they have to be exaggerated to show up on the prints. To illustrate, a common duplex receptacle is only about three inches wide. If such a receptacle were to be located on the floor plan of this building (drawn to a scale of $\frac{1}{8}$" = 1'–0"), even a small dot on the drawing would be too large to draw the receptacle exactly to scale. Therefore, the receptacle symbol is exaggerated. When such receptacles are scaled on the drawings to determine the proper location, a measurement is usually taken to the center of the symbol to determine the distance between outlets. Junction boxes, switches, and other electrical connections shown on the floor plan will be exaggerated in a similar manner. The partial floor plan drawing in *Figure 34* allows a better view of the drawing details.

7.3.1 Notes and Building Symbols

Referring again to *Figure 33*, you will notice numbers placed inside an oval symbol in each room. These numbered ovals represent the room name or type and correspond to a room schedule in the architectural drawings. For example, room number 112 is designated as the lobby in the room schedule (not shown), room number 113 is designated as office No. 1, etc. On some drawings, these room symbols are omitted and the room names are written out on the drawings.

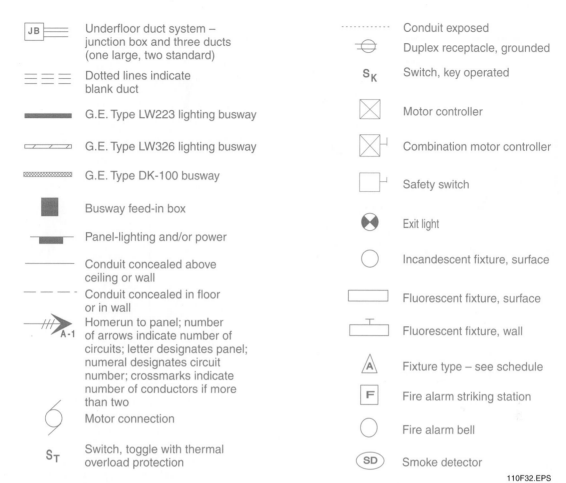

Figure 32 ◆ Sample electrical symbols list.

110F32.EPS

Understanding Contact Symbols

When a drawing shows normally open or normally closed contacts, the word normally refers to the condition of the contacts in their de-energized or shelf state.

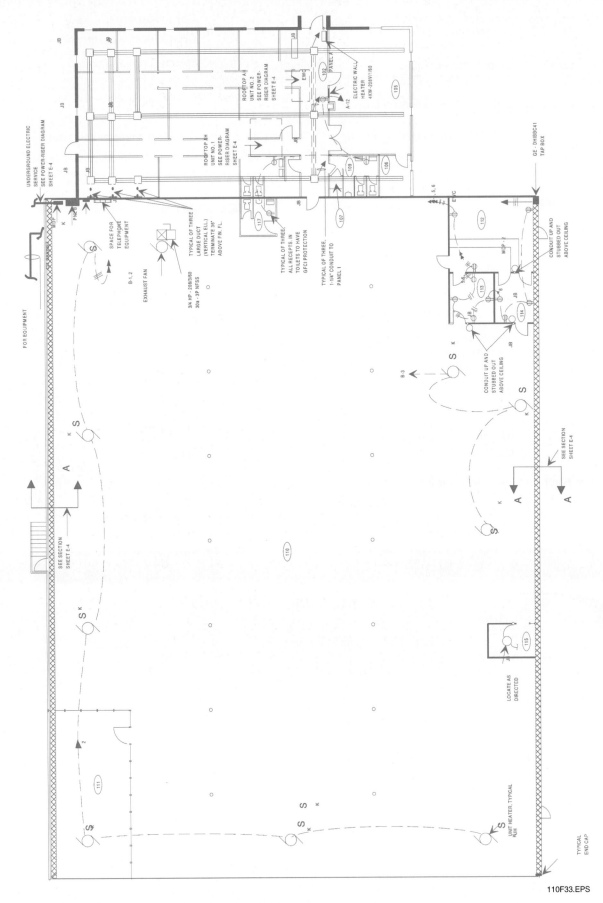

Figure 33 ◆ Power plan for an office/warehouse building.

110F33.EPS

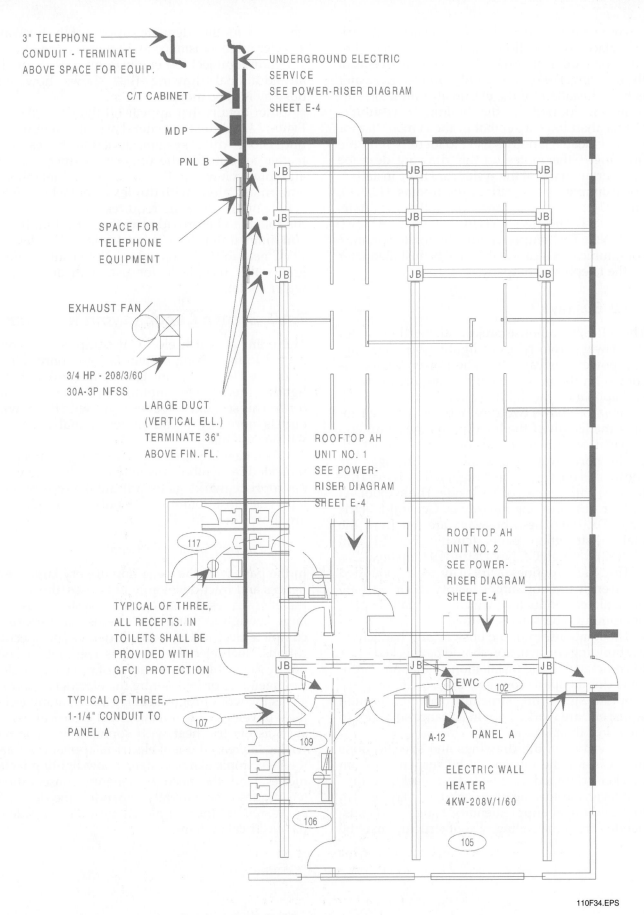

Figure 34 ◆ Partial floor plan for office/warehouse building.

110F34.EPS

There are also several notes appearing at various places on the floor plan. These notes offer additional information to clarify certain aspects of the drawing. For example, only one electric heater is to be installed by the electrical contractor; this heater is located in the building's vestibule. Rather than have a symbol in the symbol list for this one heater, a note is used to identify it on the drawing. Other notes on this drawing describe how certain parts of the system are to be installed. For example, in the office area (rooms 112, 113, and 114), you will see the following note: *CONDUIT UP AND STUBBED OUT ABOVE CEILING.* This empty conduit is for telephone/communications cables that will be installed later by the telephone company.

7.3.2 Busways

The office/warehouse project utilizes three types of busways: two types of lighting busways and one power busway. Only the power busway is shown on the power plan; the lighting busways will appear on the lighting plan.

Figure 33 shows two runs of busways: one running the length of the building on the south end (top wall on drawing), and one running the length of the north wall. The symbol list in *Figure 32* shows this busway to be designated by two parallel lines with a series of X's inside. The symbol list further describes the busway as General Electric Type DK-100. These busways are fed from the main distribution panel (circuits MDP-1 and MDP-2) through GE No. DHIBBC41 tap boxes.

The *NEC®* defines a busway as a grounded metal enclosure containing factory-mounted, bare or insulated conductors, which are usually copper or aluminum bars, rods, or tubes.

The relationship of the busway and hangers to the building construction should be checked prior to commencing the installation so that any problems due to space conflicts, inadequate or inappropriate supporting structure, openings through walls, etc. are worked out in advance so as not to incur lost time.

For example, the drawings and specifications may call for the busway to be suspended from brackets clamped or welded to steel columns. However, the spacing of the columns may be such that additional supplementary hanger rods suspended from the ceiling or roof structure may be necessary for the adequate support of the busway. To offer more assistance to workers on the office/warehouse project, the engineer may also provide an additional drawing that shows how the busway is to be mounted.

Other details that appear on the floor plan in *Figure 34* include the general arrangement of the underfloor duct system, junction boxes and feeder conduit for the underfloor duct system, and plan views of the service and telephone equipment, along with duplex receptacle outlets. A note on the drawing requires all receptacles in the toilets to be provided with ground fault circuit interrupter (GFCI) protection. The letters EWC next to the receptacle in the vestibule designate this receptacle for use with an electric water cooler.

7.4.0 Branch Circuit Layout for Power

The point at which electrical equipment is connected to the wiring system is commonly called an outlet. There are many classifications of outlets: lighting, receptacle, motor, appliance, and so forth. This section, however, deals with the power outlets normally found in residential electrical wiring systems.

When viewing an electrical drawing, outlets are indicated by symbols (usually a small circle with appropriate markings to indicate the type of outlet). The most common symbols for receptacles are shown in *Figure 35*.

7.4.1 Branch Circuit Drawings

In the past, with the exception of very large residences and tract-development houses, the size of the average residential electrical system was not large enough to justify the expense of preparing complete electrical working drawings and specifications. Such electrical systems were either laid out by the architect in the form of a sketchy outlet arrangement, or laid out by the electrician on the job as the work progressed. However, many technical developments in residential electrical use—such as electric heat with sophisticated control wiring, increased use of electrical appliances, various electronic alarm systems, new lighting techniques, and the need for energy conservation techniques—have greatly expanded the demand and extended the complexity of today's residential electrical systems.

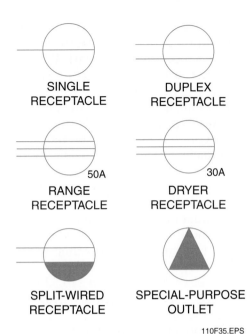

SINGLE RECEPTACLE

DUPLEX RECEPTACLE

RANGE RECEPTACLE 50A

DRYER RECEPTACLE 30A

SPLIT-WIRED RECEPTACLE

SPECIAL-PURPOSE OUTLET

110F35.EPS

Figure 35 ◆ Typical outlet symbols appearing in electrical drawings.

Each year, the number of homes with electrical systems designed by consulting engineering firms increases. Such homes are provided with complete electrical working drawings and specifications, similar to those frequently provided for commercial and industrial projects. Still, these are more the exception than the rule. Most residential projects will not have a complete set of drawings.

Circuit layout is provided on the drawings to follow for several reasons:

- They provide a visual layout of house wiring circuitry.
- They provide a sample of electrical residential drawings that are prepared by consulting engineering firms, although the number may still be limited.
- They introduce the method of showing electrical systems on working drawings so that you will have a better foundation for tackling advanced electrical systems.

Branch circuits are shown on electrical drawings by means of a single line drawn from the panelboard (or by homerun arrowheads indicating that the circuit goes to the panelboard) to the outlet or from outlet to outlet where there is more than one outlet on the circuit.

The lines indicating branch circuits can be solid to show that the conductors are to be run concealed in the ceiling or wall, dashed to show that the conductors are to be run in the floor or ceiling below, or dotted to show that the wiring is to be run exposed. *Figure 36* shows examples of these three types of branch circuit lines.

In *Figure 36*, No. 12 indicates the wire size. The slash marks shown through the circuits in *Figure 36* indicate the number of current-carrying conductors in the circuit. Although two slash marks are shown for the current-carrying conductors (along with one slash mark for the ground), in actual practice, a branch circuit containing only two conductors usually contains no slash marks;

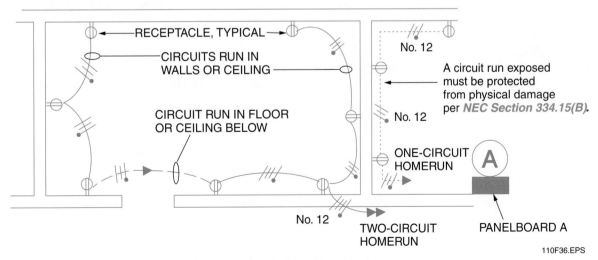

RECEPTACLE, TYPICAL

CIRCUITS RUN IN WALLS OR CEILING

CIRCUIT RUN IN FLOOR OR CEILING BELOW

No. 12

No. 12

A circuit run exposed must be protected from physical damage per *NEC Section 334.15(B)*.

ONE-CIRCUIT HOMERUN

A

No. 12

TWO-CIRCUIT HOMERUN

PANELBOARD A

110F36.EPS

Figure 36 ◆ Types of branch circuit lines shown on electrical working drawings.

that is, any circuit with no slash marks is assumed to have two conductors. However, three or more conductors are always indicated on electrical working drawings—either by slash marks for each conductor, or else by a note.

Never assume that you know the meaning of any electrical symbol. Although great efforts have been made in recent years to standardize drawing symbols, architects, consulting engineers, and electrical drafters still modify existing symbols or devise new ones to meet their own needs. Always consult the symbol list or legend on electrical working drawings for an exact interpretation of the symbols used.

7.4.2 Locating Receptacles

NEC Section 210.52 states the minimum requirements for the location of receptacles in dwelling units. It specifies that in each kitchen, family room, and dining room, receptacle outlets shall be installed so that no point along the floor line in any wall space is more than 6', measured horizontally, from an outlet in that space, including any wall space 2' or more in width and the wall space

occupied by fixed panels in exterior walls, but excluding sliding panels. This means that the outlets will be no more than 12' apart. When spaced in this manner, a 6' extension cord will reach a receptacle at any point along the wall line. Receptacle outlets shall, insofar as practicable, be spaced equal distances apart. Receptacle outlets in floors shall not be counted as part of the required number of receptacle outlets unless located within 18" of the wall.

The *NEC*® defines wall space as a wall that is unbroken along the floor line by doorways, fireplaces, or similar openings. Each wall space that is two feet or more in width must be treated individually and separately from other wall spaces within the room.

The purpose of *NEC Section 210.52* is to minimize the use of cords across doorways, fireplaces, and similar openings.

Figure 37 shows the outlets for a sample residence. In laying out these receptacle outlets, the floor line of the wall is measured (also around corners), but not across doorways, fireplaces, passageways, or other spaces where a flexible cord extended across the space would be unsuitable.

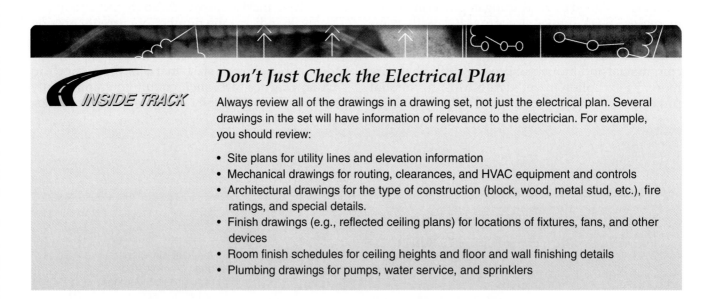

Don't Just Check the Electrical Plan

Always review all of the drawings in a drawing set, not just the electrical plan. Several drawings in the set will have information of relevance to the electrician. For example, you should review:

- Site plans for utility lines and elevation information
- Mechanical drawings for routing, clearances, and HVAC equipment and controls
- Architectural drawings for the type of construction (block, wood, metal stud, etc.), fire ratings, and special details.
- Finish drawings (e.g., reflected ceiling plans) for locations of fixtures, fans, and other devices
- Room finish schedules for ceiling heights and floor and wall finishing details
- Plumbing drawings for pumps, water service, and sprinklers

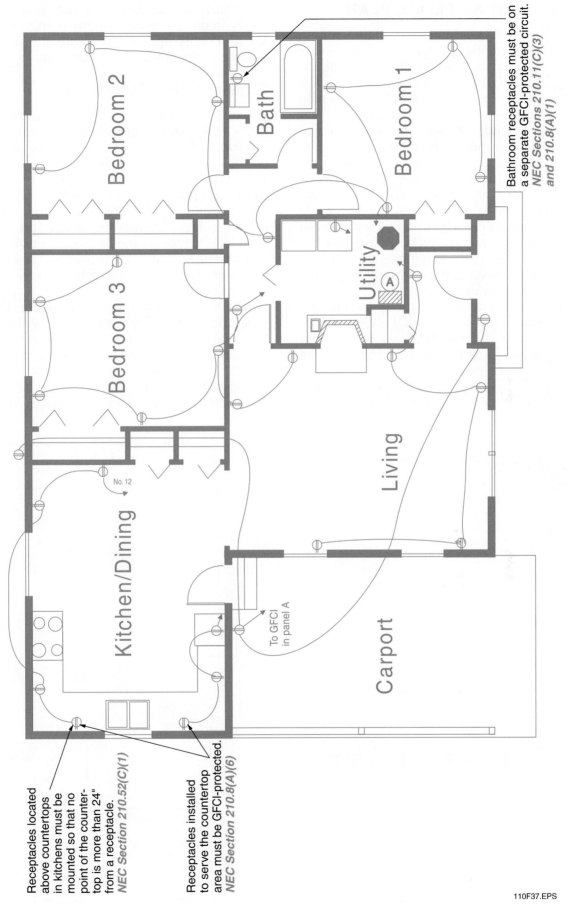

Bathroom receptacles must be on a separate GFCI-protected circuit. *NEC Sections 210.11(C)(3) and 210.8(A)(1)*

Bath

Bedroom 2

Bedroom 1

Bedroom 3

Utility

Ⓐ

Living

No. 12

Kitchen/Dining

To GFCI in panel A

Carport

Receptacles located above countertops in kitchens must be mounted so that no point of the counter-top is more than 24" from a receptacle. *NEC Section 210.52(C)(1)*

Receptacles installed to serve the countertop area must be GFCI-protected. *NEC Section 210.8(A)(6)*

110F37.EPS

Figure 37 ◆ Floor plan of a sample residence.

8.0.0 ◆ LIGHTING FLOOR PLAN

A skeleton view of a lighting floor plan is shown in *Figure 38*. Again, the architect's/engineer's title blocks appear in the lower right corner of the drawing. A key plan, as discussed previously, appears above the engineer's title block. This plan is drawn to the same scale as the power plan; that is, ⅛" = 1'–0". A lighting fixture schedule appears in the upper right corner of the drawing and some installation notes appear below the schedule.

The lighting outlet symbols found on the drawing for the office/warehouse building represent both incandescent and fluorescent types; a circle on most electrical drawings usually represents an incandescent fixture, and a rectangle represents a fluorescent one. All of these symbols are designed to indicate the physical shape of a particular fixture and are usually drawn to scale.

The type of mounting used for all lighting fixtures is usually indicated in a lighting fixture schedule, which in this case is shown on the drawings. On some projects, the schedule may be found only in the written specifications.

The type of lighting fixture is identified by a numeral placed inside a triangle near each light-ing fixture. If one type of fixture is used exclusively in one room or area, the triangular indicator need only appear once with the word ALL lettered at the bottom of the triangle.

8.1.0 Drawing Schedules

A schedule is a systematic method of presenting notes or lists of equipment on a drawing in tabular form. When properly organized and thoroughly understood, schedules are powerful timesaving devices for both those preparing the drawings and workers on the job.

For example, the lighting fixture schedule shown in *Figure 39* lists the fixture and identifies each fixture type on the drawing by number. The manufacturer and catalog number of each type are given along with the number, size, and type of lamp for each.

At times, all of the same information found in schedules will be duplicated in the written specifications, but combing through page after page of written specifications can be time consuming. Workers do not always have access to the specifications while on the job, whereas they usually do have access to the working drawings. Therefore,

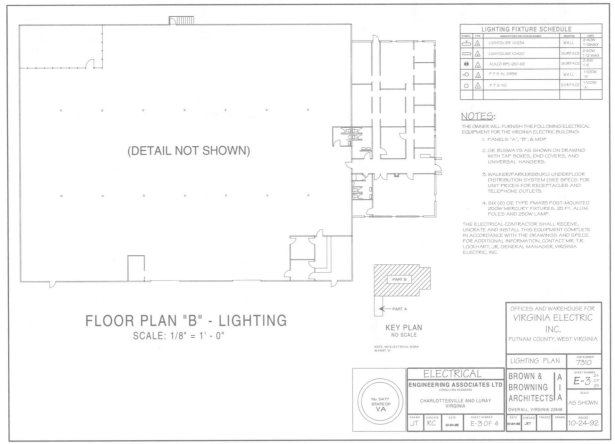

Figure 38 ◆ Sample lighting plan.

110F38.EPS

LIGHTING FIXTURE SCHEDULE

SYMBOL	TYPE	MANUFACTURER AND CATALOG NUMBER	MOUNTING	LAMPS
⊤▭	A	LIGHTOLIER 10234	WALL	2-40W T-12WWX
▭	B	LIGHTOLIER 10420	SURFACE	2-40W T-12 WWX
⊗	C	ALKCO RPC-210-6E	SURFACE	2-8W T-5
⊢○	D	P 7 S AL 2936	WALL	1-100W 'A'
○	E	P 7 S 110	SURFACE	1-100W 'A'

110F39.EPS

Figure 39 ◆ Lighting fixture schedule.

the schedule is an excellent means of providing essential information in a clear and accurate manner, allowing the workers to carry out their assignments in the least amount of time.

Other schedules that are frequently found on electrical working drawings include:

- Connected load schedule
- Panelboard schedule
- Electric heat schedule
- Kitchen equipment schedule
- Schedule of receptacle types

There are also other schedules found on electrical drawings, depending upon the type of project. However, most will deal with lists of equipment such as motors, motor controllers, and similar items.

8.2.0 Branch Circuit Layout for Lighting

A simple lighting branch circuit requires two conductors to provide a continuous path for current flow. The usual lighting branch circuit operates at either 120V or 277V; the white (grounded) circuit conductor is therefore connected to the neutral bus in the panelboard, while the black (ungrounded) circuit conductor is connected to an overcurrent protection device.

Lighting branch circuits and outlets are shown on electrical drawings by means of lines and symbols; that is, a single line is drawn from outlet to outlet and then terminated with an arrowhead to indicate a homerun to the panelboard. Several methods are used to indicate the number and size of conductors, but the most common is to indicate the number of conductors in the circuit by using slash marks through the circuit lines and then

indicate the wire size by a notation adjacent to these slash marks.

The circuits used to feed residential lighting must conform to standards established by the NEC® as well as by local and state ordinances. Most of the lighting circuits should be calculated to include the total load, although at times this is not possible because the electrician cannot be certain of the exact wattage that might be used by the homeowner. For example, an electrician may install four porcelain lampholders for the unfinished basement area, each to contain one 100-watt (100W) incandescent lamp. However, the homeowners may eventually replace the original lamps with others rated at 150W or even 200W. Thus, if the electrician initially loads the lighting circuit to full capacity, the circuit will probably become overloaded in the future.

It is recommended that no residential branch circuit be loaded to more than 80% of its rated capacity. Since most circuits used for lighting are rated at 15A, the total ampacity (in volt-amperes) for the circuit is as follows:

$$15A \times 120V = 1,800VA$$

Therefore, if the circuit is to be loaded to only 80% of its rated capacity, the maximum initial connected load should be no more than 1,440VA.

Figure 40 shows one possible lighting arrangement for the sample residence discussed earlier. All lighting fixtures are shown in their approximate physical location as they should be installed.

Electrical symbols are used to show the fixture types. Switches and lighting branch circuits are also shown by appropriate lines and symbols. The meanings of the symbols used on this drawing are explained in the symbol list in *Figure 41*.

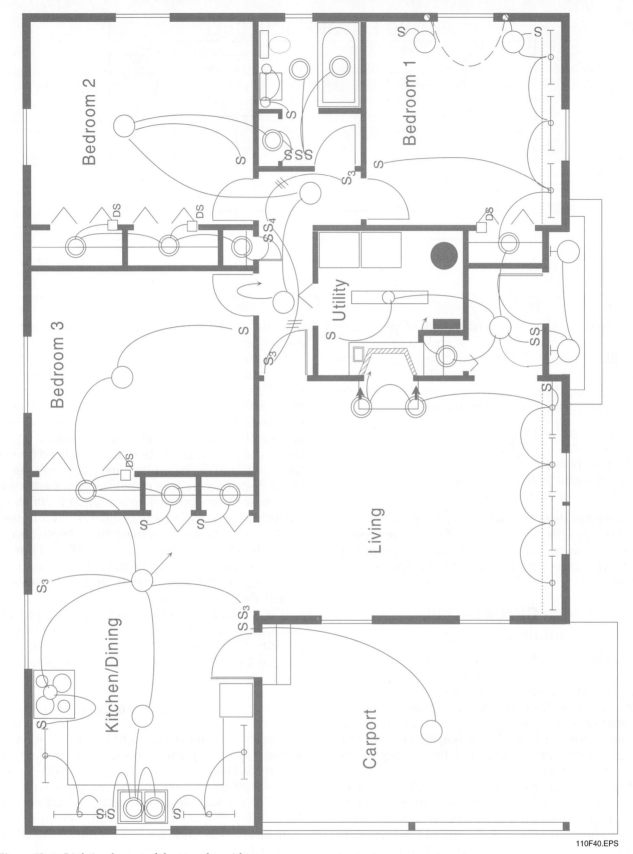

Figure 40 ◆ Lighting layout of the sample residence.

110F40.EPS

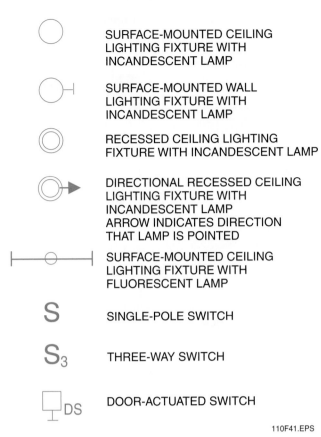

SURFACE-MOUNTED CEILING
LIGHTING FIXTURE WITH
INCANDESCENT LAMP

SURFACE-MOUNTED WALL
LIGHTING FIXTURE WITH
INCANDESCENT LAMP

RECESSED CEILING LIGHTING
FIXTURE WITH INCANDESCENT LAMP

DIRECTIONAL RECESSED CEILING
LIGHTING FIXTURE WITH
INCANDESCENT LAMP
ARROW INDICATES DIRECTION
THAT LAMP IS POINTED

SURFACE-MOUNTED CEILING
LIGHTING FIXTURE WITH
FLUORESCENT LAMP

S   SINGLE-POLE SWITCH

S_3 THREE-WAY SWITCH

DS   DOOR-ACTUATED SWITCH

110F41.EPS

Figure 41 ◆ Electrical symbols list.

In actual practice, the location of lighting fixtures and their related switches will probably be
the extent of the information shown on working
drawings. The circuits shown in *Figure 40* are
meant to illustrate how lighting circuits are
routed, not to imply that such drawings are typical for residential construction. If fixtures are used
in a closet, they must meet the requirements of
NEC Section 410.16 and be completely enclosed.

9.0.0 ◆ ELECTRICAL DETAILS AND DIAGRAMS

Electrical diagrams are drawings that are
intended to show electrical components and their
related connections. They show the electrical association of the different components, but are seldom, if ever, drawn to scale.

9.1.0 Power-Riser Diagrams

One-line (single-line) **block diagrams** are
used extensively to show the arrangement of
electric service equipment. **Power-riser diagrams** (*Figure 42*) are typical of such drawings.
These drawings show all pieces of electrical
equipment as well as the connecting lines used
to indicate service-entrance conductors and

feeders. Notes are used to identify the equipment, indicate the size of conduit necessary for
each feeder, and show the number, size, and type
of conductors in each conduit.

A panelboard schedule (*Figure 43*) is included
with the power-riser diagram to indicate the exact
components contained in each panelboard. This
panelboard schedule is for the main distribution
panel. On the actual drawings, schedules would
also be shown for the other two panels (PNL A
and PNL B).

In general, panelboard schedules usually indicate the panel number, type of cabinet (either
flush- or surface-mounted), panel mains (ampere
and voltage rating), phase (single- or three-
phase), and number of wires. A four-wire panel,
for example, indicates that a solid neutral exists in
the panel. Branches indicate the type of overcurrent protection; that is, they indicate the number
of poles, trip rating, and frame size. The items fed
by each overcurrent device are also indicated.

9.2.0 Schematic Diagrams

Complete schematic wiring diagrams are normally used only in complicated electrical systems,
such as control circuits. Components are represented by symbols, and every wire is either shown
by itself or included in an assembly of several
wires, which appear as one line on the drawing.
Each wire should be numbered when it enters an
assembly and should keep the same number when
it comes out again to be connected to some electrical component in the system. *Figure 44* shows a
complete schematic wiring diagram for a three-
phase, AC magnetic non-reversing motor starter.

Note that this diagram shows the various
devices in symbol form and indicates the actual
connections of all wires between the devices. The
three-wire supply lines are indicated by L_1, L_2, and
L_3; the motor terminals of motor M are indicated
by T_1, T_2, and T_3. Lines L_1, L_2, and L_3 each have a
thermal overload protection device (OL) con-
nected in series with normally open line contacts
C_1 and C_3, respectively, which are both controlled
by the magnetic starter coil, C. The control station,
consisting of start pushbutton 1 and stop pushbutton 2, is connected across lines L_1 and L_2. Auxiliary contacts (C_4) are connected in series with the
stop pushbutton and in parallel with the start
pushbutton. The control circuit also has normally
closed overload contacts (OC) connected in series
with the magnetic starter coil (C).

Any number of additional pushbutton stations
may be added to this control circuit similarly to
the way in which three-way and four-way
switches are added to control a lighting circuit.

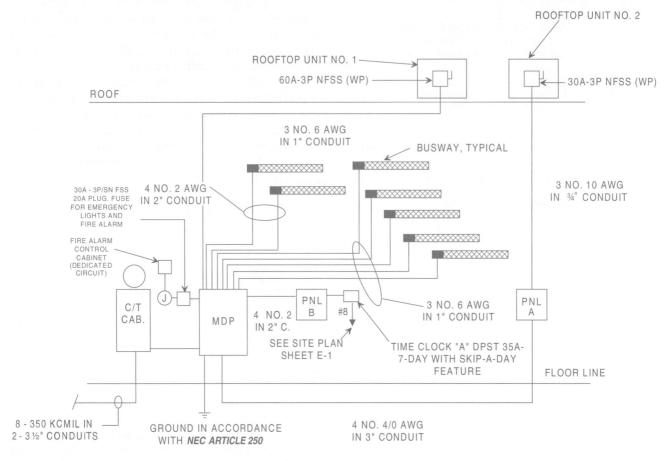

COMMERCIAL

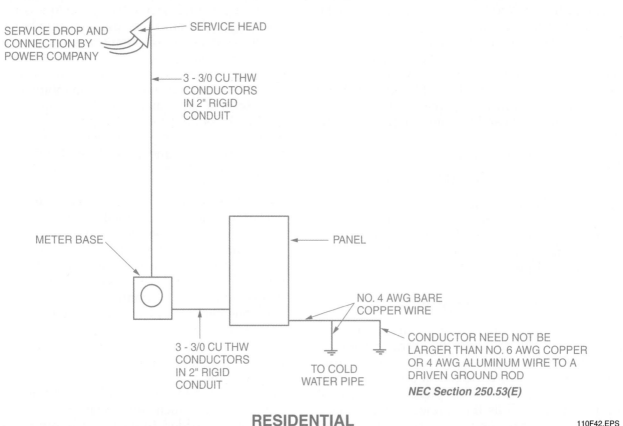

RESIDENTIAL

110F42.EPS

Figure 42 ◆ Typical power-riser diagrams.

PANELBOARD SCHEDULE

PANEL No.	CABINET TYPE	PANEL MAINS			BRANCHES					ITEMS FED OR REMARKS
		AMPS	VOLTS	PHASE	1P	2P	3P	PROT.	FRAME	
MDP	SURFACE	600A	120/208	3φ,4-W	–	–	1	225A	25,000	PANEL "A"
					–	–	1	100A	18,000	PANEL "B"
					–	–	1	100A		POWER BUSWAY
					–	–	1	60A		LIGHTING BUSWAY
					–	–	1	70A		ROOFTOP UNIT #1
					–	–	1	70A		SPARE
					–	–	1	600A	42,000	MAIN CIRCUIT BRKR

110F43.EPS

Figure 43 ◆ Typical panelboard schedule.

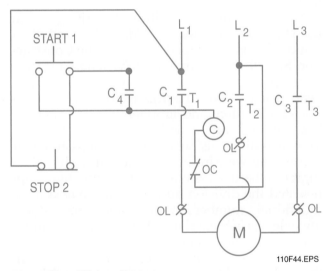

110F44.EPS

Figure 44 ◆ Wiring diagram.

When adding pushbutton stations, the stop buttons are always connected in series and the start buttons are always connected in parallel. *Figure 45* shows the same motor starter circuit in *Figure 44*, but this time it is controlled by two sets of start/stop buttons.

Schematic wiring diagrams have only been touched upon in this module; there are many other details that you will need to know to perform your work in a proficient manner. Later modules cover wiring diagrams in more detail.

9.3.0 Drawing Details

A detail drawing is a drawing of a separate item or portion of an electrical system, giving a complete and exact description of its use and all the

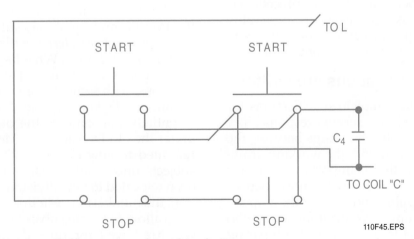

110F45.EPS

Figure 45 ◆ Circuit being controlled by two sets of start/stop buttons.

details needed to show the electrician exactly what is required for its installation. For example, the power plan for the office/warehouse has a sectional cut through the busduct. This is a good example of where an extra, detailed drawing is desirable.

A set of electrical drawings will sometimes require large-scale drawings of certain areas that are not indicated with sufficient clarity on the small-scale drawings. For example, the site plan may show exterior pole-mounted lighting fixtures that are to be installed by the contractor.

10.0.0 ◆ WRITTEN SPECIFICATIONS

The written specifications for a building or project are the written descriptions of work and duties required of the owner, architect, and consulting engineer. Together with the working drawings, these specifications form the basis of the contract requirements for the construction of the building or project. Those who use the construction drawings and specifications must always be alert to discrepancies between the working drawings and the written specifications. These are some situations where discrepancies may occur:

- Architects or engineers use standard or prototype specifications and attempt to apply them without any modification to specific working drawings.
- Previously prepared standard drawings are changed or amended by reference in the specifications only and the drawings themselves are not changed.
- Items are duplicated in both the drawings and specifications, but an item is subsequently amended in one and overlooked in the other contract document.

In such instances, the person in charge of the project has the responsibility to ascertain whether the drawings or the specifications take precedence. Such questions must be resolved, preferably before the work begins, to avoid added costs to the owner, architect/engineer, or contractor.

10.1.0 How Specifications are Written

Writing accurate and complete specifications for building construction is a serious responsibility for those who design the buildings because the specifications, combined with the working drawings, govern practically all important decisions that are made during the construction span of every project. Compiling and writing these specifications is not a simple task, even for those who have had considerable experience in preparing

such documents. A set of written specifications for a single project will usually contain thousands of products, parts, and components, and the methods of installing them, all of which must be covered in either the drawings and/or specifications. No one can memorize all of the necessary items required to describe accurately the various areas of construction. One must rely upon reference materials such as manufacturer's data, catalogs, checklists, and, most of all, a high-quality master specification.

10.2.0 Format of Specifications

For convenience in writing, speed in estimating, and ease of reference, the most suitable organization of the specifications is a series of sections dealing with the construction requirements, products, and activities that is easily understandable by the different trades. Those people who use the specifications must be able to find all the information they need without spending too much time looking for it.

The most common specification-writing format used in North America is the *MasterFormat*™. This standard was developed jointly by the Construction Specifications Institute (CSI) and Construction Specifications Canada (CSC). For many years prior to 2004, the organization of construction specifications and suppliers catalogs has been based on a standard with 16 sections, otherwise known as divisions, where the divisions and their subsections were individually identified by a five-digit numbering system. The first two digits represented the division number and the next three individual numbers represented successively lower levels of breakdown. For example, the number 13213 represents division 13, subsection 2, sub-subsection 1, and sub-sub-subsection 3. In this older version of the standard, electrical systems, including any electronic or special electrical systems, were grouped together under Division 16 – *Electrical*. Today, specifications conforming to the 16-division format may still be in use.

In 2004, the *MasterFormat*™ standard underwent a major change. What had been 16 divisions was expanded to four major groupings and 49 divisions, with some divisions reserved for future expansion (*Figure 46*). The first 14 divisions are essentially the same as the old format. Subjects under the old Division 15 – *Mechanical* have been relocated to new divisions 22 and 23. The basic subjects under old Division 16 – *Electrical* have been relocated to new divisions 26 and 27. In addition, the numbering system was changed to 6 digits to allow for more subsections in each division, which allowed for finer definition. In the new

MasterFormat™ 2004 Edition – Numbers & Titles — Division Numbers & Titles

6/8/04

Division Numbers and Titles

PROCUREMENT AND CONTRACTING REQUIREMENTS GROUP

Division 00 Procurement and Contracting Requirements

SPECIFICATIONS GROUP

GENERAL REQUIREMENTS SUBGROUP

Division 01 General Requirements

FACILITY CONSTRUCTION SUBGROUP

Division 02 Existing Conditions
Division 03 Concrete
Division 04 Masonry
Division 05 Metals
Division 06 Wood, Plastics, and Composites
Division 07 Thermal and Moisture Protection
Division 08 Openings
Division 09 Finishes
Division 10 Specialties
Division 11 Equipment
Division 12 Furnishings
Division 13 Special Construction
Division 14 Conveying Equipment
Division 15 Reserved
Division 16 Reserved
Division 17 Reserved
Division 18 Reserved
Division 19 Reserved

FACILITY SERVICES SUBGROUP

Division 20 Reserved
Division 21 Fire Suppression
Division 22 Plumbing
Division 23 Heating, Ventilating, and Air Conditioning
Division 24 Reserved
Division 25 Integrated Automation
Division 26 Electrical
Division 27 Communications
Division 28 Electronic Safety and Security
Division 29 Reserved

SITE AND INFRASTRUCTURE SUBGROUP

Division 30 Reserved
Division 31 Earthwork
Division 32 Exterior Improvements
Division 33 Utilities
Division 34 Transportation
Division 35 Waterway and Marine Construction
Division 36 Reserved
Division 37 Reserved
Division 38 Reserved
Division 39 Reserved

PROCESS EQUIPMENT SUBGROUP

Division 40 Process Integration
Division 41 Material Processing and Handling Equipment
Division 42 Process Heating, Cooling, and Drying Equipment
Division 43 Process Gas and Liquid Handling, Purification, and Storage Equipment
Division 44 Pollution Control Equipment
Division 45 Industry-Specific Manufacturing Equipment
Division 46 Reserved
Division 47 Reserved
Division 48 Electrical Power Generation
Division 49 Reserved

Div Numbers - 1

110F46.EPS

Figure 46 ◆ 2004 MasterFormat™.

Specifications

numbering system, the first two digits represent the division number. The next two digits represent subsections of the division and the two remaining digits represent the third level sub-subsection numbers. The fourth level, if required, is a decimal and number added to the end of the last two digits. For example, the number 132013.04 represents division 13, subsection 20, sub-subsection 13 and sub-sub-subsection 04. Under the new standard, the Facility Service Sub-

group contains the divisions that are most important to the electrician. These include the following divisions:

- *Division 25 – Integrated Automation*
- *Division 26 – Electrical*
- *Division 27 – Communications*
- *Division 28 – Electronic Safety and Security*

Figure 47 contains a detailed breakdown of the electrical division.

DIVISION 26 – ELECTRICAL

26 00 00   ELECTRICAL

26 01 00   Operation and Maintenance of Electrical Systems
26 01 10   Operation and Maintenance of Medium-Voltage Electrical Distribution
26 01 20   Operation and Maintenance of Low-Voltage Electrical Distribution
26 01 26   Maintenance Testing of Electrical Systems
26 01 30   Operation and Maintenance of Facility Electrical Power Generating and Storing Equipment
26 01 40   Operation and Maintenance of Electrical and Cathodic Protection Systems
26 01 50   Operation and Maintenance of Lighting
    26 01 50.51   Luminaire Relamping
    26 01 50.81   Luminaire Replacement

26 05 00   Common Work Results for Electrical
26 05 13   Medium-Voltage Cables
    26 05 13.13   Medium-Voltage Open Conductors
    26 05 13.16   Medium-Voltage, Single- and Multi-Conductor Cables
26 05 19   Low-Voltage Electrical Power Conductors and Cables
    26 05 19.13   Undercarpet Electrical Power Cables
26 05 23   Control-Voltage Electrical Power Cables
26 05 26   Grounding and Bonding for Electrical Systems
26 05 29   Hangers and Supports for Electrical Systems
26 05 33   Raceway and Boxes for Electrical Systems
26 05 36   Cable Trays for Electrical Systems
26 05 39   Underfloor Raceways for Electrical Systems
26 05 43   Underground Ducts and Raceways for Electrical Systems
26 05 46   Utility Poles for Electrical Systems
26 05 48   Vibration and Seismic Controls for Electrical Systems
26 05 53   Identification for Electrical Systems
26 05 73   Overcurrent Protective Device Coordination Study

26 06 00   Schedules for Electrical
26 06 10   Schedules for Medium-Voltage Electrical Distribution
26 06 20   Schedules for Low-Voltage Electrical Distribution
    26 06 20.13   Electrical Switchboard Schedule
    26 06 20.16   Electrical Panelboard Schedule
    26 06 20.19   Electrical Motor-Control Center Schedule
    26 06 20.23   Electrical Circuit Schedule
    26 06 20.26   Wiring Device Schedule
26 06 30   Schedules for Facility Electrical Power Generating and Storing Equipment
26 06 40   Schedules for Electrical and Cathodic Protection Systems
26 06 50   Schedules for Lighting
    26 06 50.13   Lighting Panelboard Schedule
    26 06 50.16   Lighting Fixture Schedule

26 08 00   Commissioning of Electrical Systems
26 09 00   Instrumentation and Control for Electrical Systems
26 09 13   Electrical Power Monitoring and Control
26 09 23   Lighting Control Devices
26 09 26   Lighting Control Panelboards
26 09 33   Central Dimming Controls

110F47A.EPS

Figure 47 ◆ Detailed breakdown of the electrical division (1 of 4).

26 09 33.13 Multichannel Remote-Controlled Dimmers
26 09 33.16 Remote-Controlled Dimming Stations
26 09 36 Modular Dimming Controls
26 09 36.13 Manual Modular Dimming Controls
26 09 36.16 Integrated Multipreset Modular Dimming Controls
26 09 43 Network Lighting Controls
26 09 43.13 Digital-Network Lighting Controls
26 09 43.16 Addressable Fixture Lighting Control
26 09 61 Theatrical Lighting Controls

26 10 00 MEDIUM-VOLTAGE ELECTRICAL DISTRIBUTION

26 11 00 Substations
26 11 13 Primary Unit Substations
26 11 16 Secondary Unit Substations
26 12 00 Medium-Voltage Transformers
26 12 13 Liquid-Filled, Medium-Voltage Transformers
26 12 16 Dry-Type, Medium-Voltage Transformers
26 12 19 Pad-Mounted, Liquid-Filled, Medium-Voltage Transformers
26 13 00 Medium-Voltage Switchgear
26 13 13 Medium-Voltage Circuit Breaker Switchgear
26 13 16 Medium-Voltage Fusible Interrupter Switchgear
26 13 19 Medium-Voltage Vacuum Interrupter Switchgear
26 18 00 Medium-Voltage Circuit Protection Devices
26 18 13 Medium-Voltage Cutouts
26 18 16 Medium-Voltage Fuses
26 18 19 Medium-Voltage Lightning Arresters
26 18 23 Medium-Voltage Surge Arresters
26 18 26 Medium-Voltage Reclosers
26 18 29 Medium-Voltage Enclosed Bus
26 18 33 Medium-Voltage Enclosed Fuse Cutouts
26 18 36 Medium-Voltage Enclosed Fuses
26 18 39 Medium-Voltage Motor Controllers

26 20 00 LOW-VOLTAGE ELECTRICAL TRANSMISSION

26 21 00 Low-Voltage Overhead Electrical Power Systems
26 22 00 Low-Voltage Transformers
26 22 13 Low-Voltage Distribution Transformers
26 22 16 Low-Voltage Buck-Boost Transformers
26 22 19 Control and Signal Transformers
26 23 00 Low-Voltage Switchgear
26 23 13 Paralleling Low-Voltage Switchgear
26 24 00 Switchboards and Panelboards
26 24 13 Switchboards
26 24 16 Panelboards
26 24 19 Motor-Control Centers
26 25 00 Enclosed Bus Assemblies
26 26 00 Power Distribution Units
26 27 00 Low-Voltage Distribution Equipment
26 27 13 Electricity Metering
26 27 16 Electrical Cabinets and Enclosures
26 27 19 Multi-Outlet Assemblies

26 - 2

110F47B.EPS

Figure 47 ◆ Detailed breakdown of the electrical division (2 of 4).

MasterFormat™ 2004 Edition – Numbers & Titles | Div. 26 – Electrical

6/8/04

26 27 23 Indoor Service Poles
26 27 26 Wiring Devices
26 27 73 Door Chimes

26 28 00 Low-Voltage Circuit Protective Devices

26 28 13 Fuses
26 28 16 Enclosed Switches and Circuit Breakers

26 29 00 Low-Voltage Controllers

26 29 13 Enclosed Controllers
26 29 13.13 Across-the-Line Motor Controllers
26 29 13.16 Reduced-Voltage Motor Controllers
26 29 23 Variable-Frequency Motor Controllers

26 30 00 FACILITY ELECTRICAL POWER GENERATING AND STORING EQUIPMENT

26 31 00 Photovoltaic Collectors

26 32 00 Packaged Generator Assemblies

26 32 13 Engine Generators
26 32 13.13 Diesel-Engine-Driven Generator Sets
26 32 13.16 Gas-Engine-Driven Generator Sets
26 32 16 Steam-Turbine Generators
26 32 19 Hydro-Turbine Generators
26 32 23 Wind Energy Equipment
26 32 26 Frequency Changers
26 32 29 Rotary Converters
26 32 33 Rotary Uninterruptible Power Units

26 33 00 Battery Equipment

26 33 13 Batteries
26 33 16 Battery Racks
26 33 19 Battery Units
26 33 23 Central Battery Equipment
26 33 33 Static Power Converters
26 33 43 Battery Chargers
26 33 46 Battery Monitoring
26 33 53 Static Uninterruptible Power Supply

26 35 00 Power Filters and Conditioners

26 35 13 Capacitors
26 35 16 Chokes and Inductors
26 35 23 Electromagnetic-Interference Filters
26 35 26 Harmonic Filters
26 35 33 Power Factor Correction Equipment
26 35 36 Slip Controllers
26 35 43 Static-Frequency Converters
26 35 46 Radio-Frequency-Interference Filters
26 35 53 Voltage Regulators

26 36 00 Transfer Switches

26 36 13 Manual Transfer Switches
26 36 23 Automatic Transfer Switches

26 40 00 ELECTRICAL AND CATHODIC PROTECTION

26 41 00 Facility Lightning Protection

26 41 13 Lightning Protection for Structures

26 - 3

110F47C.EPS

Figure 47 ◆ Detailed breakdown of the electrical division (3 of 4).

26 41 13.13	Lightning Protection for Buildings
26 41 16	Lightning Prevention and Dissipation
26 41 19	Early Streamer Emission Lightning Protection
26 41 23	Lightning Protection Surge Arresters and Suppressors

26 42 00   Cathodic Protection
| 26 42 13 | Passive Cathodic Protection for Underground and Submerged Piping |
| 26 42 16 | Passive Cathodic Protection for Underground Storage Tank |

26 43 00   Transient Voltage Suppression
| 26 43 13 | Transient-Voltage Suppression for Low-Voltage Electrical Power Circuits |

26 50 00   LIGHTING

26 51 00   Interior Lighting
| 26 51 13 | Interior Lighting Fixtures, Lamps, And Ballasts |

26 52 00   Emergency Lighting

26 53 00   Exit Signs

26 54 00   Classified Location Lighting

26 55 00   Special Purpose Lighting
26 55 23	Outline Lighting
26 55 29	Underwater Lighting
26 55 33	Hazard Warning Lighting
26 55 36	Obstruction Lighting
26 55 53	Security Lighting
26 55 59	Display Lighting
26 55 61	Theatrical Lighting
26 55 63	Detention Lighting
26 55 70	Healthcare Lighting

26 56 00   Exterior Lighting
26 56 13	Lighting Poles and Standards
26 56 16	Parking Lighting
26 56 19	Roadway Lighting
26 56 23	Area Lighting
26 56 26	Landscape Lighting
26 56 29	Site Lighting
26 56 33	Walkway Lighting
26 56 36	Flood Lighting
26 56 68	Exterior Athletic Lighting

26 60 00   Reserved

26 70 00   Reserved

26 80 00   Reserved

26 90 00   Reserved

110F47D.EPS

Figure 47 ◆ Detailed breakdown of the electrical division (4 of 4).

1. A section line on a drawing shows _____.
   a. the north orientation
   b. the location of the section on the plan
   c. where to locate receptacles in that section
   d. the section scale

2. An electrical drafting line with a double arrowhead represents _____.
   a. wiring concealed in the floor
   b. wiring turned down
   c. a branch circuit homerun
   d. wiring concealed in a ceiling or wall

Questions 3 through 9 refer to the seven electrical symbols shown below. In the spaces provided, place the letter corresponding to the correct answer found in the list.

3. ————o————)

4. ⊖

5. ◯ ◯—
   PC   PC

6. ⊕

7. ⊖

   Ceiling   Wall

8. ◯   ◯—

9. ————●————

110RQ01.EPS

___7___ a. Single Receptacle Outlet

___4___ b. Duplex Receptacle Outlet

___6___ c. Triplex Receptacle Outlet

___8___ d. Incandescent Fixture
          (Surface or Pendant)

___5___ e. Incandescent Fixture
          with Pull Chain
          (Surface or Pendant)

___9___ f. Head Guy

___3___ g. Sidewalk Guy

10. In dimension drawings, the dimensions written on the drawing are _____.
    a. for reference only
    b. on a larger scale
    c. inaccurate
    d. the actual dimensions

11. The architect's scale is designed so that one inch always equals one foot.
    a. True
    b. False

12. All views on a blueprint are drawn to the same scale.
    a. True
    b. False

13. The *NEC*® specifies one set of electrical drawing symbols that are used in all cases.
    a. True
    b. False

14. Dotted lines used to represent a branch circuit on a drawing mean that the wiring is to be _____.
    a. concealed in the ceiling or wall
    b. run in the floor or ceiling below
    c. exposed
    d. installed in a future building expansion

15. A branch circuit line or drawing that does *not* have slashes is assumed to have two conductors.
    a. True
    b. False

16. To meet general recommendations, a residential branch circuit rated for 2,400VA should have a connected load of no more than _____ VA.
    a. 1,680
    b. 1,920
    c. 2,040
    d. 2,160

17. Power-riser diagrams are used to show the _____.
    a. arrangement of electric service equipment
    b. branch circuit layout for power
    c. branch circuit layout for lighting
    d. panelboard schedule

18. The symbols T_1, T_2, and T_3 in a typical motor starter schematic represent _____.
    a. voltage supply lines
    b. auxiliary contacts
    c. motor terminals
    d. line contacts

19. The updated *MasterFormat*™ standard _____.
    a. is specified in the *NEC*®
    b. uses a six-digit code for division content
    c. is required by OSHA
    d. allows for fewer subsections

20. The current *MasterFormat*™ standard covering communications systems is under _____.
    a. Division 16
    b. Division 27
    c. Division 37
    d. Division 48

Summary

In this module, you learned the symbols and conventions used on architectural and engineering drawings. As an electrician, you need to know how to recognize the basic symbols used on electrical drawings and other drawings used in the building construction industry. You should also know where to find the meaning of symbols that you do not immediately recognize. Schedules, diagrams, and specifications often provide detailed information that is not included on the working drawings.

Building projects require detailed specifications. These written specifications are complex and detailed and need a unified format to be easily usable by the trades. The specification format most commonly used is the *MasterFormat*™ developed by CSI and CSC. The *MasterFormat*™ was updated in 2004 with changes to the division numbering system. Division 26 relates to electrical systems.

Reading architectural and engineering drawings takes practice and study. Now that you have the basic skills, take the time to master them.

Notes

Trade Terms Quiz

1. _____ typically include the following information: a site plan, floor plans, elevations of all exterior faces of the building, and large-scale detail drawings.

2. A(n) _____ is an exact copy or reproduction of an original drawing.

3. A simple, single-line diagram used to show electrical equipment and related connections is a(n) _____ diagram.

4. A(n) _____ shows the path of an electrical circuit or system of circuits, along with the circuit components.

5. To convey a substantial amount of detailed information to installation electricians, an engineer will use a(n) _____ drawing.

6. Shown in a separate view, a(n) _____ view is an enlarged, detailed view taken from an area of a drawing.

7. A cutaway drawing that shows the inside of an object or building is a(n) _____ drawing.

8. The sizes or measurements that are printed on a drawing are called _____.

9. The relationship between an object's size in a drawing and the object's actual size is the _____.

10. The height of the front, rear, or sides of a building is shown in a(n) _____ drawing.

11. A building's location on the site is shown in a(n) _____.

12. A drawing that has a top-down view of a building is a(n) _____ plan.

13. A drawing that has a top-down view of a single object is a(n) _____ view.

14. A(n) _____ diagram is a single-line block diagram used to indicate the electric service equipment, service conductors and feeders, and subpanels.

15. Owners, architects, and engineers use _____ to specify material and workmanship requirements.

16. A(n) _____ is a systematic way of presenting equipment lists on a drawing in tabular form.

17. Complicated circuits, such as control circuits, are shown in a(n) _____ diagram.

18. Usually developed by manufacturers, fabricators, or contractors, a(n) _____ drawing shows specific dimensions and other information about a piece of equipment and its installation methods.

Trade Terms

Architectural drawings
Block diagram
Blueprint
Detail drawing
Dimensions

Electrical drawing
Elevation drawing
Floor plan
One-line diagram
Plan view

Power-riser diagram
Scale
Schedule
Schematic diagram
Sectional view

Shop drawing
Site plan
Written specifications

Wayne Stratton

Associated Builders and Contractors

How did you choose a career in the electrical field?
Three events in my childhood created the desire to learn the electrical trade. At age six, the farmhouse we lived in was totally destroyed by fire. The cause was electrical. As a young teen, a local electrician had incorrectly wired a heating element and electrocuted several pigs. In 1973, my father hired this electrician to install a motor starter on a grain conveyor. He could not figure it out. I wanted to learn how to do this type of work and do it safely.

Tell us about your apprenticeship experience.
My education is from a technical school. I have attended several manufacturers' training sessions. I had to gain the hands-on experience after learning the trade. My observation of the apprenticeship programs is this: you get hands-on experience while you learn.

What positions have you held in the industry?
I worked as a plant industrial electrician responsible for motor control, DC motors, co-generation, and medium voltage distribution. Later, I began working for an electrical contractor who wanted to expand his business into the industrial field. I worked as a PLC

technician designing and installing control systems. In 1987, I began teaching apprenticeship classes.

What would you say is the primary factor in achieving success?
The desire to learn all that I can learn, the ability to think outside the box, and the opportunities to gain a variety of experiences. All this helps me continue to learn and share with trainees.

What does your current job involve?
I teach electrical apprenticeship levels one through four at two different locations in Iowa. My other responsibilities involve task training for electrical licensing, fire alarm, and code updates.

Do you have any advice for someone just entering the trade?
Continue to learn. Completing an apprenticeship program or acquiring an electrician's license is not the end of learning. With code changes every 3 years, there is always more to learn. If you don't understand something, ask! Observe and learn from experienced individuals.

Trade Terms
Introduced in This Module

Architectural drawings: Working drawings consisting of plans, elevations, details, and other information necessary for the construction of a building. Architectural drawings usually include:

- A site (plot) plan indicating the location of the building on the property
- Floor plans showing the walls and partitions for each floor or level
- Elevations of all exterior faces of the building
- Several vertical cross sections to indicate clearly the various floor levels and details of the footings, foundations, walls, floors, ceilings, and roof construction
- Large-scale detail drawings showing such construction details as may be required

Block diagram: A single-line diagram used to show electrical equipment and related connections. See *power-riser diagram.*

Blueprint: An exact copy or reproduction of an original drawing.

Detail drawing: An enlarged, detailed view taken from an area of a drawing and shown in a separate view.

Dimensions: Sizes or measurements printed on a drawing.

Electrical drawing: A means of conveying a large amount of exact, detailed information in an abbreviated language. Consists of lines, symbols, dimensions, and notations to accurately convey an engineer's designs to electricians who install the electrical system on a job.

Elevation drawing: An architectural drawing showing height, but not depth; usually the front, rear, and sides of a building or object.

Floor plan: A drawing of a building as if a horizontal cut were made through a building at about window level, and the top portion removed. The floor plan is what would appear if the remaining structure were viewed from above.

One-line diagram: A drawing that shows, by means of lines and symbols, the path of an electrical circuit or system of circuits along with

the various circuit components. Also called a single-line diagram.

Plan view: A drawing made as though the viewer were looking straight down (from above) on an object.

Power-riser diagram: A single-line block diagram used to indicate the electric service equipment, service conductors and feeders, and subpanels. Notes are used on power-riser diagrams to identify the equipment; indicate the size of conduit; show the number, size, and type of conductors; and list related materials. A panelboard schedule is usually included with power-riser diagrams to indicate the exact components (panel type and size), along with fuses, circuit breakers, etc., contained in each panelboard.

Scale: On a drawing, the size relationship between an object's actual size and the size it is drawn. Scale also refers to the measuring tool used to determine this relationship.

Schedule: A systematic method of presenting equipment lists on a drawing in tabular form.

Schematic diagram: A detailed diagram showing complicated circuits, such as control circuits.

Sectional view: A cutaway drawing that shows the inside of an object or building.

Shop drawing: A drawing that is usually developed by manufacturers, fabricators, or contractors to show specific dimensions and other pertinent information concerning a particular piece of equipment and its installation methods.

Site plan: A drawing showing the location of a building or buildings on the building site. Such drawings frequently show topographical lines, electrical and communication lines, water and sewer lines, sidewalks, driveways, and similar information.

Written specifications: A written description of what is required by the owner, architect, and engineer in the way of materials and workmanship. Together with working drawings, the specifications form the basis of the contract requirements for construction.

Metric Conversion Chart

METRIC CONVERSION CHART

INCHES Fractional	Decimal	METRIC mm	INCHES Fractional	Decimal	METRIC mm	INCHES Fractional	Decimal	METRIC mm
.	0.0039	0.1000	.	0.5512	14.0000	.	1.8898	48.0000
.	0.0079	0.2000	9/16	0.5625	14.2875	.	1.9291	49.0000
.	0.0118	0.3000	.	0.5709	14.5000	.	1.9685	50.0000
1/64	0.0156	0.3969	37/64	0.5781	14.6844	2	2.0000	50.8000
.	0.0157	0.4000	.	0.5906	15.0000	.	2.0079	51.0000
.	0.0197	0.5000	19/32	0.5938	15.0813	.	2.0472	52.0000
.	0.0236	0.6000	39/64	0.6094	15.4781	.	2.0866	53.0000
.	0.0276	0.7000	.	0.6102	15.5000	.	2.1260	54.0000
1/32	0.0313	0.7938	5/8	0.6250	15.8750	.	2.1654	55.0000
.	0.0315	0.8000	.	0.6299	16.0000	.	2.2047	56.0000
.	0.0354	0.9000	41/64	0.6406	16.2719	.	2.2441	57.0000
.	0.0394	1.0000	.	0.6496	16.5000	2 1/4	2.2500	57.1500
.	0.0433	1.1000	21/32	0.6563	16.6688	.	2.2835	58.0000
3/64	0.0469	1.1906	.	0.6693	17.0000	.	2.3228	59.0000
.	0.0472	1.2000	43/64	0.6719	17.0656	.	2.3622	60.0000
.	0.0512	1.3000	11/16	0.6875	17.4625	.	2.4016	61.0000
.	0.0551	1.4000	.	0.6890	17.5000	.	2.4409	62.0000
.	0.0591	1.5000	45/64	0.7031	17.8594	.	2.4803	63.0000
1/16	0.0625	1.5875	.	0.7087	18.0000	2 1/2	2.5000	63.5000
.	0.0630	1.6000	23/32	0.7188	18.2563	.	2.5197	64.0000
.	0.0669	1.7000	.	0.7283	18.5000	.	2.5591	65.0000
.	0.0709	1.8000	47/64	0.7344	18.6531	.	2.5984	66.0000
.	0.0748	1.9000	.	0.7480	19.0000	.	2.6378	67.0000
5/64	0.0781	1.9844	3/4	0.7500	19.0500	.	2.6772	68.0000
.	0.0787	2.0000	49/64	0.7656	19.4469	.	2.7165	69.0000
.	0.0827	2.1000	.	0.7677	19.5000	2 3/4	2.7500	69.8500
.	0.0866	2.2000	25/32	0.7813	19.8438	.	2.7559	70.0000
.	0.0906	2.3000	.	0.7874	20.0000	.	2.7953	71.0000
3/32	0.0938	2.3813	51/64	0.7969	20.2406	.	2.8346	72.0000
.	0.0945	2.4000	.	0.8071	20.5000	.	2.8740	73.0000
.	0.0984	2.5000	13/16	0.8125	20.6375	.	2.9134	74.0000
7/64	0.1094	2.7781	.	0.8268	21.0000	.	2.9528	75.0000
.	0.1181	3.0000	53/64	0.8281	21.0344	.	2.9921	76.0000
1/8	0.1250	3.1750	27/32	0.8438	21.4313	3	3.0000	76.2000
.	0.1378	3.5000	.	0.8465	21.5000	.	3.0315	77.0000
9/64	0.1406	3.5719	55/64	0.8594	21.8281	.	3.0709	78.0000
5/32	0.1563	3.9688	.	0.8661	22.0000	.	3.1102	79.0000
.	0.1575	4.0000	7/8	0.8750	22.2250	.	3.1496	80.0000
11/64	0.1719	4.3656	.	.8858	22.5000	.	3.1890	81.0000
.	0.1772	4.5000	57/64	.89063	22.6219	.	3.2283	82.0000
3/16	0.1875	4.7625	.	.9055	23.0000	.	3.2677	83.0000
.	0.1969	5.0000	29/32	.90625	23.0188	.	3.3071	84.0000
13/64	0.2031	5.1594	59/64	.92188	23.4156	.	3.3465	85.0000
.	0.2165	5.5000	.	.9252	23.5000	.	3.3858	86.0000
7/32	0.2188	5.5563	15/16	.93750	23.8125	.	3.4252	87.0000
15/64	0.2344	5.9531	.	.9449	24.0000	.	3.4646	88.0000
.	0.2362	6.0000	61/64	.95313	24.2094	3 1/2	3.5000	88.9000
1/4	0.2500	6.3500	.	.9646	24.5000	.	3.5039	89.0000
.	0.2559	6.5000	31/32	.96875	24.6063	.	3.5433	90.0000
17/64	0.2656	6.7469	.	.9843	25.0000	.	3.5827	91.0000
.	0.2756	7.0000	63/64	.98438	25.0031	.	3.6220	92.0000
9/32	0.2813	7.1438	1	1.000	25.40	.	3.6614	93.0000
.	0.2953	7.5000	.	1.0039	25.5000	.	3.7008	94.0000
19/64	0.2969	7.5406	.	1.0236	26.0000	.	3.7402	95.0000
5/16	0.3125	7.9375	.	1.0433	26.5000	.	3.7795	96.0000
.	0.3150	8.0000	.	1.0630	27.0000	.	3.8189	97.0000
21/64	0.3281	8.3344	.	1.0827	27.5000	.	3.8583	98.0000
.	0.3346	8.5000	.	1.1024	28.0000	.	3.8976	99.0000
11/32	0.3438	8.7313	.	1.1220	28.5000	.	3.9370	100.0000
.	0.3543	9.0000	.	1.1417	29.0000	4	4.0000	101.6000
23/64	0.3594	9.1281	.	1.1614	29.5000	.	4.3307	110.0000
.	0.3740	9.5000	.	1.1811	30.0000	4 1/2	4.5000	114.3000
3/8	0.3750	9.5250	.	1.2205	31.0000	.	4.7244	120.0000
25/64	0.3906	9.9219	1 1/4	1.2500	31.7500	5	5.0000	127.0000
.	0.3937	10.0000	.	1.2598	32.0000	.	5.1181	130.0000
13/32	0.4063	10.3188	.	1.2992	33.0000	.	5.5118	140.0000
.	0.4134	10.5000	.	1.3386	34.0000	.	5.9055	150.0000
27/64	0.4219	10.7156	.	1.3780	35.0000	6	6.0000	152.4000
.	0.4331	11.0000	.	1.4173	36.0000	.	6.2992	160.0000
7/16	0.4375	11.1125	.	1.4567	37.0000	.	6.6929	170.0000
.	0.4528	11.5000	.	1.4961	38.0000	.	7.0866	180.0000
29/64	0.4531	11.5094	1 1/2	1.5000	38.1000	.	7.4803	190.0000
15/32	0.4688	11.9063	.	1.5354	39.0000	.	7.8740	200.0000
.	0.4724	12.0000	.	1.5748	40.0000	8	8.0000	203.2000
31/64	0.4844	12.3031	.	1.6142	41.0000	.	9.8425	250.0000
.	0.4921	12.5000	.	1.6535	42.0000	.		
1/2	0.5000	12.7000	.	1.6929	43.0000	10	10.0000	254.0000
.	0.5118	13.0000	.	1.7323	44.0000	20	20.0000	508.0000
33/64	0.5156	13.0969	1 3/4	1.7500	44.4500	30	30.0000	762.0000
17/32	0.5313	13.4938	.	1.7717	45.0000	40	40.0000	1016.000
.	0.5315	13.5000	.	1.8110	46.0000	60	60.0000	1524.000
35/64	0.5469	13.8906	.	1.8504	47.0000	80	80.0000	2032.000
						100	100.0000	2540.000

TO CONVERT TO MILLIMETERS, MULTIPLY INCHES X 25.4
TO CONVERT TO INCHES, MULTIPLY MILLIMETERS X 0.03937*
*FOR SLIGHTLY GREATER ACCURACY WHEN CONVERTING TO INCHES, DIVIDE MILLIMETERS BY 25.4

110A01.EPS

This module is intended to present thorough resources for task training. The following reference works are suggested for further study. These are optional materials for continued education rather than for task training.

American Electrician's Handbook. Terrell Croft and Wilford I. Summers. New York: McGraw-Hill.

National Electrical Code® Handbook, Latest Edition. Quincy, MA: National Fire Protection Association.

Residential
Electrical Services

Phoenix Fire Station No. 50

Phoenix's Fire Station No. 50 sports many environmentally protective measures in its construction, such as a roof made of recycled aluminum cans and terra-cotta colored terrazzo flooring made by grinding down the concrete structural slab. Recycled countertops are used in the kitchen and more than 80 percent of the lighting takes advantage of natural sources to save energy. The landscape is a xeriscape design that will require no irrigation after two years.

26111-08

26111-08
Residential Electrical Services

Topics to be presented in this module include:

Overview

The first step in residential wiring is a complete review of the floor plan layout. Electrical floor plans show approximate locations of panels, switches, receptacles, lighting, and other outlets. They do not show, however, the routing of the wiring that interconnects these devices, as this task is generally left up to the electrician.

If the design engineer has not performed the load calculations, the electrician must determine the connected load for the residence, and then size the electrical service accordingly. In order to figure total connected load, certain formulas must be applied based on livable square footage of the house and other factors. Residential electricians must know how to perform load calculations accurately.

Specific wiring methods, grounding requirements, and ground fault circuit interrupting techniques for residences are strictly regulated by the *National Electrical Code®* because building occupants are constantly exposed to the hazards associated with electrical systems and devices. If the *NEC®* and local codes are not strictly followed, the inspector will not sign off on the installation. This will cause construction delays and rework expenses.

Objectives

When you have completed this module, you will be able to do the following:

1. Explain the role of the *National Electrical Code®* in residential wiring and describe how to determine electric service requirements for dwellings.
2. Explain the grounding requirements of a residential electric service.
3. Calculate and select service-entrance equipment.
4. Select the proper wiring methods for various types of residences.
5. Compute branch circuit loads and explain their installation requirements.
6. Explain the types and purposes of equipment grounding conductors.
7. Explain the purpose of ground fault circuit interrupters and tell where they must be installed.
8. Size outlet boxes and select the proper type for different wiring methods.
9. Describe rules for installing electric space heating and HVAC equipment.
10. Describe the installation rules for electrical systems around swimming pools, spas, and hot tubs.
11. Explain how wiring devices are selected and installed.
12. Describe the installation and control of lighting fixtures.

Trade Terms

Appliance
Bonding bushing
Bonding jumper
Branch circuit
Feeder
Load center
Metal-clad (MC) cable
Nonmetallic-sheathed (Type NM) cable
Romex®
Roughing in
Service drop
Service entrance
Service-entrance conductors
Service-entrance equipment
Service lateral
Switch
Switch leg

Required Trainee Materials

1. Paper and pencil
2. Copy of the latest edition of the *National Electrical Code®*
3. Appropriate personal protective equipment

Prerequisites

Before you begin this module, it is recommended that you successfully complete *Core Curriculum* and *Electrical Level One*, Modules 26101-08 through 26110-08.

This course map shows all of the modules in *Electrical Level One*. The suggested training order begins at the bottom and proceeds up. Skill levels increase as you advance on the course map. The local Training Program Sponsor may adjust the training order.

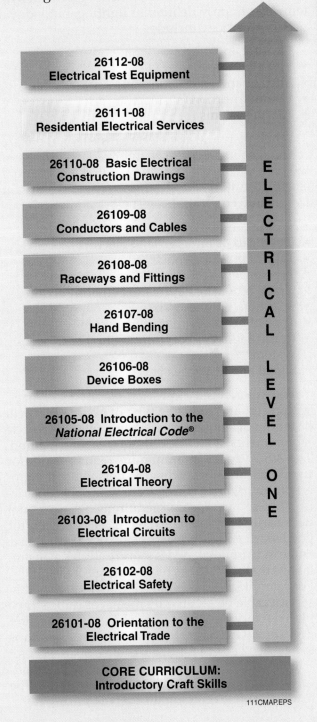

ELECTRICAL LEVEL ONE

26112-08
Electrical Test Equipment

26111-08
Residential Electrical Services

26110-08 Basic Electrical Construction Drawings

26109-08
Conductors and Cables

26108-08
Raceways and Fittings

26107-08
Hand Bending

26106-08
Device Boxes

26105-08 Introduction to the *National Electrical Code®*

26104-08
Electrical Theory

26103-08 Introduction to Electrical Circuits

26102-08
Electrical Safety

26101-08 Orientation to the Electrical Trade

CORE CURRICULUM:
Introductory Craft Skills

111CMAP.EPS

1.0.0 ◆ INTRODUCTION

The use of electricity in houses began shortly after the opening of the California Electric Light Company in 1879 and Thomas Edison's Pearl Street Station in New York City in 1882. These two companies were the first to enter the business of producing and selling electric service to the public. In 1886, the Westinghouse Electric Company secured patents that resulted in the development and introduction of alternating current; this paved the way for rapid acceleration in the use of electricity.

The primary use of early home electrical systems was to provide interior lighting, but today's uses of electricity include:

• Heating and air conditioning
• Electrical appliances
• Interior and exterior lighting
• Communications systems
• Alarm systems

When planning any electrical system, there are certain general steps to be followed, regardless of the type of construction. In planning a residential electrical system, the electrician must take certain factors into consideration. These include:

• Wiring method
• Overhead or underground electrical service
• Type of building construction
• Type of service entrance and equipment
• Grade of wiring devices and lighting fixtures
• Selection of lighting fixtures
• Type of heating and cooling system
• Control wiring for the heating and cooling system
• Signal and alarm systems

The experienced electrician readily recognizes, within certain limits, the type of system that will be required. However, always check the local code requirements when selecting a wiring method. The *NEC*® provides minimum requirements for the practical safeguarding of persons and property from hazards arising from the use of electricity. These minimum requirements are not necessarily efficient, convenient, or adequate for good service or future expansion of electrical use. Some local building codes require electrical installations that surpass the requirements of the *NEC*®. For example, *NEC Section 230.51(A)* requires that service cable be secured by means of cable straps placed every 30 inches. The electrical inspection department in one area requires these cable straps to be placed at a minimum distance of 18 inches.

If more than one wiring method may be practical, a decision as to which type to use should be made prior to beginning the installation.

NOTE

See the *Appendix* for other codes and electrical standards that apply to residential electrical installations.

In a residential occupancy, the electrician should know that a 120/240-volt (V), single-phase service entrance will invariably be provided by the utility company. The electrician knows that the service and feeders will be three-wire, that the branch circuits will be either two- or three-wire, and that the safety switches, service equipment, and panelboards will be three-wire, solid neutral. On each project, however, the electrician must consult with the local utility to determine the point of attachment for overhead connections and the location of the metering equipment.

2.0.0 ◆ SIZING THE ELECTRICAL SERVICE

It may be difficult to decide at times which comes first, the layout of the outlets or the sizing of the electric service. In many cases, the service (main disconnect, panelboard, service conductors, etc.) can be sized using the *NEC*® before the outlets are actually located. In other cases, the outlets will have to be laid out first. However, in either case, the service entrance and panelboard locations will have to be determined before the circuits can be installed—so the electrician will know in which direction (and to what points) the circuit homeruns will terminate. In this module, an actual residence will be used as a model to size the electric service according to the latest edition of the *NEC*®.

2.1.0 Floor Plans

A floor plan is a drawing that shows the length and width of a building and the rooms that it contains. A separate plan is made for each floor.

Figure 1 shows how a floor plan is developed. An imaginary cut is made through the building as shown in the view on the left. The top half of this cut is removed (bottom view), and the resulting floor plan is what the remaining structure looks like when viewed directly from above.

The floor plan for a small residence is shown in *Figure 2*. This building is constructed on a concrete slab with no basement or crawl space. There is an unfinished attic above the living area and an open carport just outside the kitchen entrance. Appliances include a 12 kilovolt-ampere (kVA) electric

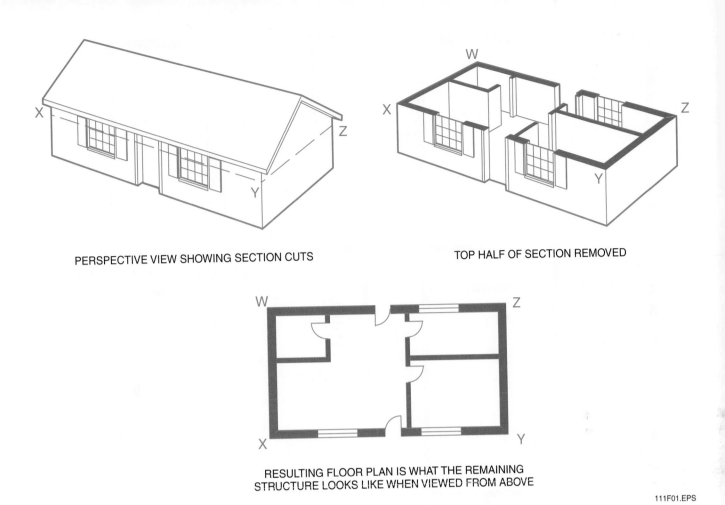

PERSPECTIVE VIEW SHOWING SECTION CUTS

TOP HALF OF SECTION REMOVED

RESULTING FLOOR PLAN IS WHAT THE REMAINING
STRUCTURE LOOKS LIKE WHEN VIEWED FROM ABOVE

111F01.EPS

Figure 1 ◆ Principles of floor plan layout.

range, a 4.5kVA water heater, a ½hp 120V disposal, and a 1.5kVA dishwasher.

There is also a washer/dryer (rated at 5.5kVA) in the utility room. A gas furnace with a ⅓hp 120V blower supplies the heating. In this module, the electrical requirements of this example building will be computed.

2.2.0 General Lighting Loads

General lighting loads are calculated on the basis of *NEC Table 220.12.* For residential occupancies, three volt-amperes (watts) per square foot of living space is the figure to use. This includes non-appliance duplex receptacles into which lamps, televisions, etc. may be connected. Therefore, the area of the building must be calculated first. If the building is under construction, the dimensions can be determined by scaling the working drawings used by the builder. If the residence is an existing building with no drawings, actual measurements will have to be made on the site.

Using the floor plan of the residence in *Figure 2* as a guide, an architect's scale is used to measure the longest width of the building (using outside dimensions). It is determined to be 33 feet. The longest length of the building is 48 feet. These two measurements multiplied together give $33 \times 48 = 1,584$ square feet of living area. However, there is an open carport on the lower left of the drawing. This carport area will have to be calculated and then deducted from 1,584 to give the true amount of living space. This open area (carport) is 12 feet wide by 19.5 feet long: $12 \times 19.5 = 234$ square feet. Subtract the carport area from 1,584 square feet: $1,584 - 234 = 1,350$ square feet of living area.

When using the square-foot method to determine lighting loads for buildings, *NEC Section 220.12* requires the floor area for each floor to be computed from the outside dimensions. When calculating lighting loads for residences, the computed floor area must not include open porches, carports, garages, or unused or unfinished spaces that are not adaptable to future use.

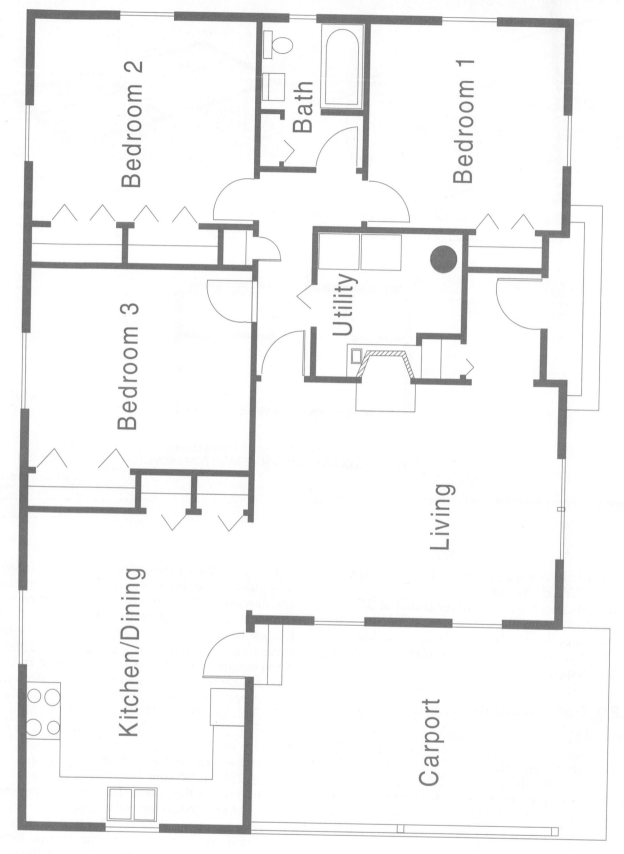

Figure 2 ◆ Floor plan of a typical residence.

111F02.EPS

2.3.0 Calculating the Electric Service Load

Figure 3 shows a standard calculation worksheet for a single-family dwelling. This form contains numbered blank spaces to be filled in while making the service calculation. Using this worksheet as a guide, the total area of our sample dwelling has been previously determined to be 1,350 square feet of living space. This figure is entered in the appropriate space (Box 1) on the form and multiplied by 3 volt-amperes (VA) for a total general lighting load of 4,050VA (Box 2).

2.3.1 Small Appliance Loads

NEC Section 210.11(C)(1) requires at least two 120V, 20A small appliance branch circuits to be installed for the small appliance loads in each kitchen area of a dwelling. Kitchen areas include the dining area, breakfast nook, pantry, and similar areas where small appliances will be used. *NEC Section 220.52* gives further requirements for residential small appliance circuits; that is, the load for those circuits is to be computed at 1,500VA each. Since our example dwelling has only one kitchen area, the number 2 is entered in Box 3 for the number of required kitchen small appliance branch circuits. Multiply the number of these circuits by 1,500 and enter the result in Box 4.

2.3.2 Laundry Circuit

NEC Section 210.11(C)(2) requires an additional 20A branch circuit to be provided for the exclusive use of the laundry area (Box 5). This circuit must not have any other outlets connected except for the laundry receptacle(s). Therefore, enter 1,500VA in Box 6 on the form.

General Lighting Load										Phase	Neutral
Square footage of the dwelling	[1]	1350	× 3VA =	[2]	4050						
Kitchen small appliance circuits	[3]	2	× 1500 =	[4]	3000						
Laundry branch circuit	[5]	1	× 1500 =	[6]	1500						
Subtotal of gen. lighting loads per *NEC Section 220.52* =				[7]	8550						
Subtract 1st 3000VA per *NEC Table 220.42*				[8]	3000	× 100% =	[9]	3000			
Remaining VA times 35% per *NEC Table 220.42*				[10]	5550	× 35% =	[11]	1943			
Total demand for general lighting loads =							[12]	4943	[13]		

Fixed Appliance Loads (Nameplate or NEC FLA of motors) per *NEC Section 220.53*						
Hot water tank, 4.5kVA, 240V	[14]	4500				
Dishwasher 1.5kVA, 120V	[15]	1500				
Disposal 1/2HP, 120V per *NEC Table 430.248* = 9.8A	[16]	1176				
Blower 1/3HP, 120V per *NEC Table 430.248* = 7.2A	[17]	864				
	[18]					
	[19]					
Subtotal of fixed appliances	[20]	8040				
If 3 or less fixed appliances take @ 100% =	[21]		[22]			
If 4 or more fixed appliances take @ 75% =	[23]	6030	[24]			

Other Loads per *NEC Section 220.14*					
Electric Range per *NEC Section 220.55* [neutral @ 70% per *NEC Section 220.61(B)*]	[25]	8000	[26]		
Electric Dryer per *NEC Section 220.54* [neutral @ 70% per *NEC Section 220.61(B)*]	[27]	5500	[28]		
Electric Heat per *NEC Section 220.51*	omit smaller load per *NEC Section 220.60*	[29]		[30]	
Air Conditioning *NEC Section 220.82(C)*					
Largest Motor = 1176	× 25% (per *NEC Section 430.24*) =	[31]	294	[32]	
Total VA Demand =		[33]	24767	[34]	
(VA divided by 240 volts) Amps =		[35]	103	[36]	
Service OCD and minimum size grounding electrode conductor		[37]	125	[38]	
AWG per *NEC Sections 310.15(B)(6)* and *Table 310.16* for neutral		[39]		[40]	

111F03.EPS

Figure 3 ◆ Calculation worksheet for residential requirements.

So far, there is enough information to complete the first portion of the service calculation form:

- General lighting   4,050VA (Box 2)
- Small appliance load   3,000VA (Box 4)
- Laundry load   1,500VA (Box 6)
- Total general lighting and appliance loads   8,550VA (Box 7)

2.3.3 Lighting Demand Factors

All residential electrical outlets are never used at one time. There may be a rare instance when all the lighting may be on for a short time every night, but even so, all the small appliances and receptacles throughout the house will never be used simultaneously. Knowing this, *NEC Section 220.42* allows a diversity or demand factor to be used when computing the general lighting load for services. Our calculation continues as follows:

- The first 3,000VA is rated at 100%   3,000VA (Box 8)
- The remaining 5,550VA (Box 10) may be rated at 35% (the allowable demand factor) Therefore, 5,550 × 0.35 =   1,943VA (Box 11)
- Net general lighting and small appliance load (rounded off)   4,943VA (Box 12)

2.3.4 Fixed Appliances

NEC Section 220.53 permits the loads for fixed appliances to be computed at 75% as long as they are not electric heating, air conditioning, electric cooking, or electric clothes dryer loads. To compute the load of the fixed appliances in this dwelling, list all the fixed appliances that meet *NEC Section 220.53*. Enter the nameplate rating of the appliance or VA for motors by using *NEC Table 430.248* to find the FLA of each motor. *NEC Section 220.5(A)* tells us to use 120V (not 115V) for calculation purposes. The fixed appliances would be as follows:

- Hot water tank   4,500VA (Box 14)
- Dishwasher   1,500VA (Box 15)
- ½hp 120V disposal (9.8A × 120V)   1,176VA (Box 16)
- Gas furnace blower (7.2A × 120V)   864VA (Box 17)
- Add the loads for the fixed appliances   8,040VA (Box 20)
- Since there are four or more fixed appliances, multiply the total in Box 20 by 75%   6,030VA (Box 23)

2.3.5 Other Loads

The remaining loads of the dwelling are now computed in the Other Loads section in *Figure 3*. *NEC Section 220.14* allows electric dryers to be computed as permitted in *NEC Table 220.54* and electric cooking appliances to be computed per *NEC Table 220.55*. For a single range rated over 8.75kVA, but not over 12kVA, Column C of *NEC Table 220.55* permits a demand of 8kVA for the range in this dwelling. Enter 8,000VA in Box 25.

The electric dryer must be computed at 5,000VA or the nameplate, whichever is greater, according to *NEC Section 220.54.* Up to four electric dryers must be taken at 100%. Enter 5,500VA in Box 27.

If this dwelling had electric space heating and/or air conditioning, it would be computed in this section using the larger of the two loads. Since they are typical noncoincidental loads, *NEC Section 220.60* permits the smaller of those loads to be omitted. There are no demand factors for either electric heating or air conditioning; therefore, the larger of the two loads would be computed at 100%.

The final step in this calculation is to add in 25% of the largest motor in the dwelling. This dwelling unit has two motors: the disposal at 9.8A and the blower at 7.2A. (See *NEC Section 430.17.*) In this case, the larger motor is the disposal; therefore, we must add 25% of the rating to meet the requirements of *NEC Section 430.24.* Enter 294VA (1,176 × 25%) in Box 31. Adding together the individual loads as computed, we have a minimum demand of 24,767VA (Box 33) for the phase conductors.

2.3.6 Required Service Size

The conventional electric service for residential use is 120/240V, three-wire, single-phase. Services are sized in amperes, and when the volt-amperes are known on single-phase services, amperes may be found by dividing the highest voltage into the total volt-amperes. For example:

$$24,767VA \div 240V = 103A \text{ (Box 35)}$$

The **service-entrance conductors** have now been calculated and must be rated at a minimum of 110A, which is a standard rating for overcurrent protection. However, this is not a typical trade size; therefore, we will use the more common rating of 125A as the size of our service.

If the demand for our dwelling unit had resulted in a load of less than 100A, *NEC Section 230.79(C)* would have required that the minimum rating of the service disconnect be 100A. *NEC Section 230.42(B)* would have required the ampacity of the service conductors to be equal to the rating of the 100A disconnect as well.

2.4.0 Demand Factors

NEC Article 220, Part III provides the rules regarding the application of demand factors to certain types of loads. Recall that a demand factor is the maximum amount of volt-amp load expected at any given time compared to the total connected load of the circuit. The maximum demand of a feeder circuit is equal to the connected load times the demand factor. The loads to which demand factors apply can be found in the *NEC®* as follows:

- Lighting loads   *NEC Table 220.42*
- Receptacle loads   *NEC Table 220.44*
- Dryer loads   *NEC Table 220.54*
- Range loads   *NEC Table 220.55*
- Kitchen equipment loads   *NEC Table 220.56*

In addition to those demand factors listed in *NEC Article 220, Part III*, alternative (optional) methods for computing loads can be found in *NEC Article 220, Part IV*. They include the following:

- Dwelling unit loads   *NEC Section 220.82*
- Existing dwelling unit
  loads   *NEC Section 220.83*
- Multi-family dwelling
  unit loads   *NEC Section 220.84*

2.5.0 General Lighting and Receptacle Load Demand Factors

NEC Table 220.42 provides the demand factors allowed for various types of lighting situations.

2.6.0 Appliance Loads

NEC Section 210.11(C) provides the number of branch circuits required for small appliances and laundry loads. Demand factors for dryers and ranges are found in *NEC Tables 220.54 and 220.55*.

2.6.1 Small Appliance Loads

The small appliance branch circuits required by *NEC Section 210.11(C)(1)* for small appliances supplied by 15A or 20A receptacles on 20A branch circuits for each kitchen area served are calculated at 1,500VA. If a dwelling has more than one kitchen area, the *NEC®* will require two small appliance branch circuits computed at 1,500VA for each kitchen area served. Where a dwelling with only one kitchen area has more than the required two small appliance branch circuits installed to serve a single kitchen area, only the first two required circuits need be computed. Additional circuits for countertops or refrigeration provide a separation of load, not additional loads. If a dwelling has two kitchen areas, then the total small appliance

branch circuits required would be four at 1,500VA each. These loads are permitted to be included with the general lighting load and subjected to the demand factors of *NEC Table 220.42*.

2.6.2 Laundry Circuit Load

A 1,500VA feeder load is added to load calculations for each two-wire laundry branch circuit installed in a home. The branch circuit is required by *NEC Section 210.11(C)(2)*. This load may also be added to the general lighting load and subjected to the same demand factors provided in *NEC Section 220.42*.

2.6.3 Dryer Load

The dryer load for each electric clothes dryer is 5,000VA or the actual nameplate value of the dryer, whichever is larger. Demand factors listed in *NEC Table 220.54* may be applied for more than one dryer in the same dwelling. If two or more single-phase dryers are supplied by a three-phase, four-wire feeder, the total load is computed by using twice the maximum number connected between any two phases.

2.6.4 Range Load

Range loads and other cooking appliances are covered under *NEC Section 220.55*. The feeder demand loads for household electric ranges, wall-mounted ovens, countertop cooking units, and other similar household appliances individually rated over 1¾kW are permitted to be computed in accordance with *NEC Table 220.55*. If two or more single-phase ranges are supplied by a three-phase, four-wire feeder, the total load is computed by using twice the maximum number connected between any two phases.

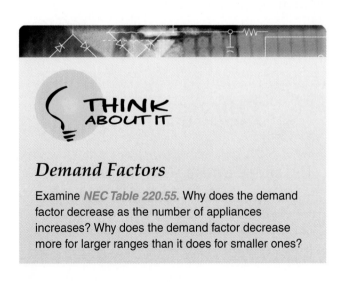

THINK ABOUT IT

Demand Factors

Examine *NEC Table 220.55*. Why does the demand factor decrease as the number of appliances increases? Why does the demand factor decrease more for larger ranges than it does for smaller ones?

2.6.5 Demand Loads for Electric Ranges

Ranges can be computed in various ways that depend on which part of *NEC Article 220* you are using and the occupancy type for the ranges involved. Note the demand factors permitted for the following occupancy types:

- Dwelling units per Part III   *NEC Section 220.55*
- Dwelling units per Part IV   *NEC Section 220.82*
- Additions to existing dwellings per Part IV   *NEC Section 220.83*
- Multi-family dwellings per Part III   *NEC Section 220.55*
- Multi-family dwellings per Part IV   *NEC Section 220.84*
- Restaurant loads per Part III   *NEC Section 220.56*
- Restaurant loads per Part IV   *NEC Section 220.88*

2.7.0 Demand Factors for Neutral Conductors

The neutral conductor of electrical systems generally carries only the maximum current imbalance of the phase conductors. For example, in a single-phase feeder circuit with one phase conductor carrying 50A and the other carrying 40A, the neutral conductor would carry 10A. Since the neutral in many cases will never be required to carry as much current as the phase conductors, the *NEC*® allows us to apply a demand factor. (See *NEC Section 220.61*.) Note that in certain circumstances such as electrical discharge lighting, data processing equipment, and other similar equipment, a demand factor cannot be applied to the neutral conductors because these types of equipment produce harmonic currents that increase the heating effect in the neutral conductor.

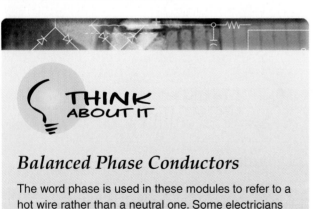

THINK ABOUT IT

Balanced Phase Conductors

The word phase is used in these modules to refer to a hot wire rather than a neutral one. Some electricians call these legs rather than phases. Why must the two phase conductors be balanced?

3.0.0 ◆ SIZING RESIDENTIAL NEUTRAL CONDUCTORS

The neutral conductor in a three-wire, single-phase service carries only the unbalanced load between the two ungrounded (hot) wires or legs. Since there are several 240V loads in the above calculations, these 240V loads will be balanced and therefore reduce the load on the service neutral conductor. Consequently, in most cases, the service neutral does not have to be as large as the ungrounded (hot) conductors.

In the previous example, the water heater does not have to be included in the neutral conductor calculation, since it is strictly 240V with no 120V loads. This takes the total number of fixed appliances on the neutral conductor down to three appliances. Therefore, each of the fixed appliance loads on the neutral must be computed at 100% (dishwasher at 1,500VA, plus disposal at 1,176VA, plus the blower at 864VA). The neutral loads of the electric range and clothes dryer are permitted by *NEC Section 220.61* to be computed at 70% of the demand for the phase conductors since these appliances have both 120V and 240V loads. In this case, the largest motor is the same for the neutral conductors as it is for the phase conductors; therefore, it is computed in the same manner. Using this information, the neutral conductor may be sized accordingly:

- Net general lighting and small appliance load   4,943VA (Box 13)
- Fixed appliance loads   3,540VA (Box 22)
- Electric range (8,000VA × 0.70)   5,600VA (Box 26)
- Clothes dryer (5,500VA × 0.70)   3,850VA (Box 28)
- Largest motor   294VA (Box 32)
- Total   18,227VA (Box 34)

To find the total phase-to-phase amperes, divide the total volt-amperes by the voltage between phases:

$$18,227VA \div 240V = 75.9A \text{ or } 76A$$

The service-entrance conductors have now been calculated and are rated at 125A with a neutral conductor rated for at least 76A. See *Figure 4* for a completed calculation form for the example residence.

In *NEC Section 310.15(B)(6)*, special consideration is given to 120/240V, single-phase residential services and feeders. Conductor sizes are shown in *NEC Table 310.15(B)(6)*. Reference to this table shows that the *NEC*® allows a No. 2 AWG copper or a 1/0 AWG aluminum conductor for a 125A service. The neutral conductor is sized per *NEC*

General Lighting Load						Phase		Neutral	
Square footage of the dwelling	[1]	1350	× 3VA =	[2]	4050				
Kitchen small appliance circuits	[3]	2	× 1500 =	[4]	3000				
Laundry branch circuit	[5]	1	× 1500 =	[6]	1500				
Subtotal of gen. lighting loads per *NEC Section 220.52* =				[7]	8550				
Subtract 1st 3000VA per *NEC Table 220.42*				[8]	3000	× 100% =	[9] 3000		
Remaining VA times 35% per *NEC Table 220.42*				[10]	5550	× 35% =	[11] 1943		
Total demand for general lighting loads =							[12] 4943	[13]	4943

Fixed Appliance Loads (Nameplate or NEC FLA of motors) per *NEC Section 220.53*			
Hot water tank, 4.5kVA, 240V		[14]	4500
Dishwasher 1.5kVA, 120V		[15]	1500
Disposal 1/2HP, 120V per *NEC Table 430.248* = 9.8A		[16]	1176
Blower 1/3HP, 120V per *NEC Table 430.248* = 7.2A		[17]	864
		[18]	
		[19]	
Subtotal of fixed appliances		[20]	8040
If 3 or less fixed appliances take @ 100% =		[21]	[22] 3540
If 4 or more fixed appliances take @ 75% =		[23] 6030	[24]

Other Loads per *NEC Section 220.14*					
Electric Range per *NEC Section 220.55* [neutral @ 70% per *NEC Section 220.61(B)*]			[25] 8000	[26]	5600
Electric Dryer per *NEC Section 220.54* [neutral @ 70% per *NEC Section 220.61(B)*]			[27] 5500	[28]	3850
Electric Heat per *NEC Section 220.51*					
Air Conditioning *NEC Section 220.82(C)*	omit smaller load per *NEC Section 220.60*		[29]	[30]	
Largest Motor = 1176	× 25% (per *NEC Section 430.24*) =		[31] 294	[32]	294

		Phase		Neutral	
Total VA Demand =		[33]	24767	[34]	18227
(VA divided by 240 volts) **Amps** =		[35]	103	[36]	76
Service OCD and minimum size grounding electrode conductor		[37]	125	[38]	8 AWG
AWG per *NEC Sections 310.15(B)(6)* and *Table 310.16* for neutral		[39]	2 AWG	[40]	4 AWG

111F04.EPS

Figure 4 ◆ Completed calculation form.

Tables 310.15(B)(6) or 310.16 using the appropriate column for the markings on the service equipment per *NEC Section 110.14(C)*. Assuming our service panel is marked as suitable for use with 75°C-rated conductors, the minimum size of the neutral would be a No. 4 AWG copper or No. 2 AWG aluminum.

When sizing the grounded conductor for services, the provisions stated in *NEC Sections 215.2, 220.61, and 230.42* must be met, along with other applicable sections.

4.0.0 ◆ SIZING THE LOAD CENTER

Each ungrounded conductor in all circuits must be provided with overcurrent protection in the form of either fuses or circuit breakers. If more than six such devices are used, a means of disconnecting the entire service must be provided using either a main disconnect switch or a main circuit breaker.

To calculate the number of fuse holders or circuit breakers required in the sample residence, look at the general lighting load first. The total general lighting load of 4,050VA can be divided by 120V to find the amperage:

$$4{,}050VA \div 120V = 33.75A$$

Either 15A or 20A circuits may be used for the lighting load. Two 20A circuits (2 × 20) equal 40A, so two 20A circuits would be adequate for the lighting. However, two 15A circuits total only 30A and 33.75A are needed. Therefore, if 15A circuits are used, three will be required for the total lighting load. In this example, three 15A circuits will be used.

In addition to the lighting circuits, the sample residence will require a minimum of two 20A

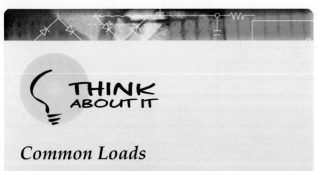

circuits for the small appliance load and one 20A circuit for the laundry. So far, the following branch circuits can be counted:

• General lighting load	Three 15A circuits
• Small appliance load	Two 20A circuits
• Laundry load	One 20A circuit
• Total	Six branch circuits

Most **load centers** and panelboards are provided with an even number of circuit breaker spaces or fuse holders (for example, four, six, eight, or ten). But before the panelboard can be selected, space must be provided for the remaining loads. Each 240V load will require two spaces. In some existing installations, you might find a two-pole fuse block containing two cartridge fuses being used to feed a residential electric range. Each 120V load will require one space each. Thus, the remaining number of circuits for this example is as follows:

• Hot water heater	One two-pole breaker
• Dishwasher	One single-pole breaker
• Disposal	One single-pole breaker
• Blower	One single-pole breaker
• Electric range	One two-pole breaker
• Electric dryer	One two-pole breaker

These additional appliances will therefore require an additional nine spaces in the load center or panelboard. *NEC Section 210.11(C)(3)* requires that a separate 20A branch circuit be provided for the bathroom receptacles. While this circuit requires extra space within a load center, it does not add to the demand on the service for a dwelling unit. Adding the nine spaces for the other loads in the dwelling, plus one for a bathroom circuit, to the six required for the general lighting and small appliance loads requires at least a 16-space load center to handle the circuits.

4.1.0 Ground Fault Circuit Interrupters

Under certain conditions, the amount of current it takes to open an overcurrent protective device can be critical. You should remember from the *Electrical Safety* module that when persons are subject to very low current values (less than one full ampere), it can be fatal. The overcurrent protection installed on services, feeders, and branch circuits protects only the conductors and equipment.

Because of this fact, the *NEC*® requires ground fault circuit interrupter (GFCI) protection for receptacle outlets and/or equipment in many locations and occupancies. The *NEC*® defines a GFCI as "a device intended for the protection of personnel that functions to de-energize a circuit or portion thereof within an established period of time when a current to ground exceeds the values established for a Class A device." Class A GFCIs trip when the current to ground has a value in the range of 4mA to 6mA.

For dwelling units, the majority of requirements to provide protection for 15A or 20A, 125V-rated receptacles can be found in *NEC Section 210.8(A)*. Further requirements for GFCI protection at dwelling units can be found in other *NEC*® articles such as *NEC Article 590* for temporary construction sites; *NEC Article 620* for special equipment such as elevators; or in *NEC Article 680* for special equipment such as swimming pools, hot tubs, and hydromassage tubs. These articles may also expand the requirements for GFCI protection to include circuits rated at more than 20A or operating at 240V.

According to *NEC Section 210.8(A)*, the 15A and 20A, 125V-rated receptacles in our dwelling that require GFCI protection will be those receptacles located in the following areas:

• Bathrooms
• Outdoor receptacles (except those provided on dedicated circuits for snow melting and de-icing equipment)
• Receptacles that serve the countertops in kitchens

Further requirements for GFCI protection at dwelling units are as follows:

• Receptacles within garages and accessory buildings, such as storage sheds or workshops, or similar uses that have a floor located at or below grade level
• Receptacles in unfinished basements
• Crawl spaces at or below grade level
• Receptacles that serve countertops and are within 6' of wet bar sinks, utility, or laundry sinks
• Boathouses

One way to provide this GFCI protection is through the use of a GFCI circuit breaker. GFCI circuit breakers require the same mounting space as standard single-pole circuit breakers and provide the same branch circuit wiring protection as standard circuit breakers. They also provide Class A ground fault protection.

Listed GFCI circuit breakers are available in single- and two-pole construction; 15A, 20A, 25A, and 30A, 50A, and 60A ratings; and have a 10,000A interrupting capacity. Single-pole units are rated at 120VAC; two-pole units are rated at 120/240VAC.

GFCI breakers can be used not only in load centers and panelboards, but they are also available factory-installed in meter pedestals and power outlet panels for recreational vehicle (RV) parks and construction sites.

The GFCI sensor continuously monitors the current balance in the ungrounded or energized (hot) load conductor and the neutral load conductor. If the current in the neutral load wire becomes less than the current in the hot load wire, then a ground fault exists, since a portion of the current is returning to the source by some means other than the neutral load wire. When a current imbalance occurs, the sensor, which is a differential current transformer, sends a signal to the solid-state circuit, which activates the ground trip solenoid mechanism and breaks the hot load connection (*Figure 5*). A current imbalance as low as four milliamps (4mA) will cause the circuit breaker to interrupt the circuit. This is indicated by the trip indicator on the front of the device.

The two-pole GFCI breaker (*Figure 6*) continuously monitors the current balance between the two hot conductors and the neutral conductor. As long as the sum of these three currents is zero, the device will not trip; that is, if the A load wire is carrying 10A of current, the neutral is carrying 5A, and the B load wire is carrying 5A, then the sensor is balanced and will not produce a signal. A current imbalance from a ground fault condition as low as 4mA will cause the sensor to produce a signal of sufficient magnitude to trip the device.

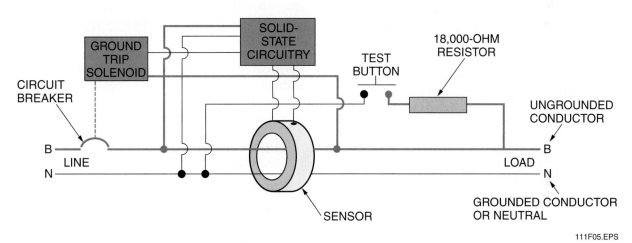

111F05.EPS

Figure 5 ◆ Operating circuitry of a typical GFCI.

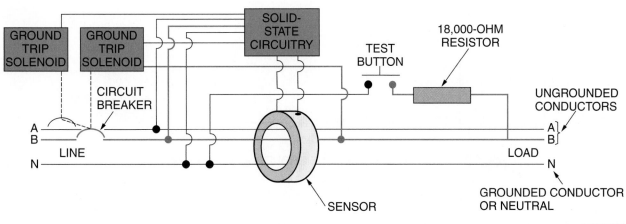

111F06.EPS

Figure 6 ◆ Operating characteristics of a two-pole GFCI.

4.1.1 Single-Pole GFCI Circuit Breakers

The single-pole GFCI breaker has two load lugs and a white wire pigtail in addition to the line side plug-on or bolt-on connector. The line side hot connection is made by installing the GFCI breaker in the panel just as any other circuit breaker is installed. The white wire pigtail is attached to the panel neutral (S/N) assembly. Both the neutral and hot wires of the branch circuit being protected are terminated in the GFCI breaker. These two load lugs are clearly marked LOAD POWER and LOAD NEUTRAL in the breaker case. Also in the case is the identifying marking for the pigtail, PANEL NEUTRAL.

> **NOTE**
>
> Single-pole GFCI circuit breakers must be installed on independent circuits. They cannot be used on multi-wire circuits.

Care should be exercised when installing GFCI breakers in existing panels. Be sure that the neutral wire for the branch circuit corresponds with the hot wire of the same circuit. Always remember that unless the current in the neutral wire is equal to that in the hot wire (within 4mA), the GFCI breaker senses this as being a possible ground fault (see *Figure 7*).

4.1.2 Two-Pole GFCI Circuit Breakers

A two-pole GFCI circuit breaker can be installed on a 120/240VAC single-phase, three-wire system; the 120/240VAC portion of a 120/240VAC three-phase, four-wire system; or the two phases and neutral of a 120/208VAC three-phase, four-wire system. Regardless of the application, the installation of the breaker is the same—connections are made to two hot buses and the panel neutral assembly. When installed on these systems, protection is provided for two-wire 240VAC or 208VAC circuits, three-wire 120/240VAC or 120/208VAC circuits, and 120VAC multiwire circuits.

The circuit in *Figure 8* illustrates the problems that are encountered when a common load neutral is used for two single-pole GFCI breakers. Either or both breakers will trip when a load is applied at the #2 duplex receptacle. The neutral current from the #2 duplex receptacle flows through breaker #1; this increase in neutral current through breaker #1 causes an imbalance in its sensor, thus causing it to produce a fault signal. At the same time, there is no neutral current flowing through breaker #2; therefore, it also senses a current imbalance. If a load is applied at the #1 duplex receptacle, and there is no load at the #2 duplex receptacle, then neither breaker will trip because neither breaker will sense a current imbalance.

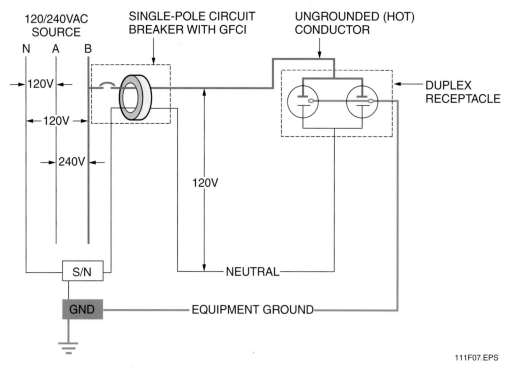

Figure 7 ◆ Operating characteristics of a single-pole circuit breaker with a GFCI.

111F07.EPS

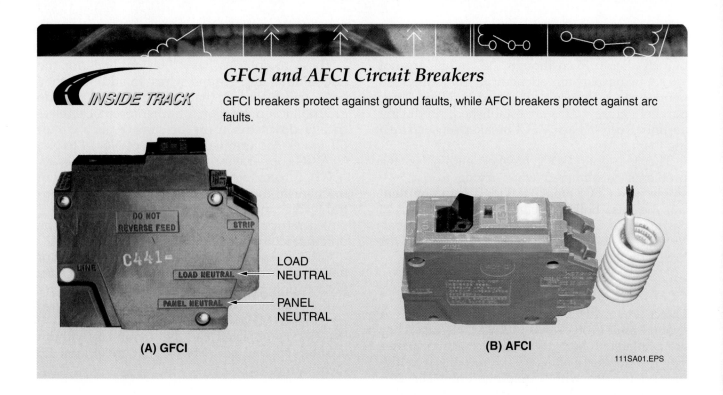

GFCI and AFCI Circuit Breakers

INSIDE TRACK

GFCI breakers protect against ground faults, while AFCI breakers protect against arc faults.

(A) GFCI

(B) AFCI

111SA01.EPS

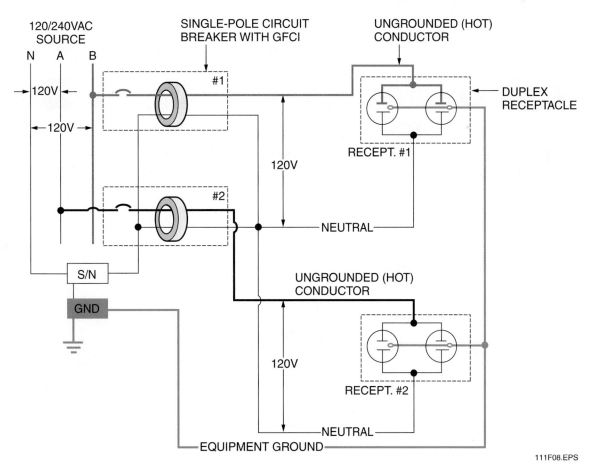

Figure 8 ◆ Circuit depicting the common load neutral.

111F08.EPS

Junction boxes can also present problems when they are used to provide taps for more than one branch circuit. Even though the circuits are not wired using a common neutral, sometimes all neutral conductors are connected together. Thus, parallel neutral paths are established, producing an imbalance in each GFCI breaker sensor, causing them to trip.

The two-pole GFCI breaker eliminates the problems encountered when trying to use two single-pole GFCI breakers with a common neutral. Because both hot currents and the neutral current pass through the same sensor, no imbalance occurs between the three currents, and the breaker will not trip.

4.1.3 Direct-Wired GFCI Receptacles

Direct-wired GFCI receptacles provide Class A ground fault protection on 120VAC circuits. They are available in both 15A and 20A arrangements. The 15A unit has a NEMA 5-15R receptacle configuration for use with 15A plugs only. The 20A device has a NEMA 5-20R receptacle configuration for use with 15A or 20A plugs. Both 15A and 20A units have a 120VAC, 20A circuit rating. This is to comply with *NEC Table 210.24*, which requires that 15A circuits use 15A receptacles but permits the use of either 15A or 20A receptacles on 20A

circuits. Therefore, GFCI receptacle units that contain a 15A receptacle may be used on 20A circuits.

These receptacles have line terminals for the hot, neutral, and ground wires. In addition, they have load terminals that can be used to provide ground fault protection for other receptacles electrically downstream on the same branch circuit (*Figure 9*). All terminals will accept No. 14 to No. 10 AWG copper wire.

GFCI receptacles have a two-pole tripping mechanism that breaks both the hot and the neutral load connections.

When tripped, the RESET button pops out. The unit is reset by pushing the button back in.

GFCI receptacles have the additional benefit of noise suppression. Noise suppression minimizes false tripping due to spurious line voltages or radio frequency (RF) signals between 10 and 500 megahertz (MHz).

GFCI receptacles can be mounted without adapters in wall outlet boxes that are at least 1.5 inches deep.

4.2.0 Arc Fault Circuit Interrupters

All branch circuits that supply the lighting and general-purpose receptacles in dwelling unit family rooms, dining rooms, living rooms, parlors,

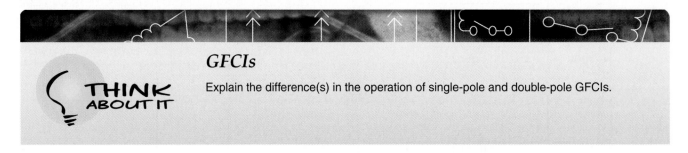

GFCIs

Explain the difference(s) in the operation of single-pole and double-pole GFCIs.

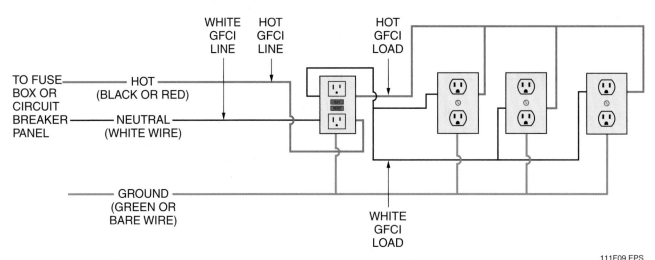

111F09.EPS

Figure 9 ◆ GFCI receptacle used to protect other outlets on the same circuit.

libraries, dens, bedrooms, sunrooms, recreation rooms, closets, hallways, or similar rooms or areas must have arc fault circuit interrupter protection to comply with *NEC Section 210.12.*

5.0.0 ◆ GROUNDING

NEC Section 250.4(A) provides the general requirements for grounding and bonding of grounded electrical systems. In order to ensure systems are properly grounded and bonded, the prescriptive requirements of *NEC Article 250* must be followed.

The grounding system is a major part of the electrical system. Its purpose is to protect people and equipment against the various electrical faults that can occur. It is sometimes possible for higher-than-normal voltages to appear at certain points in an electrical system or in the electrical equipment connected to the system. Proper grounding ensures that the electrical charges that cause these higher voltages are channeled to the earth or ground and that an effective ground fault path is provided throughout the system so that overcurrent devices will open before people are endangered or equipment is damaged.

The word ground refers to ground potential or earth ground. If a conductor is connected to the earth or some conducting body that serves in place of the earth, such as a driven ground rod (electrode), the conductor is said to be grounded. The neutral conductor in a three- or four-wire service, for example, is intentionally grounded, and therefore becomes a grounded conductor. This is the path back to the source of supply for all ground faults in an electrical system. This conductor is intended not only to carry the unbalanced loads of an installation, but also to provide the low-impedance path back to the source so that enough current will flow in the system to open the overcurrent devices. A wire that is used to connect this neutral conductor to a grounding electrode or electrodes is referred to as a grounding electrode

conductor (GEC). Note the difference in the two meanings: one is grounded, while the other provides a means for grounding.

There are two general classifications of protective grounding:

• System grounding
• Equipment grounding

The system ground relates to the **service-entrance equipment** and its interrelated and bonded components; that is, the system and circuit conductors are grounded to limit voltages due to lightning, line surges, or unintentional contact with higher voltage and to stabilize the voltage to ground during normal operation per *NEC Sections 250.4(A)(1) and (2).*

The noncurrent-carrying conductive parts of materials enclosing electrical conductors or equipment, or forming a part of such equipment, and electrically conductive materials that are likely to become energized are all connected together to the supply source in a manner that establishes an effective ground fault path per *NEC Sections 250.4(A)(3) and (4).*

NEC Section 250.4(A)(5) defines the requirements for an effective ground path. It requires that electrical equipment and wiring and other electrically conductive materials likely to become energized shall be installed in a manner that creates a permanent, low-impedance circuit capable of safely carrying the maximum ground fault current likely to be imposed on it from any point on the wiring system where a ground fault may occur to the electrical supply source. The earth shall not be used as the sole equipment grounding conductor or effective ground fault current path.

To better understand a complete grounding system, a conventional residential system will be examined, beginning at the power company's high-voltage lines and transformer, as shown in *Figure 10.* The pole-mounted transformer is fed with a two-wire, single-phase 7,200V system,

Grounding

Systematic grounding wasn't required by the *NEC®* until the mid-1950s; even then, electricians commonly grounded an outlet by wrapping an uninsulated wire around a cold-water pipe and taping it. Three-hole receptacles with grounding terminals became common in the 1960s, but in many older houses, you cannot assume that receptacles are grounded, even when you see a three-hole receptacle. Sometimes, new receptacles have simply been screwed onto old boxes where there is no equipment grounding conductor.

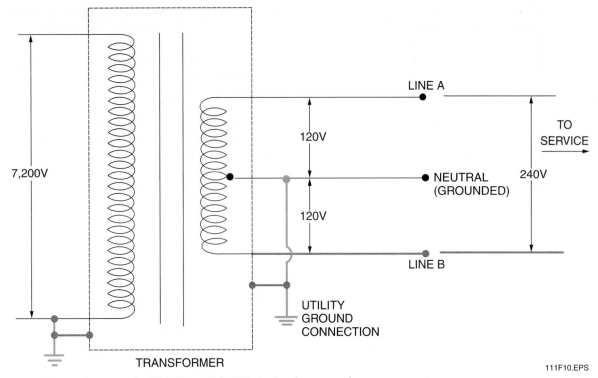

Figure 10 ◆ Wiring diagram of a 7,200V to 120/240V, single-phase transformer connection.

which is transformed and stepped down to a three-wired, 120/240V, single-phase electric service suitable for residential use. Note that the voltage between line A and line B is 240V. However, by connecting a third (neutral) wire on the secondary winding of the transformer—between the other two—the 240V is split in half, providing 120V between either line A or line B and the neutral conductor. Consequently, 240V is available for household appliances such as ranges, hot water heaters, and clothes dryers, while 120V is available for lights and small appliances.

Referring again to *Figure 10*, conductors A and B are ungrounded conductors, while the neutral is a grounded conductor. If only 240V loads were connected, the neutral (grounded conductor) would carry no current. In this instance, the neutral would be used to carry any ground fault currents from the load side of the service back to the utility instead of depending on the earth as the path back to the source. However, since 120V loads are present, the neutral will carry the unbalanced load and become a current-carrying conductor. For example, if line A carries 60A and line B carries 50A, the neutral would carry only 10A (60A – 50A = 10A). This is why the *NEC*® allows the neutral conductor in an electric service to be smaller than the ungrounded conductors. However, *NEC Section 250.24(C)(1)* requires that it must be sufficient to carry fault currents back to the source and, therefore, must not be less than the

required grounding electrode conductor using *NEC Table 250.66* for service conductors up to 1,100 kcmil and not less than 12.5% of the area of the service-entrance conductors (or equivalent) larger than 1,100 kcmil. The typical pole-mounted service drop conductors are normally routed by a messenger cable from a point on the pole to a point on the building being served, terminating at the point where service-entrance conductors exit a weatherhead. Service-entrance conductors are then typically routed through metering equipment into the service disconnecting means. This is the point where most services are grounded. See *Figure 11. NEC Section 250.24(A)(1)* requires that the grounding electrode for the structure connection to the neutral (grounded conductor) be at any accessible point from the load end of the service drop or **service lateral** to and including the terminal or bus to which the neutral (grounded service conductor) is connected to the service disconnecting means.

NOTE

Effectively grounded means intentionally connected to earth through one or more ground connection(s) of sufficiently low impedance and having sufficient current-carrying capacity to prevent the buildup of voltages that may result in a hazard to people or connected equipment.

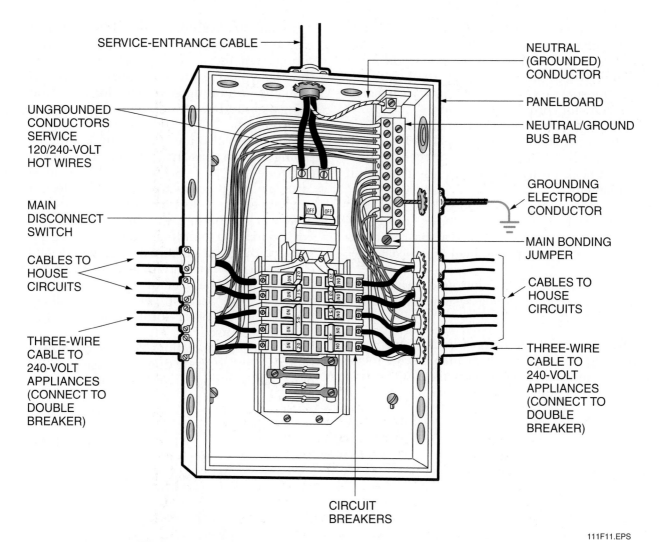

SERVICE-ENTRANCE CABLE

NEUTRAL
(GROUNDED)
CONDUCTOR

PANELBOARD

NEUTRAL/GROUND
BUS BAR

UNGROUNDED
CONDUCTORS
SERVICE
120/240-VOLT
HOT WIRES

GROUNDING
ELECTRODE
CONDUCTOR

MAIN
DISCONNECT
SWITCH

MAIN BONDING
JUMPER

CABLES TO
HOUSE
CIRCUITS

CABLES TO
HOUSE
CIRCUITS

THREE-WIRE
CABLE TO
240-VOLT
APPLIANCES
(CONNECT TO
DOUBLE
BREAKER)

THREE-WIRE
CABLE TO
240-VOLT
APPLIANCES
(CONNECT TO
DOUBLE
BREAKER)

CIRCUIT
BREAKERS

111F11.EPS

Figure 11 ◆ Interior view of panelboard showing connections.

5.1.0 Grounding Electrodes

NEC Article 250, Part III provides the requirements for connecting an electric service to the grounding electrode system of a building or structure. *NEC Section 250.50* requires, in general, that all of the electrodes described in *NEC Section 250.52(A)* be used (if present), and they must be bonded together to form the grounding electrode system. The electrodes listed in *NEC Section 250.52(A)* are as follows:

• Metal underground water pipe in direct contact with the earth for 10' or more and electrically continuous (or made electrically continuous by bonding around insulating joints or insulating pipe) to the points of connection of the grounding electrode conductor and the bonding conductors. Interior metal water piping located more than 5' from the point of entrance to the building shall not be used as part of the grounding electrode system or as a conductor to interconnect electrodes that are part of the grounding electrode system.

• Metal frame of the building or structure that complies with *NEC Section 250.52(A)(2)*.

• An electrode encased by at least 2" of concrete may be used if it is located within and near the bottom of a concrete foundation or footing that is in direct contact with the earth. The electrode must be at least 20' long and must be made of electrically conductive coated steel reinforcing bars or rods of not less than ½" in diameter, or consisting of at least 20' of bare copper conductor not smaller than No. 4 AWG wire size.

• A ground ring encircling the building or structure, in direct contact with the earth, consisting of at least 20' of bare copper conductor not smaller than No. 2 AWG.

- Rod and pipe electrodes shall not be less than 8' in length and consist of either:
  - Pipe or conduit not smaller than trade size ¾ and, where of iron or steel, shall have the outer surface galvanized or otherwise metal-coated for corrosion protection.
  - Rods of iron or steel not smaller than ⅝" in diameter. Stainless steel rods less than ⅝" in diameter, nonferrous rods, or their equivalent shall be listed and not less than ½" in diameter.
- Plate electrodes shall expose less than two square feet of surface to exterior soil. Plates made of iron or steel shall be at least ¼" thick. Nonferrous metal plates shall be at least 0.06" thick.
- Other local metal underground systems or structures such as piping systems and underground tanks.

Often in residential construction, the only grounding electrode that is available is the metal underground water piping system. *NEC Section 250.53(D)(2)* requires that whenever water piping is used as an electrode, it must be supplemented. Any of the electrodes listed above can be used to supplement the water pipe electrode. *Figure 12* shows a typical residential electric service and the available grounding electrodes for this structure using a ground rod to supplement the water pipe electrode.

This house also has a metal underground gas piping system, but this may not be used as an electrode per *NEC Section 250.52(B)*. In some cases, a water pipe electrode, building steel, and a concrete-encased electrode are not available to be used as a part of the grounding electrode system. For example, a building may be fed by plastic water piping, be constructed of wood, and an electrician may not be present at the site when the foundation for the structure is poured. When that happens, *NEC Section 250.50* requires that rod, pipe, plate, or other local metal underground structures be used.

Some local jurisdictions do not recognize water piping as an electrode due to the rise in the use of nonmetallic piping for both new and replacement

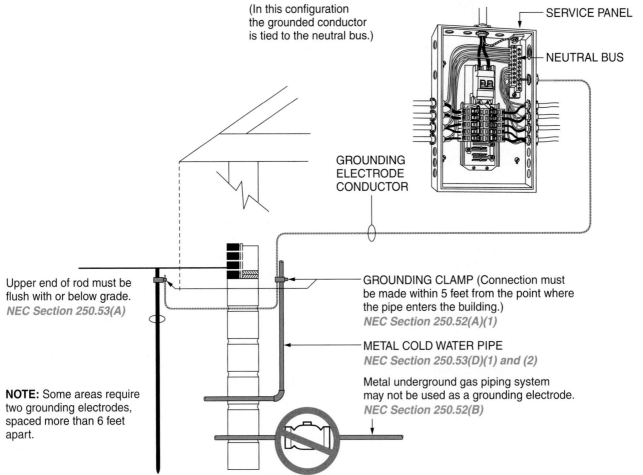

Figure 12 ◆ Components of a residential grounding system.

water systems. They do not want to rely on the maintained viability of an existing metallic water service and therefore require the use of other electrodes such as the concrete-encased or rod electrodes. This means electricians must be involved with the construction prior to the foundation being poured in order to utilize concrete-encased electrodes.

In most cases, the supplemental electrode used for a water pipe electrode will consist of either a driven rod or pipe electrode, the specifications for which are shown in *Figure 13*.

 WARNING!

A metal underground gas piping system must never be used as a grounding electrode.

5.1.1 Grounding Electrode Installations

NEC Section 250.53(A) requires that rod, pipe, and plate electrodes, where practical, be buried below the permanent moisture level and that they are free from any nonconductive coatings, such as paint or enamel. This section also requires that each electrode system used for a structure be at least 6' from other electrode systems, such as those for lightning protection.

NEC Section 250.53(G) permits a rod or pipe electrode to be driven at a 45-degree angle if rock bottom is encountered and prevents the rod or pipe from being driven vertically for at least 8'. Where driving a rod or pipe electrode at a 45-degree angle will not work, it is permitted to lay a rod or pipe horizontally in a trench that is at least 30" deep. For rod or pipe electrodes longer than 8', it is permitted to have the upper end above ground level if a suitable means of protection is provided for the grounding electrode conductor attachment; otherwise, the upper end must be flush with the earth surface.

NEC Section 250.56 requires that a single rod, pipe, or plate electrode that does not have a resistance to ground of 25Ω or less shall be augmented by one additional electrode of any of the types specified by *NEC Sections 250.52(A)(2) through (7)*. In fact, many local jurisdictions require two electrodes regardless of the resistance to ground.

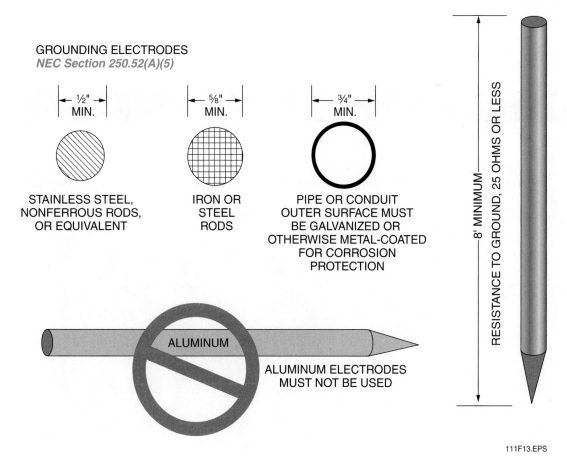

Figure 13 ◆ Specifications for rod and pipe grounding electrodes.

Always check with the local inspection authority, including the local utility, for rules that surpass the requirements of the *NEC®*.

- Where multiple rod, pipe, or plate electrodes are installed to meet the requirements of this section, they shall not be less than 6' apart.
- Plate electrodes must be buried at least 30" below the surface of the earth, according to *NEC Section 250.53(H)*.
- Where two or more electrodes are effectively bonded together, they are treated as a single electrode system.

5.1.2 Grounding Electrode Conductors (GECs)

The grounding electrode conductor (GEC) connecting the neutral (grounded conductor of the service) at the panelboard neutral bus to the grounding electrodes must meet the requirements of *NEC Section 250.62*. This requires that it be made of copper, aluminum, or copper-clad aluminum. The material selected must be suitably protected against corrosion. The GEC may be either solid or stranded, covered or bare. Note that the GEC is not an equipment grounding conductor, and thus is not required to be identified by the use of the color green or green with yellow stripes, if insulated.

5.1.3 Installation of GECs

NEC Section 250.64 provides the installation requirements for GECs and does not permit bare aluminum or copper-clad aluminum grounding conductors to be used where in direct contact with masonry, the earth, where subject to corrosive conditions, or where used outside within 18" of the earth at the termination point. Other *NEC®* requirements include the following:

- A GEC or its enclosure is required to be securely fastened to the surface on which it is carried.
- A No. 4 AWG or larger copper or aluminum GEC is required to be protected where it will be exposed to severe physical damage.
- A No. 6 AWG or larger GEC that is free from exposure to physical damage is permitted to be run along the surface of the building without metal covering or protection where it is securely fastened to the building. Otherwise, it must be installed in RMC, IMC, RNC, EMT, or a cable armor.
- GECs smaller than No. 6 AWG must be protected by RMC, IMC, RNC, EMT, or cable armor.

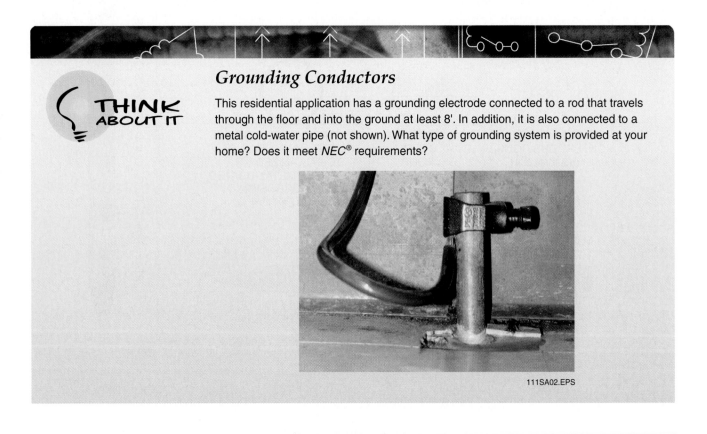

Grounding Conductors

This residential application has a grounding electrode connected to a rod that travels through the floor and into the ground at least 8'. In addition, it is also connected to a metal cold-water pipe (not shown). What type of grounding system is provided at your home? Does it meet *NEC®* requirements?

111SA02.EPS

Grounding Electrode Conductors

THINK ABOUT IT

Which *NEC*® table would you use to size the minimum GEC required for a typical residential service?

- The GEC shall be installed in one continuous length without a splice or joint, unless spliced only by irreversible compression-type connectors listed for the purpose or by an exothermic welding process. Connecting sections of busbars together to form a GEC is not considered to be a splice.
- Where a service consists of more than a single enclosure, it is permissible to connect taps to the GEC, provided each tap extends all the way into the inside of each such enclosure. The tap conductors shall be connected to the GEC in such a manner that the GEC remains without a splice.
- Ferrous metal enclosures for the GEC are required to be electrically continuous from the point of attachment to metal cabinets or metallic equipment enclosures to the GEC. They must also be securely fastened to the ground clamp or fitting.
- Ferrous metal enclosures for the GEC that are not physically continuous from a metal cabinet or metallic equipment enclosure to the grounding electrode must be made electrically continuous by bonding each to the enclosed GEC.
- GECs may be run to any convenient grounding electrode available in the grounding electrode system, or to one or more grounding electrode(s) individually. The GEC shall be sized for the largest grounding electrode conductor required among all the electrodes connected together.

5.1.4 Methods of Connecting GECs

NEC Section 250.70 requires the GEC to be connected to electrodes using exothermic welding, listed pressure connectors, listed clamps, listed lugs, or other listed means. Connections that depend on solder must never be used. To prevent corrosion, the ground clamp must be listed for the material of the grounding electrode and the GEC.

Where used on a pipe, ground rod, or other buried electrodes, the fitting must be listed for direct soil burial or concrete encasement. More than one conductor is not permitted to be connected to the grounding electrode using a single

clamp or fitting unless the clamp or fitting is specifically listed for the connection of more than one conductor.

For the connection to an electrode, you must use one of the following:

- A listed, bolted clamp of cast bronze or brass, or plain or malleable iron
- A pipe fitting, pipe plug, or other approved device that is screwed into a pipe or pipe fitting
- For indoor telecommunication purposes only, a listed sheet metal strap-type ground clamp with a rigid metal base that seats on the electrode with a strap that will not stretch during or after installation
- An equally substantial approved means

The connection of a GEC or a **bonding jumper** to a grounding electrode must be accessible unless that connection is to the concrete-encased or buried grounding electrodes permitted in *NEC Section 250.68.* Where it is necessary to ensure the grounding path for metal piping used as a grounding electrode, effective bonding shall be provided around insulated joints and around any equipment likely to be disconnected for repairs or replacement. Bonding conductors shall be of sufficient length to permit removal of such equipment while retaining the integrity of the bond. Coatings on metal piping systems must be removed to ensure that a permanent and effective grounding path is provided.

NOTE

The UL listing states that "strap-type ground clamps are not suitable for attachment of the grounding electrode conductor of an interior wiring system to a grounding electrode."

For the example house, the point of connection to the water piping is shown in *Figure 12* and would be required to be accessible after any wall coverings are installed. Any nonconductive coatings on the water piping would also have been scraped off or removed prior to installing the clamp on the water pipe.

5.1.5 Sizing GECs

Grounding electrode conductors must be sized per *NEC Section 250.66,* which uses the area of the largest service-entrance conductor (or equivalent area for paralleled conductors). Except as noted below, *NEC Table 250.66* will provide the minimum size GEC and any bonding jumpers used to interconnect grounding electrodes used.

- Where connected to rod, pipe, or plate electrodes, that portion of the GEC that is the sole connection to the grounding electrode shall not be required to be larger than No. 6 AWG copper or No. 4 AWG aluminum.
- Where connected to a concrete-encased electrode, that portion of the GEC that is the sole connection to the grounding electrode shall not be required to be larger than No. 4 AWG copper wire.
- Where the GEC is connected to a ground ring, that portion of the conductor that is the sole connection to the grounding electrode shall not be required to be larger than the conductor used for the ground ring.
- Where multiple sets of service-entrance conductors are used as permitted in *NEC Section 230.40, Exception 2,* the equivalent size of the largest service-entrance conductor is required to be determined by the largest sum of the areas of the corresponding conductors of each set.
- Where there are no service-entrance conductors, the GEC size is required to be determined by the equivalent size of the largest service-entrance conductor required for the load to be served.

For our sample dwelling unit, the size of the service-entrance conductors is No. 2 AWG. Using *NEC Table 250.66,* we can determine that the size of the conductor coming from the service panel to the water pipe (the GEC) must be at least No. 8 AWG copper. This No. 8 AWG may continue on without a splice to the ground rod, as shown in *Figure 12,* or a separate No. 6 AWG could be installed for the ground rod(s). This conductor would have to be connected to the service panel as an individual run or with a separate connector to any portion of the No. 8 AWG. It may not be connected directly to the water piping.

5.1.6 Air Terminals

Air terminal conductors and driven pipes, rods, or plate electrodes used for grounding air terminals are not permitted to be used in lieu of the grounding electrodes covered in *NEC Section 250.50* for grounding wiring systems and equipment. However, *NEC Section 250.106* requires that they be bonded to the wiring and equipment grounding electrode system for the structure.

5.2.0 Main Bonding Jumper

NEC Section 250.24(B) requires that an unspliced main bonding jumper (MBJ) shall be used to connect the equipment grounding conductor(s) and the service disconnect enclosure to the grounded conductor (neutral) of the system within the enclosure of each service disconnect.

The MBJ must be of copper or other corrosion-resistant material. An MBJ may be in the form of a wire, bus, screw, or similar suitable conductor.

Where an MBJ is in the form of a screw, it is required to be identified with a green finish so that the head of the screw is visible for inspection. An MBJ must be attached using exothermic welding, a listed pressure connector, listed clamp, or other listed means.

The MBJ cannot be smaller than the sizes given in *NEC Table 250.66* for grounding electrode conductors. See *NEC Section 250.28(D)* for service conductors that exceed 1,100 kcmil.

The MBJ is the means by which any ground fault in the branch circuits and feeders of the electrical system travels back to the source of supply at the utility. A ground fault will travel along the equipment conductors of the circuits back to the service disconnecting means. Where metallic

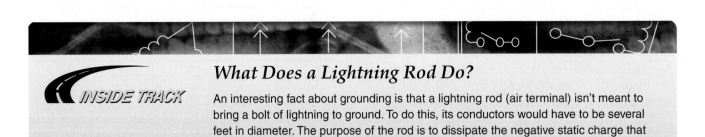

What Does a Lightning Rod Do?

An interesting fact about grounding is that a lightning rod (air terminal) isn't meant to bring a bolt of lightning to ground. To do this, its conductors would have to be several feet in diameter. The purpose of the rod is to dissipate the negative static charge that would cause the positive lightning charge to strike the house.

raceways are used as equipment grounding conductors, there will be no connection to the grounded conductor at the service. Without the MBJ, the path back to the source would be through the grounding electrode system and the earth. This does not provide a low-impedance path, and thus will not allow enough current to flow in the circuit to let the overcurrent devices open.

For example, suppose a phase conductor makes contact with the metallic housing of a 120V, 15A appliance that was wired using EMT. Further suppose that the total combined resistance of the EMT being used as the equipment grounding conductor connected to the appliance and the resistance of the metal water piping and our ground rod in the sample house is 20Ω (less than the 25Ω permitted in *NEC Section 250.56*). The amount of current that could flow back to the utility source would be 120V ÷ 20 = 6A. The smallest overcurrent device in our electrical system is 15A, and would not trip. With the MBJ installed, the path back to the utility source is through the MBJ to the grounded conductor of the service. This resistance will be much less than 1Ω, and thus would allow enough current to flow to open up the overcurrent devices within the system. The MBJ provides the path back to the source for faults that occur within the service disconnect means.

5.2.1 Bonding at the Service

Electrical continuity is required at the service per *NEC Section 250.92(A)*, which states that all of the following must be bonded:

- The service raceways, auxiliary gutters, or service cable armor or sheaths, except for underground metallic sheaths of continuously underground cables as noted in *NEC Section 250.84*
- All service enclosures containing service-entrance conductors, including meter fittings, boxes, or the like interposed in the service raceway or armor
- Any metallic raceway or armor enclosing a grounding electrode conductor as specified in *NEC Section 250.64(E)*

Bonding shall apply at each end and to all intervening raceways, boxes, and enclosures between the service equipment and the grounding electrode.

The items that typically require bonding include the mast and weatherhead, the meter enclosure, the armor of the SE cable (if it has armor), and the service disconnect.

5.2.2 Methods of Bonding at the Service

The electrical continuity of the service equipment, raceways, and enclosures will be ensured per *NEC Section 250.92(B)* through the use of the following methods:

- Bonding equipment to the grounded service conductor in a manner provided in *NEC Section 250.8*
- Connections utilizing threaded couplings or threaded bosses on enclosures where made up wrenchtight
- Threadless couplings and connectors where made up tight for metal raceways and metal-clad cables
- Other approved devices, such as bonding-type locknuts and **bonding bushings**

Bonding jumpers must be used around concentric or eccentric knockouts that are punched or otherwise formed so as to impair the electrical connection to ground. Standard locknuts or bushings shall not be the sole means for bonding.

5.2.3 Bonding and Grounding Requirements for Other Systems

An accessible means external to the service equipment enclosure is required for connecting intersystem bonding and grounding conductors and connections for the communications, radio and television (TV), community antenna television (CATV), and network-powered broadband communication system. The intersystem bonding termination must consist of at least three terminals and be installed in accordance with *NEC Section 250.94*. Any one of the following can be used:

- A set of listed terminals securely mounted to and electrically connected to the meter enclosure
- A bonding bar near the service-entrance enclosure, meter enclosure, or raceway for service conductors connected to an equipment grounding conductor in the enclosure or raceway with a minimum No. 6 AWG copper conductor
- A bonding bar near the grounding electrode conductor connected with a minimum No. 6 AWG copper conductor

5.2.4 Bonding of Water Piping Systems

Metallic water piping systems in or on a structure must be bonded as required by *NEC Section 250.104(A)*. The metallic water piping system(s) must be bonded by means of a bonding jumper sized in accordance with *NEC Table 250.66* and connected to one of the following:

- The service-entrance enclosures
- The grounded (neutral) conductor at the service
- The grounding electrode conductor where of sufficient size
- The grounding electrode(s) used

The points of attachment of the bonding jumper(s) shall be accessible. It shall be installed in accordance with *NEC Section 250.64(A), (B), and (E).* Note that while this conductor is sized in the same manner as if the water piping system is a grounding electrode, the point of attachment to the water piping is permitted to be at any convenient point on the water piping system and not just within the first 5' of where the water enters the building.

NEC Section 250.104(A)(2) states that in multi-family dwelling units (or other multiple occupancy buildings) where the metal water piping system(s) installed in or attached to a building or structure for the individual occupancies is metallically isolated from all other occupancies by use of nonmetallic water piping, the metal water piping system(s) for each occupancy shall be permitted to be bonded to the equipment grounding terminal of the panelboard or switchboard enclosure (other than service equipment) supplying that occupancy. The bonding jumper shall be sized in accordance with *NEC Table 250.122.*

5.2.5 Bonding of Other Piping Systems

NEC Section 250.104(B) requires that other piping systems, where installed in or attached to a building or structure, including gas piping, that may become energized shall be bonded to one of the following:

- The service equipment enclosure
- The grounded conductor at the service
- The grounding electrode conductor where of sufficient size
- One or more grounding electrodes used

The bonding jumper(s) shall be sized in accordance with *NEC Table 250.122* using the rating of the circuit that may energize the piping system(s). The equipment grounding conductor for the circuit that may energize the piping shall be permitted to serve as the bonding means. The points of attachment of the bonding jumper(s) shall be accessible.

NOTE

Bonding all piping and metal air ducts within the premises will provide additional safety.

6.0.0 ◆ INSTALLING THE SERVICE ENTRANCE

In practical applications, the electric service is normally one of the last components of an electrical system to be installed. However, it is one of the first considerations when laying out a residential electrical system. For instance:

- The electrician must know in which direction and to what location to route the circuit home-runs while **roughing in** the electrical wiring.
- Provisions must be made for sleeves through footings and foundations in cases where underground systems (service laterals) are used.
- The local power company must be notified as to the approximate size of service required so they may plan the best way to furnish a **service drop** to the property.

6.1.0 Service Drop Locations

The location of the service drop, electric meter, and load center should be considered first. It is always wise to consult the local power company to obtain their requirements; where you want the service drop and where they want it may not coincide. A brief meeting with the power company about the location of the service drop can prevent problems later on.

The service drop must be routed so that the service drop conductors have a clearance of not less than 3' horizontally and below windows that open, doors, porches, fire escapes, or similar locations. In addition, they must have a 10' vertical clearance that extends 3' horizontally from porches, fire escapes, balconies, and so forth, as required in *NEC Section 230.9.* Where service drop conductors pass over rooftops, driveways, yards, and so forth, they must have clearances as specified in *NEC Section 230.24.*

A plot plan (also called a site plan) is often available for new construction. The plot plan shows the entire property, with the building or buildings drawn in their proper location on the plot of land. It also shows sidewalks, driveways, streets, and existing utilities—both overhead and underground.

A plot plan of the sample residence is shown in *Figure 14*. In reviewing this drawing, you can see that the closest power pole is located across a public street from the house. By consulting with the local power company, it is learned that the service will be brought to the house from this pole by triplex cable, which will connect to the residence at a point on its left (west) end. The steel uninsulated conductor of triplex cable acts as both the grounded conductor (neutral) and as a support for the insulated (ungrounded) conductors. It is also suitable for overhead use.

When service-entrance cable is used, it will run directly from the point of attachment and service head to the meter base. However, since the carport is located on the west side of the building, a service mast *(Figure 15)* will have to be installed.

The *NEC*® requires a clearance of not less than 8' over rooftops, unless the roof has a slope of 4" in 12" or greater, in which case the clearance may be reduced to 3'. Where the service drop conductors pass over only the overhang (eaves) of a roof, the clearance may be reduced to 18" as long as no more than 6' of the conductors travel over no more than 4' of the overhang (eave). This minimum height requirement extends beyond the roof for a distance of not less than 3' in all directions, except the final portion of the span where the service drop conductors attach to the sides of a building.

6.2.0 Vertical Clearances of Service Drop

NEC Section 230.24(B) specifies the distances by which service drop conductors must clear the ground. These distances vary according to the surrounding conditions.

In general, the *NEC*® states that the vertical clearances of all service drop conductors that carry 600V or under are based on a conductor temperature of 60°F (15°C) with no wind and with the final unloaded sag in the wire, conductor, or cable. Service drop conductors must be at least 10' above the ground or other accessible surfaces at all times. More distance is required under most conditions. For example, if the service conductors pass over residential property and driveways or commercial property that is not subject to truck traffic, the conductors must be at least 15' above the ground. However, this distance may be reduced to 12' when the voltage is limited to 300V to ground.

In other areas, such as public streets, alleys, roads, parking areas subject to truck traffic, driveways on other-than-residential property, the minimum vertical distance is 18'. The conditions of the sample residence are shown in *Figure 16*.

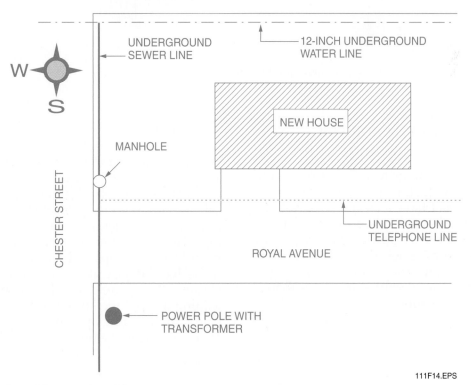

111F14.EPS

Figure 14 ◆ Plot plan of the sample residence.

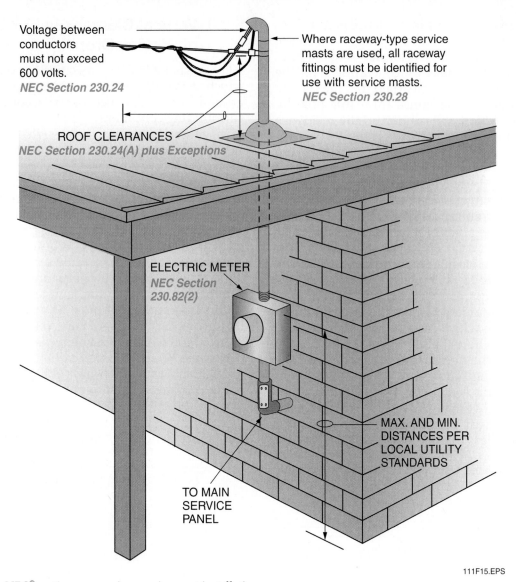

Voltage between conductors must not exceed 600 volts.
NEC Section 230.24

Where raceway-type service masts are used, all raceway fittings must be identified for use with service masts.
NEC Section 230.28

ROOF CLEARANCES
NEC Section 230.24(A) plus Exceptions

ELECTRIC METER
NEC Section 230.82(2)

MAX. AND MIN. DISTANCES PER LOCAL UTILITY STANDARDS

TO MAIN SERVICE PANEL

111F15.EPS

Figure 15 ◆ *NEC® sections governing service mast installations.*

6.3.0 Service Drop Clearances for Building Openings

Service conductors that are installed as open conductors or multiconductor cable without an overall outer jacket must have a clearance of not less than 3' from windows that are designed to be opened, doors, porches, balconies, ladders, stairs, fire escapes, or similar locations (*NEC Section 230.9*). However, conductors run above the top level of a window are permitted to be less than 3' from the window opening.

The 3' of clearance is not applicable to raceways or cable assemblies that have an overall outer jacket approved for use as a service conductor. The intention of this requirement is to protect the conductors from physical damage and/or physical contact with unprotected personnel when evacuating a structure through the window opening. The exception allows service conductors, including drip loops and service drop conductors, to be located just above the window openings because they would not interfere with ladders leaning against the structure to the right, left, or below the window opening when used to evacuate people from the building.

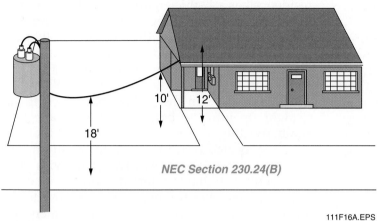

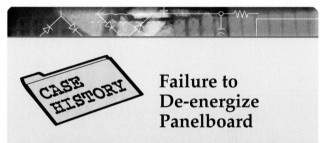

NEC Section 230.24(B)

111F16A.EPS

111F16B.EPS

Figure 16 ◆ Vertical clearances for service drop conductors.

Failure to De-energize Panelboard

A 31-year-old electrician was finishing the installation of an outdoor floodlight on a new home. He borrowed an aluminum ladder from another contractor and then proceeded with his task. He did not verify that power was removed at the panelboard, and when he used his wire strippers to remove the conductor insulation, his right thumb and index finger contacted a 110V circuit. He received a fatal shock.

The Bottom Line: Always de-energize circuits at the panelboard before beginning any electrical task, and never use aluminum ladders while working in or near electrical devices. In this case, the ladder provided a path to ground.

7.0.0 ◆ PANELBOARD LOCATION

The main service disconnect or panelboard is normally located in a portion of an unfinished basement or utility room on an outside wall so that the service cable coming from the electric meter can terminate immediately into the switch or panelboard when the cable enters the building. In the example home, however, there is no basement and the utility room is located in the center of the house with no outside walls. Consequently, a somewhat different arrangement will have to be used. A load center is a type of panelboard that is normally located at the service entrance of a residential installation. The load center usually contains a main circuit breaker, which is the main disconnect. Circuit breakers are provided for equipment such as electric water heaters, ranges, dryers, air conditioning and heating units, and breakers that feed subpanels such as lighting panels.

NEC Section 230.70 requires that the service disconnecting means be installed in a readily accessible location—either outside or inside the building. If located inside the building, it must be located nearest the point of entrance of the service conductors. In the sample home, there are at least two methods of installing the panelboard in the utility room that will comply with this *NEC®* regulation, as well as the requirements in *NEC Sections 110.26 and 240.24.*

The first method utilizes a weatherproof 100A disconnect (safety switch or circuit breaker enclosure) mounted next to the meter base on the outside of the building. With this method, service conductors are provided with overcurrent protection; the neutral conductor is also grounded at this point, as this becomes the main disconnect switch. Three-wire cable with an additional grounding wire is then routed from this main disconnect to the panelboard in the utility room. All three current-carrying conductors (two ungrounded and one neutral) must be insulated with this arrangement; the equipment ground, however, may be bare. The panelboard containing overcurrent protection devices for the branch circuits, which is located in the utility room, now becomes a subpanel. See *Figure 17.*

An alternate method utilizes conduit from the meter base that is routed under the concrete slab and then up to a main panelboard located in the utility room. *NEC Section 230.6* considers conductors to be outside of a building when they are installed under not less than 2" of concrete beneath a building or installed in a conduit not less than 18" deep beneath a building. The sample residence has a 4"-thick reinforced concrete slab—well within the *NEC®* regulations. Therefore, the service conductors from the meter base that are installed under the concrete slab in conduit are considered to be outside the house, and no disconnect is required at the meter base. When this conduit emerges in the utility room, it will run straight up into the bottom of the panelboard, again meeting the *NEC®* requirement that the panel be located nearest the point of entrance of the service conductors. Always check with your local authority having jurisdiction for specific requirements about where services are to be located. Details of this service arrangement are shown in *Figure 18.*

NOTE

Local ordinances in some areas may require a disconnect at the meter base, making the panel in the utility room a subpanel.

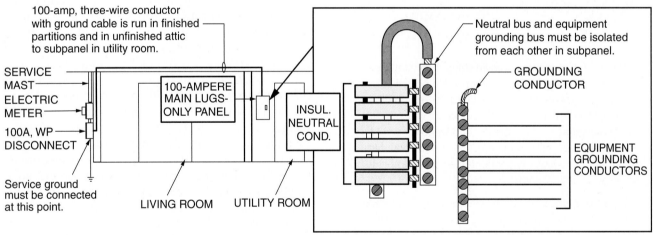

Figure 17 ◆ One method of wiring a panelboard for the sample residence.

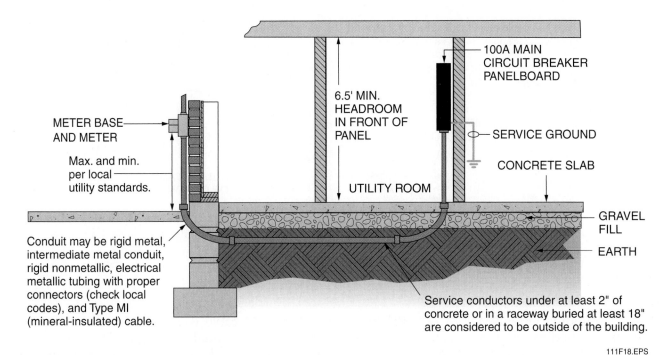

Figure 18 ◆ Alternate method of service installation for the sample residence.

8.0.0 ◆ WIRING METHODS

Branch circuits and feeders are used in residential construction to provide power wiring to operate components and equipment, and control wiring to regulate the equipment. Wiring may be further subdivided into either open or concealed wiring.

In open wiring systems, the cable and/or raceways are installed on the surface of the walls, ceilings, columns, and other areas where they are in view and are readily accessible. Open wiring is often used in areas where appearance is not important, such as in unfinished basements, attics, and garages.

Concealed wiring systems are installed inside walls, partitions, ceilings, columns, and behind baseboards or moldings where they are out of view and are not readily accessible. This type of wiring is generally used in all new construction with finished interior walls, ceilings, and floors, and it is the preferred type of wiring where appearance is important.

In general, there are two basic wiring methods used in the majority of modern residential electrical systems. They are:

- Sheathed cables of two or more conductors
- Raceway (conduit) systems

The method used on a given job is determined by the requirements of the *NEC*®, any amendments made by local authorities, the type of building construction, and the location of the wiring in the building. In most applications, either of the two methods may be used, and both methods are frequently used in combination.

8.1.0 Cable Systems

Several types of cable are used in wiring systems to feed or supply power to equipment. These include nonmetallic-sheathed cable, **metal-clad (MC) cable**, underground feeder cable, and service-entrance cable.

8.1.1 Nonmetallic-Sheathed Cable

Nonmetallic-sheathed (Type NM) cable *(NEC Article 334)* is manufactured in two- or three-wire configurations with varying sizes of conductors. In both two- and three-wire cables, conductors are color-coded: one conductor is black while the other is white in two-wire cable; in three-wire cable, the additional conductor is red. Both types also have a grounding conductor, which is usually bare, but it is sometimes covered with green plastic insulation, depending upon the manufacturer. The jacket or covering consists of rubber, plastic, or fiber. Most also have markings on this jacket giving the manufacturer's name or trademark, wire size, and number of conductors (see *Figure 19*). For example, NM 12-2 W/GRD indicates that the jacket contains two No. 12 AWG conductors along

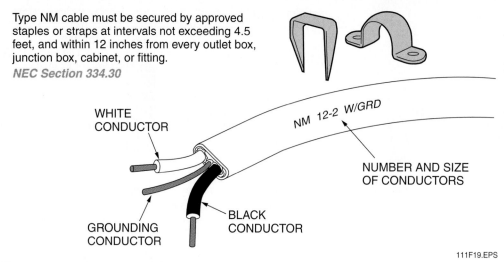

Type NM cable must be secured by approved staples or straps at intervals not exceeding 4.5 feet, and within 12 inches from every outlet box, junction box, cabinet, or fitting.
NEC Section 334.30

WHITE CONDUCTOR

NM 12-2 W/GRD

NUMBER AND SIZE OF CONDUCTORS

GROUNDING CONDUCTOR

BLACK CONDUCTOR

111F19.EPS

Figure 19 ◆ Characteristics of Type NM cable.

with a grounding wire; NM 12-3 W/GRD indicates three conductors plus a grounding wire. Type NM cable is often referred to as **Romex**®.

NEC Section 334.10 permits Type NM cable to be used in the following applications:

- One- and two-family dwelling units
- Multi-family dwellings when they are of Types III, IV, and V construction
- Other structures if concealed behind a 15-minute finish barrier and are of Types III, IV, and V construction

NEC Section 334.12 prohibits the use of Type NM cable in the following applications:

- As open runs in dropped or suspended ceilings in other than dwelling units
- As service-entrance cable
- In commercial garages that have hazardous (classified) areas
- In theaters and similar locations, except as permitted by *NEC Section 518.4*
- In motion picture studios
- In storage battery rooms
- In hoistways or on elevators or escalators
- Embedded in poured cement, concrete, or aggregate
- In hazardous (classified) areas
- Where exposed to corrosive fumes or vapors
- Embedded in masonry, adobe, fill, or plaster
- In a shallow chase in masonry, concrete, or adobe and covered with plaster, adobe, or similar finish
- Where exposed or subject to excessive moisture or dampness

Type NM cable is the most common type of cable for residential use. *Figure 20* shows additional *NEC*® regulations pertaining to the installation of Type NM cable.

8.1.2 Metal-Clad Cable

Metal-clad (MC) cable is manufactured in two-, three-, and four-wire assemblies with varying sizes of conductors, and is used in locations similar to those where Type NM cable is allowed. Unlike Type NM, it can also be used as service-entrance cable and in other locations permitted by *NEC Section 330.10*.

The metallic spiral covering on Type MC cable offers a greater degree of mechanical protection than Type NM cable and also provides a continuous grounding bond without the need for additional grounding conductors.

Type MC cable may be embedded in plaster finish, brick, or other masonry, except in damp or wet locations. It may also be run in the air voids of masonry block or tile walls, except where such walls are exposed or subject to excessive moisture or dampness. It may be used in wet locations if the conditions of *NEC Section 330.10(A)(11)* are met. It may not be used where subject to physical damage. See *Figures 21* and *22*.

NOTE

In the past, armored cable (Type AC), also called BX® cable, was commonly used in residential applications. Today, Type MC cable is used because it has a plastic wrapping to protect the conductors and does not require an insulating bushing at cable terminations.

Ampacity adjustments are required where cables are run in ambient temperatures over 86°F (such as hot attics) or where cable is embedded in insulation.
NEC Section 334.80

Where Type NM cable is run through wood joists where the edges of the bored hole is less than 1¼" from the nearest edge of the stud, or where studs are notched, a listed steel plate, or a plate not less than 1⁄16" must be used to protect the cables as shown.
NEC Sections 334.17 and 300.4(B)(1)

Where run across top of floor joists, in attic, and roof space, front edges of rafters or studs, NM cable must be protected by guard strips which are at least as high as the cable.
NEC Sections 334.23 and 320.23

Where the attic space or roof space is not accessible by permanent stairs or ladders, guard strips are required only within 6 feet of the nearest edge of the attic entrance.
NEC Sections 334.23 and 320.23

Where cable is carried along the sides of rafters, studs, or floor joists, neither guard strips nor running boards are required.
NEC Sections 334.23 and 320.23

Cables run through holes in wooden joists, rafters, or studs are considered to be supported without additional clamps or straps.
NEC Section 334.30(A)

Cable must be secured within 12" of every cabinet, box, or fitting.
NEC Section 334.30

4'-6"

NM cable must be secured in place at intervals not exceeding 4.5 feet.
NEC Section 334.30

Cables not smaller than two No. 6 AWG or three No. 8 AWG may be secured directly to the lower edges of joists in unfinished basements.
NEC Section 334.15(C)

Where run parallel to the framing members, cable may be secured to the sides of the framing members.
NEC Sections 334.17 and 300.4

Cables smaller than two No. 6 AWG that run on the bottom edge of floor joists in unfinished basements must be provided with a r unning board" and cable must be secured to it.
NEC Section 334.15(C)

Bends must not be less than five times the diameter of the cable.
NEC Section 334.24

Type NM cable may be installed in air voids in masonry block where such walls are not subject to excessive moisture or dampness.
NEC Section 334.10(A)(2)

111F20.EPS

Figure 20 ◆ *NEC*® sections governing the installation of Type NM cable.

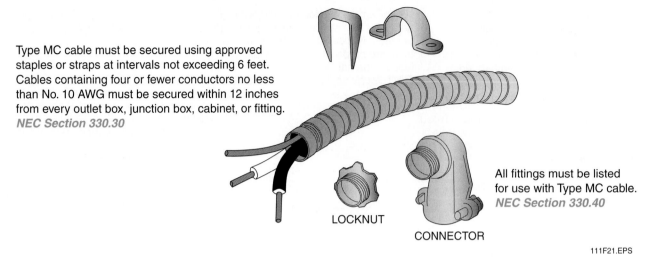

Type MC cable must be secured using approved staples or straps at intervals not exceeding 6 feet. Cables containing four or fewer conductors no less than No. 10 AWG must be secured within 12 inches from every outlet box, junction box, cabinet, or fitting. *NEC Section 330.30*

All fittings must be listed for use with Type MC cable. *NEC Section 330.40*

LOCKNUT

CONNECTOR

111F21.EPS

Figure 21 ◆ Characteristics of Type MC cable.

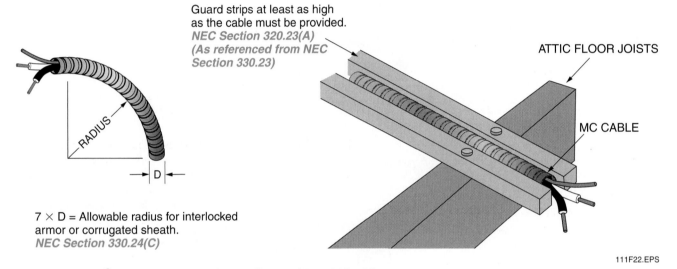

Guard strips at least as high as the cable must be provided. *NEC Section 320.23(A)* *(As referenced from NEC Section 330.23)*

ATTIC FLOOR JOISTS

MC CABLE

RADIUS

D

7 × D = Allowable radius for interlocked armor or corrugated sheath. *NEC Section 330.24(C)*

111F22.EPS

Figure 22 ◆ NEC® sections governing the installation of Type MC cable.

8.1.3 Underground Feeder Cable

Underground feeder (Type UF) cable *(NEC Article 340)* may be used underground, including direct burial in the earth, as a feeder or branch circuit cable when provided with overcurrent protection at the rated ampacity as required by the *NEC®*. When Type UF cable is used above grade where it will come in direct contact with the rays of the sun, its outer covering must be sun-resistant. Furthermore, where Type UF cable emerges from the ground, some means of mechanical protection must be provided. This protection may be in the form of conduit or guard strips. *NEC Section 300.5(D)(1)* requires that the protection extend from the minimum burial depth below grade to a point at least 8' above grade. *NEC Section 300.5(D)(4)* states that if conduit is used as protection, the permitted types are RMC, IMC, and Schedule 80 PVC, or equivalent. Type UF cable resembles Type NM cable; however, the jacket is constructed of weather-resistant material to provide the required protection for direct-burial wiring installations.

8.1.4 Service-Entrance Cable

Service-entrance (Type SE) and underground service-entrance (Type USE) cable, when used for electrical services, must be installed as specified in *NEC Articles 230 and 338*. Service-entrance cable is available with the grounded conductor bare for outside service conductors, and also with an insulated grounded conductor for interior wiring systems.

Type SE cable is permitted for use on branch circuits or feeders provided that all current-carrying conductors are insulated; this includes

Cable Stripping

Special strippers are used to remove the jackets from Type NM and Type MC cable.

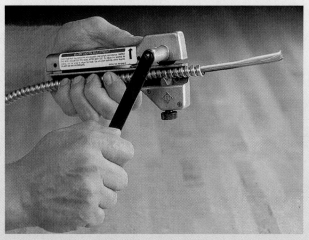

(A) NM CABLE RIPPER

(B) MC CABLE CUTTER

111SA03.EPS

the grounded or neutral conductor. Where a conductor in the cable is not insulated, it is only permitted to be used as an equipment grounding conductor for branch circuits or feeders. Where used as an interior wiring method, the installation requirements of *NEC Article 334* must be followed, except for determining the ampacity of the cable. Where installed as exterior wiring, the requirements of *NEC Article 225* must be met, with the supports for the cable in accordance with *NEC Section 334.30*.

SE Style R (SER) cable is used in residential applications for subfeeds for ranges, and it is also used for service laterals in multi-family dwellings.

Figure 23 summarizes the installation rules for Type SE cable for both exterior and interior wiring.

8.2.0 Raceways

A raceway is any channel that is designed and used solely for the purpose of holding wires, cables, or busbars. Types of raceways include rigid metal conduit, intermediate metal conduit, rigid nonmetallic conduit, flexible metallic conduit, electrical metallic tubing, and auxiliary gutters. Raceways are constructed of either metal or insulating material, such as polyvinyl chloride or PVC (plastic). Metal raceways are joined using threaded, compression, or setscrew couplings; nonmetallic raceways are joined using cement-coated couplings. Where a raceway terminates in an outlet box, junction box, or other enclosure, an approved connector must be used.

Raceways provide mechanical protection for the conductors that run in them and also prevent accidental damage to insulation and the conducting material. They also protect conductors from corrosive atmospheres and prevent fire hazards to life and property by confining arcs and flames that may occur due to faults in the wiring system. Conduits or raceways are used in residential applications for service masts, underground wiring embedded in concrete, and sometimes in unfinished basements, shops, or garage areas.

Another function of metal raceways is to provide a continuous equipment grounding system throughout the electrical system. To maintain this feature, it is extremely important that all raceway systems be securely bonded together into a continuous conductive path and properly connected to the system ground. The following section explains how this is accomplished.

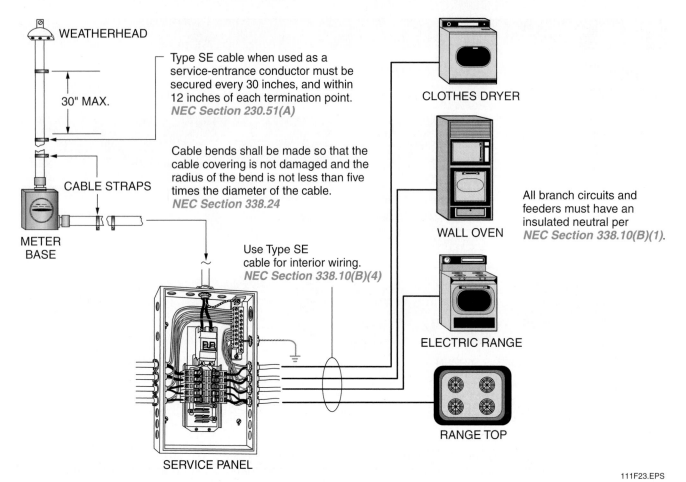

WEATHERHEAD

30" MAX.

Type SE cable when used as a service-entrance conductor must be secured every 30 inches, and within 12 inches of each termination point. *NEC Section 230.51(A)*

CABLE STRAPS

Cable bends shall be made so that the cable covering is not damaged and the radius of the bend is not less than five times the diameter of the cable. *NEC Section 338.24*

METER BASE

Use Type SE cable for interior wiring. *NEC Section 338.10(B)(4)*

CLOTHES DRYER

WALL OVEN

All branch circuits and feeders must have an insulated neutral per *NEC Section 338.10(B)(1).*

ELECTRIC RANGE

RANGE TOP

SERVICE PANEL

111F23.EPS

Figure 23 ◆ *NEC® sections governing Type SE cable.*

9.0.0 ◆ EQUIPMENT GROUNDING SYSTEM

NEC Article 250, Part IV generally requires that all metallic enclosures, raceways, and cable armor be grounded. The exceptions in *NEC Sections 250.80 and 250.86* allow metal enclosures or short sections of raceways that are used to provide support or physical protection to be ungrounded under specific conditions.

NEC Article 250, Part VI covers equipment grounding and equipment grounding conductors. This section generally requires that the exposed noncurrent-carrying metal parts of fixed equipment likely to become energized be grounded under the following conditions:

- Where within 8' vertically or 5' horizontally of ground or grounded metal objects and subject to contact by occupants or others
- Where located in wet or damp locations

- Where in electrical contact with metal
- Where in hazardous (classified) locations as covered by *NEC Articles 500 through 517*
- Where supplied by a metal-clad, metal-sheathed, or metal raceway, or other wiring method that provides an equipment ground
- Where equipment operates with any terminal at over 150V to ground

Specific equipment that is required to be grounded regardless of the voltage is listed in *NEC Section 250.112* and includes equipment such as motors, motor controllers, and light fixtures. Types of cord- and plug-connected equipment in dwelling units that are required to be grounded are found in *NEC Section 250.114* and include equipment such as refrigerators, freezers, air conditioners, information technology equipment (computers), clothes washers, clothes dryers, and dishwashing machines.

The types of equipment grounding conductors that are acceptable to be used are found in *NEC Section 250.118.* Note that among the list of wiring methods approved for use as equipment grounding conductors, both flexible metal conduit (FMC) and liquidtight flexible metal conduit (LFMC) are permitted to be used. Listed FMC is permitted to be used as an equipment grounding conductor only when the following conditions are met:

• The conduit is terminated in fittings listed for grounding.
• The circuit conductors contained in the conduit are protected by overcurrent devices rated at 20A or less.
• The combined length of FMC, FMT, and LFMC in the same ground return path does not exceed 6'.
• The conduit is not installed for flexibility.

Type LFMC is also used in dwelling units and has slightly different requirements when used as an equipment grounding conductor:

• The conduit is terminated in fittings listed for grounding.
• For trade sizes ⅜ through ½, the circuit conductors contained in the conduit are protected by overcurrent devices rated at 20A or less.
• For trade sizes ¾ through 1¼, the circuit conductors contained in the conduit are protected by overcurrent devices rated at 60A or less and there is no FMC, FMT, or LFMC in trade sizes ⅜ through ½ in the grounding path.

• The combined length of FMC, FMT, and LFMC in the same ground return path does not exceed 6'.
• The conduit is not used for flexibility.

Where external bonding jumpers are used to provide the continuity of the fault current path, *NEC Section 250.102(E)* limits the length to not more than 6', except at outside pole locations for the purposes of bonding or grounding the isolated sections of metal raceways or elbows installed in exposed risers at those pole locations. When installing an equipment grounding conductor in a raceway, *NEC Table 250.122* is used to determine the size of the equipment grounding conductor. It is permitted to install one equipment grounding conductor in a raceway that has several circuits. In that case, the size of the equipment grounding conductor is based on the rating of the largest overcurrent device protecting the circuits contained in the raceway.

NEC Section 250.148 requires that where circuit conductors are spliced within a box, or terminated on equipment within or supported by a box, separate equipment grounding conductors associated with those circuit conductors shall be spliced or joined within the box or to the box with devices suitable for the use. *Figure 24* shows several types of fittings that are suitable for this purpose.

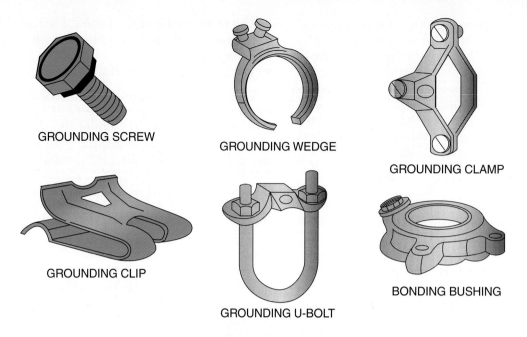

GROUNDING SCREW

GROUNDING WEDGE

GROUNDING CLAMP

GROUNDING CLIP

GROUNDING U-BOLT

BONDING BUSHING

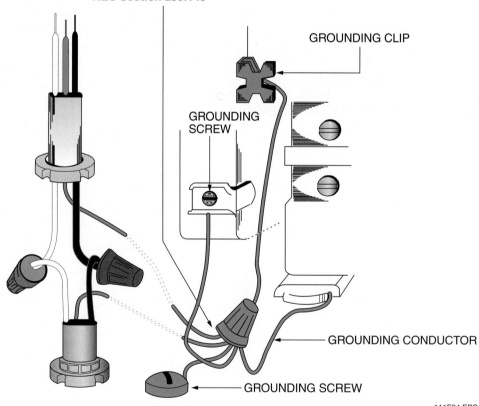

Where splices are made in
a junction box, the grounding
conductors must be spliced to
the metal junction box.
NEC Section 250.148

GROUNDING CLIP

GROUNDING
SCREW

GROUNDING CONDUCTOR

GROUNDING SCREW

111F24.EPS

Figure 24 ◆ Equipment grounding methods.

10.0.0 ◆ BRANCH CIRCUIT LAYOUT FOR POWER

The point at which electrical equipment is connected to the wiring system is commonly called an outlet. There are many classifications of outlets: lighting, receptacle, motor, appliance, and so forth. This section, however, deals with the power outlets normally found in residential electrical wiring systems.

When viewing an electrical drawing, outlets are indicated by symbols (usually a small circle with appropriate markings to indicate the type of outlet). The most common symbols for receptacles are shown in *Figure 25*.

10.1.0 Branch Circuits and Feeders

The conductors that extend from the panelboard to the various outlets are called branch circuits and are defined by the *NEC*® as the point of a wiring system that extends beyond the final overcurrent device protecting the circuit. See *Figure 26*.

A feeder consists of all conductors between the service equipment and the final overcurrent device. See *Figure 27*.

In general, the size of the branch circuit conductors varies depending upon the load requirements of the electrically operated equipment connected to the outlet. For residential use, most branch circuits consist of either No. 14 AWG, No. 12 AWG, No. 10 AWG, or No. 8 AWG conductors.

The basic branch circuit requires two wires or conductors to provide a continuous path for the flow of electric current, plus a third wire for equipment grounding. The usual receptacle branch circuit operates at 120V.

Fractional horsepower motors and small electric heaters usually operate at 120V and are connected to 120V branch circuits by means of a receptacle, junction box, or direct connection.

With the exception of very large residences and tract-development houses, the size of the average residential electrical system of the past has not been large enough to justify the expense of preparing complete electrical working drawings and specifications. Such electrical systems were usually laid out by the architect in the form of a sketchy outlet arrangement or else laid out by the electrician on the job, often only as the work progressed. However, many technical developments in residential electrical use—such as electric heat

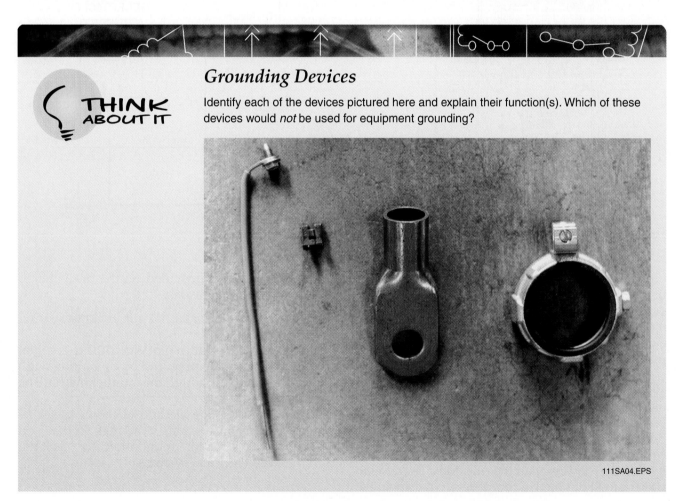

Grounding Devices

Identify each of the devices pictured here and explain their function(s). Which of these devices would *not* be used for equipment grounding?

THINK ABOUT IT

111SA04.EPS

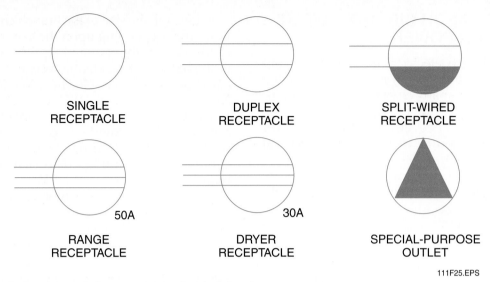

SINGLE
RECEPTACLE

DUPLEX
RECEPTACLE

SPLIT-WIRED
RECEPTACLE

50A

RANGE
RECEPTACLE

30A

DRYER
RECEPTACLE

SPECIAL-PURPOSE
OUTLET

111F25.EPS

Figure 25 ◆ Typical outlet symbols appearing in electrical drawings.

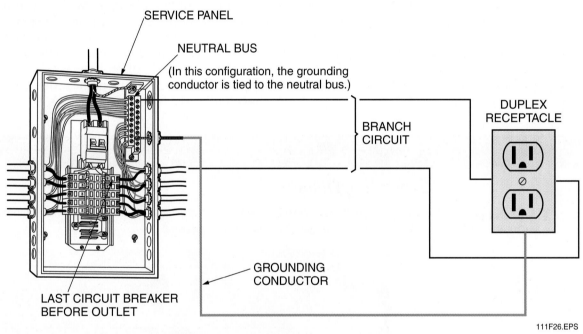

SERVICE PANEL

NEUTRAL BUS

(In this configuration, the grounding
conductor is tied to the neutral bus.)

BRANCH
CIRCUIT

DUPLEX
RECEPTACLE

GROUNDING
CONDUCTOR

LAST CIRCUIT BREAKER
BEFORE OUTLET

111F26.EPS

Figure 26 ◆ Components of a duplex receptacle branch circuit.

with sophisticated control wiring, increased use of electrical appliances, various electronic alarm systems, new lighting techniques, and the need for energy conservation techniques—have greatly expanded the demand and extended the complexity of today's residential electrical systems.

Each year, the number of homes with electrical systems designed by consulting engineering firms increases. Such homes are provided with complete electrical working drawings and specifications, similar to those frequently provided for commercial and industrial projects. Still, these are more the exception than the rule. Most residential projects will not have a complete set of drawings.

Circuit layout is provided on the drawings to follow for several reasons:

• They provide a visual layout of house wiring circuitry.
• They provide a sample of electrical residential drawings that are prepared by consulting engineering firms, although the number may still be limited.
• They introduce the method of showing electrical systems on working drawings so that you will have a better foundation for tackling advanced electrical systems.

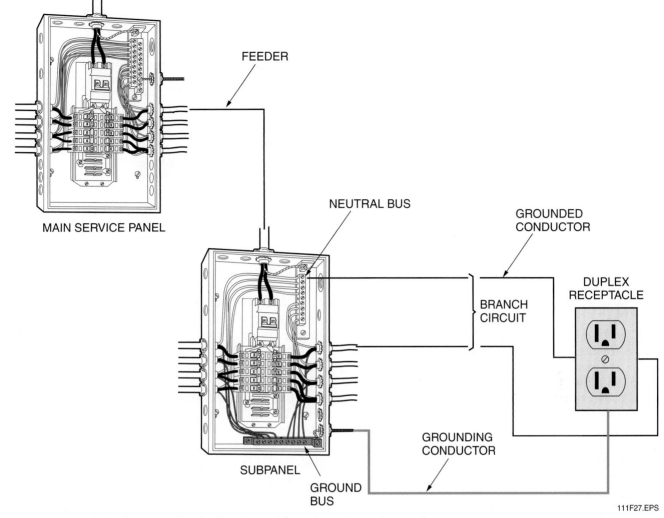

FEEDER

MAIN SERVICE PANEL

NEUTRAL BUS

GROUNDED
CONDUCTOR

BRANCH
CIRCUIT

DUPLEX
RECEPTACLE

SUBPANEL

GROUND
BUS

GROUNDING
CONDUCTOR

111F27.EPS

Figure 27 ◆ A feeder being used to feed a subpanel from the main service panel.

Branch circuits are shown on electrical drawings by means of a single line drawn from the panelboard (or by homerun arrowheads indicating that the circuit goes to the panelboard) to the outlet or from outlet to outlet where there is more than one outlet on the circuit.

The lines indicating branch circuits can be solid to show that the conductors are to be run concealed in the ceiling or wall, dashed to show that the conductors are to be run in the floor or ceiling below, or dotted to show that the wiring is to be run exposed. *Figure 28* shows examples of these three types of branch circuit lines.

In *Figure 28*, No. 12 indicates the wire size. The slash marks shown through the circuits in *Figure 28* indicate the number of current-carrying conductors in the circuit. Although two slash marks are shown, in actual practice, a branch circuit containing only two conductors usually contains no slash marks; that is, any circuit with no slash marks is assumed to have two conductors. However, three or more conductors are always indicated on electrical working drawings—either by slash marks for each conductor, or else by a note.

Never assume that you know the meaning of any electrical symbol. Although great efforts have been made in recent years to standardize drawing symbols, architects, consulting engineers, and electrical drafters still modify existing symbols or devise new ones to meet their own needs. Always consult the symbol list or legend on electrical working drawings for an exact interpretation of the symbols used.

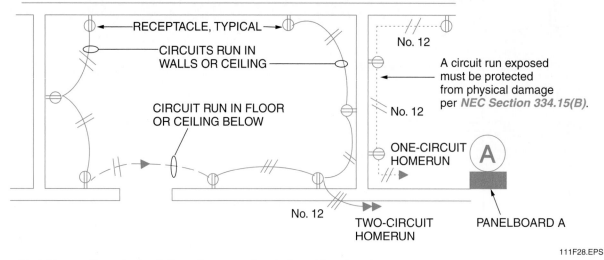

Figure 28 ◆ Types of branch circuit lines shown on electrical working drawings.

111F28.EPS

10.2.0 Locating Receptacles

NEC Section 210.52 states the minimum requirements for the location of receptacles in dwelling units. It specifies that in each kitchen, family room, and dining room, receptacle outlets shall be installed so that no point along the floor line in any wall space is more than 6', measured horizontally, from an outlet in that space, including any wall space 2' or more in width and the wall space occupied by fixed panels in exterior walls, but excluding sliding panels. Receptacle outlets shall, insofar as practicable, be spaced equal distances apart. Receptacle outlets in floors shall not be counted as part of the required number of receptacle outlets unless located within 18" of the wall.

The *NEC®* defines wall space as a wall that is unbroken along the floor line by doorways, fireplaces, or similar openings. Each wall space that is two feet or more in width must be treated individually and separately from other wall spaces within the room.

The purpose of *NEC Section 210.52* is to minimize the use of cords across doorways, fireplaces, and similar openings.

With this *NEC®* requirement in mind, outlets for our sample residence will be laid out (see *Figure 29*). In laying out these receptacle outlets, the floor line of the wall is measured (also around corners), but not across doorways, fireplaces, passageways, or other spaces where a flexible cord extended across the space would be unsuitable.

In general, duplex receptacle outlets must be no more than 12' apart. When spaced in this manner, a 6' extension cord will reach a receptacle from any point along the wall line.

Note that at no point along the wall line are any receptacles more than 12' apart or more than six

feet from any door or room opening. Where practical, no more than eight receptacles are connected to one circuit. However, this is just a design consideration since general-purpose receptacles in dwelling units are sized on the basis of 3VA per square foot of dwelling space. A 15A branch circuit is rated at 1,800VA (15A × 120V = 1,800VA) and the *NEC®* requires that for every 600 square feet (1,800VA ÷ 3VA/sq. ft. = 600 sq. ft.), a circuit to supply lighting and receptacles must be installed. Always check with the local authorities about the requirements for the number of branch circuits in a dwelling.

The utility room has at least one receptacle for the laundry on a separate circuit in order to comply with *NEC Sections 210.11(C)(2) and 210.52(F)*.

One duplex receptacle is located in the vestibule for cleaning purposes, such as feeding a portable vacuum cleaner or similar appliance. It is connected to the living room circuit. An additional duplex receptacle is required per *NEC Section 210.52(H)* in hallways of 10' or more.

Although this is not shown in the figure, the living room outlets could be split-wired (the lower half of each duplex receptacle is energized all the time, while the upper half can be switched on or off). The reason for this is that a great deal of the illumination for this area will be provided by portable table lamps, and the split-wired receptacles provide a means to control these lamps from several locations, such as at each entry to the living room, if desired. Split receptacles are discussed in more detail in the next section.

To comply with *NEC Sections 210.11(C)(1) and 210.52(B)*, the kitchen receptacles are laid out as follows. In addition to the number of branch circuits determined previously, two or more 20A small appliance branch circuits must be provided

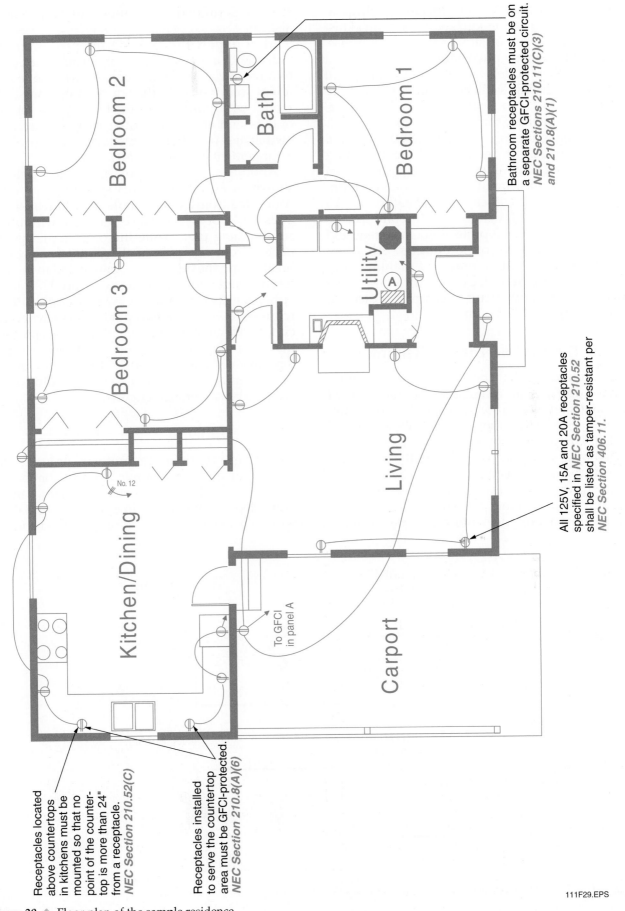

Bathroom receptacles must be on a separate GFCI-protected circuit. *NEC Sections 210.11(C)(3) and 210.8(A)(1)*

Bedroom 2

Bath

Bedroom 1

Bedroom 3

Utility

Ⓐ

Living

Kitchen/Dining

No. 12

To GFCI in panel A

Carport

All 125V, 15A and 20A receptacles specified in *NEC Section 210.52* shall be listed as tamper-resistant per *NEC Section 406.11.*

Receptacles located above countertops in kitchens must be mounted so that no point of the countertop is more than 24" from a receptacle. *NEC Section 210.52(C)*

Receptacles installed to serve the countertop area must be GFCI-protected. *NEC Section 210.8(A)(6)*

Figure 29 ◆ Floor plan of the sample residence.

111F29.EPS

to serve all receptacle outlets (including refrigeration equipment) in the kitchen, pantry, breakfast room, dining room, or similar area of the house. Such circuits, whether two or more are used, must have no other outlets connected to them. All receptacles serving a kitchen countertop require GFCI protection. No small appliance branch circuit shall serve more than one kitchen.

To comply with *NEC Sections 210.11(C)(3) and 210.52(D),* bathroom receptacle(s) must be on a separate branch circuit supplying only bathroom receptacles or on a circuit supplying a single bathroom with no loads other than that bathroom. All receptacles located within a bathroom require GFCI protection. GFCI protection is also required on garage and exterior receptacles. All other branch circuits that supply the lighting and general-purpose receptacles in dwelling units must have arc fault circuit interrupter protection to comply with *NEC Section 210.12.*

10.3.0 Split-Wired Duplex Receptacles

In modern residential construction, it is common to have duplex wall receptacles that have one of the outlets wired as a standard duplex outlet (hot all the time) and the other half controlled by a wall switch. This allows table or floor lamps to be controlled by a wall switch and leaves the other outlet available for items that are not to be switched. This wiring method is commonly referred to as a split receptacle. Note that switched receptacles are installed to provide lighting. Dimmer switches

are not permitted to be used per *NEC Section 404.14(E).*

Most duplex 15A and 20A receptacles are provided with a breakoff tab that permits each of the two receptacle outlets to be supplied from a different source or polarity. For example, one outlet would be supplied from the hot leg of a series of outlets and the other outlet supplied from the **switch leg** of a light switch. A diagram of this arrangement is shown in *Figure 30.*

Another application of split receptacles is shown in *Figure 31.* In this example, one outlet connected from a double-pole circuit breaker supplies 240V for an appliance such as a window air conditioning unit, while the other outlet is connected from one pole of the double-pole circuit breaker and the other side is connected to the neutral or grounded conductor to supply 120V for an appliance such as a lamp. *NEC Section 210.4(B)* requires the use of a two-pole breaker when two circuits are connected to one duplex receptacle so that all ungrounded conductors of the circuit are disconnected simultaneously. This circuit and the split receptacle mentioned above are both considered multiwire branch circuits.

10.4.0 Multiwire Branch Circuits

NEC Article 100 defines a multiwire branch circuit as "two or more ungrounded conductors having a potential difference between them, and a grounded conductor having equal potential difference between it and each ungrounded conduc-

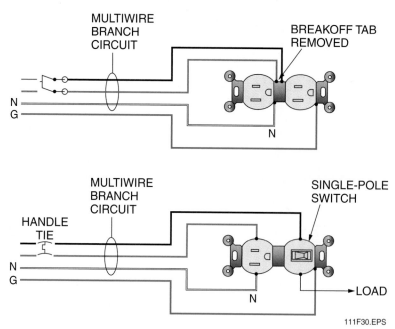

Figure 30 ◆ Two 120V receptacle outlets supplied from different sources.

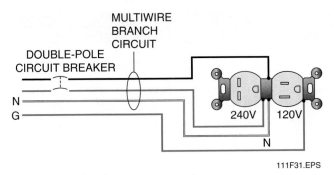

Figure 31 ◆ Combination receptacle.

111F31.EPS

tor of the circuit and that is connected to the neutral conductor of the system."

Multiwire branch circuits have many advantages, such as three wires doing the work of four (in place of two two-wire branch circuits), less raceway fill, easier balancing and phasing of a system, and less voltage drop. See *NEC Section 210.4(B)*.

10.5.0 240-Volt Circuits

The electric range, clothes dryer, and water heater in the sample residence all operate at 240VAC. Each will be fed by a separate circuit and connected to a two-pole circuit breaker of the appropriate rating in the panelboard. To determine the conductor size and overcurrent protection for the range, proceed as follows:

Step 1 Find the nameplate rating of the electric range. This has previously been determined to be 12kVA.

Step 2 Refer to *NEC Table 220.55*. Since Column A of this table applies to ranges rated at 12kVA (12kW) and under, this will be the column to use in this example.

Step 3 Under the Number of Appliances column, locate the appropriate number of appliances (one in this case), and find the maximum demand given for it in Column A. Column A states that the circuit should be sized for 8kVA (not the nameplate rating of 12kVA).

Step 4 Calculate the required conductor ampacity as follows:

$$\frac{8,000VA}{240V} = 33.33A$$

The minimum branch circuit must be rated at 40A since common residential circuit breakers are rated in steps of 15A, 20A, 30A, 40A, and so forth. A 30A circuit breaker is too small, so a 40A circuit breaker is selected. The conductors must have a current-carrying capacity that is equal to or greater than the overcurrent protection. Therefore, No. 8 AWG conductors will be used.

If a cooktop and wall oven were used instead of the electric range, the circuit would be sized similarly. The *NEC*® specifies that a branch circuit for a counter-mounted cooking unit and not more than two wall-mounted ovens, all supplied from a single branch circuit and located in the same room, is computed by adding the nameplate ratings of the individual appliances and treating this total as equivalent to one range. Therefore, two appliances of 6kVA each may be treated as a single range with a 12kVA nameplate rating.

Figure 32 shows how the electric range circuit may appear on an electrical drawing. The connection may be made directly to the range junction box, but more often a 50A range receptacle is mounted at the range location and a range cord-and-plug set is used to make the connection. This facilitates moving the appliance later for maintenance or cleaning.

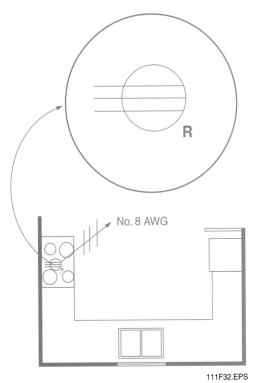

No. 8 AWG

111F32.EPS

Figure 32 ◆ Range circuit shown on an electrical drawing.

240V Circuits

THINK ABOUT IT

Calculate the ampacity required for a kitchen range with an 8kW rating. Now design the practical wiring in a labeled diagram. How will the wires be connected at the service panel and at the appliance? How will the cable be installed?

Figure 33 shows several types of receptacle configurations used in residential wiring applications. You will eventually recognize these configurations at a glance.

The branch circuit for the water heater in the sample residence must be sized for its full capacity because there is no diversity or demand factor for this appliance. Since the nameplate rating on the water heater indicates two heating elements of 4,500W each, the first inclination would be to size the circuit for a total load of 9,000W (volt-amperes). However, only one of the two elements operates at a time. See *Figure 34*. Note that each element is controlled by a separate thermostat. The lower element becomes energized when the thermostat calls for heat, and at the same time, the thermostat opens a set of contacts to prevent the upper element from operating. When the lower element's thermostat is satisfied, the lower contacts open, and at the same time, the thermostat closes the contacts for the upper element to become energized to maintain the water temperature.

With this information in hand, the circuit for the water heater may be sized as follows:

$$\frac{4{,}500\text{VA}}{240\text{V}} = 18.75\text{A} \times 1.25 = 23.44\text{A}$$

NEC Section 422.13 requires that the branch circuits that supply storage type water heaters having a capacity of 120 gallons or less be rated not less than 125% of the nameplate rating of the water heater. Our calculation shows this to be not

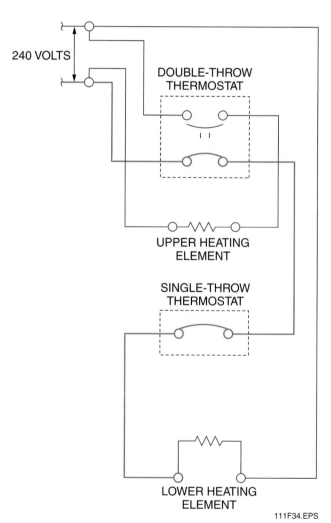

15 Amp, 125 Volts	
20 Amp, 125 Volts	
20 Amp, 250 Volts	
30 Amp, 125 Volts	
30 Amp, 250 Volts	
30 Amp, 125/250 Volts	
50 Amp, 250 Volts	
50 Amp, 125/250 Volts	

111F33.EPS

Figure 33 ◆ Residential receptacle configurations.

240 VOLTS

DOUBLE-THROW THERMOSTAT

UPPER HEATING ELEMENT

SINGLE-THROW THERMOSTAT

LOWER HEATING ELEMENT

111F34.EPS

Figure 34 ◆ Wiring diagram of water heater controls.

less than 23A. Normally, this would require a maximum rating for the branch circuit to be not more than 25A. (See standard ratings of overcurrent devices in *NEC Section 240.6.*) However, *NEC Section 422.11(E)* permits a single nonmotor-operated appliance to be protected by overcurrent devices rated up to 150% of the nameplate rating of the appliance. In this case, 4,500VA ÷ 240V = 18.75A × 150% = 28.125A. Since the next standard rating is 30A, the water heater will be wired with No. 10 AWG conductors protected by a 30A overcurrent device.

The *NEC* specifies that electric clothes dryers must be rated at 5kVA or the nameplate rating, whichever is greater. In this case, the dryer is rated at 5.5kVA, and the conductor current-carrying capacity is calculated as follows:

$$\frac{5,500VA}{240V} = 22.92A$$

A three-wire, 30A circuit will be provided (No. 10 AWG wire). It is protected by a 30A circuit breaker. The dryer may be connected directly, but a 30A dryer receptacle is normally provided for the same reasons as mentioned for the electric range.

Large appliance outlets rated at 240V are frequently shown on electrical drawings using lines and symbols to indicate the outlets and circuits. In some cases, no drawings are provided.

11.0.0 ◆ BRANCH CIRCUIT LAYOUT FOR LIGHTING

A simple lighting branch circuit requires two conductors to provide a continuous path for current flow. The usual lighting branch circuit operates at 120V; the white (grounded) circuit conductor is therefore connected to the neutral bus in the panelboard, while the black (ungrounded) circuit conductor is connected to an overcurrent protection device.

Lighting branch circuits and outlets are shown on electrical drawings by means of lines and symbols; that is, a single line is drawn from outlet to outlet and then terminated with an arrowhead to indicate a homerun to the panelboard. Several methods are used to indicate the number and size of conductors, but the most common is to indicate the number of conductors in the circuit by using slash marks through the circuit lines and then

indicate the wire size by a notation adjacent to these slash marks. For example, two slash marks indicate two conductors; three slash marks indicate three conductors. Some electrical designers omit slash marks for two-conductor circuits. In this case, the conductor size is usually indicated in the symbol list or legend.

The circuits used to feed residential lighting must conform to standards established by the *NEC®* as well as by local and state ordinances. Most of the lighting circuits should be calculated to include the total load, although at times this is not possible because the electrician cannot be certain of the exact wattage that might be used by the homeowner. For example, an electrician may install four porcelain lampholders for the unfinished basement area, each to contain one 100-watt (100W) incandescent lamp. However, the homeowners may eventually replace the original lamps with others rated at 150W or even 200W. Thus, if the electrician initially loads the lighting circuit to full capacity, the circuit will probably become overloaded in the future.

It is recommended that no residential branch circuit be loaded to more than 80% of its rated capacity. Since most circuits used for lighting are rated at 15A, the total ampacity (in volt-amperes) for the circuit is as follows:

$$15A \times 120V = 1,800VA$$

Therefore, if the circuit is to be loaded to only 80% of its rated capacity, the maximum initial connected load should be no more than 1,440VA.

Figure 35 shows one possible lighting arrangement for the sample residence. All lighting fixtures are shown in their approximate physical location as they should be installed.

Electrical symbols are used to show the fixture types. Switches and lighting branch circuits are also shown by appropriate lines and symbols. The meanings of the symbols used on this drawing are explained in the symbol list in *Figure 36.*

In actual practice, the location of lighting fixtures and their related switches will probably be the extent of the information shown on working drawings. The circuits shown in *Figure 35* are meant to illustrate how lighting circuits are routed, not to imply that such drawings are typical for residential construction. If fixtures are used in a closet, they must meet the requirements of *NEC Section 410.16.*

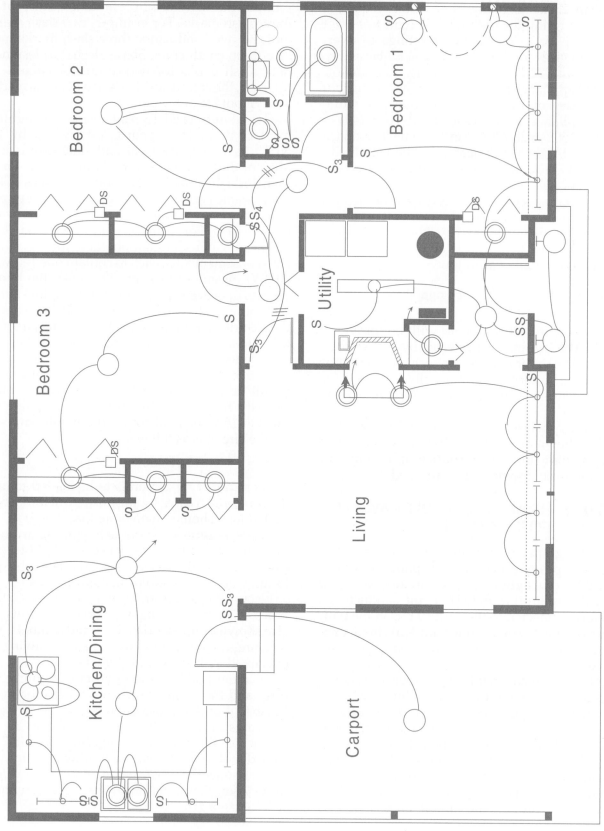

Figure 35 ◆ Lighting layout of the sample residence.

111F35.EPS

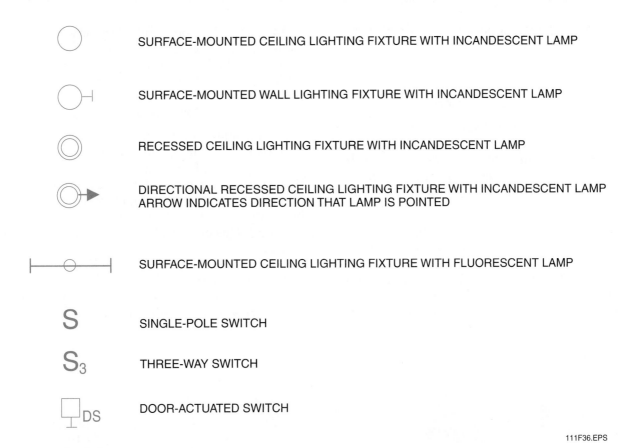

SURFACE-MOUNTED CEILING LIGHTING FIXTURE WITH INCANDESCENT LAMP

SURFACE-MOUNTED WALL LIGHTING FIXTURE WITH INCANDESCENT LAMP

RECESSED CEILING LIGHTING FIXTURE WITH INCANDESCENT LAMP

DIRECTIONAL RECESSED CEILING LIGHTING FIXTURE WITH INCANDESCENT LAMP
ARROW INDICATES DIRECTION THAT LAMP IS POINTED

SURFACE-MOUNTED CEILING LIGHTING FIXTURE WITH FLUORESCENT LAMP

S   SINGLE-POLE SWITCH

S₃   THREE-WAY SWITCH

DS   DOOR-ACTUATED SWITCH

111F36.EPS

Figure 36 ◆ Symbols.

12.0.0 ◆ OUTLET BOXES

Electricians installing residential electrical systems must be familiar with outlet box capacities, means of supporting outlet boxes, and other requirements of the *NEC®*. Boxes were discussed in detail in an earlier module, but a general review of the rules and necessary calculations is provided here.

The maximum numbers of conductors of the same size permitted in standard outlet boxes are listed in *NEC Table 314.16(A).* These figures apply where no fittings or devices such as fixture studs, cable clamps, switches, or receptacles are contained in the box and where no grounding conductors are part of the wiring within the box. Obviously, in all modern residential wiring systems there will be one or more of these items contained in every outlet box installed. Therefore, where one or more of the above-mentioned items are present, the total number of conductors will be less than that shown in the table. Also, if the box contains a looped, unbroken conductor 12" or more in length, it must be counted twice.

For example, a deduction of two conductors must be made for each strap containing a wiring device entering the box (based on the largest size conductor connected to the device) such as a switch or duplex receptacle; a further deduction of one conductor must be made for one or more equipment grounding conductors entering the box (based on the largest size grounding conductor). For instance, a 3-inch × 2-inch × 2¾-inch box is listed in the table as containing a maximum of six No. 12 wires. If the box contains cable clamps and a duplex receptacle, three wires will have to be deducted from the total of six, providing for only three No. 12 wires. If a ground wire is used, which is always the case in residential wiring, only two No. 12 wires may be used.

For example, to size a metallic outlet box for two No. 12 AWG conductors with a ground wire, cable clamp, and receptacle, proceed as follows:

Step 1 Calculate the total number of conductors and equivalents [*NEC Section 314.16(B)*]. One ground wire plus one cable clamp plus one receptacle (two wires) plus two No. 12 conductors equals a total of six No. 12 conductors.

Step 2 Determine the amount of space required for each conductor. *NEC Table 314.16(B)* gives the box volume required for each conductor. No. 12 AWG equals 2.25 cubic inches.

Step 3 Calculate the outlet box space required by multiplying the number of cubic inches required for each conductor by the total number of conductors:

6 × 2.25 = 13.5 cubic inches

Step 4 Once you have determined the required box capacity, again refer to *NEC Table 314.16(A)* and note that a 3-inch × 2-inch × 2¾-inch box comes closest to our requirements. This box is rated for 14 cubic inches.

Now, size the box for two additional conductors. Where four No. 12 conductors enter the box with two ground wires, only the two additional No. 12 conductors must be added to our previous count for a total of 8 conductors (6 + 2 = 8). Remember, any number of ground wires in a box counts as only one conductor; any number of cable clamps also counts as only one conductor. Therefore, the box size required for use with two additional No. 12 conductors may be calculated as follows:

8 × 2.25 = 18 cubic inches

Again, refer to *NEC Table 314.16(A)* and note that a 3-inch × 2-inch × 3½-inch device box with a rated capacity of 18.0 cubic inches is the closest device box that meets *NEC®* requirements. An alternative is to use a 4-inch × 1¼-inch square box with a single-gang plaster ring, as shown in *Figure 37*. This box also has a capacity of 18.0 cubic inches.

Other box sizes are calculated in a similar fashion. When sizing boxes for different size conductors, remember that the box capacity varies as shown in *NEC Table 314.16(B)*.

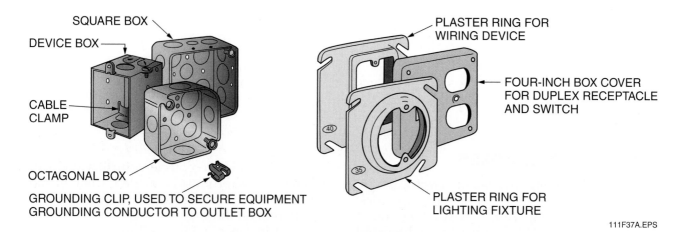

111F37A.EPS

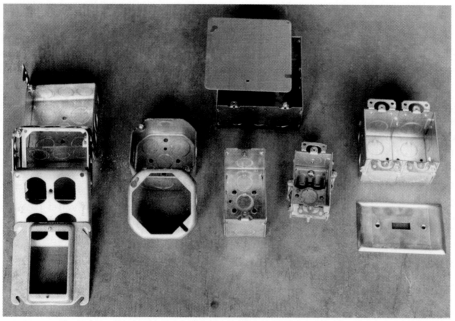

111F37B.EPS

Figure 37 ◆ Typical metallic outlet boxes with extension (plaster) rings.

Calculating Conductors

In a 4" × 4" × 1½" metal box, one 14/3 cable with ground feeds three 14/2 cables with ground wires. The red wire of the 14/3 cable feeds a receptacle, and the black wire feeds the 14/2 black wires. All of the white wires are spliced together, with one brought out to the receptacle terminal. The ground wires are all spliced, with one brought out to the grounding terminal on the receptacle and one to the ground clip on the box. All four cables are connected with box connectors rather than internal clamps. Using *NEC Section 314.16,* decide whether this wiring violates the code.

111SA05.EPS

12.1.0 Mounting Outlet Boxes

Outlet box configurations are almost endless, and if you research the various methods of mounting these boxes, you will be astonished. In this section, some common outlet boxes and their mounting considerations will be reviewed.

The conventional metallic device box, which is used for residential duplex receptacles and switches for lighting control, may be mounted to wall studs using 16d (penny) nails placed through the round mounting holes passing through the interior of the box. The nails are then driven into the wall stud. When nails are used for mounting outlet boxes in this manner, the nails must be located within ¼" of the back or ends of the enclosure.

Nonmetallic boxes normally have mounting nails fitted to the box for mounting. Other boxes have mounting brackets. When mounting outlet boxes with brackets, use either wide-head roofing nails or box nails about 1¼" in length. *Figure 38* shows various methods of mounting outlet boxes.

Before mounting any boxes during the rough wiring process, first find out what type and thickness of finish will be used on the walls. This will dictate the depth to which the boxes must be mounted to comply with *NEC®* regulations. For example, the finish on plastered walls or ceilings is normally ½" thick; gypsum board or drywall is either ½" or ⅝" thick; and wood paneling is normally only ¼" thick. (Some tongue-and-groove wood paneling is ½" to ⅝" thick.)

The *NEC®* specifies the amount of space permitted from the edge of the outlet box to the finished wall. When a noncombustible wall finish (such as plaster, masonry, or tile) is used, the box may be recessed ¼". However, when combustible finishes are used (such as wood paneling), the box must be flush (even) with the finished wall or ceiling. See *Figure 39* and *NEC Section 314.20.*

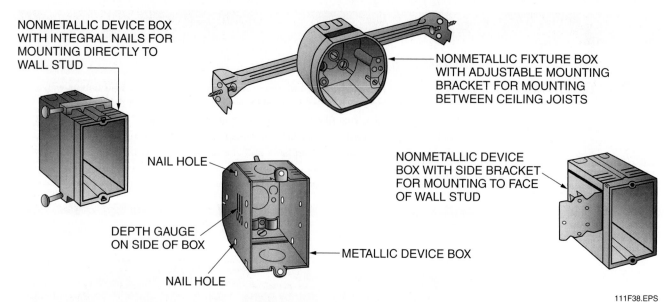

NONMETALLIC DEVICE BOX WITH INTEGRAL NAILS FOR MOUNTING DIRECTLY TO WALL STUD

NONMETALLIC FIXTURE BOX WITH ADJUSTABLE MOUNTING BRACKET FOR MOUNTING BETWEEN CEILING JOISTS

NAIL HOLE

NONMETALLIC DEVICE BOX WITH SIDE BRACKET FOR MOUNTING TO FACE OF WALL STUD

DEPTH GAUGE ON SIDE OF BOX

METALLIC DEVICE BOX

NAIL HOLE

111F38.EPS

Figure 38 ◆ Several methods of mounting outlet boxes.

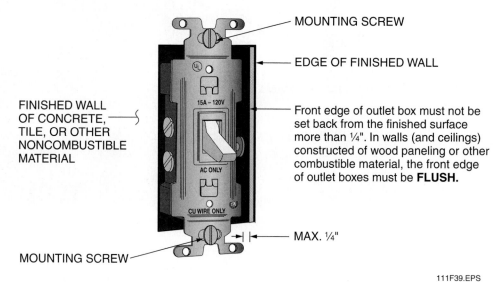

MOUNTING SCREW

EDGE OF FINISHED WALL

FINISHED WALL OF CONCRETE, TILE, OR OTHER NONCOMBUSTIBLE MATERIAL

15A – 120V

AC ONLY

CU WIRE ONLY

Front edge of outlet box must not be set back from the finished surface more than ¼". In walls (and ceilings) constructed of wood paneling or other combustible material, the front edge of outlet boxes must be **FLUSH.**

MAX. ¼"

MOUNTING SCREW

111F39.EPS

Figure 39 ◆ Outlet box installation.

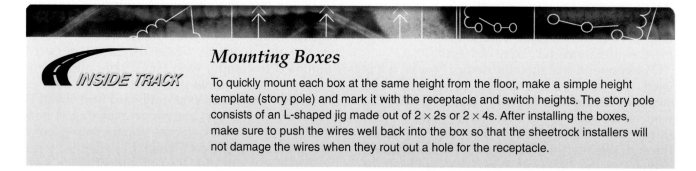

Mounting Boxes

To quickly mount each box at the same height from the floor, make a simple height template (story pole) and mark it with the receptacle and switch heights. The story pole consists of an L-shaped jig made out of 2 × 2s or 2 × 4s. After installing the boxes, make sure to push the wires well back into the box so that the sheetrock installers will not damage the wires when they rout out a hole for the receptacle.

When Type NM cable is used in either metallic or nonmetallic outlet boxes, the cable assembly, including the sheath, must extend into the box by not less than ¼" [*NEC Section 314.17(C)*]. In all instances, all permitted wiring methods must be secured to the boxes by means of either cable clamps or approved connectors. The one exception to this rule is where Type NM cable is used with 2¼-inch × 4-inch (or smaller) nonmetallic boxes where the cable is fastened within eight inches of the box. In this case, the cable does not have to be secured to the box. See *NEC Section 314.17(C), Exception.*

13.0.0 ◆ WIRING DEVICES

Wiring devices include various types of receptacles and switches, the latter being used for lighting control. Switches are covered in *NEC Article 404*, while regulations for receptacles may be found in *NEC Article 406*.

13.1.0 Receptacles

Receptacles are rated by voltage and amperage capacity. *NEC Section 406.3* requires that receptacles connected to a 15A or 20A circuit have the correct voltage and current rating for the application, and be of the grounding type. *NEC Section 406.11* requires that all 15A and 20A, 125V receptacles installed in dwelling units be listed as tamper-resistant.

Where there is only one outlet on a circuit, the receptacle's rating must be equal to or greater than the capacity of the conductors feeding it per *NEC Section 210.21(B)(1)*. For example, if one receptacle is connected to a 20A residential laundry circuit, the receptacle must be rated at 20A or more. When more than one outlet is on a circuit, the total connected load must be equal to or less than the capacity of the branch circuit conductors feeding the receptacles.

Refer to *Figure 40* for some of the characteristics of a standard 125V, 15A duplex receptacle. Note that the terminals are color coded as follows:

- *Green* – Connection for the equipment grounding conductor
- *Silver* – Connection for the neutral or grounded conductor
- *Brass* – Connection for the ungrounded conductor

A standard 125V, 15A receptacle is also typically imprinted with the following symbols:

- *UL* – Underwriters Laboratories, Inc., listing
- *CSA* – Canadian Standards Association
- *CO/ALR* – Designed for use with both copper and aluminum wire
- *15A* – Receptacle rated for a maximum of 15A
- *125V* – Receptacle rated for a maximum of 125V

The UL label means that the receptacle has undergone testing by Underwriters Laboratories, Inc., and meets minimum safety requirements. Underwriters Laboratories, Inc., was created by the National Board of Fire Underwriters to test electrical devices and materials. The UL label is a safety rating only and does not mean that the device or equipment meets any type of quality standard. The CSA label means that the receptacle is approved by the Canadian Standards Association, the Canadian equivalent to Underwriters Laboratories, Inc. The CSA label means that the receptacle is acceptable for use in Canada.

The CO/ALR symbol means that the device is suitable for use with copper, aluminum, or copper-clad aluminum wire. The CO in the symbol stands for copper while ALR stands for aluminum revised. The CO/ALR symbol replaces the earlier CU/AL mark, which appeared on

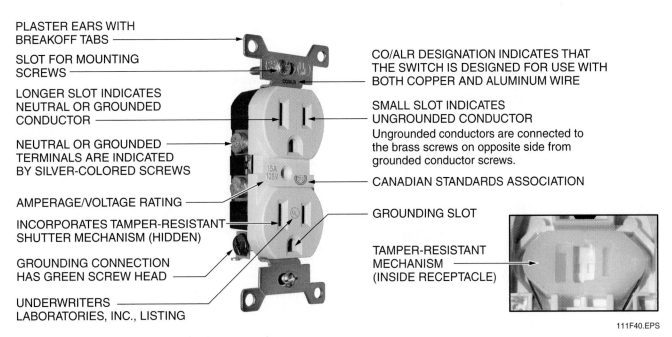

Figure 40 ◆ Standard 125V, 15A duplex receptacle.

wiring devices that were later found to be inadequate for use with aluminum wire in the 15A to 20A range. Therefore, any receptacle or wall switch marked with the CU/AL configuration or anything other than CO/ALR should be used only for copper wire.

These same configurations also apply to wall switches used for lighting control. These will be discussed next.

14.0.0 ◆ LIGHTING CONTROL

There are many types of lighting control devices. These devices have been designed to make the best use of the lighting equipment provided by the lighting industry. They include:

* Automatic timing devices for outdoor lighting
* Dimmers for residential lighting
* Common single-pole, three-way, and four-way switches

For the purposes of this module, a switch is defined as a device that is used on branch circuits to control lighting. Switches fall into the following basic categories:

* Snap-action switches
* Quiet switches

A single-pole snap-action switch consists of a device containing two stationary current-carrying elements, a moving current-carrying element, a toggle handle, a spring, and a housing. When the contacts are open, as shown in *Figure 41*, the circuit is broken and no current flows. When the moving element is closed by manually flipping the toggle handle, the contacts complete the circuit and the lamp is energized. See *Figure 42*.

The quiet switch (*Figure 43*) is the most common switch for use in lighting applications. Its operation is much quieter than the snap-action switch.

The quiet switch consists of a stationary contact and a moving contact that are close together when the switch is open. Only a short, gentle movement is required to open and close the switch, producing very little noise. This type of switch may be used only on alternating current.

Quiet switches are common for loads from 10A to 20A, and are available in single-pole, three-way, and four-way configurations.

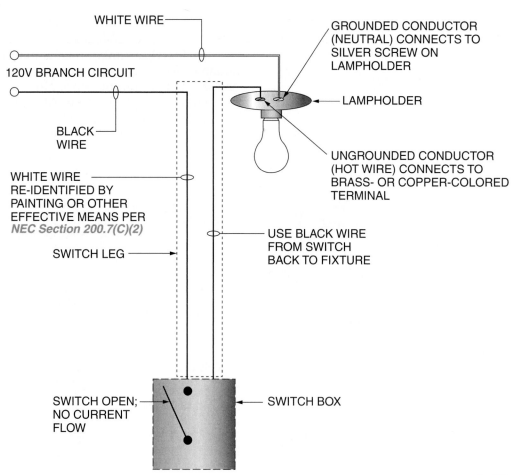

Figure 41 ◆ Switch operation, contacts open.

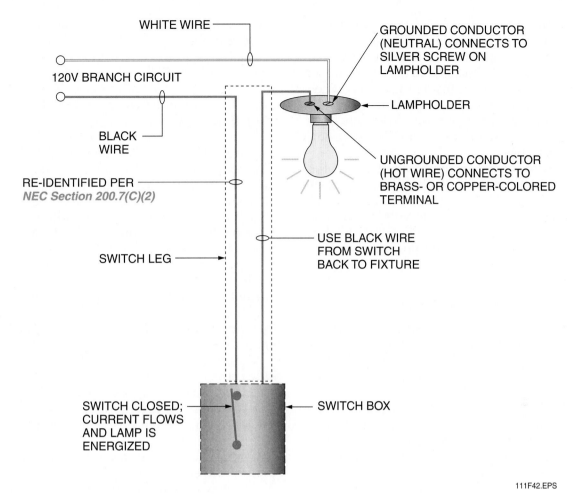

WHITE WIRE

GROUNDED CONDUCTOR (NEUTRAL) CONNECTS TO SILVER SCREW ON LAMPHOLDER

120V BRANCH CIRCUIT

LAMPHOLDER

BLACK WIRE

UNGROUNDED CONDUCTOR (HOT WIRE) CONNECTS TO BRASS- OR COPPER-COLORED TERMINAL

RE-IDENTIFIED PER *NEC Section 200.7(C)(2)*

USE BLACK WIRE FROM SWITCH BACK TO FIXTURE

SWITCH LEG

SWITCH CLOSED; CURRENT FLOWS AND LAMP IS ENERGIZED

SWITCH BOX

111F42.EPS

Figure 42 ◆ Switch operation, contacts closed.

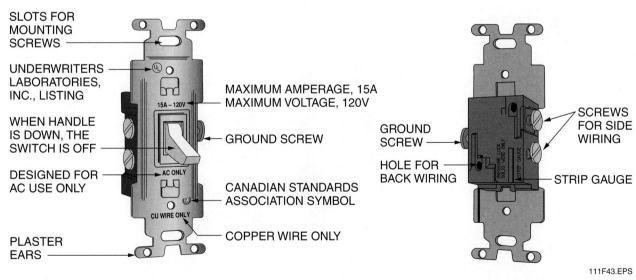

SLOTS FOR MOUNTING SCREWS

UNDERWRITERS LABORATORIES, INC., LISTING

MAXIMUM AMPERAGE, 15A
MAXIMUM VOLTAGE, 120V

WHEN HANDLE IS DOWN, THE SWITCH IS OFF

GROUND SCREW

DESIGNED FOR AC USE ONLY

CANADIAN STANDARDS ASSOCIATION SYMBOL

COPPER WIRE ONLY

PLASTER EARS

SCREWS FOR SIDE WIRING

GROUND SCREW

HOLE FOR BACK WIRING

STRIP GAUGE

111F43.EPS

Figure 43 ◆ Characteristics of a single-pole quiet switch.

Many other types of switches are available for lighting control. One type of switch used mainly in residential occupancies is the door-actuated switch. It is generally installed in the door jamb of a closet to control a light inside the closet. When the door is open, the light comes on; when the door is closed, the light goes out. Most refrigerator and oven lights are also controlled by door-actuated switches.

Combination switch/indicator light assemblies are available for use where the light cannot be seen from the switch location, such as an attic or garage. Switches are also made with small neon lamps in the handle that light when the switch is off. These low-current-consuming lamps make the switches easy to find in the dark.

14.1.0 Three-Way Switches

Three-way switches are used to control one or more lamps from two different locations, such as at the top and bottom of stairways, in a room that has two entrances, etc. A typical three-way switch is shown in *Figure 44*.

A three-way switch has three terminals. The single terminal at one end of the switch is called the common or hinge point. This terminal is easily identified because it is darker than the other two terminals. The feeder (hot wire) or switch leg is always connected to the common dark or black terminal. The two remaining terminals are called traveler terminals. These terminals are used to connect three-way switches together.

The connection of two three-way switches is shown in *Figure 45*. By means of the two switches, it is possible to control the lamp from two locations. By tracing the circuit, it may be seen how these three-way switches operate.

A 120V circuit emerges from the left side of the drawing. The white or neutral wire connects directly to the neutral terminal of the lamp. The hot wire carries current, in the direction of the arrows, to the common terminal of the three-way switch on the left. Since the handle is in the up position, the current continues to the top traveler terminal and is carried by this traveler to the other three-way switch. Note that the handle is also in the up position on this switch; this picks up the current flow and carries it to the common point, which continues to the ungrounded terminal of the lamp to make a complete circuit. The lamp is energized.

Moving the handle to a different position on either three-way switch will break the circuit, which in turn de-energizes the lamp. For example, let's say a person leaves the room at the point of the three-way switch on the left, and the switch handle is flipped down, as shown in *Figure 46*. Note that the current flow is now directed to the bottom traveler terminal, but since the handle of the three-way switch on the right is still in the up position, no current will flow to the lamp.

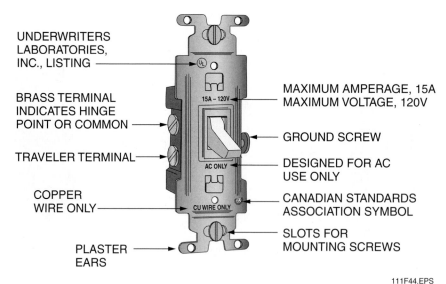

111F44.EPS

Figure 44 ◆ Typical three-way switch.

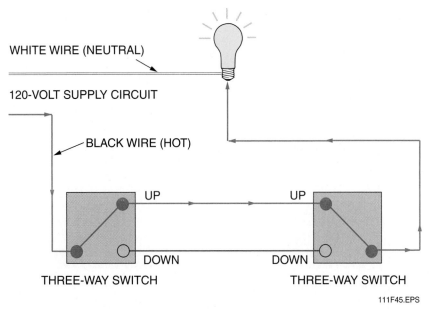

WHITE WIRE (NEUTRAL)

120-VOLT SUPPLY CIRCUIT

BLACK WIRE (HOT)

UP   UP

DOWN   DOWN

THREE-WAY SWITCH   THREE-WAY SWITCH

111F45.EPS

Figure 45 ◆ Three-way switches in the ON position; both handles are up.

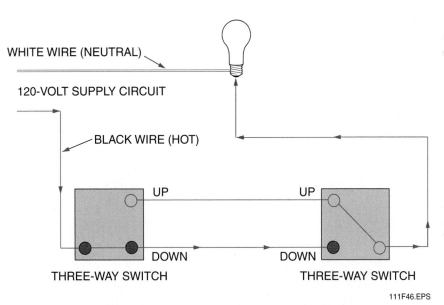

WHITE WIRE (NEUTRAL)

120-VOLT SUPPLY CIRCUIT

BLACK WIRE (HOT)

UP   UP

DOWN   DOWN

THREE-WAY SWITCH   THREE-WAY SWITCH

111F46.EPS

Figure 46 ◆ Three-way switches in the OFF position; one handle is down, one handle is up.

If another person enters the room at the location of the three-way switch on the right, and the handle is flipped downward, as shown in *Figure 47*, this change provides a complete circuit to the lamp, which causes it to be energized. In this example, current flow is on the bottom traveler. Again, changing the position of the switch handle (pivot point) on either three-way switch will de-energize the lamp.

In actual practice, the exact wiring of the two three-way switches to control the operation of a lamp will be slightly different from the routing shown in these three diagrams. There are several ways that two three-way switches may be connected. One solution is shown in *Figure 48*. In this case, two-wire, Type NM cable is fed to the three-way switch on the left.

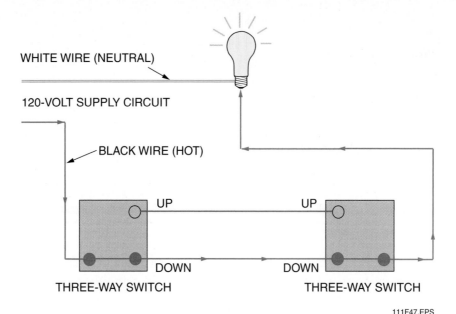

WHITE WIRE (NEUTRAL)

120-VOLT SUPPLY CIRCUIT

BLACK WIRE (HOT)

UP   UP

DOWN   DOWN

THREE-WAY SWITCH   THREE-WAY SWITCH

111F47.EPS

Figure 47 ◆ Three-way switches with both handles down; the light is energized.

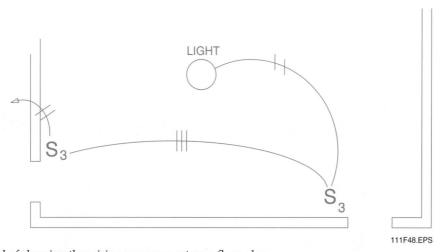

LIGHT

S_3

S_3

111F48.EPS

Figure 48 ◆ Method of showing the wiring arrangement on a floor plan.

The black or hot conductor is connected to the common terminal on the switch, while the white or neutral conductor is spliced to the white conductor of the three-wire, Type NM cable leaving the switch. This three-wire cable is necessary to carry the two travelers plus the neutral to the three-way switch on the right. At this point, the black and red wires connect to the two traveler terminals, respectively. The white or neutral wire is again spliced—this time to the white wire of another two-wire, Type NM cable. The neutral wire is never connected to the switch itself. The black wire of the two-wire, Type NM cable connects to the common terminal on the three-way switch. This cable, carrying the hot and neutral conductors, is routed to the lighting fixture outlet for connection to the fixture.

Another solution is to feed the lighting fixture outlet with two-wire cable. Run another two-wire cable carrying the hot and neutral conductors to one of the three-way switches. A three-wire cable is pulled between the two three-way switches, and then another two-wire cable is routed from the other three-way switch to the lighting fixture outlet.

Some electricians use a shortcut method that eliminates one of the two-wire cables in the preceding method. In this case, a two-wire cable is run from the lighting fixture outlet to one three-way switch. Three-wire cable is pulled between the two three-way switches—two of the wires for travelers and the third for the common point return. This method is shown in *Figure 49.*

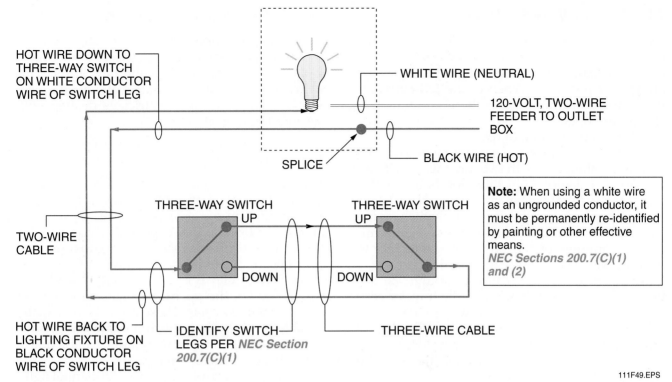

HOT WIRE DOWN TO
THREE-WAY SWITCH
ON WHITE CONDUCTOR
WIRE OF SWITCH LEG

WHITE WIRE (NEUTRAL)

120-VOLT, TWO-WIRE
FEEDER TO OUTLET
BOX

SPLICE

BLACK WIRE (HOT)

THREE-WAY SWITCH
UP

THREE-WAY SWITCH
UP

Note: When using a white wire
as an ungrounded conductor, it
must be permanently re-identified
by painting or other effective
means.
*NEC Sections 200.7(C)(1)
and (2)*

TWO-WIRE
CABLE

DOWN

DOWN

HOT WIRE BACK TO
LIGHTING FIXTURE ON
BLACK CONDUCTOR
WIRE OF SWITCH LEG

IDENTIFY SWITCH
LEGS PER *NEC Section
200.7(C)(1)*

THREE-WIRE CABLE

111F49.EPS

Figure 49 ◆ One way to connect a pair of three-way switches to control one lighting fixture.

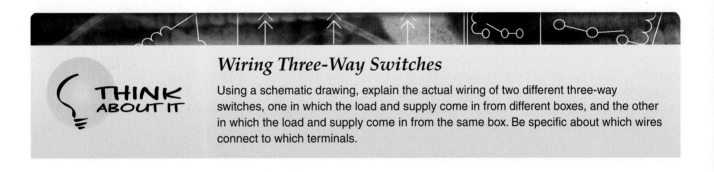

Wiring Three-Way Switches

Using a schematic drawing, explain the actual wiring of two different three-way
switches, one in which the load and supply come in from different boxes, and the other
in which the load and supply come in from the same box. Be specific about which wires
connect to which terminals.

14.2.0 Four-Way Switches

Two three-way switches may be used in conjunction with any number of four-way switches to control a lamp, or a series of lamps, from any number of positions. When connected correctly, the actuation of any one of these switches will change the operating condition of the lamp (i.e., turn the lamp either on or off).

Figure 50 shows how a four-way switch may be used in combination with two three-way switches to control a device from three locations. In this example, note that the hot wire is connected to the common terminal on the three-way switch on the left. Current then travels to the top traveler terminal and continues on the top traveler conductor to the four-way switch. Since the handle is up on the four-way switch, current flows through the top terminals of the switch and into the traveler conductor going to the other three-way switch.

Again, the switch is in the up position. Therefore, current is carried from the top traveler terminal to the common terminal and then to the lighting fixture to energize it.

If the position of any one of the three switch handles is changed, the circuit will be broken and no current will flow to the lamp. For example, assume that the four-way switch handle is flipped downward. The circuit will now appear as shown in *Figure 51*, and the light will be out.

Remember, any number of four-way switches may be used in combination with two three-way switches, but two three-way switches are always necessary for the correct operation of one or more four-way switches.

14.3.0 Photoelectric Switch

The chief application of the photoelectric switch is to control outdoor lighting, especially the dusk-to-dawn lights found in suburban areas. This switch has an endless number of possible uses and is a great tool for electricians dealing with outdoor lighting situations.

14.4.0 Relays

Next to switches, relays play the most important part in the control of light. However, the design and application of relays is a study in itself, and they are far beyond the scope of this module. Still, a brief mention of relays is necessary to round out your knowledge of lighting controls.

An electric relay is a device whereby an electric current causes the opening or closing of one or more pairs of contacts. These contacts are usually capable of controlling much more power than is necessary to operate the relay itself. This is one of the main advantages of relays.

One popular use of the relay in residential lighting systems is that of remote control lighting. In this type of system, all relays are designed to operate on a 24V circuit and are used to control 120V lighting circuits. They are rated at 20A, which is sufficient to control the full load of a normal lighting branch circuit, if desired.

Remote control switching makes it possible to install a switch wherever it is convenient and practical to do so or wherever there is an obvious need for a switch, no matter how remote it is from the lamp or lamps it is to control. This method enables

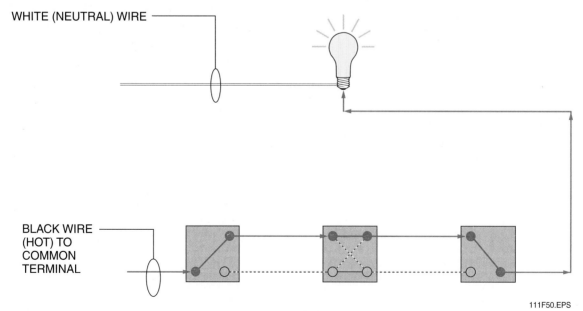

111F50.EPS

Figure 50 ◆ Three- and four-way switches used in combination; the light is on.

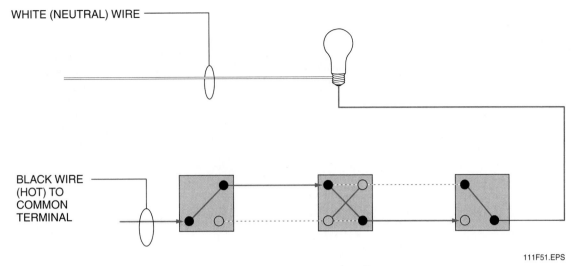

WHITE (NEUTRAL) WIRE

BLACK WIRE
(HOT) TO
COMMON
TERMINAL

111F51.EPS

Figure 51 ◆ Three- and four-way switches used in combination; the light is off.

lighting designs to achieve new advances in lighting control convenience at a reasonable cost. Remote control switching is also ideal for rewiring existing homes with finished walls and ceilings.

One relay is required for each fixture or each group of fixtures that are controlled together. Switch locations for remote control follow the same rules as for conventional direct switching. However, since it is easy to add switches to control a given relay, no opportunities should be overlooked for adding a switch to improve the convenience of control.

Remote control lighting also has the advantage of using selector switches at central locations. For example, selector switches located in the master bedroom or in the kitchen of a home enable the owner to control every lighting fixture on the property from this location. For example, the selector switch may be used to control outside or basement lights that might otherwise be left on inadvertently.

14.5.0 Dimmers

Dimming a lighting system provides control of the quantity of illumination. It may be done to create certain moods or to blend the lighting from different sources for various lighting effects.

For example, in homes with formal dining rooms, a chandelier mounted directly above the dining table and controlled by a dimmer switch becomes the centerpiece of the room while providing general illumination. The dimmer adds versatility since it can set the mood for the activity—low brilliance (candlelight effect) for formal dining or bright for an evening of playing cards. When chandeliers with exposed lamps are used, the dimmer

is essential to avoid a garish and uncomfortable atmosphere. The chandelier should be sized in proportion to the dining area.

NOTE

It is very important that dimmers be matched to the wattage of the application. Check the manufacturer's data.

14.6.0 Switch Locations

Although the location of wall switches is usually provided for convenience, the *NEC®* also stipulates certain mandatory locations for lighting fixtures and wall switches. See *NEC Section 210.70(A)* for specific switch locations in dwelling units. These locations are deemed necessary for added safety in the home for both the occupants and service personnel.

For example, the *NEC®* requires adequate light in areas where heating, ventilating, and air conditioning (HVAC) equipment is placed. Furthermore, these lights must be conveniently controlled so that homeowners and service personnel do not have to enter a dark area where they might come in contact with dangerous equipment. Three-way switches are required under certain conditions. The *NEC®* also specifies regulations governing lighting fixtures in clothes closets, along with those governing lighting fixtures that may be mounted directly to the outlet box without further support. *Figure 52* summarizes some of the *NEC®* requirements for light and switch placement in the home. For further details, refer to the appropriate sections in the *NEC®*.

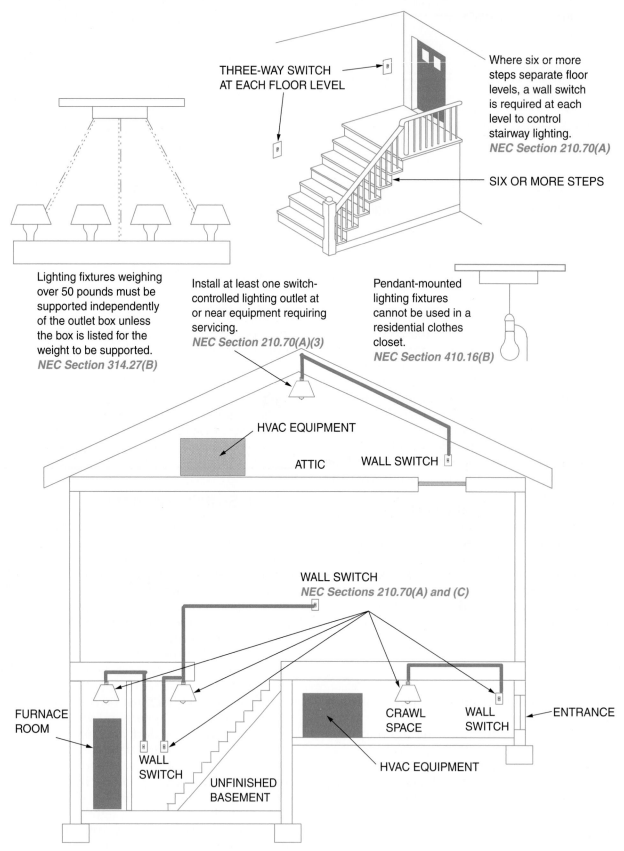

THREE-WAY SWITCH AT EACH FLOOR LEVEL

Where six or more steps separate floor levels, a wall switch is required at each level to control stairway lighting. *NEC Section 210.70(A)*

SIX OR MORE STEPS

Lighting fixtures weighing over 50 pounds must be supported independently of the outlet box unless the box is listed for the weight to be supported. *NEC Section 314.27(B)*

Install at least one switch-controlled lighting outlet at or near equipment requiring servicing. *NEC Section 210.70(A)(3)*

Pendant-mounted lighting fixtures cannot be used in a residential clothes closet. *NEC Section 410.16(B)*

HVAC EQUIPMENT

ATTIC

WALL SWITCH

WALL SWITCH
NEC Sections 210.70(A) and (C)

FURNACE ROOM

WALL SWITCH

UNFINISHED BASEMENT

HVAC EQUIPMENT

CRAWL SPACE

WALL SWITCH

ENTRANCE

111F52.EPS

Figure 52 ◆ *NEC*® requirements for light and switch placement.

14.7.0 Low-Voltage Electrical Systems

Conventional lighting systems operate and are controlled by the same system voltage, generally 120V in residential lighting circuits. The *NEC®* permits the use of low-voltage systems to control lighting circuits. There are some advantages to low-voltage systems. One advantage is that the control of lighting from several different locations is more easily accomplished, such as with the remote control system discussed earlier. For example, outside flood lighting can be controlled from several different rooms in a house. The cost of the control wiring is less in that it is rated for a lower voltage and only carries a minimum amount of current compared to a standard lighting system. When extensive or complex lighting control is required, low-voltage systems are preferred. Also, since these circuits are low-energy circuits, circuit protection is not required.

14.7.1 NEC® Requirements for Low-Voltage Systems

NEC Article 725 governs the installation of low-voltage system wiring. These provisions apply to remote control circuits, low-voltage relay switching, low-energy power circuits, and low-voltage circuits. The *NEC®* divides these circuits into three categories:

- Remote control
- Signaling
- Power-limited circuits

As mentioned earlier, circuit protection of the low-voltage circuit is not required; however, the high-voltage side of the transformer that supplies the low-voltage system must be protected. *NEC Chapter 9, Tables 11(A) and 11(B)* cover circuits that are inherently limited in power output and therefore require no overcurrent protection or are limited by a combination of power source and overcurrent protection.

There are a number of requirements of the power systems described in *NEC Chapter 9, Tables 11(A) and 11(B)* and the notes preceding the tables. You should read and study all applicable portions of the *NEC®* before installing low-voltage power systems.

Low-voltage systems are described in more detail in later modules.

15.0.0 ◆ ELECTRIC HEATING

The use of electric heating in residential occupancies has risen tremendously over the past decade or so, and the practice will no doubt continue. This is due to the following advantages of electric heat over most other heating systems:

- Electric heat is noncombustible and is therefore safer than combustible fuels.
- It requires no storage space, fuel tanks, or chimneys.
- It requires little maintenance.
- The initial installation cost is relatively inexpensive when compared to other types of heating systems.
- The comfort level may be improved since each room may be controlled separately by its own thermostat.

There are also some disadvantages to using electric baseboard heat, especially in northern climates. Some of these disadvantages include:

- Electric heat is often more expensive to operate than other types of fuels.
- Receptacles must not be installed above electric baseboard heaters.
- Electric baseboard heaters tend to discolor the wall area immediately above the heater, especially if there are smokers in the home.

NOTE
It is very important to calculate the extra electric load of an electric heater installation (especially in an add-on situation). Ensure that the extra load does not exceed the maximum amperage draw of either the circuit or the panel.

The type of electric heating system used for a given residence will usually depend on the structural conditions, the kind of room, and the activities for which the room will be used. The homeowner's preference will also enter into the final decision.

Electric heating equipment is available in baseboard, wall, ceiling, kick space, and floor units; in resistance cable embedded in the ceiling or concrete floor; in forced-air duct systems similar to conventional oil- or gas-fired hot air systems; and in electric boilers for hot water baseboard heat.

Electric heat pumps have also become popular for HVAC systems in certain parts of the country. The term heat pump, as applied to a year-round air conditioning system, commonly denotes a system in which refrigeration equipment is used in such a manner that heat is taken from a heat source and transferred to the conditioned space when heating is desired; heat is removed from the space and discharged to a heat sink when cooling and dehumidification are desired.

A heat pump has the unique ability to furnish more energy than it consumes. This is due to the fact that under certain outdoor conditions, electrical energy is required only to move the refrigerant and run the fan; thus, a heat pump can attain a heating efficiency of two or more to one; that is, it will put out an equivalent of two or three watts of heat for every watt consumed. For this reason, its use is highly desirable for the conservation of energy.

In general, electric baseboard heating equipment should be located on the outside wall near the areas where the greatest heat loss will occur, such as under windows, etc. The controls for wall-mounted thermostats should be located on an interior wall, about 50 inches above the floor to sense the average room temperature. *Figure 53* shows an electric heating arrangement for the sample residence. *NEC®* regulations governing the installation of these units are also noted.

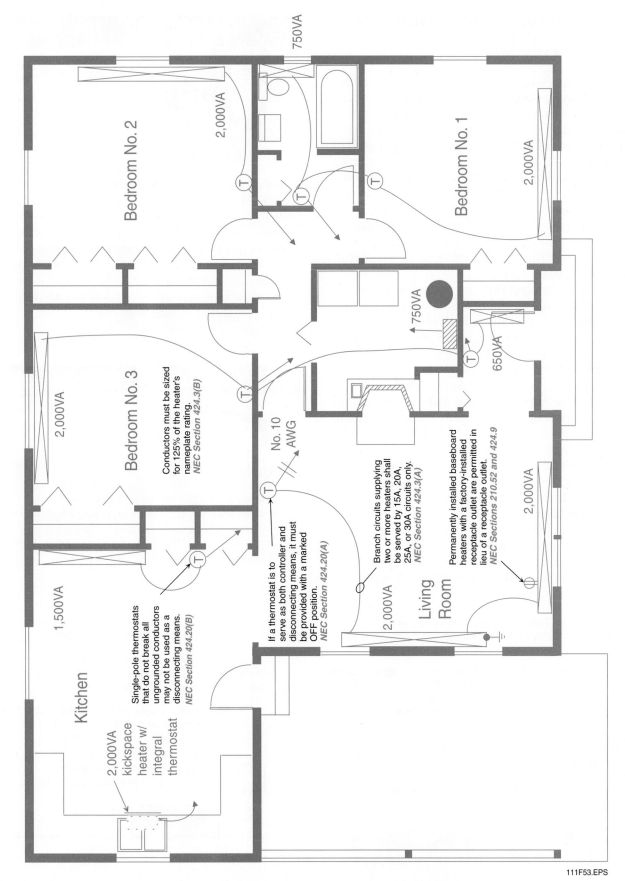

750VA

Bedroom No. 2

2,000VA

Bedroom No. 1

2,000VA

750VA

650VA

2,000VA

Conductors must be sized for 125% of the heater's nameplate rating.
NEC Section 424.3(B)

Bedroom No. 3

2,000VA

No. 10 AWG

If a thermostat is to serve as both controller and disconnecting means, it must be provided with a marked OFF position.
NEC Section 424.20(A)

Branch circuits supplying two or more heaters shall be served by 15A, 20A, 25A, or 30A circuits only.
NEC Section 424.3(A)

Permanently installed baseboard heaters with a factory-installed receptacle outlet are permitted in lieu of a receptacle outlet.
NEC Sections 210.52 and 424.9

2,000VA

Living Room

2,000VA

Kitchen

1,500VA

Single-pole thermostats that do not break all ungrounded conductors may not be used as a disconnecting means.
NEC Section 424.20(B)

2,000VA kickspace heater w/ integral thermostat

111F53.EPS

Figure 53 ◆ Electric heating arrangement for the sample residence.

16.0.0 ◆ RESIDENTIAL SWIMMING POOLS, SPAS, AND HOT TUBS

The *NEC®* recognizes the potential danger of electric shock to persons in swimming pools, wading pools, and therapeutic pools, or near decorative pools or fountains. This shock could occur from electric potential in the water itself or as a result of a person in the water or a wet area touching an enclosure that is not at ground potential. Accordingly, the *NEC®* provides rules for the safe installation of electrical equipment and wiring in or adjacent to swimming pools and similar locations. *NEC Article 680* covers the specific rules governing the installation and maintenance of swimming pools, spas, and hot tubs.

The electrical installation procedures for hot tubs and swimming pools are too vast to be covered in detail in this module. However, the general requirements for the installation of outlets, overhead fans and lighting fixtures, and other items are summarized in *Figure 54.*

Besides *NEC Article 680,* another good source for learning more about electrical installations in and around swimming pools is from manufacturers of swimming pool equipment, including those who manufacture and distribute underwater lighting fixtures. Many of these manufacturers offer pamphlets detailing the installation of their equipment with helpful illustrations, code explanations, and similar details. This literature is usually available at little or no cost to qualified personnel. You can write directly to manufacturers to request information about available literature, or contact your local electrical supplier or contractor who specializes in installing residential swimming pools. See *Figure 55.*

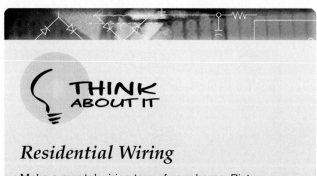

THINK ABOUT IT

Residential Wiring

Make a mental wiring tour of your home. Picture several rooms, including the kitchen and utility/laundry room. How is each device connected to the power source, and what is the probable amperage and overcurrent protection? What other devices might or might not be included in the circuit? How many branch circuits serve each room? Later, examine the panelboard and exposed wiring to see how accurately you identified the branch circuits.

INSIDE TRACK

Pools

Pools are a common site for homeowners to add lights, receptacles, or heaters, and code violations are common. If you are called in for a service problem, ask the homeowner about any do-it-yourself wiring.

Lighting fixtures, lighting outlets, and ceiling fans located over the hot tub or within 5 feet from its inside walls shall be a minimum of 7 feet 6 inches above the maximum water level and shall be GFCI-protected [*NEC Section 680.43(B)*].

At least one receptacle must be located at a minimum of 6 feet and no more than 10 feet from the inside wall of the hot tub [*NEC Section 680.43(A)*]. Also, all receptacles must be located at least 6 feet from the inside wall of the hot tub per *NEC Section 680.43(A)(1)* and all 125-volt receptacles located within 10 feet of the inside wall of the hot tub must be GFCI-protected [*NEC Section 680.43(A)(2)*]. Wall switches must be located at least 5 feet from the hot tub per *NEC Section 680.43(C).*

Maintenance disconnect must be accessible and within sight of the hot tub *(NEC Section 680.12)* and located at least 5 feet from the inside wall of the hot tub.

All electrical equipment associated with the circulating system of the hot tub must be grounded [*NEC Section 680.43(F)*].

Any outlet that supplies a hot tub shall be GFCI-protected [*NEC Section 680.43(A)(3)*].

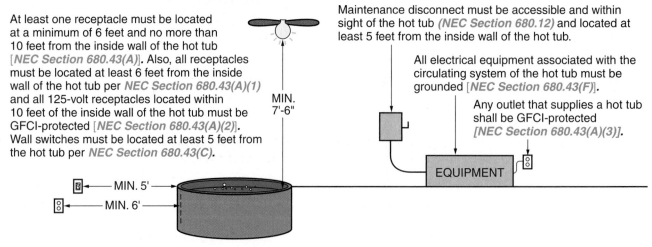

111F54.EPS

Figure 54 ◆ *NEC®* requirements for packaged indoor hot tubs.

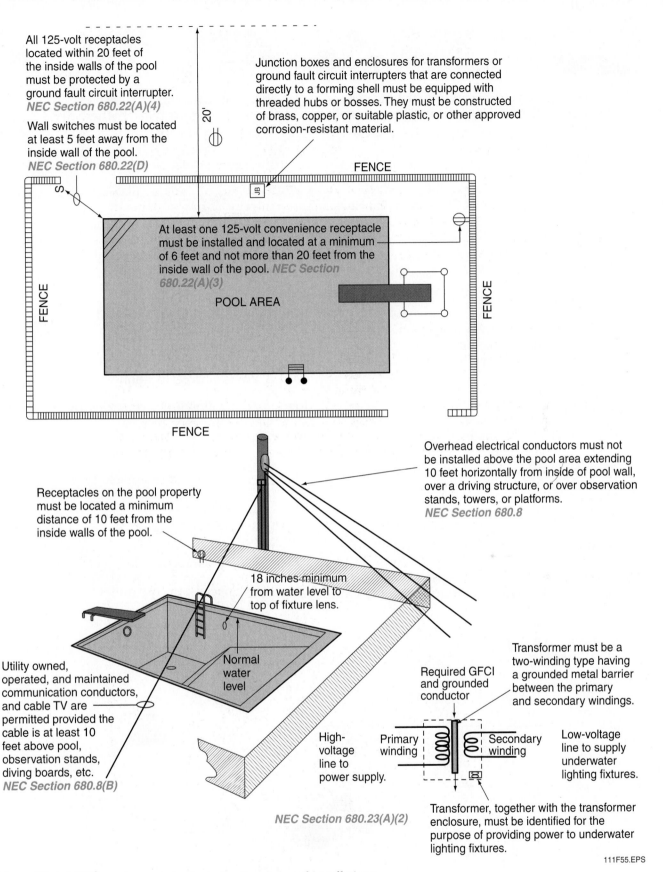

All 125-volt receptacles located within 20 feet of the inside walls of the pool must be protected by a ground fault circuit interrupter. *NEC Section 680.22(A)(4)*

Wall switches must be located at least 5 feet away from the inside wall of the pool. *NEC Section 680.22(D)*

Junction boxes and enclosures for transformers or ground fault circuit interrupters that are connected directly to a forming shell must be equipped with threaded hubs or bosses. They must be constructed of brass, copper, or suitable plastic, or other approved corrosion-resistant material.

20'

FENCE

S

At least one 125-volt convenience receptacle must be installed and located at a minimum of 6 feet and not more than 20 feet from the inside wall of the pool. *NEC Section 680.22(A)(3)*

FENCE

FENCE

POOL AREA

FENCE

Receptacles on the pool property must be located a minimum distance of 10 feet from the inside walls of the pool.

Overhead electrical conductors must not be installed above the pool area extending 10 feet horizontally from inside of pool wall, over a driving structure, or over observation stands, towers, or platforms. *NEC Section 680.8*

18 inches minimum from water level to top of fixture lens.

Normal water level

Utility owned, operated, and maintained communication conductors, and cable TV are permitted provided the cable is at least 10 feet above pool, observation stands, diving boards, etc. *NEC Section 680.8(B)*

High-voltage line to power supply.

Required GFCI and grounded conductor

Primary winding

Secondary winding

Transformer must be a two-winding type having a grounded metal barrier between the primary and secondary windings.

Low-voltage line to supply underwater lighting fixtures.

NEC Section 680.23(A)(2)

Transformer, together with the transformer enclosure, must be identified for the purpose of providing power to underwater lighting fixtures.

111F55.EPS

Figure 55 ◆ *NEC*® requirements for typical swimming pool installations.

1. When sizing electrical services, at what percentage is the first 3,000VA rated?
   a. 20%
   b. 45%
   c. 80%
   d. 100%

2. What section of the *NEC*® requires that fittings be identified for use with service masts?
   a. *NEC Section 230.28*
   b. *NEC Section 230.40*
   c. *NEC Section 250.46*
   d. *NEC Section 250.83*

3. A service conductor without an overall jacket must have a clearance of not less than _____ above a window that can be opened.
   a. two feet
   b. three feet
   c. eight feet
   d. ten feet

4. *NEC Section 230.6* considers conductors installed under at least _____ inch(es) of concrete to be outside the building.
   a. one
   b. two
   c. four
   d. five

5. Type NM cable may *not* be used in _____.
   a. shallow chases of masonry, concrete, or adobe
   b. the framework of a building
   c. protective strips
   d. attic spaces

6. Type MC cable may *not* be used _____.
   a. in concrete or plaster where dry
   b. in dry masonry
   c. in attic spaces
   d. where subject to physical damage

7. Type SE cable is available with _____ for interior wiring systems.
   a. a non-insulated ground or neutral conductor
   b. an insulated grounded conductor
   c. no ground conductor
   d. guard strips

8. Type SER cable may be used _____.
   a. in overhead applications
   b. underground
   c. as a subfeed under certain conditions
   d. in hazardous locations

Identify the following receptacles by numbers shown below for each receptacle.

9. _____ Single receptacle

10. _____ Split-wired receptacle

11. _____ Dryer receptacle

12. _____ Range receptacle

13. _____ Duplex receptacle

14. _____ Special-purpose outlet

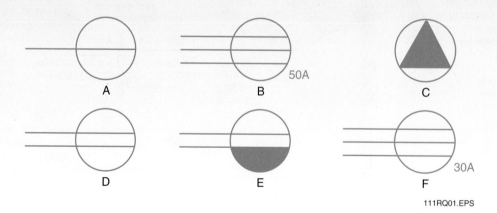

A   B   50A   C

D   E   F   30A

111RQ01.EPS

Review Questions

15. Using *NEC Table 314.16(A),* calculate the cubic inches required for the receptacle outlets shown in the table below. Then, indicate the size of the metallic box that should be used.

Number and Size of Conductors in Box	Free Space within Box for Each Conductor	Total Cubic Inches of Box Space Required	What Size Metallic Box May Be Used?
A. Six No. 12 conductors and three ground wires	2.25	_____	_____
B. Seven No. 12 conductors and three ground wires with one receptacle	2.25	_____	_____
C. Two No. 14 conductors and one ground wire	2.0	_____	_____
D. Four No. 14 conductors and two ground wires	2.0	_____	_____
E. Six No. 14 conductors and three ground wires with one receptacle	2.0	_____	_____

Summary

This module covered the basics of residential wiring, including load calculations and wiring devices.

Residential electrical system design begins with the floor plan. The square footage of the home is used to calculate the lighting loads. The lighting loads, along with small appliance loads, fixed appliance loads, and other loads must be calculated to determine the service size. Demand factors are applied and the neutral conductor is sized per *NEC Article 220*. GFCI and AFCI break-ers must be installed in all required locations, and the system must be grounded and bonded per *NEC Article 250*. Most residences use Type NM or MC cable for branch circuits. Switches must be installed in all locations required by the *NEC®*. Special consideration must be given to electric heating units and circuits supplying swimming pools and hot tubs.

A thorough knowledge of the *NEC®* is essential to the safe and successful installation of residential wiring systems.

Notes

Trade Terms Quiz

1. _____ consists of wires with a spiral-wound, flexible steel outer jacketing.

2. A type of cable that is popular in residential and small commercial wiring systems, _____ may be used for both exposed and concealed work in normally dry locations; this type of cable is also referred to as _____.

3. A(n) _____ is a piece of equipment that has been designed for a particular purpose.

4. A(n) _____ is used for turning an electrical circuit on and off.

5. The circuit that is routed to a switch box for controlling electric lights is known as a(n) _____.

6. A(n) _____ is equipped with a conductor terminal to take a bonding jumper.

7. A(n) _____ is a bare or green insulated conductor used to ensure conductivity between metal parts that are required to be electrically connected.

8. The _____ is comprised of the conductors that extend from the last power company pole to the point of connection at the service facilities.

9. The _____ is the point where power is supplied to a building.

10. _____ lie between the point of termination of the overhead service drop or underground service lateral and the main disconnecting device in the building.

11. _____ mainly provides overcurrent protection to the feeder and service conductors.

12. A(n) _____ is comprised of the underground conductors through which service is supplied between the power company's distribution facilities and their first point of connection to the building.

13. The portion of a wiring system that extends beyond the final overcurrent device is the _____.

14. A(n) _____ is a circuit that carries current from the service equipment to a subpanel or a branch circuit panel or to some point in the wiring system.

15. Normally located at the service entrance of a residential installation, a(n) _____ usually contains the main disconnect.

16. Raceway, cable, wires, boxes, and other equipment are installed during _____.

Trade Terms

Appliance
Bonding bushing
Bonding jumper
Branch circuit
Feeder
Load center

Metal-clad (Type MC)
  cable
Nonmetallic-sheathed
  (Type NM) cable
Romex®

Roughing in
Service drop
Service entrance
Service-entrance
  conductors

Service-entrance
  equipment
Service lateral
Switch
Switch leg

Dan Lamphear
Associated Builders and Contractors, Inc.

Like many other people, Dan Lamphear just fell into his career as an electrician. But once he discovered the electrical trade, he knew he had found a home. Since then, he has progressed from a helper to an apprentice, a journeyman, an independent contractor, an inventor, and finally, a teacher.

It was as much luck as anything else that led Dan toward a career as a professional electrician more than two decades ago. He wasn't sure what he wanted to do with his life after graduating from high school. However, after watching an electrician perform a commercial wiring job at a friend's business—and providing a helping hand—he was hooked. "It seemed like a challenging career, and I was curious to learn more about how electricity works," he recalls.

Dan was hired by that same electrician, under whom he apprenticed for several years before hearing about the NCCER program. He jumped at the chance to further his skills through the program. "Like they say, knowledge is money," he smiles.

After graduating from the program, Dan struck out on his own as an independent electrician,

specializing in plant maintenance, industrial, and commercial work. His ability to diagnose and repair electrical problems in factory machinery soon made him a valuable contractor in Milwaukee's industrial sector.

He also discovered his knack for invention, and he has designed and built specialized machinery for a company that hired him as its full-time electrical maintenance supervisor. "Knowing the electrical side of machinery allowed me to understand how they operate mechanically," he says of his work as an inventor.

Dan later returned to the Associated Builders and Contractors (ABC) chapter, which trains out of a local community college, to repay the favor that helped him embark on his career. He teaches Electrical Level 2 courses for students who represent the next generation of professional electricians.

"Knowing how to use test instruments is perhaps the most important aspect of the job," he notes. "I still have some of the same meters I started out with."

David Lewis

Instructor
Putnam Career & Technical Center

David Lewis started his career working in coal mines. After a few years he opened his own electrical business. Now he is an electrical instructor and works with the State Department of Education on curriculum development. He also serves on the NCCER revision team for the Electrical curriculum.

How did you first get interested in the field?
After graduating from high school in 1972 and attending college for a while, I decided that I wanted to work in the coal mines. Electricity interested me, so I became a maintenance foreman/electrician. Eventually, I started my own business.

What kind of training have you been through?
While working in the coal mines, I attended several electrical training classes and obtained my underground electrical license. While in business for myself I got my Master Electrician license and attended many update classes. Since I have started teaching, I have attended classes on PLCs and other topics. I also went back to college and obtained a bachelor of science degree in Career and Technical Education.

What work have you done in your career?
In the coal mines I worked on all types of mining equipment. After starting my own business, I worked mainly in residential and light commercial wiring.

Tell us about your present job and what you like about it.
I enjoy being an instructor in Electrical Technology. I work mostly with high school students and during the two years they are with me, it is great to see them grasp the knowledge of electricity.

What factors have contributed most to your success?
Hard work and the willingness to learn from experienced electricians.

What advice would you give to those new to the field?
Try to learn all you can. Work with an experienced electrician and learn from them. Attend any training or classes that you can. There is always something new to learn.

Trade Terms
Introduced in This Module

Appliance: Equipment designed for a particular purpose (for example, using electricity to produce heat, light, or mechanical motion). Appliances are usually self-contained, are generally available for applications other than industrial use, and are normally produced in standard sizes or types.

Bonding bushing: A special conduit bushing equipped with a conductor terminal to take a bonding jumper. It also has a screw or other sharp device to bite into the enclosure wall to bond the conduit to the enclosure without a jumper when there are no concentric knockouts left in the wall of the enclosure.

Bonding jumper: A bare or green insulated conductor used to ensure the required electrical conductivity between metal parts required to be electrically connected. Bonding jumpers are frequently used from a bonding bushing to the service-equipment enclosure to provide a path around concentric knockouts in an enclosure wall, and they may also be used to bond one raceway to another.

Branch circuit: The portion of a wiring system extending beyond the final overcurrent device protecting a circuit.

Feeder: A circuit, such as conductors in conduit or a cable run, that carries current from the service equipment to a subpanel or a branch circuit panel or to some point in the wiring system.

Load center: A type of panelboard that is normally located at the service entrance of a residential installation. It usually contains the main disconnect.

Metal-clad (Type MC) cable: A type of cable with a metal jacket that is popular for use in residential and small commercial wiring systems. In general, it may be used for both exposed and concealed work in normally dry locations. It may also be used as service-entrance cable and in other locations as permitted by the *NEC*®.

Nonmetallic-sheathed (Type NM) cable: A type of cable that is popular for use in residential and small commercial wiring systems. In general, it may be used for both exposed and concealed work in normally dry locations. *See Romex*®.

Romex®*:* General Cable's trade name for Type NM cable; however, it is often used generically to refer to any nonmetallic-sheathed cable.

Roughing in: The first stage of an electrical installation, when the raceway, cable, wires, boxes, and other equipment are installed. This is the electrical work that must be done before any finishing work can be done.

Service drop: The overhead conductors, through which electrical service is supplied, between the last power company pole and the point of their connection to the service facilities located at the building.

Service entrance: The point where power is supplied to a building (including the equipment used for this purpose). The service entrance includes the service main switch or panelboard, metering devices, overcurrent protective devices, and conductors/raceways for connecting to the power company's conductors.

Service-entrance conductors: The conductors between the point of termination of the overhead service drop or underground service lateral and the main disconnecting device in the building.

Service-entrance equipment: Equipment that provides overcurrent protection to the feeder and service conductors, a means of disconnecting the feeders from energized service conductors, and a means of measuring the energy used.

Service lateral: The underground conductors through which service is supplied between the power company's distribution facilities and the first point of their connection to the building or area service facilities located at the building.

Switch: A mechanical device used for turning an electrical circuit on and off.

Switch leg: A circuit routed to a switch box for controlling electric lights.

Other Codes and Standards That Apply to Electrical Installations

Until 2000, there were three model building codes. These included the following:

- *Standard Building Code (SBC)* – Published by the Southern Building Code Congress International.
- *BOCA National Building Code (NBC)* – Published by the Building Officials and Code Administrators.
- *Uniform Building Code (UBC)* – Published by the International Conference of Building Officials.

The three code writing groups, SBCCI, BOCA, and UBC, combined into one organization called the International Code Council with the purpose of writing one nationally accepted family of building and fire codes. The first edition of the *International Building Code* was published in 2000 and the second edition in 2003, and updated again in 2006. It is intended to continue on a three-year cycle.

The International Residential Code (IRC) is adopted as part of the electrical code requirements in many areas of the country. The IRC covers one- and two-family dwellings of three stories or less. The IRC includes requirements for such things as ventilating fans for bathrooms, requirements for smoke detectors, and other items not specified by the *NEC®*.

The IRC covers all trades, including building, plumbing, mechanical, gas, energy, and electrical.

In 2002, the NFPA published its own building code, *NFPA 5000*. There are now two nationally recognized codes competing for adoption by the 50 states.

To be thoroughly competent in the electrical trade, you should become familiar with the contents of these codes and the terminology used in them.

NOTE

Always refer to the latest editions of codes in effect in your area.

This module is intended to present thorough resources for task training. The following reference work is suggested for further study. This is optional material for continued education rather than for task training.

National Electrical Code® Handbook, Latest Edition. Quincy, MA: National Fire Protection Association.

NCCER makes every effort to keep these textbooks up-to-date and free of technical errors. We appreciate your help in this process. If you have an idea for improving this textbook, or if you find an error, a typographical mistake, or an inaccuracy in NCCER's Contren® textbooks, please write us, using this form or a photocopy. Be sure to include the exact module number, page number, a detailed description, and the correction, if applicable. Your input will be brought to the attention of the Technical Review Committee. Thank you for your assistance.

Instructors – If you found that additional materials were necessary in order to teach this module effectively, please let us know so that we may include them in the Equipment/Materials list in the Annotated Instructor's Guide.

Write: Product Development and Revision
National Center for Construction Education and Research
3600 NW 43rd St., Bldg. G, Gainesville, FL 32606

Fax: 352-334-0932

E-mail: curriculum@nccer.org

Craft _____ Module Name _____

Copyright Date _____ Module Number _____ Page Number(s) _____

Description _____

_____

_____

_____

_____

(Optional) Correction _____

_____

_____

(Optional) Your Name and Address _____

_____

_____

Electrical Test Equipment

The Biodesign Institute at Arizona State University

The Biodesign Institute is the centerpiece of new growth at Arizona State University (ASU). This unique complex was built to meet the stringent demands posed by experimental programs in biotechnology and nanotechnology, and to enhance communication and collaboration between researchers with an open, shared lab design and a central atrium linking all floors.

26112-08

26112-08
Electrical Test Equipment

Topics to be presented in this module include:

Overview

Electrical systems must be tested for proper operation when they are installed and periodically for maintenance purposes. If a system develops a problem after installation, the cause of the problem must be located and repaired. The best way to test and troubleshoot an electrical system is to take electrical measurements at specific points in the circuit. These measurements are taken using various types of electrical test equipment.

Electricians must be able to select the right test equipment for the application. Older test equipment relied on analog meter movements and scales. Many electrical meters now use digital technology with digital readouts. Electricians must keep up with changes in testing technology and learn how to use various types of test equipment.

Note: *National Electrical Code®* and *NEC®* are registered trademarks of the National Fire Protection Association, Inc., Quincy, MA 02269. All *National Electrical Code®* and *NEC®* references in this module refer to the 2008 edition of the *National Electrical Code®*.

Objectives

When you have completed this module, you will be able to do the following:

1. Explain the operation of and describe the following pieces of test equipment:
   - Voltmeter
   - Ohmmeter
   - Clamp-on ammeter
   - Multimeter
   - Megohmmeter
   - Motor and phase rotation testers
2. Select the appropriate meter for a given work environment based on category ratings.
3. Identify the safety hazards associated with various types of test equipment.

Trade Terms

Coil
Continuity
d'Arsonval meter movement
Frequency

Required Trainee Materials

1. Pencil and paper
2. Copy of the latest edition of the *National Electrical Code®*
3. Appropriate personal protective equipment

Prerequisites

Before you begin this module, it is recommended that you successfully complete *Core Curriculum* and *Electrical Level One*, Modules 26101-08 through 26111-08.

This course map shows all of the modules in *Electrical Level One*. The suggested training order begins at the bottom and proceeds up. Skill levels increase as you advance on the course map. The local Training Program Sponsor may adjust the training order.

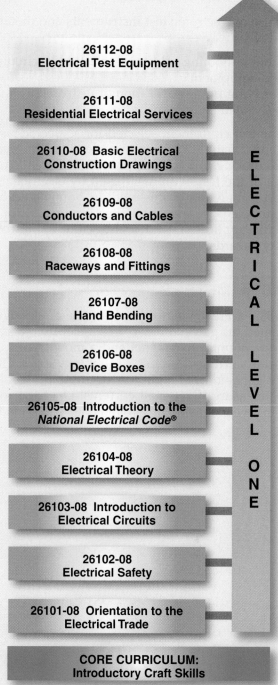

26112-08
Electrical Test Equipment

26111-08
Residential Electrical Services

26110-08 Basic Electrical
Construction Drawings

26109-08
Conductors and Cables

26108-08
Raceways and Fittings

26107-08
Hand Bending

26106-08
Device Boxes

26105-08 Introduction to the
National Electrical Code®

26104-08
Electrical Theory

26103-08 Introduction to
Electrical Circuits

26102-08
Electrical Safety

26101-08 Orientation to the
Electrical Trade

CORE CURRICULUM:
Introductory Craft Skills

ELECTRICAL LEVEL ONE

112CMAP.EPS

1.0.0 ◆ INTRODUCTION

The use of electronic test instruments and meters generally involves:

- Troubleshooting electrical/electronic circuits and equipment
- Verifying proper operation of instruments and associated equipment

The test equipment selected for a specific task depends on the type of measurement and the level of accuracy required. This module will focus on some of the test equipment used by electricians. Upon completion of this module, you should be able to select the appropriate test equipment for a specific application and identify the applicable safety hazards.

2.0.0 ◆ METERS

In 1882, a Frenchman named Arsene d'Arsonval invented the galvanometer. This meter used a stationary permanent magnet and a moving coil to indicate current flow on a calibrated scale. The early galvanometer was very accurate but could only measure very small currents. Over the following years, many improvements were made that extended the range of the meter and increased its ruggedness. The d'Arsonval meter movement (*Figure 1*) is the basis for analog meters.

A moving-coil meter movement operates on the electromagnetic principle. In its simplest form, the moving-coil meter uses a coil of very fine wire wound on a light aluminum frame. A permanent magnet surrounds the coil. The aluminum frame is mounted on pivots to allow it and the coil to rotate freely between the poles of the permanent magnet. When current flows through the coil, it becomes magnetized, and the polarity of the coil is repelled by the field of the permanent magnet. This causes the coil frame to rotate on its pivots, and the distance it rotates is determined by the amount of current that flows through the coil. By attaching a pointer to the coil frame and adding a calibrated scale, the amount of current flowing through the meter can be measured. Multiplier resistors are used to extend the range of the meter movement for voltage measurements, while shunt resistors are used to extend the range of the meter movement for current measurements.

Today, most meters are solid-state digital systems; they are easier to read than mechanical (analog) meters and have no meter movement or moving parts.

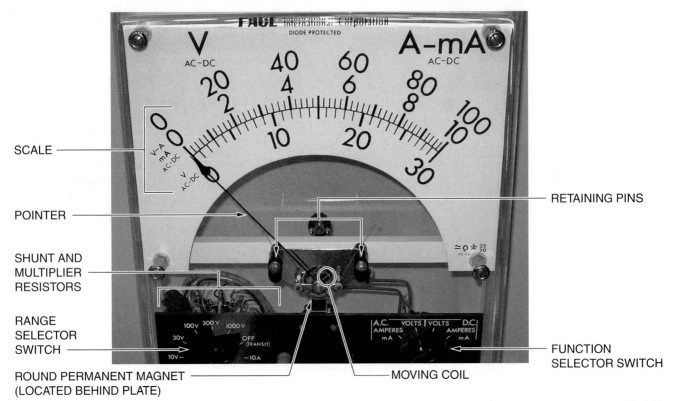

Figure 1 ◆ d'Arsonval meter movement.

112F01.EPS

Phantom Readings

The sensitivity of a digital meter can sometimes produce a low reading known as a phantom or ghost reading. This is due to the induction from the electrical field around the energized conductors in close proximity to the meter.

2.1.0 Voltmeter

A voltmeter is used to measure voltage, also known as potential difference or electromotive force (emf). It is connected in parallel with the circuit or component being measured. An analog meter uses the basic d'Arsonval meter movement with internally switched resistors to measure different voltage ranges. A digital meter uses an analog-to-digital converter chip to convert the sensed values into a digital or graphic display.

Many digital voltmeters are autoranging, which means that the meter will automatically search for the correct scale. When using a voltmeter that is not autoranging, always start with the highest voltage range and work down until the indication reads somewhere between half and three-quarter scale. This will provide a more accurate reading and prevent damage to the meter. On many meters, a DC value is indicated by a straight line with three dashes beneath it, while an AC value is indicated by a sine wave.

 WARNING!

Measuring voltages above 50V exposes the technician to potentially life-threatening hazards. Follow all applicable safety procedures as found in *NFPA 70E*, OSHA standards, and company/institutional policies and standards.

A voltmeter is used when the exact value of the voltage is required. However, electricians are often concerned with identifying only whether voltage is present, and if so, the general range of the voltage. In other words, is it energized, and if so, is it at 120V, 240V, or 480V? In these cases, a voltage tester is used. The range of voltage and the type of current (AC and/or DC) that a voltage tester is capable of measuring are usually indicated on the scales that display the reading. See *Figure 2*.

Advanced voltage testers offer additional features, such as a digital readout, GFCI test capability, and even the ability to switch between use as a contact and noncontact detector. See *Figure 3*.

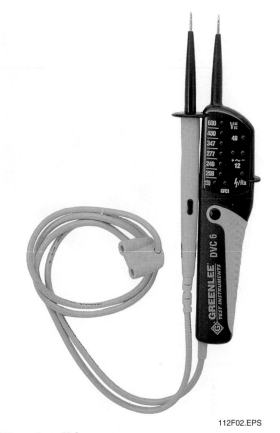

112F02.EPS

Figure 2 ◆ Voltage tester.

A voltage tester must be checked before each use to make sure that it is in good condition and is operating correctly. The external check of the tester should include a careful inspection of the insulation on the leads for cracks or frayed areas. Faulty leads constitute a safety hazard, so they must be replaced. As a check to make sure that the voltage tester is operating correctly, the probes of the tester are first connected to a known energized source. The voltage indicated on the tester should match the voltage of the source. If there is no indication, the voltage tester is not operating correctly, and it must be repaired or replaced. It must also be repaired or replaced if it indicates a voltage different from the known voltage of the source.

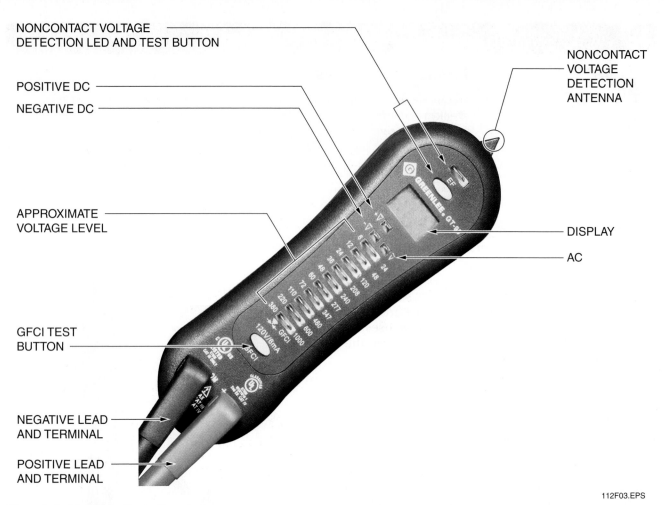

NONCONTACT VOLTAGE
DETECTION LED AND TEST BUTTON

NONCONTACT
VOLTAGE
DETECTION
ANTENNA

POSITIVE DC

NEGATIVE DC

APPROXIMATE
VOLTAGE LEVEL

DISPLAY

AC

GFCI TEST
BUTTON

NEGATIVE LEAD
AND TERMINAL

POSITIVE LEAD
AND TERMINAL

112F03.EPS

Figure 3 ◆ Multi-function voltage tester.

CAUTION

Care should be taken when placing the probes of the tester across the voltage source. Some voltage testers are designed to take quick readings and may be damaged if left in contact with the voltage source for too long.

Voltage testers are used to make sure that voltage is available when it is needed and to ensure that power has been cut off when it should have been. In a troubleshooting situation, it might be necessary to verify that power is available in order to be sure that lack of power is not the problem. For example, if there were a problem with a power tool, such as a drill, a voltage tester might be used to make sure that power is available to run the drill. A voltage tester might also be used to verify that there is power available to a three-phase motor that will not start.

WARNING!

When testing for voltage to verify that a circuit is de-energized as part of an electrical lockout, always perform a live-dead-live test. This involves first verifying operation of the test equipment on a known energized (live) source, then de-energizing and testing the target circuit to ensure it is dead, then again testing the known energized (live) source before making contact with the de-energized circuit. This is an OSHA requirement for systems above 600V but is a good practice at all voltage levels.

2.2.0 Ohmmeter

An ohmmeter measures the resistance of a circuit or component. It can also be used to locate open circuits or shorted circuits. An ohmmeter consists of a DC current meter movement, a low-voltage DC power source (usually a battery), and current-limiting resistors, all of which are connected in series with the meter (*Figure 4*).

Voltage Detectors

Simple noncontact (proximity) voltage detectors can also be used to indicate the presence of voltage within their specified range rating, but do not discriminate between ranges of values in the same way as a voltage tester. They are handy for quickly scanning for the presence of voltage in junction boxes or termination cabinets, and can even be used to trace circuits through walls. The voltage detector shown here glows in the presence of voltages between 50V and 1,000V.

112SA01.EPS

the leads. Some analog ohmmeters have a zero adjustment knob. With the leads connected together, turn the adjustment knob until the meter registers zero ohms. This adjustment must be made each time a different range is selected.

Many digital ohmmeters are autoranging. The correct scale is internally selected and the reading will indicate the range (ohms, K-ohms, or M-ohms). Analog ohmmeters require that you select the desired range. Most digital meters also have an audible tone when the measured value is very low or at zero ohms. This indicates a closed circuit and is useful when using the meter as a continuity tester. A continuity test is used to determine if a circuit is complete.

Figure 4 ◆ Ohmmeter schematic.

112F04.EPS

WARNING!

Prior to taking a reading with an ohmmeter, verify that both sides of the circuit are de-energized by using a voltmeter. If the circuit were energized, its voltage could cause a damaging current to flow through the meter. This can damage the meter and/or circuit and cause personal injury.

Before measuring the resistance of an unknown resistor or electrical circuit, connect the test leads together. This zeroes out or nulls the resistance of

When making resistance measurements in circuits, each component in the circuit can be tested individually by removing the component from the circuit and connecting the ohmmeter leads across it. However, the component does not have to be totally removed from the circuit. Usually, the part can be effectively isolated by disconnecting one of its leads from the circuit. Note that this method can still be somewhat time consuming.

2.3.0 Ammeter

A clamp-on ammeter, also known as a clamp meter, can measure current without having to make contact with uninsulated wires (*Figure 5*). This type of meter operates by sensing the strength of the electromagnetic field around the wire(s).

Clamp-on ammeters measure current by using simple transformer principles. The conductor(s) being measured would be the primary and the jaws (clamp) of the meter would be the secondary. The current in the primary winding induces a current in the secondary winding. If the ratio of the primary winding to the secondary winding is 1,000, then the secondary current is $\frac{1}{1000}$ of the current flowing in the primary. The smaller secondary current is connected to the meter's input. For example, a 1A current in the conductor will produce 0.001A (1mA) in the meter.

To measure the current, open the jaws of the meter and close them around the conductor(s) that are to be measured. Make sure that the jaws are clean and close tightly. Then read the magnitude of the current on the meter display.

Many meters have a HOLD function. This is useful in tight locations when it is hard to read the

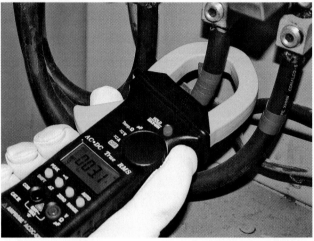

112F05.EPS

Figure 5 ◆ Clamp-on ammeter.

meter while it is clamped around the conductor(s). Just press the HOLD button when the value is measured, then remove and read the meter.

CAUTION

When using a clamp-on ammeter, make sure that the range of the meter is at least as high as the current to be measured. If the meter is digital and the current is too high, the display will read OL (overload). This means that the meter has been overloaded. If the meter is an analog meter, the indicating needle will peg (move) above the maximum limit on the scale, which might damage the meter.

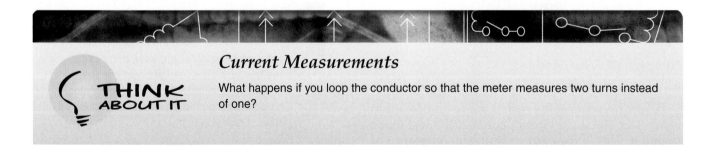

Some meters have a MIN/MAX or peak function. This allows the technician to record the maximum inrush, as with a motor start.

WARNING!

Using a clamp-on ammeter may expose you to energized systems and equipment. Never use this type of meter unless you are qualified and are following NFPA, OSHA, and company/institutional safety procedures.

2.4.0 Multimeter

The multimeter is also known as a volt-ohm-milliammeter (VOM). An analog VOM is shown in *Figure 6*. It is a multi-purpose instrument that combines the three previous meters discussed. When using an analog meter, you must select the proper voltage (DC or AC) and range (in volts, amps, or ohms). When using a digital VOM (*Figure 7*), you must select the proper voltage. Most have an autoranging feature for the magnitude.

WARNING!

Before use, a VOM must be checked on a known power source.

112F06.EPS

Figure 6 ◆ Analog VOM.

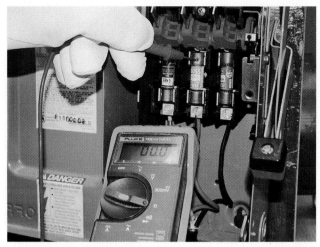

112F07.EPS

Figure 7 ◆ Digital VOM.

Use of a VOM is the same as using the individual voltmeter, ohmmeter, and clamp-on ammeter. Current clamps can be used with most multimeters for measuring AC and DC currents above the milliamp level. These can be plugged into either the amp or voltage test lead connections, depending on the current clamp. Always refer to the manufacturer's instructions for the device in use. Current clamps are available in various current ranges, from 50A to several thousand amps. The jaws are available in different shapes and sizes to suit various applications, from round to rectangular, and even flexible.

Some multimeters can also measure the frequency of an AC waveform—this function is useful for diagnosing harmonic problems in an electrical distribution system.

Another additional feature is the MIN/MAX memory function. It will record the minimum and maximum readings over the time period selected. Other common multimeter functions include capacitance measurement, diode and transistor testers, temperature measurement, and true RMS measurement for accurate voltage and current readings at different frequencies. Refer to the manufacturer's instructions for the meter in use. In addition to the current clamps used with standard multimeters, clamp-on multimeters are also available (*Figure 8*). They are used to measure AC current, AC and DC voltage, resistance, and other values.

2.5.0 Megohmmeter

An ordinary ohmmeter cannot be used for measuring resistances of several million ohms, such as those found in conductor insulation or between motor or transformer windings. The instrument used to measure very high resistances is known as a megohmmeter. Megohmmeters are also called

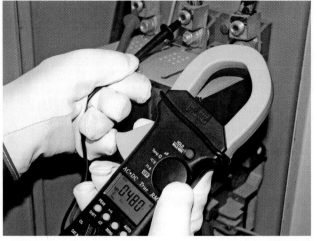

Figure 8 ◆ Clamp-on multimeter.

112F08.EPS

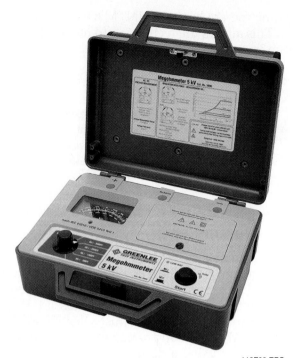

112F09.EPS

Figure 9 ◆ Battery-operated megohmmeter.

Meggers®, or insulation resistance testers. They can be powered by alternating current, battery (*Figure 9*), or hand cranking (*Figure 10*).

When using a megohmmeter, you could be injured or cause damage to the equipment you are working on if the following minimum safety precautions are not followed:

- High voltages are present when using a megohmmeter. For example, in 600V class systems, applied megohmmeter voltages are typically 500V and 1,000V. Only qualified individuals may use this equipment. Always wear appropriate personal protective equipment when approaching energized parts.
- De-energize and verify the de-energization of the circuit before connecting the meter. Make sure all capacitors are discharged.
- If possible, disconnect the item being checked from the other circuit components before using the meter.
- Do not exceed the manufacturer's recommended voltage test levels for the cable or equipment under test. Many manufacturers have different test levels based on the age of the cable or equipment being tested.
- Never touch the test leads when the meter is energized or powered. Meggers generate high voltage, and touching the leads could result in injury or electrical shock.
- After the test, discharge any energy that may be left in the circuit by grounding the conductor or equipment for a period of time equal to the duration of the test.
- When megging cables or busducts where you have exposed parts that are remote from your testing position, safely secure or barricade the exposed end to protect others from inadvertent contact with the test voltage.

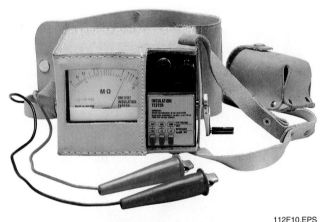

112F10.EPS

Figure 10 ◆ Hand-crank megohmmeter.

CAUTION

If a megohmmeter is used to test switchgear, all of the electronics must be disconnected prior to testing the switchgear. The voltage produced by the megohmmeter may damage electronic equipment.

Meter manufacturers supply detailed manuals for testing various devices and equipment. Always follow these instructions.

Meter Care

Like all meters, a megohmmeter is a sensitive instrument. Treat it with care and keep it in its case when not in use.

112SA02.EPS

2.6.0 Motor and Phase Rotation Testers

Before connecting a three-phase motor to a circuit, you must first match the legs or windings of the motor (T1, T2, and T3) to the phases of the circuit (L1, L2, and L3). This ensures that the motor rotates in the proper direction. A motor rotation tester can be used to identify the legs of the motor (*Figure 11*), while a phase rotation tester can be used to identify the phases of the circuit (*Figure 12*).

 WARNING!
Do not connect a motor rotation tester to energized equipment. This can result in injury and equipment damage.

A motor rotation tester is a passive device; it operates on residual magnetism present in the motor after it has been run or tested by the manufacturer prior to shipping. To use a motor rotation tester, connect the three motor wires to the T1, T2, and T3 leads on the tester, then rotate the motor shaft a half-turn while pressing the TEST button

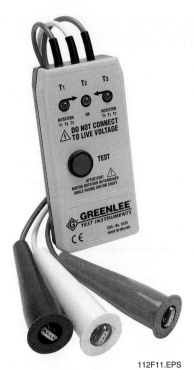

112F11.EPS

Figure 11 ◆ Motor rotation tester.

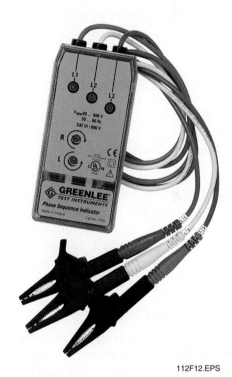

112F12.EPS

Figure 12 ◆ Phase rotation tester.

(the direction of rotation depends on the tester in use; always follow the manufacturer's instructions). Either the clockwise or counterclockwise LED will light up. If the required rotation is clockwise and the clockwise LED lights up, tag the motor wires to correspond to the motor rotation leads. If the required rotation is clockwise and the

Phase Rotation Tester

The leads for a phase rotation tester may not correspond to typical circuit color coding. A good idea is to put phase tape on your meter leads to correspond to the circuit color coding. This will help to ensure correct connections.

counterclockwise LED lights up, switch a pair of leads and retest.

A phase rotation tester, also called a phase sequence indicator, is used on three-phase electrical systems to indicate the phase sequence rotation of the voltages. These testers typically have LEDs to indicate the phase rotation. A phase sequence is measured as clockwise or counterclockwise rotation.

WARNING!

Phase rotation testers may only be used by qualified individuals. Always wear appropriate personal protective equipment when approaching energized parts.

A phase rotation tester is used when it is necessary to ensure the same phase rotation throughout a facility. To test phase rotation, de-energize and lock out power to the circuit, then connect the three leads of the tester to the phase conductors in the circuit. Next, safely energize the circuit and observe the meter. Make note of the color scheme of the connected leads to the system, along with the phase sequences as indicated on the meter. This is necessary to ensure that the added equipment follows the same phase rotation. De-energize and lock out the circuit before disconnecting the leads.

2.7.0 Recording Instruments

The term recording instrument describes many instruments that make a permanent record of measured quantities over a period of time.

Recording instruments use either a paper strip or electronic accessible memory. Those using electronic memory are usually called data loggers. These instruments record electrical quantities, including potential difference, current, power, resistance, and frequency. They can also record nonelectrical quantities by electrical means, such as a temperature recorder that uses a potentiometer system to record thermocouple output.

It is often necessary to know the conditions that exist in an electrical circuit over a period of time to determine such things as peak loads, voltage fluctuations, and so on. An automatic recording instrument can be connected to take readings at specified intervals for later review and analysis. Some meters can upload data to a PC for real-time data logging and graphing. See *Figure 13*.

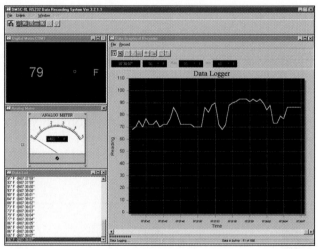

112F13.EPS

Figure 13 ◆ Data recording system.

3.0.0 ◆ CATEGORY RATINGS

Distribution systems and loads are becoming more complex, increasing the risk of transient power spikes. Lightning strikes on outdoor transmission lines and switching surges from normal switching operations can also produce dangerous high-energy transients. Motors, capacitors, variable speed drives, and power conversion equipment can also generate power spikes.

Safety systems are built into test equipment to protect electricians from transient power spikes. The International Electrotechnical Commission (IEC) developed a new safety standard, *IEC 1010*, for test equipment that was adapted as *UL Standard UL3111-1*. These standards define four overvoltage installation categories, often abbreviated as CAT I, CAT II, CAT III, and CAT IV (*Table 1*). These categories identify the hazards posed by transients; the higher the category number, the greater the risk to the electrician. A higher category number refers to an installation with higher power available and higher-energy transients.

When selecting a meter, choose a meter rated for the highest category you will be working in. Then select the appropriate voltage level. In addition, make sure that your test leads are rated as high as your meter. Choose meters that are independently tested and certified by UL, CSA, or another recognized testing organization. Certified meters are marked with the category rating on the meter housing (*Figure 14*).

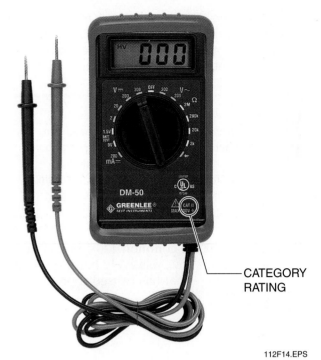

CATEGORY RATING

112F14.EPS

Figure 14 ◆ Category rating on a typical meter.

4.0.0 ◆ SAFETY

Safety must be the primary responsibility of all personnel on a job site. The safe installation, maintenance, and operation of electrical equipment requires strict adherence to local and national codes and safety standards, as well as facility and company safety policies. Carelessness can result in serious injury or death due to electrical shock, burns, falls, flying objects, etc. After an accident has occurred, investigation almost invariably shows that it could have been prevented by the exercise of simple safety precautions and procedures. It is your personal responsibility to identify and eliminate unsafe conditions and unsafe acts that cause accidents.

You must bear in mind that de-energizing main supply circuits by opening supply switches will not necessarily de-energize all circuits in a given piece of equipment. A source of danger that has often been neglected or ignored, sometimes with tragic results, is the input to electrical equipment from other sources, such as backfeeds. The rescue of a victim shocked by the power input from a backfeed is often hampered because of the time required to determine the source of power and

Table 1	Overvoltage Installation Categories
Overvoltage Category	**Installation Examples**
CAT I	Electronic equipment and circuitry
CAT II	Single-phase loads such as small appliances and tools, outlets at more than 30 feet from a CAT III source or 60 feet from a CAT IV source
CAT III	Three-phase motors, single-phase commercial or industrial lighting, switchgear, busduct and feeders in industrial plants
CAT IV	Three-phase power at meter, service-entrance, or utility connection, any outdoor conductors

isolate it. Always turn off all power inputs before working on equipment and lock out and tag, then check with an operating voltage tester to be sure that the equipment is safe to work on.

 WARNING!

When performing lockout/tagout procedures, remember that other forms of energy may be present and must also be locked out and tagged. These include water pressure, steam, springs, gravity, and other forms of energy.

Remember that the common 120V power supply voltage is not a low, relatively harmless voltage but is a voltage that has caused more deaths than any other.

Safety can never be stressed enough. There are times when your life literally depends on it. Always observe the following precautions:

- Thoroughly inspect all test equipment before each use. Check for broken leads or knobs, damaged plugs, or frayed cords. Do not use equipment that is wet or damaged.
- Make sure the rating of any leads or accessories meets or exceeds the rating of the meter.
- Do not work with energized equipment unless you are both qualified and approved by your supervisor.
- Never shortcut safety; strictly adhere to all energized work policies and procedures.
- When testing circuits, test at higher ranges first, then work your way down to lower ranges.
- Always have a standby person present during hot work; he or she should know whom to contact in case of emergency and how to disconnect the power.

Putting It All Together

What kind of measuring device would you select or need for the following tasks, and how would you apply the device?

- Identify a short circuit in house wiring
- Measure the secondary voltage of an AC transformer
- Identify a blown fuse in a circuit
- Identify the contact configuration of a three-way switch or multi-pole relay

1. Digital meters do *not* have a(n) _____.
   a. red lead wire
   b. battery
   c. meter movement
   d. autoranging setting

2. A voltmeter is used to test _____.
   a. exact voltages
   b. voltage ranges
   c. power
   d. sine waves

3. In order to ensure safety, before measuring low voltages, you should first test for _____.
   a. resistance
   b. current
   c. vibration
   d. higher voltages

4. An ammeter is used to measure _____.
   a. current
   b. voltage
   c. resistance
   d. insulation value

5. Clamp-on ammeters operate by _____.
   a. using d'Arsonval meter movement
   b. sensing the strength of the electromagnetic field around the wire
   c. measuring the high resistance end of a power transformer
   d. using a resistive shunt

6. An insulation tester is another name for a(n) _____.
   a. megohmmeter
   b. ammeter
   c. multimeter
   d. continuity tester

7. A motor rotation tester _____.
   a. tests an energized motor
   b. gets connected to the motor supply conductors
   c. works on residual magnetism
   d. works only on clockwise rotation

8. The highest level of protection is provided by _____ rated instruments.
   a. CAT I
   b. CAT II
   c. CAT III
   d. CAT IV

9. The International Electrotechnical Commission (IEC) developed a new safety standard for overvoltage installation categories of CAT I, CAT II, CAT III, and CAT IV for _____.
   a. wiring
   b. electrical equipment
   c. test equipment
   d. signaling circuits

10. De-energizing the main supply circuit will always de-energize all circuits in a given piece of equipment.
    a. True
    b. False

Summary

Meters and other devices are used to test and troubleshoot circuits and electrical equipment. One of the most important tests you will perform is verifying the absence of voltage before working on a device or circuit. This is often done using a voltage tester. Other common test equipment includes multimeters, clamp-on ammeters, megohmmeters, and motor/phase rotation testers. You must understand the operation of and safety precautions for each piece of test equipment. In addition, you must be able to select the appropriate test equipment based on the task and category rating of the environment in which the work is to be performed. Always inspect and verify the operation of all test equipment before using it. Your life may depend on it.

Notes

Trade Terms Quiz

1. Used for electromagnetic effects or for providing electrical resistance, a _____ is a number of turns of wire.

2. A _____ uses a permanent magnet and moving coil arrangement to move a pointer across a scale.

3. Usually expressed in hertz, _____ is the number of cycles completed each second by a given AC voltage.

4. _____ is an uninterrupted electrical path for current flow.

Trade Terms

Coil
Continuity
d'Arsonval meter movement
Frequency

Ed Cockrell
All Star Electric

How did you choose a career in the electrical field?
I was looking for a job that would hold my interest for more than a year. I studied electrical engineering at Louisiana State University for two years. I was given an opportunity to work as an electrical helper for a summer and found it very rewarding.

Tell us about your apprenticeship experience.
I entered an electrical apprenticeship program with very little idea that it would lead to the position that I now have. I started work with a very tough ex-Marine as my first journeyman. The first day on the job I learned some valuable lessons. It left me thinking that I had chosen the wrong field of work. I had very good instructors in the classroom, but I found that there was much to learn in the field. I lucked out and found an excellent mentor who changed my perception of the electrical field, my classes, and my possible future. He taught me that entering into this field would be an open door to many other options that I had never thought about. These were to be a foreman, superintendent, project manager, and potentially an owner of my own business.

What positions have you held in the industry?
I started as a summer helper; then apprentice, journeyman, foreman, field superintendent, electrical designer, project manager, electrical instructor, lead electrical instructor, and Director of Education for ABC.

What would you say is the primary factor in achieving success?
Education, luck, and the perseverance to stick with something that I enjoyed doing.

What does your current job involve?
My job is directing the efforts of some very good craftpersons (instructors) who are trying to help our students become craft professionals. I also create the budgets and beg for contributions to the school. I believe that a very important part of my job is to instill in our students that they determine where they end up in life by their actions. Becoming a craft professional is one point along the way, but there are many more options available to them if they truly wish to take steps to achieve them.

Do you have any advice for someone just entering the trade?
Stick with it and don't fall into the thought process that we must only learn from our mistakes. Take time to learn from your successes by noting the path to success. Life is like a grindstone—those who learn to align themselves with it become sharper, and those who push head first with little regard for the right way to do things tend to lose their edge.

Trade Terms
Introduced in This Module

Coil: A number of turns of wire, especially in spiral form, used for electromagnetic effects or for providing electrical resistance.

Continuity: An electrical term used to describe a complete (unbroken) circuit that is capable of conducting current. Such a circuit is also said to be closed.

d'Arsonval meter movement: A meter movement that uses a permanent magnet and moving coil arrangement to move a pointer across a scale.

Frequency: The number of cycles completed each second by a given AC voltage; usually expressed in hertz. One hertz equals one cycle per second.

This module is intended to be a thorough resource for task training. The following reference works are suggested for further study. These are optional materials for continued education rather than for task training.

ABCs of Multimeter Safety, Everett, WA: Fluke Corporation.

ABCs of DMMs, Multimeter Features and Functions Explained, Everett, WA: Fluke Corporation.

Clamp Meter ABCs, Everett, WA: Fluke Corporation.

Electronics Fundamentals: Circuits, Devices, and Applications, Thomas L. Floyd. New York: Prentice Hall.

Power Quality Analyzer Uses for Electricians, Everett, WA: Fluke Corporation.

Principles of Electric Circuits, Thomas L. Floyd. New York: Prentice Hall.

NCCER makes every effort to keep these textbooks up-to-date and free of technical errors. We appreciate your help in this process. If you have an idea for improving this textbook, or if you find an error, a typographical mistake, or an inaccuracy in NCCER's Contren® textbooks, please write us, using this form or a photocopy. Be sure to include the exact module number, page number, a detailed description, and the correction, if applicable. Your input will be brought to the attention of the Technical Review Committee. Thank you for your assistance.

Instructors – If you found that additional materials were necessary in order to teach this module effectively, please let us know so that we may include them in the Equipment/Materials list in the Annotated Instructor's Guide.

Write: Product Development and Revision
           National Center for Construction Education and Research
           3600 NW 43rd St., Bldg. G, Gainesville, FL 32606

Fax: 352-334-0932

E-mail: curriculum@nccer.org

Craft _____ Module Name _____

Copyright Date _____ Module Number _____ Page Number(s) _____

Description _____

_____

_____

_____

(Optional) Correction _____

_____

_____

(Optional) Your Name and Address _____

_____

_____

Glossary

90° bend: A bend that changes the direction of the conduit by 90°.

Accessible: Able to be reached, as for service or repair.

Ammeter: An instrument for measuring electrical current.

Ampacity: The current in amperes a conductor can carry continuously under the conditions of use without exceeding its temperature rating.

Ampere (A): A unit of electrical current. For example, one volt across one ohm of resistance causes a current flow of one ampere.

Appliance: Equipment designed for a particular purpose (for example, using electricity to produce heat, light, or mechanical motion). Appliances are usually self-contained, are generally available for applications other than industrial use, and are normally produced in standard sizes or types.

Approved: Meeting the requirements of an appropriate regulatory agency.

Architectural drawings: Working drawings consisting of plans, elevations, details, and other information necessary for the construction of a building. Architectural drawings usually include:

- A site (plot) plan indicating the location of the building on the property
- Floor plans showing the walls and partitions for each floor or level
- Elevations of all exterior faces of the building
- Several vertical cross sections to indicate clearly the various floor levels and details of the footings, foundations, walls, floors, ceilings, and roof construction
- Large-scale detail drawings showing such construction details as may be required

Articles: The articles are the main topics of the *NEC®*, beginning with *NEC Article 90*, Introduction, and ending with *NEC Article 830*, Network-Powered Broadband Communications Systems.

Atom: The smallest particle to which an element may be divided and still retain the properties of the element.

Back-to-back bend: Any bend formed by two 90° bends with a straight section of conduit between the bends.

Battery: A DC voltage source consisting of two or more cells that convert chemical energy into electrical energy.

Block diagram: A single-line diagram used to show electrical equipment and related connections. See power-riser diagram.

Blueprint: An exact copy or reproduction of an original drawing.

Bonding bushing: A special conduit bushing equipped with a conductor terminal to take a bonding jumper. It also has a screw or other sharp device to bite into the enclosure wall to bond the conduit to the enclosure without a jumper when there are no concentric knockouts left in the wall of the enclosure.

Bonding jumper: A bare or green insulated conductor used to ensure the required electrical conductivity between metal parts required to be electrically connected. Bonding jumpers are frequently used from a bonding bushing to the service-equipment enclosure to provide a path around concentric knockouts in an enclosure wall, and they may also be used to bond one raceway to another.

Bonding wire: A wire used to make a continuous grounding path between equipment and ground.

Branch circuit: The portion of a wiring system extending beyond the final overcurrent device protecting a circuit.

Cable trays: Rigid structures used to support electrical conductors.

Capstan: The turning drum of the cable puller on which the rope is wrapped and pulled.

Chapters: Nine chapters form the broad structure of the *NEC®*.

Circuit: A complete path for current flow.

Coil: A number of turns of wire, especially in spiral form, used for electromagnetic effects or for providing electrical resistance.

Concentric bends: 90° bends made in two or more parallel runs of conduit with the radius of each bend increasing from the inside of the run toward the outside.

Conductor: A material through which it is relatively easy to maintain an electric current.

Conduit: A round raceway, similar to pipe, that houses conductors.

Connector: Device used to physically connect conduit or cable to an outlet box, cabinet, or other enclosure.

Continuity: An electrical term used to describe a complete (unbroken) circuit that is capable of conducting current. Such a circuit is also said to be closed.

Coulomb: An electrical charge equal to 6.25×10^{18} electrons or 6,250,000,000,000,000,000 electrons. A coulomb is the common unit of quantity used for specifying the size of a given charge.

Current: The movement, or flow, of electrons in a circuit. Current (I) is measured in amperes.

d'Arsonval meter movement: A meter movement that uses a permanent magnet and moving coil arrangement to move a pointer across a scale.

Detail drawing: An enlarged, detailed view taken from an area of a drawing and shown in a separate view.

Developed length: The actual length of the conduit that will be bent.

Dimensions: Sizes or measurements printed on a drawing.

Double-insulated/ungrounded tool: An electrical tool that is constructed so that the case is insulated from electrical energy. The case is made of a nonconductive material.

Electrical drawing: A means of conveying a large amount of exact, detailed information in an abbreviated language. Consists of lines, symbols, dimensions, and notations to accurately convey an engineer's designs to electricians who install the electrical system on a job.

Electrical service: The electrical components that are used to connect the commercial power to the premises wiring system.

Electron: A negatively charged particle that orbits the nucleus of an atom.

Elevation drawing: An architectural drawing showing height, but not depth; usually the front, rear, and sides of a building or object.

Exceptions: Exceptions follow the applicable sections of the *NEC*® and allow alternative methods to be used under specific conditions.

Explosion-proof: Designed and constructed to withstand an internal explosion without creating an external explosion or fire.

Exposed location: Not permanently closed in by the structure or finish of a building; able to be installed or removed without damage to the structure.

Feeder: A circuit, such as conductors in conduit or a cable run, that carries current from the service equipment to a subpanel or a branch circuit panel or to some point in the wiring system.

Fibrillation: Very rapid irregular contractions of the muscle fibers of the heart that result in the heartbeat and pulse going out of rhythm with each other.

Fine print note (FPN): Explanatory material that follows specific *NEC*® sections.

Fish tape: A hand device used to pull a wire through a conduit run.

Floor plan: A drawing of a building as if a horizontal cut were made through a building at about window level, and the top portion removed. The floor plan is what would appear if the remaining structure were viewed from above.

Frequency: The number of cycles completed each second by a given AC voltage; usually expressed in hertz. One hertz equals one cycle per second.

Gain: Because a conduit bends in a radius and not at right angles, the length of conduit needed for a bend will not equal the total determined length. Gain is the distance saved by the arc of a 90° bend.

Ground fault circuit interrupter (GFCI): A protective device that functions to de-energize a circuit or portion thereof within an established period of time when a current to ground exceeds some predetermined value. This value is less than that required to operate the overcurrent protective device of the supply circuit.

Grounded tool: An electrical tool with a three-prong plug at the end of its power cord or some other means to ensure that stray current travels to ground without passing through the body of the user. The ground plug is bonded to the conductive frame of the tool.

Handy box: Single-gang outlet box used for surface mounting to enclose receptacles or wall switches on concrete or concrete block construction of industrial and commercial buildings; nongangable; also made for recessed mounting; also known as a utility box.

Insulator: A material through which it is difficult to conduct an electric current.

Joule (J): A unit of measurement that represents one newton-meter (Nm), which is a unit of measure for doing work.

Junction box: An enclosure where one or more raceways or cables enter, and in which electrical conductors can be, or are, spliced.

Kick: A bend in a piece of conduit, usually less than 45°, made to change the direction of the conduit.

Kilo: A prefix used to indicate one thousand; for example, one kilowatt is equal to one thousand watts.

Kirchhoff's current law: The statement that the total amount of current flowing through a parallel circuit is equal to the sum of the amounts of current flowing through each current path.

Kirchhoff's voltage law: The statement that the sum of all the voltage drops in a circuit is equal to the source voltage of the circuit.

Load center: A type of panelboard that is normally located at the service entrance of a residential installation. It usually contains the main disconnect.

Matter: Any substance that has mass and occupies space.

Mega: A prefix used to indicate one million; for example, one megawatt is equal to one million watts.

Metal-clad (Type MC) cable: A type of cable with a metal jacket that is popular for use in residential and small commercial wiring systems. In general, it may be used for both exposed and concealed work in normally dry locations. It may also be used as service-entrance cable and in other locations as permitted by the *NEC*®.

Mouse: A cylinder of foam rubber that fits inside the conduit and is then propelled by compressed air or vacuumed through the conduit run, pulling a line or tape.

National Electrical Manufacturers Association (NEMA): The association that maintains and improves the quality and reliability of electrical products.

National Fire Protection Association (NFPA): The publishers of the *NEC*®. The NFPA develops standards to minimize the possibility and effects of fire.

Nationally Recognized Testing Laboratories (NRTLs): Product safety certification laboratories that are responsible for testing and certifying electrical equipment.

Neutrons: Electrically neutral particles (neither positive nor negative) that have the same mass as a proton and are found in the nucleus of an atom.

Nonmetallic-sheathed (Type NM) cable: A type of cable that is popular for use in residential and small commercial wiring systems. In general, it may be used for both exposed and concealed work in normally dry locations. See Romex®.

Nucleus: The center of an atom. It contains the protons and neutrons of the atom.

Occupational Safety and Health Administration (OSHA): The federal government agency established to ensure a safe and healthy environment in the workplace.

Offsets: An offset (kick) is two bends placed in a piece of conduit to change elevation to go over or under obstructions or for proper entry into boxes, cabinets, etc.

Ohm (Ω): The basic unit of measurement for resistance.

Ohm's law: A statement of the relationships among current, voltage, and resistance in an electrical circuit: current (I) equals voltage (E) divided by resistance (R). Generally expressed as a mathematical formula: $I = E/R$.

Ohmmeter: An instrument used for measuring resistance.

One-line diagram: A drawing that shows, by means of lines and symbols, the path of an electrical circuit or system of circuits along with the various circuit components. Also called a single-line diagram.

On-the-job training (OJT): Job-related learning acquired while working.

Outlet box: A metallic or nonmetallic box installed in an electrical wiring system from which current is taken to supply some apparatus or device.

Parallel circuits: Circuits containing two or more parallel paths through which current can flow.

Parts: Certain articles in the *NEC®* are subdivided into parts. Parts have letter designations (e.g., Part A).

Plan view: A drawing made as though the viewer were looking straight down (from above) on an object.

Polychlorinated biphenyls (PCBs): Toxic chemicals that may be contained in liquids used to cool certain types of large transformers and capacitors.

Power: The rate of doing work or the rate at which energy is used or dissipated. Electrical power is the rate of doing electrical work. Electrical power is measured in watts.

Power-riser diagram: A single-line block diagram used to indicate the electric service equipment, service conductors and feeders, and subpanels. Notes are used on power-riser diagrams to identify the equipment; indicate the size of conduit; show the number, size, and type of conductors; and list related materials. A panelboard schedule is usually included with power-riser diagrams to indicate the exact components (panel type and size), along with fuses, circuit breakers, etc., contained in each panelboard.

Protons: The smallest positively charged particles of an atom. Protons are contained in the nucleus of an atom.

Pull box: A sheet metal box-like enclosure used in conduit runs to facilitate the pulling of cables from point to point in long runs, or to provide for the installation of conduit support bushings needed to support the weight of long riser cables, or to provide for turns in multiple-conduit runs.

Raceway systems: Conduit, fittings, boxes, and enclosures that house the conductors in an electrical system.

Raceways: Enclosed channels designed expressly for holding wires, cables, or busbars, with additional functions as permitted in the *NEC®*.

Raintight: Constructed or protected so that exposure to a beating rain will not result in the entrance of water under specified test conditions.

Relay: An electromechanical device consisting of a coil and one or more sets of contacts. Used as a switching device.

Resistance: An electrical property that opposes the flow of current through a circuit. Resistance (R) is measured in ohms.

Resistor: Any device in a circuit that resists the flow of electrons.

Rise: The length of the bent section of conduit measured from the bottom, centerline, or top of the straight section to the end of the bent section.

Romex®: General Cable's trade name for Type NM cable; however, it is often used generically to refer to any nonmetallic-sheathed cable.

Rough-in: The installation of the raceway system (including conduit, boxes, and enclosures), wiring, or cable.

Roughing in: The first stage of an electrical installation, when the raceway, cable, wires, boxes, and other equipment are installed. This is the electrical work that must be done before any finishing work can be done.

Scale: On a drawing, the size relationship between an object's actual size and the size it is drawn. Scale also refers to the measuring tool used to determine this relationship.

Schedule: A systematic method of presenting equipment lists on a drawing in tabular form.

Schematic: A type of drawing in which symbols are used to represent the components in a system.

Schematic diagram: A detailed diagram showing complicated circuits, such as control circuits.

Sectional view: A cutaway drawing that shows the inside of an object or building.

Sections: Parts and articles are subdivided into sections. Sections have numeric designations that follow the article number and are preceded by a period (e.g., 501.4).

Segment bend: A large bend formed by multiple short bends or shots.

Series circuit: A circuit with only one path for current flow.

Series circuits: Circuits with only one path for current flow.

Series-parallel circuits: Circuits that contain both series and parallel current paths.

Service drop: The overhead conductors, through which electrical service is supplied, between the last power company pole and the point of their connection to the service facilities located at the building.

Service entrance: The point where power is supplied to a building (including the equipment used for this purpose). The service entrance includes the service main switch or panelboard, metering devices, overcurrent protective devices, and conductors/raceways for connecting to the power company's conductors.

Service lateral: The underground conductors through which service is supplied between the power company's distribution facilities and the first point of their connection to the building or area service facilities located at the building.

Service-entrance conductors: The conductors between the point of termination of the overhead service drop or underground service lateral and the main disconnecting device in the building.

Service-entrance equipment: Equipment that provides overcurrent protection to the feeder and service conductors, a means of disconnecting the feeders from energized service conductors, and a means of measuring the energy used.

Shop drawing: A drawing that is usually developed by manufacturers, fabricators, or contractors to show specific dimensions and other pertinent information concerning a particular piece of equipment and its installation methods.

Site plan: A drawing showing the location of a building or buildings on the building site. Such drawings frequently show topographical lines, electrical and communication lines, water and sewer lines, sidewalks, driveways, and similar information.

Solenoid: An electromagnetic coil used to control a mechanical device such as a valve.

Splice: Connection of two or more conductors.

Stub-up: Another name for the rise in a section of conduit. Also, a term used for conduit penetrating a slab or the ground.

Switch: A mechanical device used for turning an electrical circuit on and off.

Switch leg: A circuit routed to a switch box for controlling electric lights.

Tap: Intermediate point on a main circuit where another wire is connected to supply electrical current to another circuit.

Transformer: A device consisting of one or more coils of wire wrapped around a common core. It is commonly used to step voltage up or down.

Trim-out: After rough-in, the installation and termination of devices and fixtures.

Trough: A long, narrow box used to house electrical connections that could be exposed to the environment.

Underwriters Laboratories, Inc. (UL): An agency that evaluates and approves electrical components and equipment.

Valence shell: The outermost ring of electrons that orbit about the nucleus of an atom.

Volt (V): The unit of measurement for voltage (electromotive force or difference of potential). One volt is equivalent to the force required to produce a current of one ampere through a resistance of one ohm.

Voltage: The driving force that makes current flow in a circuit. Voltage (E) is also referred to as electromotive force or difference of potential.

Voltage drop: The change in voltage across a component that is caused by the current flowing through it and the amount of resistance opposing it.

Voltmeter: An instrument for measuring voltage. The resistance of the voltmeter is fixed. When the voltmeter is connected to a circuit, the current passing through the meter will be directly proportional to the voltage at the connection points.

Waterproof: Constructed so that moisture will not interfere with successful operation.

Watertight: Constructed so that water will not enter the enclosure under specified test conditions.

Watt (W): The basic unit of measurement for electrical power.

Weatherproof: Constructed or protected so that exposure to the weather will not interfere with successful operation.

Wire grip: A device used to link pulling rope to cable during a pull.

Wireways: Steel troughs designed to carry electrical wire and cable.

Written specifications: A written description of what is required by the owner, architect, and engineer in the way of materials and workmanship. Together with working drawings, the specifications form the basis of the contract requirements for construction.

Module 26101-08

Associated Builders and Contractors, Inc.,
    Module divider
Tim Dean, 101F01A
Topaz Publications, Inc., 101F01B–101F01E,
    101F02A, 101F03–101F08
Progress Lighting, 101F01F
Mike Powers, 101F02B, 101F02C
Notifier, 101F02D
Hubbell Lighting, 101F02E
Kim Lighting, 101F02F
Dumond Chemicals, Inc., 101SA01
Pepsi Center, Denver, Colorado, Module
    overview, 101SA02

Module 26102-08

TIC, The Industrial Company, Module divider
Topaz Publications, Inc., 102F02, 102F05,
    102F06A
Panduit Corp., 102F06B
Mike Powers, 102F08–102F10, 102F14,
    102F16–102F20, 102SA01, 102SA03–102SA10,
    102SA12–102SA14
The Stanley Works, 102F12A, 102F13
Greenlee Textron, Inc., a subsidiary of Textron
    Inc., 102F12B, 102F12D
RIDGID®, 102F12C
Tim Ely, 102SA02
Datum Filing Systems, Inc., 102SA11
Occupational Safety and Health Administration,
    102T01, 102T02

Module 26103-08

AGC of America, Module divider
Topaz Publications, Inc., 103F13B (Bottom),
    103F18, 103F19, 103F25, 103SA02, 103SA04,
    103SA05
Tim Dean, Module overview, 103F20

Amprobe Instruments, 103F24
U.S. Army Corps of Engineers, 103SA01
Extech Instruments, Inc., 103SA06

Module 26104-08

AGC of America, Module divider
Topaz Publications, Inc., 104SA01

Module 26105-08

Associated Builders and Contractors, Inc.,
    Module divider
John Traister, 105F03
Topaz Publications, Inc., 105SA01, 105SA02,
    105SA04
Tim Ely, 105SA03

Module 26106-08

Associated Builders and Contractors, Inc.,
    Module divider
Topaz Publications, Inc., Module overview,
    106F01, 106F04, 106F06, 106F17–106F20,
    106SA01
John Traister, 106F09–106F12, 106F14
John Autrey, 106SA02

Module 26107-08

Associated Builders and Contractors, Inc.,
    Module divider
Topaz Publications, Inc., Module overview,
    107F01, 107F03–107F05, 107SA01–107SA03,
    107F11B, 107F24B, 107F25, 107F26B, 107F27,
    107SA07–107SA09
John Traister, 107T01, 107F08–107F10
Greenlee Textron, Inc., a subsidiary of Textron
    Inc., 107SA04
Tim Ely, 107SA05
Klein Tools, Inc., 107SA06

Module 26108-08

Associated Builders and Contractors, Inc., Module divider
Veronica Westfall, Module overview
Topaz Publications, Inc., 108F03, 108F04A, 108F08,108F12–108F19, 108F25, 108F35B, 108F42, 108F62, 108F78, 108SA01–108SA09, 108SA11, 108SA12, 108SA15
Tim Dean, 108F04B, 108F61, 108F77, 108SA10, 108SA14
Wiremold/Legrand, 108F58, 108F59
Panduit Corp., 108F60
Cooper B-Line, 108F65
Jim Mitchem, 108F69, 108F70, 108SA13
Greenlee Textron, Inc., a subsidiary of Textron Inc., 108F74, 108F75
VP Buildings, 108F79
Nucor Corporation—Vulcraft Group, 108F80
Reprinted with permission from National Fire Protection Association 70-2008. National Electrical Code, Copyright ©2007, National Fire Protection Association, Quincy, MA 02269. This reprinted material is not the complete and official position of the NFPA on the referenced subject, which is represented only by the standard in its entirety, 108T01.

Module 26109-08

AGC of America, Module divider
Tim Dean, 109F04, 109F16
Topaz Publications, Inc., 109F07B, 109F08B, 109F15, 109F17B, 109F18, 109F19, 109F23, 109F24, 109F26, 109SA01–109SA03
Jim Mitchem, 109F09
The Okonite Company, 109F10
General Cable, Module overview, 109F12
Greenlee Textron, Inc., a subsidiary of Textron Inc., 109F22, 109F25

Module 26110-08

AGC of America, Module divider
John Traister, 110F06–110F18, 110F20–110F23, 110F25, 110F27, 110F29–110F31, 110F33–110F40, 110F43–110F45

The Number and Titles used in this textbook are from *MasterFormat*™ 2004, published by the Construction Specifications Institute (CSI) and Construction Specifications Canada (CSC), and are used with permission from CSI. For those interested in a more in-depth explanation of *MasterFormat*™ 2004 and its use in the construction industry visit www.csinet.org/masterformat or contact:
The Construction Specifications Institute (CSI)
99 Canal Center Plaza, Suite 300
Alexandria, VA 22314
800-689-2900; 703-684-0300
http://www.csinet.org, 110F46, 110F47
Topaz Publications, Inc., Module overview, 110SA01
Staedtler USA, 110SA02
Scalex Corporation, 110SA03

Module 26111-08

Associated Builders and Contractors, Inc., Module divider
Veronica Westfall, Module overview
John Traister, 111F02, 111F05–111F08, 111F10–111F13, 111F19–111F23, 111F25–111F29, 111F35, 111F43, 111F44, 111F53, 111F55
Topaz Publications, Inc., 111F16B, 111F37B, 111SA01A, 111SA02, 111SA04, 111SA05
Tim Dean, 111F40, 111SA01B
Greenlee Textron, Inc., a subsidiary of Textron Inc., 111SA03

Module 26112-08

AGC of America, Module divider
Tim Dean, 112F01, 112F05, 112F07, 112F08
Greenlee Textron, Inc., a subsidiary of Textron Inc., Module overview, 112F02, 112F03, 112F06, 112F09–112F14, 112SA01
Topaz Publications, Inc., 112SA02

Index

A

Above finished floor (AFF), 10.35
Absenteeism, 1.13–1.14
AC. *See* Alternating current (AC)
Accessibility, 8.9
Accessible, defined, 8.60
Acids, safety, 2.33
ADA (Americans with Disabilities Act of 1990), 1.14
Adjustable resistors. *See* Variable resistors
AFCI (arc fault circuit interrupters), 11.13, 11.14–11.15
Affected employee, defined, 2.17
Air-purifying respirators, 2.32
Air terminals, 11.22
Alternating current (AC)
  frequency of, 12.7
  measuring voltage of, 3.2, 3.10, 3.19, 3.19*f*
Aluminum, conductors and, 2.2, 9.3, 9.3*f*
Aluminum boxes, 6.3
Aluminum conduit, 8.5, 8.6
Aluminum wire
  grounding electrode conductors and, 11.20, 11.22
  terminating, 9.4, 9.5
American National Standards Institute (ANSI)
  classification system for rubber protective equipment, 2.7
  electrical symbols and, 10.16, 10.17*f*
  standards for electrical safety, 2.20
American Society for Testing and Materials (ASTM) International
  classification system for rubber protective equipment, 2.7
  standards for electrical safety, 2.20
Americans with Disabilities Act of 1990 (ADA), 1.14
American Wire Gauge (AWG) system, 9.2, 9.2*f*
Ammeters
  clamp-on, 3.18–3.19, 3.18*f*, 3.19*f*, 12.6–12.7, 12.6*f*
  combination multimeter/clamp-on ammeters, 3.21
  defined, 3.31
  overload protection, 12.6
  safety and, 12.6, 12.7
Ampacity
  conductors and, 9.3–9.4
  defined, 9.28
Ampere-hours, 3.9
Amperes (A)
  defined, 3.31
  electrical power and, 3.22–3.23
  electrical theory and, 3.7–3.8
Analog multimeters, 3.18, 3.18*f*, 12.2, 12.3, 12.7, 12.7*f*
Anchorage devices, personal fall arrest systems, 2.37
Anchors
  ceiling installations, 8.30, 8.31
  drilling anchor holes, 8.27–8.28, 8.28*f*
  epoxy anchoring systems, 8.31–8.32, 8.31*f*
  mechanical anchors, 8.24–8.31

Angles
  calculating with trigonometry, 7.8*f*, 7.9, 7.25–7.27
  right triangles, 7.4, 7.4*f*
Annexes A through M, National Fire Protection Association (NFPA), 2.20
ANSI. *See* American National Standards Institute (ANSI)
Appliance loads, 11.7–11.8
Appliances
  defined, 11.72
  electric heating appliances and, 11.61–11.62
  equipment grounding system and, 11.34
  fixed appliance circuits and, 11.6
  planning residential electrical systems and, 11.2
  small appliance loads, 11.5, 11.7
Appointed authorized employee, defined, 2.17
Apprenticeships
  Department of Labor's Office of Apprenticeship, 1.9
  sample apprentice training recognitions, 1.21–1.24
  standards for, 1.10
  Youth Apprenticeship Program, 1.11
Approval blocks, 10.11, 10.11*f*, 10.13, 10.13*f*
Approved, 8.8
  defined, 8.60
Arc, 2.5
Arc blast, 2.5, 2.6
Arc burns, 2.5
Arc fault circuit interrupters (AFCI), 11.13, 11.14–11.15
Arc flash hoods, 1.15, 1.15*f*
Architect's scales, 10.27, 10.28–10.31, 10.28*f*, 10.29*f*, 10.30*f*, 10.31*f*
Architectural drawings, 10.2, 10.4
  defined, 10.64
Armored cable (Type AC), 9.11, 11.30
Articles
  defined, 5.20
  National Electric Code and, 5.3, 5.5*f*, 5.6
Asbestos
  duct materials and, 8.48
  insulation and, 2.33
  safety and, 2.32–2.33
Asbestos containing products, 2.33, 8.48
Asbestosis, 2.32–2.33
ASTM. *See* American Society for Testing and Materials (ASTM) International
Atoms, defined, 3.31
Attachment plugs (plug caps), 2.15, 5.9
Audio continuity testers, 3.20
Authorized employee, defined, 2.17
Authorized supervisor, defined, 2.17
Automatic timing devices, 11.52
Autoranging ohmmeters, 12.5
Autoranging voltmeters, 12.3
Auxiliary gutters, as wireways, 8.34